Texts in Applied Mathematics 7

Editors
J.E. Marsden
L. Sirovich
M. Golubitsky

Advisors
G. Iooss
P. Holmes
D. Barkley
M. Dellnitz
P. Newton

Springer
New York
Berlin
Heidelberg
Hong Kong
London
Milan
Paris
Tokyo

Texts in Applied Mathematics

1. *Sirovich:* Introduction to Applied Mathematics.
2. *Wiggins:* Introduction to Applied Nonlinear Dynamical Systems and Chaos, 2nd ed.
3. *Hale/Koçak:* Dynamics and Bifurcations.
4. *Chorin/Marsden:* A Mathematical Introduction to Fluid Mechanics, 3rd ed.
5. *Hubbard/West:* Differential Equations: A Dynamical Systems Approach: Ordinary Differential Equations.
6. *Sontag:* Mathematical Control Theory: Deterministic Finite Dimensional Systems, 2nd ed.
7. *Perko:* Differential Equations and Dynamical Systems, 3rd ed.
8. *Seaborn:* Hypergeometric Functions and Their Applications.
9. *Pipkin:* A Course on Integral Equations.
10. *Hoppensteadt/Peskin:* Modeling and Simulation in Medicine and the Life Sciences, 2nd ed.
11. *Braun:* Differential Equations and Their Applications, 4th ed.
12. *Stoer/Bulirsch:* Introduction to Numerical Analysis, 3rd ed.
13. *Renardy/Rogers:* An Introduction to Partial Differential Equations.
14. *Banks:* Growth and Diffusion Phenomena: Mathematical Frameworks and Applications.
15. *Brenner/Scott:* The Mathematical Theory of Finite Element Methods, 2nd ed.
16. *Van de Velde:* Concurrent Scientific Computing.
17. *Marsden/Ratiu:* Introduction to Mechanics and Symmetry, 2nd ed.
18. *Hubbard/West:* Differential Equations: A Dynamical Systems Approach: Higher-Dimensional Systems.
19. *Kaplan/Glass:* Understanding Nonlinear Dynamics.
20. *Holmes:* Introduction to Perturbation Methods.
21. *Curtain/Zwart:* An Introduction to Infinite-Dimensional Linear Systems Theory.
22. *Thomas:* Numerical Partial Differential Equations: Finite Difference Methods.
23. *Taylor:* Partial Differential Equations: Basic Theory.
24. *Merkin:* Introduction to the Theory of Stability of Motion.
25. *Naber:* Topology, Geometry, and Gauge Fields: Foundations.
26. *Polderman/Willems:* Introduction to Mathematical Systems Theory: A Behavioral Approach.
27. *Reddy:* Introductory Functional Analysis with Applications to Boundary-Value Problems and Finite Elements.
28. *Gustafson/Wilcox:* Analytical and Computational Methods of Advanced Engineering Mathematics.
29. *Tveito/Winther:* Introduction to Partial Differential Equations: A Computational Approach.
30. *Gasquet/Witomski:* Fourier Analysis and Applications: Filtering, Numerical Computation, Wavelets.

(continued after index)

Lawrence Perko

Differential Equations and Dynamical Systems

Third Edition

With 241 Illustrations

 Springer

Lawrence Perko
Department of Mathematics
Northern Arizona University
Flagstaff, AZ 86011
USA
Lawrence.Perko@nau.edu

Series Editors
J.E. Marsden
Control and Dynamical Systems, 107-81
California Institute of Technology
Pasadena, CA 91125
USA

L. Sirovich
Department of Applied Mathematics
Brown University
Providence, RI 02912
USA

M. Golubitsky
Department of Mathematics
University of Houston
Houston, TX 77204-3476
USA

Mathematics Subject Classification (2000): 34A34, 34C35, 58F14, 58F21

Library of Congress Cataloging-in-Publication Data
Perko, Lawrence.
 Differential equations and dynamical systems / Lawrence Perko.—3rd. ed.
   p. cm. — (Texts in applied mathematics ; 7)
 Includes bibliographical references and index.
 ISBN 0-387-95116-4 (alk. paper)
   1. Differential equations, Nonlinear. 2. Differentiable dynamical systems. I. Title.
 II. Series.
 QA372.P47 2000
 515.353—dc21   00-058305

ISBN 0-387-95116-4   Printed on acid-free paper.

© 2001, 1996, 1991 Springer-Verlag, New York, Inc.
All rights reserved. This work may not be translated or copied in whole or in part without the written permission of the publisher (Springer-Verlag New York, Inc., 175 Fifth Avenue, New York, NY 10010, USA), except for brief excerpts in connection with reviews or scholarly analysis. Use in connection with any form of information storage and retrieval, electronic adaptation, computer software, or by similar or dissimilar methodology now know or hereafter developed is forbidden.
The use in this publication of trade names, trademarks, service marks, and similar terms, even if they are not identified as such, is not to be taken as an expression of opinion as to whether or not they are subject to proprietary rights.

Printed in the United States of America

9 8 7 6 5 4 3   SPIN 10956625

www.springer-ny.com

Springer-Verlag New York Berlin Heidelberg
A member of BertelsmannSpringer Science+Business Media GmbH

To my wife, Kathy, and children, Mary, Mike, Vince, Jenny, and John, for all the joy they bring to my life.

Series Preface

Mathematics is playing an ever more important role in the physical and biological sciences, provoking a blurring of boundaries between scientific disciplines and a resurgence of interest in the modern as well as the classical techniques of applied mathematics. This renewal of interest, both in research and teaching, has led to the establishment of the series: *Texts in Applied Mathematics (TAM)*.

The development of new courses is a natural consequence of a high level of excitement on the research frontier as newer techniques, such as numerical and symbolic computer systems, dynamical systems, and chaos, mix with and reinforce the traditional methods of applied mathematics. Thus, the purpose of this textbook series is to meet the current and future needs of these advances and encourage the teaching of new courses.

TAM will publish textbooks suitable for use in advanced undergraduate and beginning graduate courses, and will complement the *Applied Mathematical Sciences (AMS)* series, which will focus on advanced textbooks and research level monographs.

Pasadena, California	J.E. Marsden
Providence, Rhode Island	L. Sirovich
Houston, Texas	M. Golubitsky

Preface to the Third Edition

This book covers those topics necessary for a clear understanding of the qualitative theory of ordinary differential equations and the concept of a dynamical system. It is written for advanced undergraduates and for beginning graduate students. It begins with a study of linear systems of ordinary differential equations, a topic already familiar to the student who has completed a first course in differential equations. An efficient method for solving any linear system of ordinary differential equations is presented in Chapter 1.

The major part of this book is devoted to a study of nonlinear systems of ordinary differential equations and dynamical systems. Since most nonlinear differential equations cannot be solved, this book focuses on the qualitative or geometrical theory of nonlinear systems of differential equations originated by Henri Poincaré in his work on differential equations at the end of the nineteenth century as well as on the functional properties inherent in the solution set of a system of nonlinear differential equations embodied in the more recent concept of a dynamical system. Our primary goal is to describe the qualitative behavior of the solution set of a given system of differential equations including the invariant sets and limiting behavior of the dynamical system or flow defined by the system of differential equations. In order to achieve this goal, it is first necessary to develop the local theory for nonlinear systems. This is done in Chapter 2 which includes the fundamental local existence–uniqueness theorem, the Hartman–Grobman Theorem and the Stable Manifold Theorem. These latter two theorems establish that the qualitative behavior of the solution set of a nonlinear system of ordinary differential equations near an equilibrium point is typically the same as the qualitative behavior of the solution set of the corresponding linearized system near the equilibrium point.

After developing the local theory, we turn to the global theory in Chapter 3. This includes a study of limit sets of trajectories and the behavior of trajectories at infinity. Some unresolved problems of current research interest are also presented in Chapter 3. For example, the Poincaré–Bendixson Theorem, established in Chapter 3, describes the limit sets of trajectories of two-dimensional systems; however, the limit sets of trajectories of three-dimensional (and higher dimensional) systems can be much more complicated and establishing the nature of these limit sets is a topic of current research interest in mathematics. In particular, higher dimensional systems can exhibit strange attractors and chaotic dynamics. All of the preliminary material necessary for studying these more advance topics is contained in this textbook. This book can therefore serve as a springboard for those students interested in continuing their study of ordinary differential equations and dynamical systems and doing research in these areas. Chapter 3 ends with a technique for constructing the global phase portrait of a dynamical system. The global phase portrait describes the qualitative behavior of the solution set for all time. In general, this is as close as we can come to "solving" nonlinear systems.

In Chapter 4, we study systems of differential equations depending on parameters. The question of particular interest is: For what parameter values does the global phase portrait of a dynamical system change its qualitative structure? The answer to this question forms the subject matter of bifurcation theory. An introduction to bifurcation theory is presented in Chapter 4 where we discuss bifurcations at nonhyperbolic equilibrium points and periodic orbits as well as Hopf bifurcations. Chapter 4 ends with a discussion of homoclinic loop and Takens–Bogdanov bifurcations for planar systems and an introduction to tangential homoclinic bifurcations and the resulting chaotic dynamics that can occur in higher dimensional systems.

The prerequisites for studying differential equations and dynamical systems using this book are courses in linear algebra and real analysis. For example, the student should know how to find the eigenvalues and eigenvectors of a linear transformation represented by a square matrix and should be familiar with the notion of uniform convergence and related concepts. In using this book, the author hopes that the student will develop an appreciation for just how useful the concepts of linear algebra, real analysis and geometry are in developing the theory of ordinary differential equations and dynamical systems. The heart of the geometrical theory of nonlinear differential equations is contained in Chapters 2–4 of this book and in order to cover the main ideas in those chapters in a one semester course, it is necessary to cover Chapter 1 as quickly as possible.

In addition to the new sections on center manifold and normal form theory, higher codimension bifurcations, higher order Melnikov theory, the Takens–Bogdanov bifurcation and bounded quadratic systems in $\mathbf{R}^2$ that were added to the second edition of this book, the third edition contains two new sections, Section 4.12 on Françoise's algorithm for higher order Melnikov functions and Section 4.15 on the higher codimension bifurcations that occur in the class of bounded quadratic systems. Also, some new results on the structural stability of polynomial systems on $\mathbf{R}^2$ have been added at the end of Section 4.1, some recent results for determining the order of a weak focus of a planar quadratic system have been added at the end of Section 4.4, and several new problems have been interspersed throughout the book.

Preface to the Third Edition

A solutions manual for this book has been prepared by the author and is now available under separate cover from Springer-Verlag at no additional cost.

I would like to express my sincere appreciation to my colleagues Freddy Dumortier, Iliya Iliev, Doug Shafer and especially to Terence Blows and Jim Swift for their many helpful suggestions which substantially improved this book. I would also like to thank Louella Holter for her patience and precision in typing the original manuscript for this book.

Flagstaff, Arizona   Lawrence Perko

Contents

Series Preface		vii
Preface to the Third Edition		ix

1 Linear Systems — 1

1.1	Uncoupled Linear Systems	1
1.2	Diagonalization	6
1.3	Exponentials of Operators	10
1.4	The Fundamental Theorem for Linear Systems	16
1.5	Linear Systems in $\mathbf{R}^2$	20
1.6	Complex Eigenvalues	28
1.7	Multiple Eigenvalues	32
1.8	Jordan Forms	39
1.9	Stability Theory	51
1.10	Nonhomogeneous Linear Systems	60

2 Nonlinear Systems: Local Theory — 65

2.1	Some Preliminary Concepts and Definitions	65
2.2	The Fundamental Existence-Uniqueness Theorem	70
2.3	Dependence on Initial Conditions and Parameters	79
2.4	The Maximal Interval of Existence	87
2.5	The Flow Defined by a Differential Equation	95
2.6	Linearization	101
2.7	The Stable Manifold Theorem	105
2.8	The Hartman–Grobman Theorem	119
2.9	Stability and Liapunov Functions	129
2.10	Saddles, Nodes, Foci and Centers	136
2.11	Nonhyperbolic Critical Points in $\mathbf{R}^2$	147
2.12	Center Manifold Theory	154
2.13	Normal Form Theory	163
2.14	Gradient and Hamiltonian Systems	171

3 Nonlinear Systems: Global Theory — 181

- 3.1 Dynamical Systems and Global Existence Theorems — 182
- 3.2 Limit Sets and Attractors — 191
- 3.3 Periodic Orbits, Limit Cycles and Separatrix Cycles — 202
- 3.4 The Poincaré Map — 211
- 3.5 The Stable Manifold Theorem for Periodic Orbits — 220
- 3.6 Hamiltonian Systems with Two Degrees of Freedom — 234
- 3.7 The Poincaré–Bendixson Theory in $\mathbf{R}^2$ — 244
- 3.8 Lienard Systems — 253
- 3.9 Bendixson's Criteria — 264
- 3.10 The Poincaré Sphere and the Behavior at Infinity — 267
- 3.11 Global Phase Portraits and Separatrix Configurations — 293
- 3.12 Index Theory — 298

4 Nonlinear Systems: Bifurcation Theory — 315

- 4.1 Structural Stability and Peixoto's Theorem — 316
- 4.2 Bifurcations at Nonhyperbolic Equilibrium Points — 334
- 4.3 Higher Codimension Bifurcations at Nonhyperbolic Equilibrium Points — 343
- 4.4 Hopf Bifurcations and Bifurcations of Limit Cycles from a Multiple Focus — 349
- 4.5 Bifurcations at Nonhyperbolic Periodic Orbits — 362
- 4.6 One-Parameter Families of Rotated Vector Fields — 383
- 4.7 The Global Behavior of One-Parameter Families of Periodic Orbits — 395
- 4.8 Homoclinic Bifurcations — 401
- 4.9 Melnikov's Method — 415
- 4.10 Global Bifurcations of Systems in $\mathbf{R}^2$ — 431
- 4.11 Second and Higher Order Melnikov Theory — 452
- 4.12 Françoise's Algorithm for Higher Order Melnikov Functions — 466
- 4.13 The Takens–Bogdanov Bifurcation — 477
- 4.14 Coppel's Problem for Bounded Quadratic Systems — 487
- 4.15 Finite Codimension Bifurcations in the Class of Bounded Quadratic Systems — 528

References — 541

Additional References — 543

Index — 549

1
Linear Systems

This chapter presents a study of linear systems of ordinary differential equations:
$$\dot{\mathbf{x}} = A\mathbf{x} \qquad (1)$$
where $\mathbf{x} \in \mathbf{R}^n$, A is an $n \times n$ matrix and
$$\dot{\mathbf{x}} = \frac{d\mathbf{x}}{dt} = \begin{bmatrix} \frac{dx_1}{dt} \\ \vdots \\ \frac{dx_n}{dt} \end{bmatrix}.$$

It is shown that the solution of the linear system (1) together with the initial condition $\mathbf{x}(0) = \mathbf{x}_0$ is given by
$$\mathbf{x}(t) = e^{At}\mathbf{x}_0$$
where e^{At} is an $n \times n$ matrix function defined by its Taylor series. A good portion of this chapter is concerned with the computation of the matrix e^{At} in terms of the eigenvalues and eigenvectors of the square matrix A. Throughout this book all vectors will be written as column vectors and A^T will denote the transpose of the matrix A.

1.1 Uncoupled Linear Systems

The method of separation of variables can be used to solve the first-order linear differential equation
$$\dot{x} = ax.$$
The general solution is given by
$$x(t) = ce^{at}$$
where the constant $c = x(0)$, the value of the function $x(t)$ at time $t = 0$.
Now consider the uncoupled linear system
$$\dot{x}_1 = -x_1$$
$$\dot{x}_2 = 2x_2.$$

This system can be written in matrix form as

$$\dot{\mathbf{x}} = A\mathbf{x} \qquad (1)$$

where

$$A = \begin{bmatrix} -1 & 0 \\ 0 & 2 \end{bmatrix}.$$

Note that in this case A is a diagonal matrix, $A = \mathrm{diag}[-1, 2]$, and in general whenever A is a diagonal matrix, the system (1) reduces to an uncoupled linear system. The general solution of the above uncoupled linear system can once again be found by the method of separation of variables. It is given by

$$\begin{aligned} x_1(t) &= c_1 e^{-t} \\ x_2(t) &= c_2 e^{2t} \end{aligned} \qquad (2)$$

or equivalently by

$$\mathbf{x}(t) = \begin{bmatrix} e^{-t} & 0 \\ 0 & e^{2t} \end{bmatrix} \mathbf{c} \qquad (2')$$

where $\mathbf{c} = \mathbf{x}(0)$. Note that the solution curves (2) lie on the algebraic curves $y = k/x^2$ where the constant $k = c_1^2 c_2$. The solution (2) or (2') defines a motion along these curves; i.e., each point $\mathbf{c} \in \mathbf{R}^2$ moves to the point $\mathbf{x}(t) \in \mathbf{R}^2$ given by (2') after time t. This motion can be described geometrically by drawing the solution curves (2) in the x_1, x_2 plane, referred to as the *phase plane*, and by using arrows to indicate the direction of the motion along these curves with increasing time t; cf. Figure 1. For $c_1 = c_2 = 0$, $x_1(t) = 0$ and $x_2(t) = 0$ for all $t \in \mathbf{R}$ and the origin is referred to as an *equilibrium point* in this example. Note that solutions starting on the x_1-axis approach the origin as $t \to \infty$ and that solutions starting on the x_2-axis approach the origin as $t \to -\infty$.

The *phase portrait* of a system of differential equations such as (1) with $\mathbf{x} \in \mathbf{R}^n$ is the set of all solution curves of (1) in the phase space $\mathbf{R}^n$. Figure 1 gives a geometrical representation of the phase portrait of the uncoupled linear system considered above. The *dynamical system* defined by the linear system (1) in this example is simply the mapping $\phi \colon \mathbf{R} \times \mathbf{R}^2 \to \mathbf{R}^2$ defined by the solution $\mathbf{x}(t, \mathbf{c})$ given by (2'); i.e., the dynamical system for this example is given by

$$\phi(t, \mathbf{c}) = \begin{bmatrix} e^{-t} & 0 \\ 0 & e^{2t} \end{bmatrix} \mathbf{c}.$$

Geometrically, the dynamical system describes the motion of the points in phase space along the solution curves defined by the system of differential equations.

The function

$$\mathbf{f}(\mathbf{x}) = A\mathbf{x}$$

on the right-hand side of (1) defines a mapping $\mathbf{f} \colon \mathbf{R}^2 \to \mathbf{R}^2$ (linear in this case). This mapping (which need not be linear) defines a *vector field on*

1.1. Uncoupled Linear Systems

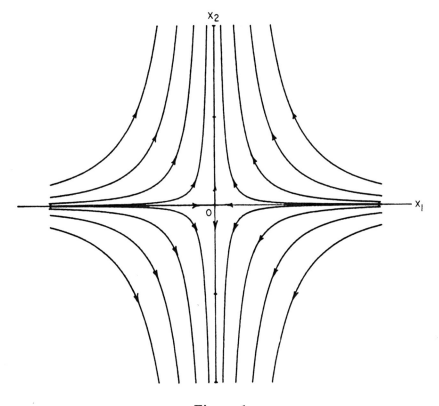

Figure 1

$\mathbf{R}^2$; i.e., to each point $\mathbf{x} \in \mathbf{R}^2$, the mapping $\mathbf{f}$ assigns a vector $\mathbf{f}(\mathbf{x})$. If we draw each vector $\mathbf{f}(\mathbf{x})$ with its initial point at the point $\mathbf{x} \in \mathbf{R}^2$, we obtain a geometrical representation of the vector field as shown in Figure 2. Note that at each point $\mathbf{x}$ in the phase space $\mathbf{R}^2$, the solution curves (2) are tangent to the vectors in the vector field $A\mathbf{x}$. This follows since at time $t = t_0$, the velocity vector $\mathbf{v}_0 = \dot{\mathbf{x}}(t_0)$ is tangent to the curve $\mathbf{x} = \mathbf{x}(t)$ at the point $\mathbf{x}_0 = \mathbf{x}(t_0)$ and since $\dot{\mathbf{x}} = A\mathbf{x}$ along the solution curves.

Consider the following uncoupled linear system in $\mathbf{R}^3$:

$$\begin{aligned} \dot{x}_1 &= x_1 \\ \dot{x}_2 &= x_2 \\ \dot{x}_3 &= -x_3 \end{aligned} \qquad (3)$$

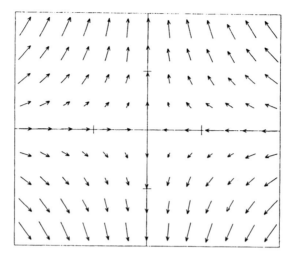

Figure 2

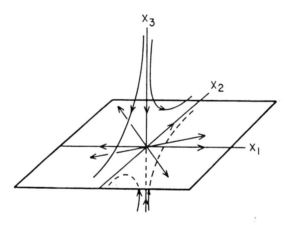

Figure 3

The general solution is given by

$$x_1(t) = c_1 e^t$$
$$x_2(t) = c_2 e^t$$
$$x_3(t) = c_3 e^{-t}.$$

And the phase portrait for this system is shown in Figure 3 above. The x_1, x_2 plane is referred to as the *unstable subspace* of the system (3) and

1.1. Uncoupled Linear Systems

the x_3 axis is called the *stable subspace* of the system (3). Precise definitions of the stable and unstable subspaces of a linear system will be given in the next section.

PROBLEM SET 1

1. Find the general solution and draw the phase portrait for the following linear systems:

   (a) $\begin{aligned} \dot{x}_1 &= x_1 \\ \dot{x}_2 &= x_2 \end{aligned}$

   (b) $\begin{aligned} \dot{x}_1 &= x_1 \\ \dot{x}_2 &= 2x_2 \end{aligned}$

   (c) $\begin{aligned} \dot{x}_1 &= x_1 \\ \dot{x}_2 &= 3x_2 \end{aligned}$

   (d) $\begin{aligned} \dot{x}_1 &= -x_2 \\ \dot{x}_2 &= x_1 \end{aligned}$

   (e) $\begin{aligned} \dot{x}_1 &= -x_1 + x_2 \\ \dot{x}_2 &= -x_2 \end{aligned}$

   Hint: Write (d) as a second-order linear differential equation with constant coefficients, solve it by standard methods, and note that $x_1^2 + x_2^2 =$ constant on the solution curves. In (e), find $x_2(t) = c_2 e^{-t}$ and then the x_1-equation becomes a first order linear differential equation.

2. Find the general solution and draw the phase portraits for the following three-dimensional linear systems:

   (a) $\begin{aligned} \dot{x}_1 &= x_1 \\ \dot{x}_2 &= x_2 \\ \dot{x}_3 &= x_3 \end{aligned}$

   (b) $\begin{aligned} \dot{x}_1 &= -x_1 \\ \dot{x}_2 &= -x_2 \\ \dot{x}_3 &= x_3 \end{aligned}$

   (c) $\begin{aligned} \dot{x}_1 &= -x_2 \\ \dot{x}_2 &= x_1 \\ \dot{x}_3 &= -x_3 \end{aligned}$

   Hint: In (c), show that the solution curves lie on right circular cylinders perpendicular to the x_1, x_2 plane. Identify the stable and unstable subspaces in (a) and (b). The x_3-axis is the stable subspace in (c) and the x_1, x_2 plane is called the center subspace in (c); cf. Section 1.9.

3. Find the general solution of the linear system

$$\dot{x}_1 = x_1$$
$$\dot{x}_2 = ax_2$$

where a is a constant. Sketch the phase portraits for $a = -1, a = 0$ and $a = 1$ and notice that the qualitative structure of the phase portrait is the same for all $a < 0$ as well as for all $a > 0$, but that it changes at the parameter value $a = 0$ called a bifurcation value.

4. Find the general solution of the linear system (1) when A is the $n \times n$ diagonal matrix $A = \text{diag}[\lambda_1, \lambda_2, \ldots, \lambda_n]$. What condition on the eigenvalues $\lambda_1, \ldots, \lambda_n$ will guarantee that $\lim_{t \to \infty} \mathbf{x}(t) = \mathbf{0}$ for all solutions $\mathbf{x}(t)$ of (1)?

5. What is the relationship between the vector fields defined by

$$\dot{\mathbf{x}} = A\mathbf{x}$$

and

$$\dot{\mathbf{x}} = kA\mathbf{x}$$

where k is a non-zero constant? (Describe this relationship both for k positive and k negative.)

6. (a) If $\mathbf{u}(t)$ and $\mathbf{v}(t)$ are solutions of the linear system (1), prove that for any constants a and b, $\mathbf{w}(t) = a\mathbf{u}(t) + b\mathbf{v}(t)$ is a solution.

   (b) For

$$A = \begin{bmatrix} 1 & 0 \\ 0 & -2 \end{bmatrix},$$

   find solutions $\mathbf{u}(t)$ and $\mathbf{v}(t)$ of $\dot{\mathbf{x}} = A\mathbf{x}$ such that every solution is a linear combination of $\mathbf{u}(t)$ and $\mathbf{v}(t)$.

1.2 Diagonalization

The algebraic technique of diagonalizing a square matrix A can be used to reduce the linear system

$$\dot{\mathbf{x}} = A\mathbf{x} \qquad (1)$$

to an uncoupled linear system. We first consider the case when A has real, distinct eigenvalues. The following theorem from linear algebra then allows us to solve the linear system (1).

Theorem. *If the eigenvalues $\lambda_1, \lambda_2, \ldots, \lambda_n$ of an $n \times n$ matrix A are real and distinct, then any set of corresponding eigenvectors $\{\mathbf{v}_1, \mathbf{v}_2, \ldots, \mathbf{v}_n\}$ forms a basis for $\mathbf{R}^n$, the matrix $P = [\mathbf{v}_1 \ \mathbf{v}_2 \ \cdots \ \mathbf{v}_n]$ is invertible and*

$$P^{-1}AP = \text{diag}[\lambda_1, \ldots, \lambda_n].$$

1.2. Diagonalization

This theorem says that if a linear transformation $T\colon \mathbf{R}^n \to \mathbf{R}^n$ is represented by the $n \times n$ matrix A with respect to the standard basis $\{\mathbf{e}_1, \mathbf{e}_2, \ldots, \mathbf{e}_n\}$ for $\mathbf{R}^n$, then with respect to any basis of eigenvectors $\{\mathbf{v}_1, \mathbf{v}_2, \ldots, \mathbf{v}_n\}$, T is represented by the diagonal matrix of eigenvalues, $\mathrm{diag}[\lambda_1, \lambda_2, \ldots, \lambda_n]$. A proof of this theorem can be found, for example, in Lowenthal [Lo].

In order to reduce the system (1) to an uncoupled linear system using the above theorem, define the linear transformation of coordinates

$$\mathbf{y} = P^{-1}\mathbf{x}$$

where P is the invertible matrix defined in the theorem. Then

$$\mathbf{x} = P\mathbf{y},$$
$$\dot{\mathbf{y}} = P^{-1}\dot{\mathbf{x}} = P^{-1}A\mathbf{x} = P^{-1}AP\mathbf{y}$$

and, according to the above theorem, we obtain the uncoupled linear system

$$\dot{\mathbf{y}} = \mathrm{diag}[\lambda_1, \ldots, \lambda_n]\mathbf{y}.$$

This uncoupled linear system has the solution

$$\mathbf{y}(t) = \mathrm{diag}[e^{\lambda_1 t}, \ldots, e^{\lambda_n t}]\mathbf{y}(0).$$

(Cf. problem 4 in Problem Set 1.) And then since $\mathbf{y}(0) = P^{-1}\mathbf{x}(0)$ and $\mathbf{x}(t) = P\mathbf{y}(t)$, it follows that (1) has the solution

$$\mathbf{x}(t) = PE(t)P^{-1}\mathbf{x}(0). \tag{2}$$

where $E(t)$ is the diagonal matrix

$$E(t) = \mathrm{diag}[e^{\lambda_1 t}, \ldots, e^{\lambda_n t}].$$

Corollary. *Under the hypotheses of the above theorem, the solution of the linear system (1) is given by the function $\mathbf{x}(t)$ defined by (2).*

Example. Consider the linear system

$$\dot{x}_1 = -x_1 - 3x_2$$
$$\dot{x}_2 = 2x_2$$

which can be written in the form (1) with the matrix

$$A = \begin{bmatrix} -1 & -3 \\ 0 & 2 \end{bmatrix}.$$

The eigenvalues of A are $\lambda_1 = -1$ and $\lambda_2 = 2$. A pair of corresponding eigenvectors is given by

$$\mathbf{v}_1 = \begin{bmatrix} 1 \\ 0 \end{bmatrix}, \quad \mathbf{v}_2 = \begin{bmatrix} -1 \\ 1 \end{bmatrix}.$$

The matrix P and its inverse are then given by

$$P = \begin{bmatrix} 1 & -1 \\ 0 & 1 \end{bmatrix} \quad \text{and} \quad P^{-1} = \begin{bmatrix} 1 & 1 \\ 0 & 1 \end{bmatrix}.$$

The student should verify that

$$P^{-1}AP = \begin{bmatrix} -1 & 0 \\ 0 & 2 \end{bmatrix}.$$

Then under the coordinate transformation $\mathbf{y} = P^{-1}\mathbf{x}$, we obtain the uncoupled linear system

$$\dot{y}_1 = -y_1$$
$$\dot{y}_2 = 2y_2$$

which has the general solution $y_1(t) = c_1 e^{-t}$, $y_2(t) = c_2 e^{2t}$. The phase portrait for this system is given in Figure 1 in Section 1.1 which is reproduced below. And according to the above corollary, the general solution to the original linear system of this example is given by

$$\mathbf{x}(t) = P \begin{bmatrix} e^{-t} & 0 \\ 0 & e^{2t} \end{bmatrix} P^{-1} \mathbf{c}$$

where $\mathbf{c} = \mathbf{x}(0)$, or equivalently by

$$\begin{aligned} x_1(t) &= c_1 e^{-t} + c_2(e^{-t} - e^{2t}) \\ x_2(t) &= c_2 e^{2t}. \end{aligned} \tag{3}$$

The phase portrait for the linear system of this example can be found by sketching the solution curves defined by (3). It is shown in Figure 2. The

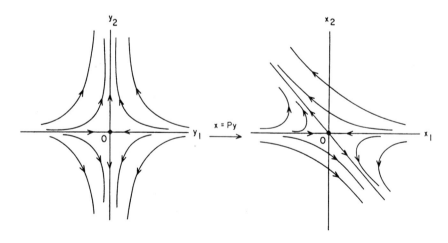

Figure 1   **Figure 2**

1.2. Diagonalization

phase portrait in Figure 2 can also be obtained from the phase portrait in Figure 1 by applying the linear transformation of coordinates $\mathbf{x} = P\mathbf{y}$. Note that the subspaces spanned by the eigenvectors $\mathbf{v}_1$ and $\mathbf{v}_2$ of the matrix A determine the stable and unstable subspaces of the linear system (1) according to the following definition:

Suppose that the $n \times n$ matrix A has k negative eigenvalues $\lambda_1, \ldots, \lambda_k$ and $n - k$ positive eigenvalues $\lambda_{k+1}, \ldots, \lambda_n$ and that these eigenvalues are distinct. Let $\{\mathbf{v}_1, \ldots, \mathbf{v}_n\}$ be a corresponding set of eigenvectors. Then the *stable and unstable subspaces of the linear system* (1), E^s and E^u, are the linear subspaces spanned by $\{\mathbf{v}_1, \ldots, \mathbf{v}_k\}$ and $\{\mathbf{v}_{k+1}, \ldots, \mathbf{v}_n\}$ respectively; i.e.,

$$E^s = \mathrm{Span}\{\mathbf{v}_1, \ldots, \mathbf{v}_k\}$$
$$E^u = \mathrm{Span}\{\mathbf{v}_{k+1}, \ldots, \mathbf{v}_n\}.$$

If the matrix A has pure imaginary eigenvalues, then there is also a center subspace E^c; cf. Problem 2(c) in Section 1.1. The stable, unstable and center subspaces are defined for the general case in Section 1.9.

Problem Set 2

1. Find the eigenvalues and eigenvectors of the matrix A and show that $B = P^{-1}AP$ is a diagonal matrix. Solve the linear system $\dot{\mathbf{y}} = B\mathbf{y}$ and then solve $\dot{\mathbf{x}} = A\mathbf{x}$ using the above corollary. And then sketch the phase portraits in both the $\mathbf{x}$ plane and $\mathbf{y}$ plane.

   (a) $A = \begin{bmatrix} 3 & 1 \\ 1 & 3 \end{bmatrix}$

   (b) $A = \begin{bmatrix} 1 & 3 \\ 3 & 1 \end{bmatrix}$

   (c) $A = \begin{bmatrix} -1 & 1 \\ 1 & -1 \end{bmatrix}$.

2. Find the eigenvalues and eigenvectors for the matrix A, solve the linear system $\dot{\mathbf{x}} = A\mathbf{x}$, determine the stable and unstable subspaces for the linear system, and sketch the phase portrait for

$$\dot{\mathbf{x}} = \begin{bmatrix} 1 & 0 & 0 \\ 1 & 2 & 0 \\ 1 & 0 & -1 \end{bmatrix} \mathbf{x}.$$

3. Write the following linear differential equations with constant coefficients in the form of the linear system (1) and solve:

   (a) $\ddot{x} + \dot{x} - 2x = 0$
   (b) $\ddot{x} + x = 0$

(c) $\dddot{x} - 2\ddot{x} - \dot{x} + 2x = 0$

Hint: Let $x_1 = x, x_2 = \dot{x}_1$, etc.

4. Using the corollary of this section solve the initial value problem
$$\dot{x} = Ax$$
$$x(0) = x_0$$
   (a) with A given by 1(a) above and $x_0 = (1,2)^T$
   (b) with A given in problem 2 above and $x_0 = (1,2,3)^T$.

5. Let the $n \times n$ matrix A have real, distinct eigenvalues. Find conditions on the eigenvalues that are necessary and sufficient for $\lim_{t \to \infty} x(t) = 0$ where $x(t)$ is any solution of $\dot{x} = Ax$.

6. Let the $n \times n$ matrix A have real, distinct eigenvalues. Let $\phi(t, x_0)$ be the solution of the initial value problem
$$\dot{x} = Ax$$
$$x(0) = x_0.$$
Show that for each fixed $t \in \mathbf{R}$,
$$\lim_{y_0 \to x_0} \phi(t, y_0) = \phi(t, x_0).$$
This shows that the solution $\phi(t, x_0)$ is a continuous function of the initial condition.

7. Let the 2×2 matrix A have real, distinct eigenvalues λ and μ. Suppose that an eigenvector of λ is $(1,0)^T$ and an eigenvector of μ is $(-1,1)^T$. Sketch the phase portraits of $\dot{x} = Ax$ for the following cases:

   (a) $0 < \lambda < \mu$ (b) $0 < \mu < \lambda$ (c) $\lambda < \mu < 0$
   (d) $\lambda < 0 < \mu$ (e) $\mu < 0 < \lambda$ (f) $\lambda = 0, \mu > 0$.

1.3 Exponentials of Operators

In order to define the exponential of a linear operator $T: \mathbf{R}^n \to \mathbf{R}^n$, it is necessary to define the concept of convergence in the linear space $L(\mathbf{R}^n)$ of linear operators on $\mathbf{R}^n$. This is done using the *operator norm of T* defined by
$$\|T\| = \max_{|x| \le 1} |T(x)|$$
where $|x|$ denotes the Euclidean norm of $x \in \mathbf{R}^n$; i.e.,
$$|x| = \sqrt{x_1^2 + \cdots + x_n^2}.$$
The operator norm has all of the usual properties of a norm, namely, for $S, T \in L(\mathbf{R}^n)$

1.3. Exponentials of Operators

(a) $\|T\| \geq 0$ and $\|T\| = 0$ iff $T = 0$

(b) $\|kT\| = |k|\,\|T\|$ for $k \in \mathbf{R}$

(c) $\|S + T\| \leq \|S\| + \|T\|$.

It follows from the Cauchy–Schwarz inequality that if $T \in L(\mathbf{R}^n)$ is represented by the matrix A with respect to the standard basis for $\mathbf{R}^n$, then $\|A\| \leq \sqrt{n}\ell$ where ℓ is the maximum length of the rows of A.

The convergence of a sequence of operators $T_k \in L(\mathbf{R}^n)$ is then defined in terms of the operator norm as follows:

Definition 1. A sequence of linear operators $T_k \in L(\mathbf{R}^n)$ is said to converge to a linear operator $T \in L(\mathbf{R}^n)$ as $k \to \infty$, i.e.,

$$\lim_{k \to \infty} T_k = T,$$

if for all $\varepsilon > 0$ there exists an N such that for $k \geq N$, $\|T - T_k\| < \varepsilon$.

Lemma. *For $S, T \in L(\mathbf{R}^n)$ and $\mathbf{x} \in \mathbf{R}^n$,*

(1) $|T(\mathbf{x})| \leq \|T\|\,|\mathbf{x}|$

(2) $\|TS\| \leq \|T\|\,\|S\|$

(3) $\|T^k\| \leq \|T\|^k$ *for $k = 0, 1, 2, \ldots$.*

Proof. (1) is obviously true for $\mathbf{x} = 0$. For $\mathbf{x} \neq 0$ define the unit vector $\mathbf{y} = \mathbf{x}/|\mathbf{x}|$. Then from the definition of the operator norm,

$$\|T\| \geq |T(\mathbf{y})| = \frac{1}{|\mathbf{x}|}|T(\mathbf{x})|.$$

(2) For $|\mathbf{x}| \leq 1$, it follows from (1) that

$$|T(S(\mathbf{x}))| \leq \|T\|\,|S(\mathbf{x})|$$
$$\leq \|T\|\,\|S\|\,|\mathbf{x}|$$
$$\leq \|T\|\,\|S\|.$$

Therefore,

$$\|TS\| = \max_{|\mathbf{x}| \leq 1} |TS(\mathbf{x})| \leq \|T\|\,\|S\|$$

and (3) is an immediate consequence of (2).

Theorem. *Given $T \in L(\mathbf{R}^n)$ and $t_0 > 0$, the series*

$$\sum_{k=0}^{\infty} \frac{T^k t^k}{k!}$$

is absolutely and uniformly convergent for all $|t| \leq t_0$.

Proof. Let $\|T\| = a$. It then follows from the above lemma that for $|t| \leq t_0$,

$$\left\|\frac{T^k t^k}{k!}\right\| \leq \frac{\|T\|^k |t|^k}{k!} \leq \frac{a^k t_0^k}{k!}.$$

But

$$\sum_{k=0}^{\infty} \frac{a^k t_0^k}{k!} = e^{at_0}.$$

It therefore follows from the Weierstrass M-Test that the series

$$\sum_{k=0}^{\infty} \frac{T^k t^k}{k!}$$

is absolutely and uniformly convergent for all $|t| \leq t_0$; cf. [R], p. 148.

The exponential of the linear operator T is then defined by the absolutely convergent series

$$e^T = \sum_{k=0}^{\infty} \frac{T^k}{k!}.$$

It follows from properties of limits that e^T is a linear operator on $\mathbf{R}^n$ and it follows as in the proof of the above theorem that $\|e^T\| \leq e^{\|T\|}$.

Since our main interest in this chapter is the solution of linear systems of the form

$$\dot{\mathbf{x}} = A\mathbf{x},$$

we shall assume that the linear transformation T on $\mathbf{R}^n$ is represented by the $n \times n$ matrix A with respect to the standard basis for $\mathbf{R}^n$ and define the exponential e^{At}.

Definition 2. Let A be an $n \times n$ matrix. Then for $t \in \mathbf{R}$,

$$e^{At} = \sum_{k=0}^{\infty} \frac{A^k t^k}{k!}.$$

For an $n \times n$ matrix A, e^{At} is an $n \times n$ matrix which can be computed in terms of the eigenvalues and eigenvectors of A. This will be carried out

1.3. Exponentials of Operators

in the remainder of this chapter. As in the proof of the above theorem $\|e^{At}\| \leq e^{\|A\| |t|}$ where $\|A\| = \|T\|$ and T is the linear transformation $T(\mathbf{x}) = A\mathbf{x}$.

We next establish some basic properties of the linear transformation e^T in order to facilitate the computation of e^T or of the $n \times n$ matrix e^A.

Proposition 1. *If P and T are linear transformations on $\mathbf{R}^n$ and $S = PTP^{-1}$, then $e^S = Pe^T P^{-1}$.*

Proof. It follows from the definition of e^S that

$$e^S = \lim_{n \to \infty} \sum_{k=0}^{n} \frac{(PTP^{-1})^k}{k!} = P \lim_{n \to \infty} \sum_{k=0}^{n} \frac{T^k}{k!} P^{-1} = Pe^T P^{-1}.$$

The next result follows directly from Proposition 1 and Definition 2.

Corollary 1. *If $P^{-1}AP = \operatorname{diag}[\lambda_j]$ then $e^{At} = P\operatorname{diag}[e^{\lambda_j t}]P^{-1}$.*

Proposition 2. *If S and T are linear transformations on $\mathbf{R}^n$ which commute, i.e., which satisfy $ST = TS$, then $e^{S+T} = e^S e^T$.*

Proof. If $ST = TS$, then by the binomial theorem

$$(S+T)^n = n! \sum_{j+k=n} \frac{S^j T^k}{j!k!}.$$

Therefore,

$$e^{S+T} = \sum_{n=0}^{\infty} \sum_{j+k=n} \frac{S^j T^k}{j!k!} = \sum_{j=0}^{\infty} \frac{S^j}{j!} \sum_{k=0}^{\infty} \frac{T^k}{k!} = e^S e^T.$$

We have used the fact that the product of two absolutely convergent series is an absolutely convergent series which is given by its Cauchy product; cf. [R], p. 74.

Upon setting $S = -T$ in Proposition 2, we obtain

Corollary 2. *If T is a linear transformation on $\mathbf{R}^n$, the inverse of the linear transformation e^T is given by $(e^T)^{-1} = e^{-T}$.*

Corollary 3. *If*

$$A = \begin{bmatrix} a & -b \\ b & a \end{bmatrix}$$

then

$$e^A = e^a \begin{bmatrix} \cos b & -\sin b \\ \sin b & \cos b \end{bmatrix}.$$

Proof. If $\lambda = a + ib$, it follows by induction that

$$\begin{bmatrix} a & -b \\ b & a \end{bmatrix}^k = \begin{bmatrix} \mathrm{Re}(\lambda^k) & -\mathrm{Im}(\lambda^k) \\ \mathrm{Im}(\lambda^k) & \mathrm{Re}(\lambda^k) \end{bmatrix}$$

where Re and Im denote the real and imaginary parts of the complex number λ respectively. Thus,

$$\begin{aligned} e^A &= \sum_{k=0}^{\infty} \begin{bmatrix} \mathrm{Re}\left(\frac{\lambda^k}{k!}\right) & -\mathrm{Im}\left(\frac{\lambda^k}{k!}\right) \\ \mathrm{Im}\left(\frac{\lambda^k}{k!}\right) & \mathrm{Re}\left(\frac{\lambda^k}{k!}\right) \end{bmatrix} \\ &= \begin{bmatrix} \mathrm{Re}(e^\lambda) & -\mathrm{Im}(e^\lambda) \\ \mathrm{Im}(e^\lambda) & \mathrm{Re}(e^\lambda) \end{bmatrix} \\ &= e^a \begin{bmatrix} \cos b & -\sin b \\ \sin b & \cos b \end{bmatrix}. \end{aligned}$$

Note that if $a = 0$ in Corollary 3, then e^A is simply a rotation through b radians.

Corollary 4. *If*

$$A = \begin{bmatrix} a & b \\ 0 & a \end{bmatrix}$$

then

$$e^A = e^a \begin{bmatrix} 1 & b \\ 0 & 1 \end{bmatrix}.$$

Proof. Write $A = aI + B$ where

$$B = \begin{bmatrix} 0 & b \\ 0 & 0 \end{bmatrix}.$$

Then aI commutes with B and by Proposition 2,

$$e^A = e^{aI} e^B = e^a e^B.$$

And from the definition

$$e^B = I + B + B^2/2! + \cdots = I + B$$

since by direct computation $B^2 = B^3 = \cdots = 0$.

We can now compute the matrix e^{At} for any 2×2 matrix A. In Section 1.8 of this chapter it is shown that there is an invertible 2×2 matrix P (whose columns consist of generalized eigenvectors of A) such that the matrix

$$B = P^{-1}AP$$

has one of the following forms

$$B = \begin{bmatrix} \lambda & 0 \\ 0 & \mu \end{bmatrix}, \quad B = \begin{bmatrix} \lambda & 1 \\ 0 & \lambda \end{bmatrix} \quad \text{or} \quad B = \begin{bmatrix} a & -b \\ b & a \end{bmatrix}.$$

1.3. Exponentials of Operators

It then follows from the above corollaries and Definition 2 that

$$e^{Bt} = \begin{bmatrix} e^{\lambda t} & 0 \\ 0 & e^{\mu t} \end{bmatrix}, \quad e^{Bt} = e^{\lambda t}\begin{bmatrix} 1 & t \\ 0 & 1 \end{bmatrix} \quad \text{or} \quad e^{Bt} = e^{at}\begin{bmatrix} \cos bt & -\sin bt \\ \sin bt & \cos bt \end{bmatrix}$$

respectively. And by Proposition 1, the matrix e^{At} is then given by

$$e^{At} = Pe^{Bt}P^{-1}.$$

As we shall see in Section 1.4, finding the matrix e^{At} is equivalent to solving the linear system (1) in Section 1.1.

Problem Set 3

1. Compute the operator norm of the linear transformation defined by the following matrices:

   (a) $\begin{bmatrix} 2 & 0 \\ 0 & -3 \end{bmatrix}$

   (b) $\begin{bmatrix} 1 & 2 \\ 0 & -1 \end{bmatrix}$

   (c) $\begin{bmatrix} 1 & 0 \\ 5 & 1 \end{bmatrix}.$

   Hint: In (c) maximize $|A\mathbf{x}|^2 = 26x_1^2 + 10x_1x_2 + x_2^2$ subject to the constraint $x_1^2 + x_2^2 = 1$ and use the result of Problem 2; or use the fact that $\|A\| = [\text{Max eigenvalue of } A^T A]^{1/2}$. Follow this same hint for (b).

2. Show that the operator norm of a linear transformation T on $\mathbf{R}^n$ satisfies

$$\|T\| = \max_{|\mathbf{x}|=1} |T(\mathbf{x})| = \sup_{\mathbf{x}\neq 0} \frac{|T(\mathbf{x})|}{|\mathbf{x}|}.$$

3. Use the lemma in this section to show that if T is an invertible linear transformation then $\|T\| > 0$ and

$$\|T^{-1}\| \geq \frac{1}{\|T\|}.$$

4. If T is a linear transformation on $\mathbf{R}^n$ with $\|T - I\| < 1$, prove that T is invertible and that the series $\sum_{k=0}^{\infty}(I - T)^k$ converges absolutely to T^{-1}.

   Hint: Use the geometric series.

5. Compute the exponentials of the following matrices:

   (a) $\begin{bmatrix} 2 & 0 \\ 0 & -3 \end{bmatrix}$ (b) $\begin{bmatrix} 1 & 2 \\ 0 & -1 \end{bmatrix}$ (c) $\begin{bmatrix} 1 & 0 \\ 5 & 1 \end{bmatrix}$

(d) $\begin{bmatrix} 5 & -6 \\ 3 & -4 \end{bmatrix}$ (e) $\begin{bmatrix} 2 & -1 \\ 1 & 2 \end{bmatrix}$ (f) $\begin{bmatrix} 0 & 1 \\ 1 & 0 \end{bmatrix}$.

6. (a) For each matrix in Problem 5 find the eigenvalues of e^A.

   (b) Show that if $\mathbf{x}$ is an eigenvector of A corresponding to the eigenvalue λ, then $\mathbf{x}$ is also an eigenvector of e^A corresponding to the eigenvalue e^λ.

   (c) If $A = P \operatorname{diag}[\lambda_j] P^{-1}$, use Corollary 1 to show that
   $$\det e^A = e^{\operatorname{trace} A}.$$
   Also, using the results in the last paragraph of this section, show that this formula holds for any 2×2 matrix A.

7. Compute the exponentials of the following matrices:

   (a) $\begin{bmatrix} 1 & 0 & 0 \\ 0 & 2 & 0 \\ 0 & 0 & 3 \end{bmatrix}$ (b) $\begin{bmatrix} 1 & 0 & 0 \\ 0 & 2 & 1 \\ 0 & 0 & 2 \end{bmatrix}$ (c) $\begin{bmatrix} 2 & 0 & 0 \\ 1 & 2 & 0 \\ 0 & 1 & 2 \end{bmatrix}$.

   Hint: Write the matrices in (b) and (c) as a diagonal matrix S plus a matrix N. Show that S and N commute and compute e^S as in part (a) and e^N by using the definition.

8. Find 2×2 matrices A and B such that $e^{A+B} \neq e^A e^B$.

9. Let T be a linear operator on $\mathbf{R}^n$ that leaves a subspace $E \subset \mathbf{R}^n$ invariant; i.e., for all $\mathbf{x} \in E$, $T(\mathbf{x}) \in E$. Show that e^T also leaves E invariant.

1.4 The Fundamental Theorem for Linear Systems

Let A be an $n \times n$ matrix. In this section we establish the fundamental fact that for $\mathbf{x}_0 \in \mathbf{R}^n$ the initial value problem

$$\dot{\mathbf{x}} = A\mathbf{x}$$
$$\mathbf{x}(0) = \mathbf{x}_0 \tag{1}$$

has a unique solution for all $t \in \mathbf{R}$ which is given by

$$\mathbf{x}(t) = e^{At} \mathbf{x}_0. \tag{2}$$

Notice the similarity in the form of the solution (2) and the solution $x(t) = e^{at} x_0$ of the elementary first-order differential equation $\dot{x} = ax$ and initial condition $x(0) = x_0$.

1.4. The Fundamental Theorem for Linear Systems

In order to prove this theorem, we first compute the derivative of the exponential function e^{At} using the basic fact from analysis that two convergent limit processes can be interchanged if one of them converges uniformly. This is referred to as Moore's Theorem; cf. Graves [G], p. 100 or Rudin [R], p. 149.

Lemma. *Let A be a square matrix, then*

$$\frac{d}{dt}e^{At} = Ae^{At}.$$

Proof. Since A commutes with itself, it follows from Proposition 2 and Definition 2 in Section 3 that

$$\frac{d}{dt}e^{At} = \lim_{h \to 0} \frac{e^{A(t+h)} - e^{At}}{h}$$

$$= \lim_{h \to 0} e^{At} \frac{(e^{Ah} - I)}{h}$$

$$= e^{At} \lim_{h \to 0} \lim_{k \to \infty} \left(A + \frac{A^2 h}{2!} + \cdots + \frac{A^k h^{k-1}}{k!}\right)$$

$$= Ae^{At}.$$

The last equality follows since by the theorem in Section 1.3 the series defining e^{Ah} converges uniformly for $|h| \leq 1$ and we can therefore interchange the two limits.

Theorem (The Fundamental Theorem for Linear Systems). *Let A be an $n \times n$ matrix. Then for a given $\mathbf{x}_0 \in \mathbf{R}^n$, the initial value problem*

$$\dot{\mathbf{x}} = A\mathbf{x}$$
$$\mathbf{x}(0) = \mathbf{x}_0 \tag{1}$$

has a unique solution given by

$$\mathbf{x}(t) = e^{At}\mathbf{x}_0. \tag{2}$$

Proof. By the preceding lemma, if $\mathbf{x}(t) = e^{At}\mathbf{x}_0$, then

$$\mathbf{x}'(t) = \frac{d}{dt}e^{At}\mathbf{x}_0 = Ae^{At}\mathbf{x}_0 = A\mathbf{x}(t)$$

for all $t \in \mathbf{R}$. Also, $\mathbf{x}(0) = I\mathbf{x}_0 = \mathbf{x}_0$. Thus $\mathbf{x}(t) = e^{At}\mathbf{x}_0$ is a solution. To see that this is the only solution, let $\mathbf{x}(t)$ be any solution of the initial value problem (1) and set

$$\mathbf{y}(t) = e^{-At}\mathbf{x}(t).$$

Then from the above lemma and the fact that $\mathbf{x}(t)$ is a solution of (1)
$$\begin{aligned} \mathbf{y}'(t) &= -Ae^{-At}\mathbf{x}(t) + e^{-At}\mathbf{x}'(t) \\ &= -Ae^{-At}\mathbf{x}(t) + e^{-At}A\mathbf{x}(t) \\ &= 0 \end{aligned}$$

for all $t \in \mathbf{R}$ since e^{-At} and A commute. Thus, $\mathbf{y}(t)$ is a constant. Setting $t = 0$ shows that $\mathbf{y}(t) = \mathbf{x}_0$ and therefore any solution of the initial value problem (1) is given by $\mathbf{x}(t) = e^{At}\mathbf{y}(t) = e^{At}\mathbf{x}_0$. This completes the proof of the theorem.

Example. Solve the initial value problem
$$\dot{\mathbf{x}} = A\mathbf{x}$$
$$\mathbf{x}(0) = \begin{bmatrix} 1 \\ 0 \end{bmatrix}$$

for
$$A = \begin{bmatrix} -2 & -1 \\ 1 & -2 \end{bmatrix}$$

and sketch the solution curve in the phase plane $\mathbf{R}^2$. By the above theorem and Corollary 3 of the last section, the solution is given by

$$\mathbf{x}(t) = e^{At}\mathbf{x}_0 = e^{-2t}\begin{bmatrix} \cos t & -\sin t \\ \sin t & \cos t \end{bmatrix}\begin{bmatrix} 1 \\ 0 \end{bmatrix} = e^{-2t}\begin{bmatrix} \cos t \\ \sin t \end{bmatrix}.$$

It follows that $|\mathbf{x}(t)| = e^{-2t}$ and that the angle $\theta(t) = \tan^{-1}x_2(t)/x_1(t) = t$. The solution curve therefore spirals into the origin as shown in Figure 1 below.

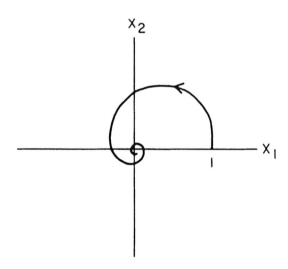

Figure 1

1.4. The Fundamental Theorem for Linear Systems

PROBLEM SET 4

1. Use the forms of the matrix e^{Bt} computed in Section 1.3 and the theorem in this section to solve the linear system $\dot{x} = Bx$ for

   (a) $B = \begin{bmatrix} \lambda & 0 \\ 0 & \mu \end{bmatrix}$

   (b) $B = \begin{bmatrix} \lambda & 1 \\ 0 & \lambda \end{bmatrix}$

   (c) $B = \begin{bmatrix} a & -b \\ b & a \end{bmatrix}$.

2. Solve the following linear system and sketch its phase portrait

   $$\dot{x} = \begin{bmatrix} -1 & -1 \\ 1 & -1 \end{bmatrix} x.$$

   The origin is called a stable focus for this system.

3. Find e^{At} and solve the linear system $\dot{x} = Ax$ for

   (a) $A = \begin{bmatrix} 3 & 1 \\ 1 & 3 \end{bmatrix}$

   (b) $A = \begin{bmatrix} 1 & 3 \\ 3 & 1 \end{bmatrix}$.

   Cf. Problem 1 in Problem Set 2.

4. Given

   $$A = \begin{bmatrix} 1 & 0 & 0 \\ 1 & 2 & 0 \\ 1 & 0 & -1 \end{bmatrix}.$$

   Compute the 3×3 matrix e^{At} and solve $\dot{x} = Ax$. Cf. Problem 2 in Problem Set 2.

5. Find the solution of the linear system $\dot{x} = Ax$ where

   (a) $A = \begin{bmatrix} 2 & -1 \\ 0 & 2 \end{bmatrix}$

   (b) $A = \begin{bmatrix} 2 & -1 \\ 1 & 2 \end{bmatrix}$

   (c) $A = \begin{bmatrix} 0 & 1 \\ 1 & 0 \end{bmatrix}$

   (d) $A = \begin{bmatrix} -2 & 0 & 0 \\ 1 & -2 & 0 \\ 0 & 1 & -2 \end{bmatrix}$.

6. Let T be a linear transformation on $\mathbf{R}^n$ that leaves a subspace $E \subset \mathbf{R}^n$ invariant (i.e., for all $\mathbf{x} \in E$, $T(\mathbf{x}) \in E$) and let $T(\mathbf{x}) = A\mathbf{x}$ with respect to the standard basis for $\mathbf{R}^n$. Show that if $\mathbf{x}(t)$ is the solution of the initial value problem

$$\dot{\mathbf{x}} = A\mathbf{x}$$
$$\mathbf{x}(0) = \mathbf{x}_0$$

with $\mathbf{x}_0 \in E$, then $\mathbf{x}(t) \in E$ for all $t \in \mathbf{R}$.

7. Suppose that the square matrix A has a negative eigenvalue. Show that the linear system $\dot{\mathbf{x}} = A\mathbf{x}$ has at least one nontrivial solution $\mathbf{x}(t)$ that satisfies

$$\lim_{t \to \infty} \mathbf{x}(t) = 0.$$

8. (Continuity with respect to initial conditions.) Let $\phi(t, \mathbf{x}_0)$ be the solution of the initial value problem (1). Use the Fundamental Theorem to show that for each fixed $t \in \mathbf{R}$

$$\lim_{\mathbf{y} \to \mathbf{x}_0} \phi(t, \mathbf{y}) = \phi(t, \mathbf{x}_0).$$

1.5 Linear Systems in $\mathbf{R}^2$

In this section we discuss the various phase portraits that are possible for the linear system

$$\dot{\mathbf{x}} = A\mathbf{x} \qquad (1)$$

when $\mathbf{x} \in \mathbf{R}^2$ and A is a 2×2 matrix. We begin by describing the phase portraits for the linear system

$$\dot{\mathbf{x}} = B\mathbf{x} \qquad (2)$$

where the matrix $B = P^{-1}AP$ has one of the forms given at the end of Section 1.3. The phase portrait for the linear system (1) above is then obtained from the phase portrait for (2) under the linear transformation of coordinates $\mathbf{x} = P\mathbf{y}$ as in Figures 1 and 2 in Section 1.2.

First of all, if

$$B = \begin{bmatrix} \lambda & 0 \\ 0 & \mu \end{bmatrix}, \quad B = \begin{bmatrix} \lambda & 1 \\ 0 & \lambda \end{bmatrix}, \quad \text{or} \quad B = \begin{bmatrix} a & -b \\ b & a \end{bmatrix},$$

it follows from the fundamental theorem in Section 1.4 and the form of the matrix e^{Bt} computed in Section 1.3 that the solution of the initial value problem (2) with $\mathbf{x}(0) = \mathbf{x}_0$ is given by

$$\mathbf{x}(t) = \begin{bmatrix} e^{\lambda t} & 0 \\ 0 & e^{\mu t} \end{bmatrix} \mathbf{x}_0, \quad \mathbf{x}(t) = e^{\lambda t} \begin{bmatrix} 1 & t \\ 0 & 1 \end{bmatrix} \mathbf{x}_0,$$

1.5. Linear Systems in $\mathbf{R}^2$

or
$$\mathbf{x}(t) = e^{at} \begin{bmatrix} \cos bt & -\sin bt \\ \sin bt & \cos bt \end{bmatrix} \mathbf{x}_0$$

respectively. We now list the various phase portraits that result from these solutions, grouped according to their topological type with a finer classification of sources and sinks into various types of unstable and stable nodes and foci:

Case I. $B = \begin{bmatrix} \lambda & 0 \\ 0 & \mu \end{bmatrix}$ with $\lambda < 0 < \mu$.

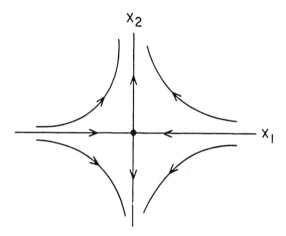

Figure 1. A saddle at the origin.

The phase portrait for the linear system (2) in this case is given in Figure 1. See the first example in Section 1.1. The system (2) is said to have a *saddle at the origin* in this case. If $\mu < 0 < \lambda$, the arrows in Figure 1 are reversed. Whenever A has two real eigenvalues of opposite sign, $\lambda < 0 < \mu$, the phase portrait for the linear system (1) is linearly equivalent to the phase portrait shown in Figure 1; i.e., it is obtained from Figure 1 by a linear transformation of coordinates; and the stable and unstable subspaces of (1) are determined by the eigenvectors of A as in the Example in Section 1.2. The four non-zero trajectories or solution curves that approach the equilibrium point at the origin as $t \to \pm\infty$ are called *separatrices* of the system.

Case II. $B = \begin{bmatrix} \lambda & 0 \\ 0 & \mu \end{bmatrix}$ with $\lambda \leq \mu < 0$ or $B = \begin{bmatrix} \lambda & 1 \\ 0 & \lambda \end{bmatrix}$ with $\lambda < 0$.

The phase portraits for the linear system (2) in these cases are given in Figure 2. Cf. the phase portraits in Problems 1(a), (b) and (c) of Problem Set 1 respectively. The origin is referred to as a *stable node* in each of these cases. It is called a proper node in the first case with $\lambda = \mu$ and an improper node in the other two cases. If $\lambda \geq \mu > 0$ or if $\lambda > 0$ in Case II, the arrows in Figure 2 are reversed and the origin is referred to as an unstable node. Whenever A has two negative eigenvalues $\lambda \leq \mu < 0$, the phase portrait of the linear system (1) is linearly equivalent to one of the phase portraits shown in Figure 2. The stability of the node is determined by the sign of the eigenvalues: *stable* if $\lambda \leq \mu < 0$ and *unstable* if $\lambda \geq \mu > 0$. Note that each trajectory in Figure 2 approaches the equilibrium point at the origin along a well-defined tangent line $\theta = \theta_0$, determined by an eigenvector of A, as $t \to \infty$.

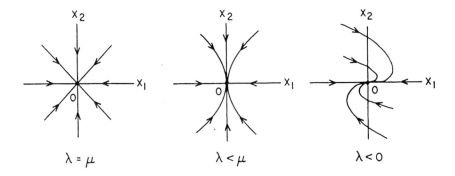

Figure 2. A stable node at the origin.

Case III. $B = \begin{bmatrix} a & -b \\ b & a \end{bmatrix}$ with $a < 0$.

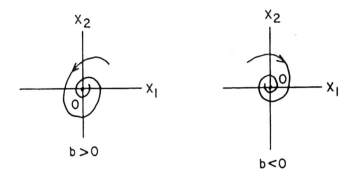

Figure 3. A stable focus at the origin.

The phase portrait for the linear system (2) in this case is given in Figure 3. Cf. Problem 9. The origin is referred to as a *stable focus* in these cases. If $a > 0$, the trajectories spiral away from the origin with increasing

1.5. Linear Systems in $\mathbf{R}^2$

t and the origin is called an unstable focus. Whenever A has a pair of complex conjugate eigenvalues with nonzero real part, $a \pm ib$, with $a < 0$, the phase portraits for the system (1) is linearly equivalent to one of the phase portraits shown in Figure 3. Note that the trajectories in Figure 3 do not approach the origin along well defined tangent lines; i.e., the angle $\theta(t)$ that the vector $\mathbf{x}(t)$ makes with the x_1-axis does not approach a constant θ_0 as $t \to \infty$, but rather $|\theta(t)| \to \infty$ as $t \to \infty$ and $|\mathbf{x}(t)| \to 0$ as $t \to \infty$ in this case.

Case IV. $B = \begin{bmatrix} 0 & -b \\ b & 0 \end{bmatrix}$

The phase portrait for the linear system (2) in this case is given in Figure 4. Cf. Problem 1(d) in Problem Set 1. The system (2) is said to have a *center at the origin* in this case. Whenever A has a pair of pure imaginary complex conjugate eigenvalues, $\pm ib$, the phase portrait of the linear system (1) is linearly equivalent to one of the phase portraits shown in Figure 4. Note that the trajectories or solution curves in Figure 4 lie on circles $|\mathbf{x}(t)|$ = constant. In general, the trajectories of the system (1) will lie on ellipses and the solution $\mathbf{x}(t)$ of (1) will satisfy $m \leq |\mathbf{x}(t)| \leq M$ for all $t \in \mathbf{R}$; cf. the following Example. The angle $\theta(t)$ also satisfies $|\theta(t)| \to \infty$ as $t \to \infty$ in this case.

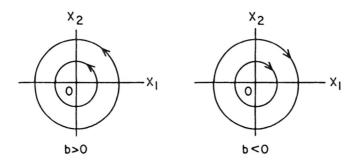

Figure 4. A center at the origin.

If one (or both) of the eigenvalues of A is zero, i.e., if $\det A = 0$, the origin is called a *degenerate equilibrium point of* (1). The various portraits for the linear system (1) are determined in Problem 4 in this case.

Example (A linear system with a center at the origin). The linear system

$$\dot{\mathbf{x}} = A\mathbf{x}$$

with

$$A = \begin{bmatrix} 0 & -4 \\ 1 & 0 \end{bmatrix}$$

has a center at the origin since the matrix A has eigenvalues $\lambda = \pm 2i$. According to the theorem in Section 1.6, the invertible matrix

$$P = \begin{bmatrix} 2 & 0 \\ 0 & 1 \end{bmatrix} \text{ with } P^{-1} = \begin{bmatrix} 1/2 & 0 \\ 0 & 1 \end{bmatrix}$$

reduces A to the matrix

$$B = P^{-1}AP = \begin{bmatrix} 0 & -2 \\ 2 & 0 \end{bmatrix}.$$

The student should verify the calculation.

The solution to the linear system $\dot{\mathbf{x}} = A\mathbf{x}$, as determined by Sections 1.3 and 1.4, is then given by

$$\mathbf{x}(t) = P \begin{bmatrix} \cos 2t & -\sin 2t \\ \sin 2t & \cos 2t \end{bmatrix} P^{-1}\mathbf{c} = \begin{bmatrix} \cos 2t & -2\sin 2t \\ 1/2 \sin 2t & \cos 2t \end{bmatrix} \mathbf{c}$$

where $\mathbf{c} = \mathbf{x}(0)$, or equivalently by

$$x_1(t) = c_1 \cos 2t - 2c_2 \sin 2t$$
$$x_2(t) = 1/2 c_1 \sin 2t + c_2 \cos 2t.$$

It is then easily shown that the solutions satisfy

$$x_1^2(t) + 4x_2^2(t) = c_1^2 + 4c_2^2$$

for all $t \in \mathbf{R}$; i.e., the trajectories of this system lie on ellipses as shown in Figure 5.

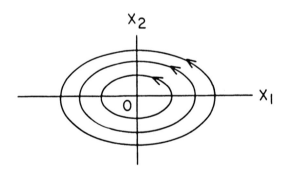

Figure 5. A center at the origin.

Definition 1. The linear system (1) is said to have *a saddle, a node, a focus* or *a center at the origin* if the matrix A is similar to one of the matrices B in Cases I, II, III or IV respectively, i.e., if its phase portrait

1.5. Linear Systems in $\mathbf{R}^2$

is linearly equivalent to one of the phase portraits in Figures 1, 2, 3 or 4 respectively.

Remark. If the matrix A is similar to the matrix B, i.e., if there is a nonsingular matrix P such that $P^{-1}AP = B$, then the system (1) is transformed into the system (2) by the linear transformation of coordinates $\mathbf{x} = P\mathbf{y}$. If B has the form III, then the phase portrait for the system (2) consists of either a counterclockwise motion (if $b > 0$) or a clockwise motion (if $b < 0$) on either circles (if $a = 0$) or spirals (if $a \neq 0$). Furthermore, the direction of rotation of trajectories in the phase portraits for the systems (1) and (2) will be the same if $\det P > 0$ (i.e., if P is orientation preserving) and it will be opposite if $\det P < 0$ (i.e., if P is orientation reversing). In either case, the two systems (1) and (2) are topologically equivalent in the sense of Definition 1 in Section 2.8 of Chapter 2.

For $\det A \neq 0$ there is an easy method for determining if the linear system has a saddle, node, focus or center at the origin. This is given in the next theorem. Note that if $\det A \neq 0$ then $A\mathbf{x} = \mathbf{0}$ iff $\mathbf{x} = \mathbf{0}$; i.e., the origin is the only equilibrium point of the linear system (1) when $\det A \neq 0$. If the origin is a focus or a center, the sign σ of $\dot{x}_2$ for $x_2 = 0$ (and for small $x_1 > 0$) can be used to determine whether the motion is counterclockwise (if $\sigma > 0$) or clockwise (if $\sigma < 0$).

Theorem. *Let* $\delta = \det A$ *and* $\tau = \text{trace } A$ *and consider the linear system*

$$\dot{\mathbf{x}} = A\mathbf{x}. \quad (1)$$

(a) *If* $\delta < 0$ *then* (1) *has a saddle at the origin.*

(b) *If* $\delta > 0$ *and* $\tau^2 - 4\delta \geq 0$ *then* (1) *has a node at the origin; it is stable if* $\tau < 0$ *and unstable if* $\tau > 0$.

(c) *If* $\delta > 0$, $\tau^2 - 4\delta < 0$, *and* $\tau \neq 0$ *then* (1) *has a focus at the origin; it is stable if* $\tau < 0$ *and unstable if* $\tau > 0$.

(d) *If* $\delta > 0$ *and* $\tau = 0$ *then* (1) *has a center at the origin.*

Note that in case (b), $\tau^2 \geq 4|\delta| > 0$; i.e., $\tau \neq 0$.

Proof. The eigenvalues of the matrix A are given by

$$\lambda = \frac{\tau \pm \sqrt{\tau^2 - 4\delta}}{2}$$

Thus (a) if $\delta < 0$ there are two real eigenvalues of opposite sign.

(b) If $\delta > 0$ and $\tau^2 - 4\delta \geq 0$ then there are two real eigenvalues of the same sign as τ;

(c) if $\delta > 0$, $\tau^2 - 4\delta < 0$ and $\tau \neq 0$ then there are two complex conjugate eigenvalues $\lambda = a \pm ib$ and, as will be shown in Section 1.6, A is similar to the matrix B in Case III above with $a = \tau/2$; and

(d) if $\delta > 0$ and $\tau = 0$ then there are two pure imaginary complex conjugate eigenvalues. Thus, cases a, b, c and d correspond to the Cases I, II, III and IV discussed above and we have a saddle, node, focus or center respectively.

Definition 2. A stable node or focus of (1) is called a *sink* of the linear system and an unstable node or focus of (1) is called a *source* of the linear system.

The above results can be summarized in a "bifurcation diagram," shown in Figure 6, which separates the (τ, δ)-plane into three components in which the solutions of the linear system (1) have the same "qualitative structure" (defined in Section 2.8 of Chapter 2). In describing the topological behavior or qualitative structure of the solution set of a linear system, we do not distinguish between nodes and foci, but only if they are stable or unstable. There are eight different topological types of behavior that are possible for a linear system according to whether $\delta \neq 0$ and it has a source, a sink, a center or a saddle or whether $\delta = 0$ and it has one of the four types of behavior determined in Problem 4.

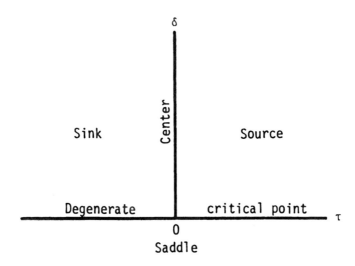

Figure 6. A bifurcation diagram for the linear system (1).

PROBLEM SET 5

1. Use the theorem in this section to determine if the linear system $\dot{\mathbf{x}} = A\mathbf{x}$ has a saddle, node, focus or center at the origin and determine the stability of each node or focus:

(a) $A = \begin{bmatrix} 1 & 2 \\ 3 & 4 \end{bmatrix}$ (d) $A = \begin{bmatrix} 1 & -1 \\ 2 & 3 \end{bmatrix}$

1.5. Linear Systems in $\mathbf{R}^2$

(b) $A = \begin{bmatrix} 3 & 1 \\ 1 & 3 \end{bmatrix}$ (e) $A = \begin{bmatrix} \lambda & -2 \\ 1 & \lambda \end{bmatrix}$

(c) $A = \begin{bmatrix} 0 & -1 \\ 2 & 0 \end{bmatrix}$ (f) $A = \begin{bmatrix} \lambda & 2 \\ 1 & \lambda \end{bmatrix}$.

2. Solve the linear system $\dot{\mathbf{x}} = A\mathbf{x}$ and sketch the phase portrait for

(a) $A = \begin{bmatrix} 3 & 0 \\ 0 & 3 \end{bmatrix}$ (c) $A = \begin{bmatrix} 1 & 0 \\ 0 & 3 \end{bmatrix}$.

(b) $A = \begin{bmatrix} 3 & 0 \\ 0 & 1 \end{bmatrix}$ (d) $A = \begin{bmatrix} 1 & 1 \\ 0 & 1 \end{bmatrix}$.

3. For what values of the parameters a and b does the linear system $\dot{\mathbf{x}} = A\mathbf{x}$ have a sink at the origin?

$$A = \begin{bmatrix} a & -b \\ b & 2 \end{bmatrix}.$$

4. If $\det A = 0$, then the origin is a degenerate critical point of $\dot{\mathbf{x}} = A\mathbf{x}$. Determine the solution and the corresponding phase portraits for the linear system with

(a) $A = \begin{bmatrix} \lambda & 0 \\ 0 & 0 \end{bmatrix}$

(b) $A = \begin{bmatrix} 0 & 1 \\ 0 & 0 \end{bmatrix}$

(c) $A = \begin{bmatrix} 0 & 0 \\ 0 & 0 \end{bmatrix}$.

Note that the origin is not an isolated equilibrium point in these cases. The four different phase portraits determined in (a) with $\lambda > 0$ or $\lambda < 0$, (b) and (c) above, together with the sources, sinks, centers and saddles discussed in this section, illustrate the eight different types of qualitative behavior that are possible for a linear system.

5. Write the second-order differential equation

$$\ddot{x} + a\dot{x} + bx = 0$$

as a system in $\mathbf{R}^2$ and determine the nature of the equilibrium point at the origin.

6. Find the general solution and draw the phase portrait for the linear system

$$\dot{x}_1 = x_1$$
$$\dot{x}_2 = -x_1 + 2x_2.$$

What role do the eigenvectors of the matrix A play in determining the phase portrait? Cf. Case II.

7. Describe the separatrices for the linear system

$$\dot{x}_1 = x_1 + 2x_2$$
$$\dot{x}_2 = 3x_1 + 4x_2.$$

Hint: Find the eigenspaces for A.

8. Determine the functions $r(t) = |\mathbf{x}(t)|$ and $\theta(t) = \tan^{-1} x_2(t)/x_1(t)$ for the linear system

$$\dot{x}_1 = -x_2$$
$$\dot{x}_2 = x_1$$

9. (*Polar Coordinates*) Given the linear system

$$\dot{x}_1 = ax_1 - bx_2$$
$$\dot{x}_2 = bx_1 + ax_2.$$

Differentiate the equations $r^2 = x_1^2 + x_2^2$ and $\theta = \tan^{-1}(x_2/x_1)$ with respect to t in order to obtain

$$\dot{r} = \frac{x_1 \dot{x}_1 + x_2 \dot{x}_2}{r} \quad \text{and} \quad \dot{\theta} = \frac{x_1 \dot{x}_2 - x_2 \dot{x}_1}{r^2}$$

for $r \neq 0$. For the linear system given above, show that these equations reduce to

$$\dot{r} = ar \quad \text{and} \quad \dot{\theta} = b.$$

Solve these equations with the initial conditions $r(0) = r_0$ and $\theta(0) = \theta_0$ and show that the phase portraits in Figures 3 and 4 follow immediately from your solution. (Polar coordinates are discussed more thoroughly in Section 2.10 of Chapter 2).

1.6 Complex Eigenvalues

If the $2n \times 2n$ real matrix A has complex eigenvalues, then they occur in complex conjugate pairs and if A has $2n$ distinct complex eigenvalues, the following theorem from linear algebra proved in Hirsch and Smale [H/S] allows us to solve the linear system

$$\dot{\mathbf{x}} = A\mathbf{x}.$$

Theorem. *If the $2n \times 2n$ real matrix A has $2n$ distinct complex eigenvalues $\lambda_j = a_j + ib_j$ and $\bar{\lambda}_j = a_j - ib_j$ and corresponding complex eigenvectors*

1.6. Complex Eigenvalues

$\mathbf{w}_j = \mathbf{u}_j + i\mathbf{v}_j$ and $\bar{\mathbf{w}}_j = \mathbf{u}_j - i\mathbf{v}_j, j = 1, \ldots, n$, then $\{\mathbf{u}_1, \mathbf{v}_1, \ldots, \mathbf{u}_n, \mathbf{v}_n\}$ is a basis for $\mathbf{R}^{2n}$, the matrix

$$P = [\mathbf{v}_1 \quad \mathbf{u}_1 \quad \mathbf{v}_2 \quad \mathbf{u}_2 \quad \cdots \quad \mathbf{v}_n \quad \mathbf{u}_n]$$

is invertible and

$$P^{-1}AP = \operatorname{diag}\begin{bmatrix} a_j & -b_j \\ b_j & a_j \end{bmatrix},$$

a real $2n \times 2n$ matrix with 2×2 blocks along the diagonal.

Remark. Note that if instead of the matrix P we use the invertible matrix

$$Q = [\mathbf{u}_1 \quad \mathbf{v}_1 \quad \mathbf{u}_2 \quad \mathbf{v}_2 \quad \cdots \quad \mathbf{u}_n \quad \mathbf{v}_n]$$

then

$$Q^{-1}AQ = \operatorname{diag}\begin{bmatrix} a_j & b_j \\ -b_j & a_j \end{bmatrix}.$$

The next corollary then follows from the above theorem and the fundamental theorem in Section 1.4.

Corollary. *Under the hypotheses of the above theorem, the solution of the initial value problem*

$$\dot{\mathbf{x}} = A\mathbf{x} \qquad (1)$$
$$\mathbf{x}(0) = \mathbf{x}_0$$

is given by

$$\mathbf{x}(t) = P \operatorname{diag} e^{a_j t} \begin{bmatrix} \cos b_j t & -\sin b_j t \\ \sin b_j t & \cos b_j t \end{bmatrix} P^{-1}\mathbf{x}_0.$$

Note that the matrix

$$R = \begin{bmatrix} \cos bt & -\sin bt \\ \sin bt & \cos bt \end{bmatrix}$$

represents a rotation through bt radians.

Example. Solve the initial value problem (1) for

$$A = \begin{bmatrix} 1 & -1 & 0 & 0 \\ 1 & 1 & 0 & 0 \\ 0 & 0 & 3 & -2 \\ 0 & 0 & 1 & 1 \end{bmatrix}.$$

The matrix A has the complex eigenvalues $\lambda_1 = 1+i$ and $\lambda_2 = 2+i$ (as well as $\bar{\lambda}_1 = 1-i$ and $\bar{\lambda}_2 = 2-i$). A corresponding pair of complex eigenvectors is

$$\mathbf{w}_1 = \mathbf{u}_1 + i\mathbf{v}_1 = \begin{bmatrix} i \\ 1 \\ 0 \\ 0 \end{bmatrix} \text{ and } \mathbf{w}_2 = \mathbf{u}_2 + i\mathbf{v}_2 = \begin{bmatrix} 0 \\ 0 \\ 1+i \\ 1 \end{bmatrix}.$$

The matrix
$$P = [\mathbf{v}_1 \ \mathbf{u}_1 \ \mathbf{v}_2 \ \mathbf{u}_2] = \begin{bmatrix} 1 & 0 & 0 & 0 \\ 0 & 1 & 0 & 0 \\ 0 & 0 & 1 & 1 \\ 0 & 0 & 0 & 1 \end{bmatrix}$$
is invertible,
$$P^{-1} = \begin{bmatrix} 1 & 0 & 0 & 0 \\ 0 & 1 & 0 & 0 \\ 0 & 0 & 1 & -1 \\ 0 & 0 & 0 & 1 \end{bmatrix},$$
and
$$P^{-1}AP = \begin{bmatrix} 1 & -1 & 0 & 0 \\ 1 & 1 & 0 & 0 \\ 0 & 0 & 2 & -1 \\ 0 & 0 & 1 & 2 \end{bmatrix}.$$

The solution to the initial value problem (1) is given by

$$\mathbf{x}(t) = P \begin{bmatrix} e^t \cos t & -e^t \sin t & 0 & 0 \\ e^t \sin t & e^t \cos t & 0 & 0 \\ 0 & 0 & e^{2t} \cos t & -e^{2t} \sin t \\ 0 & 0 & e^{2t} \sin t & e^{2t} \cos t \end{bmatrix} P^{-1} \mathbf{x}_0$$

$$= \begin{bmatrix} e^t \cos t & -e^t \sin t & 0 & 0 \\ e^t \sin t & e^t \cos t & 0 & 0 \\ 0 & 0 & e^{2t}(\cos t + \sin t) & -2e^{2t} \sin t \\ 0 & 0 & e^{2t} \sin t & e^{2t}(\cos t - \sin t) \end{bmatrix} \mathbf{x}_0.$$

In case A has both real and complex eigenvalues and they are distinct, we have the following result: If A has distinct real eigenvalues λ_j and corresponding eigenvectors $\mathbf{v}_j, j = 1, \ldots, k$ and distinct complex eigenvalues $\lambda_j = a_j + ib_j$ and $\bar{\lambda}_j = a_j - ib_j$ and corresponding eigenvectors $\mathbf{w}_j = \mathbf{u}_j + i\mathbf{v}_j$ and $\bar{\mathbf{w}}_j = \mathbf{u}_j - i\mathbf{v}_j, j = k+1, \ldots, n$, then the matrix

$$P = [\mathbf{v}_1 \ \cdots \ \mathbf{v}_k \ \mathbf{v}_{k+1} \ \mathbf{u}_{k+1} \ \cdots \ \mathbf{v}_n \ \mathbf{u}_n]$$

is invertible and

$$P^{-1}AP = \operatorname{diag}[\lambda_1, \ldots, \lambda_k, B_{k+1}, \ldots, B_n]$$

where the 2×2 blocks

$$B_j = \begin{bmatrix} a_j & -b_j \\ b_j & a_j \end{bmatrix}$$

for $j = k+1, \ldots, n$. We illustrate this result with an example.

Example. The matrix
$$A = \begin{bmatrix} -3 & 0 & 0 \\ 0 & 3 & -2 \\ 0 & 1 & 1 \end{bmatrix}$$

1.6. Complex Eigenvalues

has eigenvalues $\lambda_1 = -3, \lambda_2 = 2 + i$ (and $\bar{\lambda}_2 = 2 - i$). The corresponding eigenvectors

$$\mathbf{v}_1 = \begin{bmatrix} 1 \\ 0 \\ 0 \end{bmatrix} \quad \text{and} \quad \mathbf{w}_2 = \mathbf{u}_2 + i\mathbf{v}_2 = \begin{bmatrix} 0 \\ 1 + i \\ 1 \end{bmatrix}.$$

Thus

$$P = \begin{bmatrix} 1 & 0 & 0 \\ 0 & 1 & 1 \\ 0 & 0 & 1 \end{bmatrix}, \quad P^{-1} = \begin{bmatrix} 1 & 0 & 0 \\ 0 & 1 & -1 \\ 0 & 0 & 1 \end{bmatrix}$$

and

$$P^{-1}AP = \begin{bmatrix} -3 & 0 & 0 \\ 0 & 2 & -1 \\ 0 & 1 & 2 \end{bmatrix}.$$

The solution of the initial value problem (1) is given by

$$\mathbf{x}(t) = P \begin{bmatrix} e^{-3t} & 0 & 0 \\ 0 & e^{2t}\cos t & -e^{2t}\sin t \\ 0 & e^{2t}\sin t & e^{2t}\cos t \end{bmatrix} P^{-1}\mathbf{x}_0$$

$$= \begin{bmatrix} e^{-3t} & 0 & 0 \\ 0 & e^{2t}(\cos t + \sin t) & -2e^{2t}\sin t \\ 0 & e^{2t}\sin t & e^{2t}(\cos t - \sin t) \end{bmatrix} \mathbf{x}_0.$$

The stable subspace E^s is the x_1-axis and the unstable subspace E^u is the x_2, x_3 plane. The phase portrait is given in Figure 1.

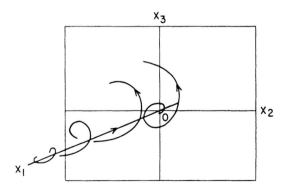

Figure 1

Problem Set 6

1. Solve the initial value problem (1) with
$$A = \begin{bmatrix} 3 & -2 \\ 1 & 1 \end{bmatrix}.$$

2. Solve the initial value problem (1) with
$$A = \begin{bmatrix} 0 & -2 & 0 \\ 1 & 2 & 0 \\ 0 & 0 & -2 \end{bmatrix}.$$

Determine the stable and unstable subspaces and sketch the phase portrait.

3. Solve the initial value problem (1) with
$$A = \begin{bmatrix} 1 & 0 & 0 \\ 0 & 2 & -3 \\ 1 & 3 & 2 \end{bmatrix}.$$

4. Solve the initial value problem (1) with
$$A = \begin{bmatrix} -1 & -1 & 0 & 0 \\ 1 & -1 & 0 & 0 \\ 0 & 0 & 0 & -2 \\ 0 & 0 & 1 & 2 \end{bmatrix}.$$

1.7 Multiple Eigenvalues

The fundamental theorem for linear systems in Section 1.4 tells us that the solution of the linear system
$$\dot{\mathbf{x}} = A\mathbf{x} \tag{1}$$
together with the initial condition $\mathbf{x}(0) = \mathbf{x}_0$ is given by
$$\mathbf{x}(t) = e^{At}\mathbf{x}_0.$$

We have seen how to find the $n \times n$ matrix e^{At} when A has distinct eigenvalues. We now complete the picture by showing how to find e^{At}, i.e., how to solve the linear system (1), when A has multiple eigenvalues.

Definition 1. Let λ be an eigenvalue of the $n \times n$ matrix A of multiplicity $m \leq n$. Then for $k = 1, \ldots, m$, any nonzero solution $\mathbf{v}$ of
$$(A - \lambda I)^k \mathbf{v} = 0$$
is called a *generalized eigenvector of A*.

1.7. Multiple Eigenvalues

Definition 2. An $n \times n$ matrix N is said to be *nilpotent of order k* if $N^{k-1} \neq 0$ and $N^k = 0$.

The following theorem is proved, for example, in Appendix III of Hirsch and Smale [II/S].

Theorem 1. *Let A be a real $n \times n$ matrix with real eigenvalues $\lambda_1, \ldots, \lambda_n$ repeated according to their multiplicity. Then there exists a basis of generalized eigenvectors for $\mathbf{R}^n$. And if $\{\mathbf{v}_1, \ldots, \mathbf{v}_n\}$ is any basis of generalized eigenvectors for $\mathbf{R}^n$, the matrix $P = [\mathbf{v}_1 \cdots \mathbf{v}_n]$ is invertible,*

$$A = S + N$$

where

$$P^{-1}SP = \operatorname{diag}[\lambda_j],$$

the matrix $N = A - S$ is nilpotent of order $k \leq n$, and S and N commute, i.e., $SN = NS$.

This theorem together with the propositions in Section 1.3 and the fundamental theorem in Section 1.4 then lead to the following result:

Corollary 1. *Under the hypotheses of the above theorem, the linear system (1), together with the initial condition $\mathbf{x}(0) = \mathbf{x}_0$, has the solution*

$$\mathbf{x}(t) = P \operatorname{diag}[e^{\lambda_j t}] P^{-1} \left[I + Nt + \cdots + \frac{N^{k-1} t^{k-1}}{(k-1)!} \right] \mathbf{x}_0.$$

If λ is an eigenvalue of multiplicity n of an $n \times n$ matrix A, then the above results are particularly easy to apply since in this case

$$S = \operatorname{diag}[\lambda]$$

with respect to the usual basis for $\mathbf{R}^n$ and

$$N = A - S.$$

The solution to the initial value problem (1) together with $\mathbf{x}(0) = \mathbf{x}_0$ is therefore given by

$$\mathbf{x}(t) = e^{\lambda t} \left[I + Nt + \cdots + \frac{N^k t^k}{k!} \right] \mathbf{x}_0.$$

Let us consider two examples where the $n \times n$ matrix A has an eigenvalue of multiplicity n. In these examples, we do not need to compute a basis of generalized eigenvectors to solve the initial value problem!

Example 1. Solve the initial value problem for (1) with

$$A = \begin{bmatrix} 3 & 1 \\ -1 & 1 \end{bmatrix}.$$

It is easy to determine that A has an eigenvalue $\lambda = 2$ of multiplicity 2; i.e., $\lambda_1 = \lambda_2 = 2$. Thus,

$$S = \begin{bmatrix} 2 & 0 \\ 0 & 2 \end{bmatrix}$$

and

$$N = A - S = \begin{bmatrix} 1 & 1 \\ -1 & -1 \end{bmatrix}.$$

It is easy to compute $N^2 = 0$ and the solution of the initial value problem for (1) is therefore given by

$$\mathbf{x}(t) = e^{At}\mathbf{x}_0 = e^{2t}[I + Nt]\mathbf{x}_0$$
$$= e^{2t}\begin{bmatrix} 1+t & t \\ -t & 1-t \end{bmatrix}\mathbf{x}_0.$$

Example 2. Solve the initial value problem for (1) with

$$A = \begin{bmatrix} 0 & -2 & -1 & -1 \\ 1 & 2 & 1 & 1 \\ 0 & 1 & 1 & 0 \\ 0 & 0 & 0 & 1 \end{bmatrix}.$$

In this case, the matrix A has an eigenvalue $\lambda = 1$ of multiplicity 4. Thus, $S = I_4$,

$$N = A - S = \begin{bmatrix} -1 & -2 & -1 & -1 \\ 1 & 1 & 1 & 1 \\ 0 & 1 & 0 & 0 \\ 0 & 0 & 0 & 0 \end{bmatrix}$$

and it is easy to compute

$$N^2 = \begin{bmatrix} -1 & -1 & -1 & -1 \\ 0 & 0 & 0 & 0 \\ 1 & 1 & 1 & 1 \\ 0 & 0 & 0 & 0 \end{bmatrix}$$

and $N^3 = 0$; i.e., N is nilpotent of order 3. The solution of the initial value problem for (1) is therefore given by

$$\mathbf{x}(t) = e^t[I + Nt + N^2 t^2/2]\mathbf{x}_0$$
$$= e^t \begin{bmatrix} 1 - t - t^2/2 & -2t - t^2/2 & -t - t^2/2 & -t - t^2/2 \\ t & 1+t & t & t \\ t^2/2 & t + t^2/2 & 1 + t^2/2 & t^2/2 \\ 0 & 0 & 0 & 1 \end{bmatrix}\mathbf{x}_0.$$

1.7. Multiple Eigenvalues

In the general case, we must first determine a basis of generalized eigenvectors for $\mathbf{R}^n$, then compute $S = P\operatorname{diag}[\lambda_j]P^{-1}$ and $N = A - S$ according to the formulas in the above theorem, and then find the solution of the initial value problem for (1) as in the above corollary.

Example 3. Solve the initial value problem for (1) when

$$A = \begin{bmatrix} 1 & 0 & 0 \\ -1 & 2 & 0 \\ 1 & 1 & 2 \end{bmatrix}.$$

It is easy to see that A has the eigenvalues $\lambda_1 = 1, \lambda_2 = \lambda_3 = 2$. And it is not difficult to find the corresponding eigenvectors

$$\mathbf{v}_1 = \begin{bmatrix} 1 \\ 1 \\ -2 \end{bmatrix} \quad \text{and} \quad \mathbf{v}_2 = \begin{bmatrix} 0 \\ 0 \\ 1 \end{bmatrix}.$$

Nonzero multiples of these eigenvectors are the only eigenvectors of A corresponding to $\lambda_1 = 1$ and $\lambda_2 = \lambda_3 = 2$ respectively. We therefore must find one generalized eigenvector corresponding to $\lambda = 2$ and independent of $\mathbf{v}_2$ by solving

$$(A - 2I)^2 \mathbf{v} = \begin{bmatrix} 1 & 0 & 0 \\ 1 & 0 & 0 \\ -2 & 0 & 0 \end{bmatrix} \mathbf{v} = 0.$$

We see that we can choose $\mathbf{v}_3 = (0, 1, 0)^T$. Thus,

$$P = \begin{bmatrix} 1 & 0 & 0 \\ 1 & 0 & 1 \\ -2 & 1 & 0 \end{bmatrix} \quad \text{and} \quad P^{-1} = \begin{bmatrix} 1 & 0 & 0 \\ 2 & 0 & 1 \\ -1 & 1 & 0 \end{bmatrix}.$$

We then compute

$$S = P \begin{bmatrix} 1 & 0 & 0 \\ 0 & 2 & 0 \\ 0 & 0 & 2 \end{bmatrix} P^{-1} = \begin{bmatrix} 1 & 0 & 0 \\ -1 & 2 & 0 \\ 2 & 0 & 2 \end{bmatrix},$$

$$N = A - S = \begin{bmatrix} 0 & 0 & 0 \\ 0 & 0 & 0 \\ -1 & 1 & 0 \end{bmatrix},$$

and $N^2 = 0$. The solution is then given by

$$\mathbf{x}(t) = P \begin{bmatrix} e^t & 0 & 0 \\ 0 & e^{2t} & 0 \\ 0 & 0 & e^{2t} \end{bmatrix} P^{-1}[I + Nt]\mathbf{x}_0$$

$$= \begin{bmatrix} e^t & 0 & 0 \\ e^t - e^{2t} & e^{2t} & 0 \\ -2e^t + (2-t)e^{2t} & te^{2t} & e^{2t} \end{bmatrix} \mathbf{x}_0.$$

In the case of multiple complex eigenvalues, we have the following theorem also proved in Appendix III of Hirsch and Smale [H/S]:

Theorem 2. *Let A be a real $2n \times 2n$ matrix with complex eigenvalues $\lambda_j = a_j + ib_j$ and $\bar{\lambda}_j = a_j - ib_j$, $j = 1, \ldots, n$. Then there exists generalized complex eigenvectors $\mathbf{w}_j = \mathbf{u}_j + i\mathbf{v}_j$ and $\bar{\mathbf{w}}_j = \mathbf{u}_j - i\mathbf{v}_j$, $i = 1, \ldots, n$ such that $\{\mathbf{u}_1, \mathbf{v}_1, \ldots, \mathbf{u}_n, \mathbf{v}_n\}$ is a basis for $\mathbf{R}^{2n}$. For any such basis, the matrix $P = [\mathbf{v}_1 \ \mathbf{u}_1 \ \cdots \ \mathbf{v}_n \ \mathbf{u}_n]$ is invertible,*

$$A = S + N$$

where

$$P^{-1}SP = \mathrm{diag} \begin{bmatrix} a_j & -b_j \\ b_j & a_j \end{bmatrix},$$

the matrix $N = A - S$ is nilpotent of order $k \leq 2n$, and S and N commute.

The next corollary follows from the fundamental theorem in Section 1.4 and the results in Section 1.3:

Corollary 2. *Under the hypotheses of the above theorem, the solution of the initial value problem (1), together with $\mathbf{x}(0) = \mathbf{x}_0$, is given by*

$$\mathbf{x}(t) = P \,\mathrm{diag}\, e^{a_j t} \begin{bmatrix} \cos b_j t & -\sin b_j t \\ \sin b_j t & \cos b_j t \end{bmatrix} P^{-1} \left[I + \cdots + \frac{N^k t^k}{k!} \right] \mathbf{x}_0.$$

We illustrate these results with an example.

Example 4. Solve the initial value problem for (1) with

$$A = \begin{bmatrix} 0 & -1 & 0 & 0 \\ 1 & 0 & 0 & 0 \\ 0 & 0 & 0 & -1 \\ 2 & 0 & 1 & 0 \end{bmatrix}.$$

The matrix A has eigenvalues $\lambda = i$ and $\bar{\lambda} = -i$ of multiplicity 2. The equation

$$(A - \lambda I)\mathbf{w} = \begin{bmatrix} -i & -1 & 0 & 0 \\ 1 & -i & 0 & 0 \\ 0 & 0 & -i & -1 \\ 2 & 0 & 1 & -i \end{bmatrix} \begin{bmatrix} z_1 \\ z_2 \\ z_3 \\ z_4 \end{bmatrix} = 0$$

is equivalent to $z_1 = z_2 = 0$ and $z_3 = iz_4$. Thus, we have one eigenvector $\mathbf{w}_1 = (0, 0, i, 1)^T$. Also, the equation

$$(A - \lambda I)^2 \mathbf{w} = \begin{bmatrix} -2 & 2i & 0 & 0 \\ -2i & -2 & 0 & 0 \\ -2 & 0 & -2 & 2i \\ -4i & -2 & -2i & -2 \end{bmatrix} \begin{bmatrix} z_1 \\ z_2 \\ z_3 \\ z_4 \end{bmatrix} = 0$$

1.7. Multiple Eigenvalues

is equivalent to $z_1 = iz_2$ and $z_3 = iz_4 - z_1$. We therefore choose the generalized eigenvector $\mathbf{w}_2 = (i, 1, 0, 1)$. Then $\mathbf{u}_1 = (0, 0, 0, 1)^T$, $\mathbf{v}_1 = (0, 0, 1, 0)^T$, $\mathbf{u}_2 = (0, 1, 0, 1)^T$, $\mathbf{v}_2 = (1, 0, 0, 0)^T$, and according to the above theorem,

$$P = \begin{bmatrix} 0 & 0 & 1 & 0 \\ 0 & 0 & 0 & 1 \\ 1 & 0 & 0 & 0 \\ 0 & 1 & 0 & 1 \end{bmatrix}, \quad P^{-1} = \begin{bmatrix} 0 & 0 & 1 & 0 \\ 0 & -1 & 0 & 1 \\ 1 & 0 & 0 & 0 \\ 0 & 1 & 0 & 0 \end{bmatrix},$$

$$S = P \begin{bmatrix} 0 & -1 & 0 & 0 \\ 1 & 0 & 0 & 0 \\ 0 & 0 & 0 & -1 \\ 0 & 0 & 1 & 0 \end{bmatrix} P^{-1} = \begin{bmatrix} 0 & -1 & 0 & 0 \\ 1 & 0 & 0 & 0 \\ 0 & 0 & 1 & 0 & -1 \\ 1 & 0 & 1 & 0 \end{bmatrix},$$

$$N = A - S = \begin{bmatrix} 0 & 0 & 0 & 0 \\ 0 & 0 & 0 & 0 \\ 0 & -1 & 0 & 0 \\ 1 & 0 & 0 & 0 \end{bmatrix}$$

and $N^2 = 0$. Thus, the solution to the initial value problem

$$\mathbf{x}(t) = \begin{bmatrix} \cos t & -\sin t & 0 & 0 \\ \sin t & \cos t & 0 & 0 \\ 0 & 0 & \cos t & -\sin t \\ 0 & 0 & \sin t & \cos t \end{bmatrix} P^{-1}[I + Nt]\mathbf{x}_0$$

$$= \begin{bmatrix} \cos t & -\sin t & 0 & 0 \\ \sin t & \cos t & 0 & 0 \\ -t\sin t & \sin t - t\cos t & \cos t & -\sin t \\ \sin t + t\cos t & -t\sin t & \sin t & \cos t \end{bmatrix} \mathbf{x}_0.$$

Remark. If A has both real and complex repeated eigenvalues, a combination of the above two theorems can be used as in the result and example at the end of Section 1.6.

PROBLEM SET 7

1. Solve the initial value problem (1) with the matrix

   (a) $A = \begin{bmatrix} 0 & 1 \\ -1 & 2 \end{bmatrix}$

   (b) $A = \begin{bmatrix} 1 & -1 \\ 1 & 3 \end{bmatrix}$

   (c) $A = \begin{bmatrix} 1 & 1 \\ 0 & 1 \end{bmatrix}$

   (d) $A = \begin{bmatrix} 1 & 1 \\ 0 & -1 \end{bmatrix}$

2. Solve the initial value problem (1) with the matrix

(a) $A = \begin{bmatrix} 1 & 0 & 0 \\ 2 & 1 & 0 \\ 3 & 2 & 1 \end{bmatrix}$

(b) $A = \begin{bmatrix} -1 & 1 & -2 \\ 0 & -1 & 4 \\ 0 & 0 & 1 \end{bmatrix}$

(c) $A = \begin{bmatrix} 1 & 0 & 0 \\ -1 & 2 & 0 \\ 1 & 0 & 2 \end{bmatrix}$

(d) $A = \begin{bmatrix} 2 & 1 & 1 \\ 0 & 2 & 2 \\ 0 & 0 & 2 \end{bmatrix}$.

3. Solve the initial value problem (1) with the matrix

(a) $A = \begin{bmatrix} 0 & 0 & 0 & 0 \\ 1 & 0 & 0 & 1 \\ 1 & 0 & 0 & 1 \\ 0 & -1 & 1 & 0 \end{bmatrix}$

(b) $A = \begin{bmatrix} 2 & 0 & 0 & 0 \\ 1 & 2 & 0 & 0 \\ 0 & 1 & 2 & 0 \\ 0 & 0 & 1 & 2 \end{bmatrix}$

(c) $A = \begin{bmatrix} 1 & 1 & 1 & 1 \\ 2 & 2 & 2 & 2 \\ 3 & 3 & 3 & 3 \\ 4 & 4 & 4 & 4 \end{bmatrix}$

(d) $A = \begin{bmatrix} 0 & 1 & 0 & 0 \\ 1 & 0 & 0 & 0 \\ 0 & 0 & 1 & -1 \\ 0 & 0 & 1 & 1 \end{bmatrix}$

(e) $A = \begin{bmatrix} 1 & -1 & 0 & 0 \\ 1 & 1 & 0 & 0 \\ 0 & 0 & 1 & -1 \\ 0 & 0 & 1 & 1 \end{bmatrix}$

(f) $A = \begin{bmatrix} 1 & -1 & 1 & 0 \\ 1 & 1 & 0 & 1 \\ 0 & 0 & 1 & -1 \\ 0 & 0 & 1 & 1 \end{bmatrix}$.

4. The "Putzer Algorithm" given below is another method for computing e^{At} when we have multiple eigenvalues; cf. [W], p. 49.

$$e^{At} = r_1(t)I + r_2(t)P_1 + \cdots + r_n(t)P_{n-1}$$

where

$$P_1 = (A - \lambda_1 I), \quad P_2 = (A - \lambda_1 I)(A - \lambda_2 I), \ldots,$$
$$P_n = (A - \lambda_1 I) \cdots (A - \lambda_n I)$$

and $r_j(t)$, $j = 1, \ldots, n$, are the solutions of the first-order linear differential equations and initial conditions

$$r'_1 = \lambda_1 r_1 \text{ with } r_1(0) = 1,$$
$$r'_2 = \lambda_2 r_2 + r_1 \text{ with } r_2(0) = 0$$
$$\vdots$$
$$r'_n = \lambda_n r_n + r_{n-1} \text{ with } r_n(0) = 0.$$

Use the Putzer Algorithm to compute e^{At} for the matrix A given in

(a) Example 1

(b) Example 3

(c) Problem 2(c)

(d) Problem 3(b).

1.8 Jordan Forms

The Jordan canonical form of a matrix gives some insight into the form of the solution of a linear system of differential equations and it is used in proving some theorems later in the book. Finding the Jordan canonical form of a matrix A is not necessarily the best method for solving the related linear system since finding a basis of generalized eigenvectors which reduces A to its Jordan canonical form may be difficult. On the other hand, *any* basis of generalized eigenvectors can be used in the method described in the previous section. The Jordan canonical form, described in the next theorem, does result in a particularly simple form for the nilpotent part N of the matrix A and it is therefore useful in the theory of ordinary differential equations.

Theorem (The Jordan Canonical Form). *Let A be a real matrix with real eigenvalues λ_j, $j = 1, \ldots, k$ and complex eigenvalues $\lambda_j = a_j + ib_j$ and $\bar{\lambda}_j = a_j - ib_j$, $j = k+1, \ldots, n$. Then there exists a basis $\{\mathbf{v}_1, \ldots, \mathbf{v}_k, \mathbf{v}_{k+1}, \mathbf{u}_{k+1}, \ldots, \mathbf{v}_n, \mathbf{u}_n\}$ for $\mathbf{R}^{2n-k}$, where $\mathbf{v}_j$, $j = 1, \ldots, k$ and $\mathbf{w}_j$, $j =$*

$k+1,\ldots,n$ are generalized eigenvectors of A, $\mathbf{u}_j = \mathrm{Re}(\mathbf{w}_j)$ and $\mathbf{v}_j = \mathrm{Im}(\mathbf{w}_j)$ for $j = k+1,\ldots,n$, such that the matrix $P = [\mathbf{v}_1 \cdots \mathbf{v}_k \ \mathbf{v}_{k+1} \ \mathbf{u}_{k+1} \cdots \mathbf{v}_n \ \mathbf{u}_n]$ is invertible and

$$P^{-1}AP = \begin{bmatrix} B_1 & & \\ & \ddots & \\ & & B_r \end{bmatrix} \qquad (1)$$

where the elementary Jordan blocks $B = B_j, j = 1,\ldots,r$ are either of the form

$$B = \begin{bmatrix} \lambda & 1 & 0 & \cdots & 0 \\ 0 & \lambda & 1 & \cdots & 0 \\ \cdots & & & & \\ 0 & \cdots & & \lambda & 1 \\ 0 & \cdots & & 0 & \lambda \end{bmatrix} \qquad (2)$$

for λ one of the real eigenvalues of A or of the form

$$B = \begin{bmatrix} D & I_2 & 0 & \cdots & 0 \\ 0 & D & I_2 & \cdots & 0 \\ \cdots & & & & \\ 0 & \cdots & & D & I_2 \\ 0 & \cdots & & 0 & D \end{bmatrix} \qquad (3)$$

with

$$D = \begin{bmatrix} a & -b \\ b & a \end{bmatrix}, \quad I_2 = \begin{bmatrix} 1 & 0 \\ 0 & 1 \end{bmatrix} \quad \text{and} \quad 0 = \begin{bmatrix} 0 & 0 \\ 0 & 0 \end{bmatrix}$$

for $\lambda = a + ib$ one of the complex eigenvalues of A.

This theorem is proved in Coddington and Levinson [C/L] or in Hirsch and Smale [H/S]. The Jordan canonical form of a given $n \times n$ matrix A is unique except for the order of the elementary Jordan blocks in (1) and for the fact that the 1's in the elementary blocks (2) or the I_2's in the elementary blocks (3) may appear either above or below the diagonal. We shall refer to (1) with the B_j given by (2) or (3) as the upper Jordan canonical form of A.

The Jordan canonical form of A yields some explicit information about the form of the solution of the initial value problem

$$\dot{\mathbf{x}} = A\mathbf{x}$$
$$\mathbf{x}(0) = \mathbf{x}_0 \qquad (4)$$

which, according to the Fundamental Theorem for Linear Systems in Section 1.4, is given by

$$\mathbf{x}(t) = P\,\mathrm{diag}[e^{B_j t}]P^{-1}\mathbf{x}_0. \qquad (5)$$

1.8. Jordan Forms

If $B_j = B$ is an $m \times m$ matrix of the form (2) and λ is a real eigenvalue of A then $B = \lambda I + N$ and

$$e^{Bt} = e^{\lambda t} e^{Nt} = e^{\lambda t} \begin{bmatrix} 1 & t & t^2/2! & \cdots & t^{m-1}/(m-1)! \\ 0 & 1 & t & \cdots & t^{m-2}/(m-2)! \\ 0 & 0 & 1 & \cdots & t^{m-3}/(m-3)! \\ \cdots & & & & \\ 0 & \cdots & & 1 & t \\ 0 & \cdots & & 0 & 1 \end{bmatrix}$$

since the $m \times m$ matrix

$$N = \begin{bmatrix} 0 & 1 & 0 & \cdots & 0 \\ 0 & 0 & 1 & \cdots & 0 \\ \cdots & & & & \\ 0 & \cdots & & 0 & 1 \\ 0 & \cdots & & 0 & 0 \end{bmatrix}$$

is nilpotent of order m and

$$N^2 = \begin{bmatrix} 0 & 0 & 1 & 0 & \cdots & 0 \\ 0 & 0 & 0 & 1 & \cdots & 0 \\ & & \cdots & & & \\ 0 & \cdots & & & & 0 \end{bmatrix}, \ldots N^{m-1} = \begin{bmatrix} 0 & 0 & \cdots & 0 & 1 \\ 0 & 0 & \cdots & 0 & 0 \\ & & \cdots & & \\ 0 & & \cdots & & 0 \end{bmatrix}.$$

Similarly, if $B_j = B$ is a $2m \times 2m$ matrix of the form (3) and $\lambda = a + ib$ is a complex eigenvalue of A, then

$$e^{Bt} = e^{at} \begin{bmatrix} R & Rt & Rt^2/2! & \cdots & Rt^{m-1}/(m-1)! \\ 0 & R & Rt & \cdots & Rt^{m-2}/(m-2)! \\ 0 & 0 & R & \cdots & Rt^{m-3}/(m-3)! \\ \cdots & & & & \\ 0 & \cdots & & R & Rt \\ 0 & \cdots & & 0 & R \end{bmatrix}$$

where the rotation matrix

$$R = \begin{bmatrix} \cos bt & -\sin bt \\ \sin bt & \cos bt \end{bmatrix}$$

since the $2m \times 2m$ matrix

$$B = \begin{bmatrix} 0 & I_2 & 0 & \cdots & 0 \\ 0 & 0 & I_2 & \cdots & 0 \\ & & \cdots & & \\ 0 & \cdots & & 0 & I_2 \\ 0 & \cdots & & 0 & 0 \end{bmatrix}$$

is nilpotent of order m and

$$N^2 = \begin{bmatrix} 0 & 0 & I_2 & \cdots & 0 \\ 0 & 0 & 0 & I_2 & \cdots & 0 \\ & & \cdots & & & \\ 0 & & \cdots & & & 0 \end{bmatrix}, \ldots N^{m-1} = \begin{bmatrix} 0 & 0 & \cdots & 0 & I_2 \\ 0 & & \cdots & & 0 \\ & & \cdots & & \\ 0 & & \cdots & & 0 \end{bmatrix}.$$

The above form of the solution (5) of the initial value problem (4) then leads to the following result:

Corollary. *Each coordinate in the solution* $\mathbf{x}(t)$ *of the initial value problem (4) is a linear combination of functions of the form*

$$t^k e^{at} \cos bt \quad \text{or} \quad t^k e^{at} \sin bt$$

where $\lambda = a + ib$ *is an eigenvalue of the matrix* A *and* $0 \leq k \leq n-1$.

We next describe a method for finding a basis which reduces A to its Jordan canonical form. But first we need the following definitions:

Definition. Let λ be an eigenvalue of the $n \times n$ matrix A of multiplicity n. The *deficiency indices*

$$\delta_k = \dim \mathrm{Ker}(A - \lambda I)^k.$$

The kernel of a linear operator $T \colon \mathbf{R}^n \to \mathbf{R}^n$

$$\mathrm{Ker}(T) = \{\mathbf{x} \in \mathbf{R}^n \mid T(\mathbf{x}) = \mathbf{0}\}.$$

The deficiency indices δ_k can be found by Gaussian reduction; in fact, δ_k is the number of rows of zeros in the reduced row echelon form of $(A - \lambda I)^k$. Clearly

$$\delta_1 \leq \delta_2 \leq \cdots \leq \delta_n = n.$$

Let ν_k be the number of elementary Jordan blocks of size $k \times k$ in the Jordan canonical form (1) of the matrix A. Then it follows from the above theorem and the definition of δ_k that

$$\begin{aligned}
\delta_1 &= \nu_1 + \nu_2 + \cdots + \nu_n \\
\delta_2 &= \nu_1 + 2\nu_2 + \cdots + 2\nu_n \\
\delta_3 &= \nu_1 + 2\nu_2 + 3\nu_3 + \cdots + 3\nu_n \\
&\cdots \\
\delta_{n-1} &= \nu_1 + 2\nu_2 + 3\nu_3 + \cdots + (n-1)\nu_{n-1} + (n-1)\nu_n \\
\delta_n &= \nu_1 + 2\nu_2 + 3\nu_3 + \cdots + (n-1)\nu_{n-1} + n\nu_n.
\end{aligned}$$

Cf. Hirsch and Smale [H/S], p. 124. These equations can then be solved for

$$\begin{aligned}
\nu_1 &= 2\delta_1 - \delta_2 \\
\nu_2 &= 2\delta_2 - \delta_3 - \delta_1 \\
&\cdots \\
\nu_k &= 2\delta_k - \delta_{k+1} - \delta_{k-1} \text{ for } 1 < k < n \\
&\cdots \\
\nu_n &= \delta_n - \delta_{n-1}.
\end{aligned}$$

1.8. Jordan Forms

Example 1. The only upper Jordan canonical forms for a 2×2 matrix with a real eigenvalue λ of multiplicity 2 and the corresponding deficiency indices are given by

$$\begin{bmatrix} \lambda & 0 \\ 0 & \lambda \end{bmatrix} \quad \text{and} \quad \begin{bmatrix} \lambda & 1 \\ 0 & \lambda \end{bmatrix}$$
$$\delta_1 = \delta_2 = 2 \qquad \delta_1 = 1, \delta_2 = 2.$$

Example 2. The (upper) Jordan canonical forms for a 3×3 matrix with a real eigenvalue λ of multiplicity 3 and the corresponding deficiency indices are given by

$$\begin{bmatrix} \lambda & 0 & 0 \\ 0 & \lambda & 0 \\ 0 & 0 & \lambda \end{bmatrix} \quad \begin{bmatrix} \lambda & 1 & 0 \\ 0 & \lambda & 0 \\ 0 & 0 & \lambda \end{bmatrix} \quad \begin{bmatrix} \lambda & 1 & 0 \\ 0 & \lambda & 1 \\ 0 & 0 & \lambda \end{bmatrix}$$
$$\delta_1 = \delta_2 = \delta_3 = 3 \quad \delta_1 = 2, \delta_2 = \delta_3 = 3 \quad \delta_1 = 1, \delta_2 = 2, \delta_3 = 3.$$

We next give an algorithm for finding a basis B of generalized eigenvectors such that the $n \times n$ matrix A with a real eigenvalue λ of multiplicity n assumes its Jordan canonical form J with respect to the basis B; cf. Curtis [Cu]:

1. Find a basis $\{\mathbf{v}_j^{(1)}\}_{j=1}^{\delta_1}$ for $\text{Ker}(A-\lambda I)$; i.e., find a linearly independent set of eigenvectors of A corresponding to the eigenvalue λ.

2. If $\delta_2 > \delta_1$, choose a basis $\{\mathbf{V}_j^{(1)}\}_{j=1}^{\delta_1}$ for $\text{Ker}(A - \lambda I)$ such that

$$(A - \lambda I)\mathbf{v}_j^{(2)} = \mathbf{V}_j^{(1)}$$

has $\delta_2 - \delta_1$ linearly independent solutions $\mathbf{v}_j^{(2)}$, $j = 1, \ldots, \delta_2 - \delta_1$. Then $\{\mathbf{v}_j^{(2)}\}_{j=1}^{\delta_2} = \{\mathbf{V}_j^{(1)}\}_{j=1}^{\delta_1} \cup \{\mathbf{v}_j^{(2)}\}_{j=1}^{\delta_2-\delta_1}$ is a basis for $\text{Ker}(A - \lambda I)^2$.

3. If $\delta_3 > \delta_2$, choose a basis $\{\mathbf{V}_j^{(2)}\}_{j=1}^{\delta_2}$ for $\text{Ker}(A - \lambda I)^2$ with $\mathbf{V}_j^{(2)} \in \text{span}\{\mathbf{v}_j^{(2)}\}_{j=1}^{\delta_2-\delta_1}$ for $j = 1, \ldots, \delta_2 - \delta_1$ such that

$$(A - \lambda I)\mathbf{v}_j^{(3)} = \mathbf{V}_j^{(2)}$$

has $\delta_3 - \delta_2$ linearly independent solutions $\mathbf{v}_j^{(3)}$, $j = 1, \ldots, \delta_3 - \delta_2$. If for $j = 1, \ldots, \delta_2 - \delta_1$, $\mathbf{V}_j^{(2)} = \sum_{i=1}^{\delta_2-\delta_1} c_i \mathbf{v}_i^{(2)}$, let $\tilde{\mathbf{V}}_j^{(1)} = \sum_{i=1}^{\delta_2-\delta_1} c_i \mathbf{V}_i^{(1)}$ and $\tilde{\mathbf{V}}_j^{(1)} = \mathbf{V}_j^{(1)}$ for $j = \delta_2 - \delta_1 + 1, \ldots, \delta_1$. Then

$$\{\mathbf{v}_j^{(3)}\}_{j=1}^{\delta_3} = \{\tilde{\mathbf{V}}_j^{(1)}\}_{j=1}^{\delta_1} \cup \{\mathbf{V}_j^{(2)}\}_{j=1}^{\delta_2-\delta_1} \cup \{\mathbf{v}_j^{(3)}\}_{j=1}^{\delta_3-\delta_2}$$

is a basis for $\text{Ker}(A - \lambda I)^3$.

4. Continue this process untill the kth step when $\delta_k = n$ to obtain a basis $B = \{\mathbf{v}_j^{(k)}\}_{j=1}^n$ for $\mathbf{R}^n$. The matrix A will then assume its Jordan canonical form with respect to this basis.

The diagonalizing matrix $P = [\mathbf{v}_1 \cdots \mathbf{v}_n]$ in the above theorem which satisfies $P^{-1}AP = J$ is then obtained by an appropriate ordering of the basis B. The manner in which the matrix P is obtained from the basis B is indicated in the following examples. Roughly speaking, each generalized eigenvector $\mathbf{v}_j^{(i)}$ satisfying $(A - \lambda I)\mathbf{v}_j^{(i)} = \mathbf{V}_j^{(i-1)}$ is listed immediately following the generalized eigenvector $\mathbf{V}_j^{(i-1)}$.

Example 3. Find a basis for $\mathbf{R}^3$ which reduces

$$A = \begin{bmatrix} 2 & 1 & 0 \\ 0 & 2 & 0 \\ 0 & -1 & 2 \end{bmatrix}$$

to its Jordan canonical form. It is easy to find that $\lambda = 2$ is an eigenvalue of multiplicity 3 and that

$$A - \lambda I = \begin{bmatrix} 0 & 1 & 0 \\ 0 & 0 & 0 \\ 0 & -1 & 0 \end{bmatrix}.$$

Thus, $\delta_1 = 2$ and $(A - \lambda I)\mathbf{v} = 0$ is equivalent to $x_2 = 0$. We therefore choose

$$\mathbf{v}_1^{(1)} = \begin{bmatrix} 1 \\ 0 \\ 0 \end{bmatrix} \quad \text{and} \quad \mathbf{v}_2^{(1)} = \begin{bmatrix} 0 \\ 0 \\ 1 \end{bmatrix}$$

as a basis for $\mathrm{Ker}(A - \lambda I)$. We next solve

$$\begin{bmatrix} 0 & 1 & 0 \\ 0 & 0 & 0 \\ 0 & -1 & 0 \end{bmatrix} \mathbf{v} = c_1 \mathbf{v}_1^{(1)} + c_2 \mathbf{v}_2^{(1)} = \begin{bmatrix} c_1 \\ 0 \\ c_2 \end{bmatrix}.$$

This is equivalent to $x_2 = c_1$ and $-x_2 = c_2$; i.e., $c_1 = -c_2$. We choose

$$\mathbf{V}_1^{(1)} = \begin{bmatrix} 1 \\ 0 \\ -1 \end{bmatrix}, \mathbf{v}_1^{(2)} = \begin{bmatrix} 0 \\ 1 \\ 0 \end{bmatrix} \quad \text{and} \quad \mathbf{V}_2^{(1)} = \begin{bmatrix} 1 \\ 0 \\ 0 \end{bmatrix}.$$

These three vectors which we re-label as $\mathbf{v}_1, \mathbf{v}_2$ and $\mathbf{v}_3$ respectively are then a basis for $\mathrm{Ker}(A - \lambda I)^2 = \mathbf{R}^3$. (Note that we could also choose $\mathbf{V}_2^{(1)} = \mathbf{v}_3 = (0,0,1)^T$ and obtain the same result.) The matrix $P = [\mathbf{v}_1, \mathbf{v}_2, \mathbf{v}_3]$ and its inverse are then given by

$$P = \begin{bmatrix} 1 & 0 & 1 \\ 0 & 1 & 0 \\ -1 & 0 & 0 \end{bmatrix} \text{ and } P^{-1} = \begin{bmatrix} 0 & 0 & -1 \\ 0 & 1 & 0 \\ 1 & 0 & 1 \end{bmatrix}$$

1.8. Jordan Forms

respectively. The student should verify that

$$P^{-1}AP = \begin{bmatrix} 2 & 1 & 0 \\ 0 & 2 & 0 \\ 0 & 0 & 2 \end{bmatrix}.$$

Example 4. Find a basis for $\mathbf{R}^4$ which reduces

$$A = \begin{bmatrix} 0 & -1 & -2 & -1 \\ 1 & 2 & 1 & 1 \\ 0 & 0 & 1 & 0 \\ 0 & 0 & 1 & 1 \end{bmatrix}$$

to its Jordan canonical form. We find $\lambda = 1$ is an eigenvalue of multiplicity 4 and

$$A - \lambda I = \begin{bmatrix} -1 & -1 & -2 & -1 \\ 1 & 1 & 1 & 1 \\ 0 & 0 & 0 & 0 \\ 0 & 0 & 1 & 0 \end{bmatrix}.$$

Using Gaussian reduction, we find $\delta_1 = 2$ and that the following two vectors

$$\mathbf{v}_1^{(1)} = \begin{bmatrix} -1 \\ 1 \\ 0 \\ 0 \end{bmatrix} \quad \mathbf{v}_2^{(1)} = \begin{bmatrix} -1 \\ 0 \\ 0 \\ 1 \end{bmatrix}$$

span Ker$(A - \lambda I)$. We next solve

$$(A - \lambda I)\mathbf{v} = c_1 \mathbf{v}_1^{(1)} + c_2 \mathbf{v}_2^{(1)}.$$

These equations are equivalent to $x_3 = c_2$ and $x_1 + x_2 + x_3 + x_4 = c_1$. We can therefore choose $c_1 = 1$, $c_2 = 0$, $x_1 = 1$, $x_2 = x_3 = x_4 = 0$ and find

$$\mathbf{v}_1^{(2)} = (1, 0, 0, 0)^T$$

(with $\mathbf{V}_1^{(1)} = (-1, 1, 0, 0)^T$); and we can choose $c_1 = 0$, $c_2 = 1 = x_3$, $x_1 = -1$, $x_2 = x_4 = 0$ and find

$$\mathbf{v}_2^{(2)} = (-1, 0, 1, 0)^T$$

(with $\mathbf{V}_2^{(1)} = (-1, 0, 0, 1)^T$). Thus the vectors $\mathbf{V}_1^{(1)}, \mathbf{v}_1^{(2)}, \mathbf{V}_2^{(1)}, \mathbf{v}_2^{(2)}$, which we re-label as $\mathbf{v}_1, \mathbf{v}_2, \mathbf{v}_3$ and $\mathbf{v}_4$ respectively, form a basis B for $\mathbf{R}^4$. The matrix $P = [\mathbf{v}_1 \quad \cdots \quad \mathbf{v}_4]$ and its inverse are then given by

$$P = \begin{bmatrix} -1 & 1 & -1 & -1 \\ 1 & 0 & 0 & 0 \\ 0 & 0 & 0 & 1 \\ 0 & 0 & 1 & 0 \end{bmatrix}, \quad P^{-1} = \begin{bmatrix} 0 & 1 & 0 & 0 \\ 1 & 1 & 1 & 1 \\ 0 & 0 & 0 & 1 \\ 0 & 0 & 1 & 0 \end{bmatrix}$$

and
$$P^{-1}AP = \begin{bmatrix} 1 & 1 & 0 & 0 \\ 0 & 1 & 0 & 0 \\ 0 & 0 & 1 & 1 \\ 0 & 0 & 0 & 1 \end{bmatrix}.$$

In this case we have $\delta_1 = 2$, $\delta_2 = \delta_3 = \delta_4 = 4$, $\nu_1 = 2\delta_1 - \delta_2 = 0$, $\nu_2 = 2\delta_2 - \delta_3 - \delta_1 = 2$ and $\nu_3 = \nu_4 = 0$.

Example 5. Find a basis for $\mathbf{R}^4$ which reduces
$$A = \begin{bmatrix} 0 & -2 & -1 & -1 \\ 1 & 2 & 1 & 1 \\ 0 & 1 & 1 & 0 \\ 0 & 0 & 0 & 1 \end{bmatrix}$$

to its Jordan canonical form. We find $\lambda = 1$ is an eigenvalue of multiplicity 4 and
$$A - \lambda I = \begin{bmatrix} -1 & -2 & -1 & -1 \\ 1 & 1 & 1 & 1 \\ 0 & 1 & 0 & 0 \\ 0 & 0 & 0 & 0 \end{bmatrix}.$$

Using Gaussian reduction, we find $\delta_1 = 2$ and that the following two vectors
$$\mathbf{v}_1^{(1)} = \begin{bmatrix} -1 \\ 0 \\ 1 \\ 0 \end{bmatrix}, \quad \mathbf{v}_2^{(1)} = \begin{bmatrix} -1 \\ 0 \\ 0 \\ 1 \end{bmatrix}$$

span $\mathrm{Ker}(A - \lambda I)$. We next solve
$$(A - \lambda I)\mathbf{v} = c_1 \mathbf{v}_1^{(1)} + c_2 \mathbf{v}_2^{(1)}.$$

The last row implies that $c_2 = 0$ and the third row implies that $x_2 = 1$. The remaining equations are then equivalent to $x_1 + x_2 + x_3 + x_4 = 0$. Thus, $\mathbf{V}_1^{(1)} = \mathbf{v}_1^{(1)}$ and we choose
$$\mathbf{v}_1^{(2)} = (-1, 1, 0, 0)^T.$$

Using Gaussian reduction, we next find that $\delta_2 = 3$ and $\{\mathbf{V}_1^{(1)}, \mathbf{v}_1^{(2)}, \mathbf{V}_2^{(1)}\}$ with $\mathbf{V}_2^{(1)} = \mathbf{v}_2^{(1)}$ spans $\mathrm{Ker}(A - \lambda I)^2$. Similarly we find $\delta_3 = 4$ and we must find $\delta_3 - \delta_2 = 1$ solution of
$$(A - \lambda I)\mathbf{v} = \mathbf{V}_1^{(2)}.$$

where $\mathbf{V}_1^{(2)} = \mathbf{v}_1^{(2)}$. The third row of this equation implies that $x_2 = 0$ and the remaining equations are then equivalent to $x_1 + x_3 + x_4 = 0$. We choose
$$\mathbf{v}_1^{(3)} = (1, 0, 0, 0)^T.$$

1.8. Jordan Forms

Then $B = \{\mathbf{v}_1^{(3)}, \mathbf{v}_2^{(3)}, \mathbf{v}_3^{(3)}, \mathbf{v}_4^{(3)}\} = \{\mathbf{V}_1^{(1)}, \mathbf{V}_1^{(2)}, \mathbf{v}_1^{(3)}, \mathbf{V}_2^{(1)}\}$ is a basis for $\text{Ker}(A - \lambda I)^3 = \mathbf{R}^4$. The matrix $P = [\mathbf{v}_1 \cdots \mathbf{v}_4]$, with $\mathbf{v}_1 = \mathbf{V}_1^{(1)}, \mathbf{v}_2 = \mathbf{V}_1^{(2)}, \mathbf{v}_3 = \mathbf{v}_1^{(3)}$ and $\mathbf{v}_4 = \mathbf{V}_2^{(1)}$, and its inverse are then given by

$$P = \begin{bmatrix} 1 & 1 & 1 & -1 \\ 0 & 1 & 0 & 0 \\ 1 & 0 & 0 & 0 \\ 0 & 0 & 0 & 1 \end{bmatrix} \quad \text{and} \quad P^{-1} = \begin{bmatrix} 0 & 0 & 1 & 0 \\ 0 & 1 & 0 & 0 \\ 1 & 1 & 1 & 1 \\ 0 & 0 & 0 & 1 \end{bmatrix}$$

respectively. And then we obtain

$$P^{-1}AP = \begin{bmatrix} 1 & 1 & 0 & 0 \\ 0 & 1 & 1 & 0 \\ 0 & 0 & 1 & 0 \\ 0 & 0 & 0 & 1 \end{bmatrix}.$$

In this case we have $\delta_1 = 2$, $\delta_2 = 3$, $\delta_3 = \delta_4 = 4$, $\nu_1 = 2\delta_1 - \delta_2 = 1$, $\nu_2 = 2\delta_2 - \delta_3 - \delta_1 = 0$, $\nu_3 = 2\delta_3 - \delta_4 - \delta_2 = 1$ and $\nu_4 = \delta_4 - \delta_3 = 0$.

It is worth mentioning that the solution to the initial value problem (4) for this last example is given by

$$\mathbf{x}(t) = e^{At}\mathbf{x}_0 = Pe^{Jt}P^{-1}\mathbf{x}_0 = Pe^t \begin{bmatrix} 1 & t & t^2/2 & 0 \\ 0 & 1 & t & 0 \\ 0 & 0 & 1 & 0 \\ 0 & 0 & 0 & 1 \end{bmatrix} P^{-1}\mathbf{x}_0$$

$$= e^t \begin{bmatrix} 1-t-t^2/2 & -2t-t^2/2 & -t-t^2/2 & -t-t^2/2 \\ t & 1+t & t & t \\ t^2/2 & t+t^2/2 & 1+t^2/2 & t^2/2 \\ 0 & 0 & 0 & 1 \end{bmatrix} \mathbf{x}_0.$$

Problem Set 8

1. Find the Jordan canonical forms for the following matrices

(a) $A = \begin{bmatrix} 1 & 0 \\ 0 & 1 \end{bmatrix}$

(b) $A = \begin{bmatrix} 1 & 1 \\ 0 & 1 \end{bmatrix}$

(c) $A = \begin{bmatrix} 0 & 1 \\ 1 & 0 \end{bmatrix}$

(d) $A = \begin{bmatrix} 1 & 0 \\ 0 & -1 \end{bmatrix}$

(e) $A = \begin{bmatrix} 1 & 1 \\ 0 & -1 \end{bmatrix}$

(f) $A = \begin{bmatrix} 0 & -1 \\ 1 & 0 \end{bmatrix}$

(g) $A = \begin{bmatrix} 1 & 1 \\ 1 & 1 \end{bmatrix}$

(h) $A = \begin{bmatrix} 1 & 1 \\ -1 & 1 \end{bmatrix}$

(i) $A = \begin{bmatrix} 1 & -1 \\ 0 & 1 \end{bmatrix}$.

2. Find the Jordan canonical forms for the following matrices

(a) $A = \begin{bmatrix} 1 & 1 & 0 \\ 0 & 1 & 0 \\ 0 & 0 & 1 \end{bmatrix}$

(b) $A = \begin{bmatrix} 1 & 1 & 0 \\ 0 & 1 & 1 \\ 0 & 0 & 1 \end{bmatrix}$

(c) $A = \begin{bmatrix} 1 & 0 & 0 \\ 0 & 0 & -1 \\ 0 & 1 & 0 \end{bmatrix}$

(d) $A = \begin{bmatrix} 1 & 1 & 0 \\ 0 & 1 & 0 \\ 0 & 0 & -1 \end{bmatrix}$

(e) $A = \begin{bmatrix} 1 & 0 & 0 \\ 0 & 1 & 1 \\ 0 & 0 & -1 \end{bmatrix}$

(f) $A = \begin{bmatrix} 1 & 0 & 0 \\ 0 & 0 & 1 \\ 0 & 1 & 0 \end{bmatrix}$.

3. (a) List the five upper Jordan canonical forms for a 4×4 matrix A with a real eigenvalue λ of multiplicity 4 and give the corresponding deficiency indices in each case.

   (b) What is the form of the solution of the initial value problem (4) in each of these cases?

4. (a) What are the four upper Jordan canonical forms for a 4×4 matrix A having complex eigenvalues?

   (b) What is the form of the solution of the initial value problem (4) in each of these cases?

5. (a) List the seven upper Jordan canonical forms for a 5×5 matrix A with a real eigenvalue λ of multiplicity 5 and give the corresponding deficiency indices in each case.

1.8. Jordan Forms

(b) What is the form of the solution of the initial value problem (4) in each of these cases?

6. Find the Jordan canonical forms for the following matrices

(a) $A = \begin{bmatrix} 1 & 0 & 0 \\ 1 & 2 & 0 \\ 1 & 2 & 3 \end{bmatrix}$

(b) $A = \begin{bmatrix} 1 & 0 & 0 \\ -1 & 2 & 0 \\ 1 & 0 & 2 \end{bmatrix}$

(c) $A = \begin{bmatrix} 1 & 1 & 2 \\ 0 & 2 & 1 \\ 0 & 0 & 2 \end{bmatrix}$

(d) $A = \begin{bmatrix} 2 & 1 & 2 \\ 0 & 2 & 1 \\ 0 & 0 & 2 \end{bmatrix}$

(e) $A = \begin{bmatrix} 1 & 0 & 0 & 0 \\ 1 & 2 & 0 & 0 \\ 1 & 2 & 3 & 0 \\ 1 & 2 & 3 & 4 \end{bmatrix}$

(f) $A = \begin{bmatrix} 1 & 0 & 0 & 0 \\ 1 & 2 & 0 & 0 \\ 1 & 0 & 2 & 0 \\ 1 & 1 & 0 & 2 \end{bmatrix}$

(g) $A = \begin{bmatrix} 2 & 1 & 4 & 0 \\ 0 & 2 & 1 & 0 \\ 0 & 0 & 2 & 0 \\ 0 & 0 & 0 & 2 \end{bmatrix}$

(h) $A = \begin{bmatrix} 2 & 1 & 4 & 0 \\ 0 & 2 & 1 & -1 \\ 0 & 0 & 2 & 1 \\ 0 & 0 & 0 & 2 \end{bmatrix}$.

Find the solution of the initial value problem (4) for each of these matrices.

7. Suppose that B is an $m \times m$ matrix given by equation (2) and that $Q = \text{diag}[1, \varepsilon, \varepsilon^2, \ldots, \varepsilon^{m-1}]$. Note that B can be written in the form

$$B = \lambda I + N$$

where N is nilpotent of order m and show that for $\varepsilon > 0$

$$Q^{-1}BQ = \lambda I + \varepsilon N.$$

This shows that the ones above the diagonal in the upper Jordan canonical form of a matrix can be replaced by any $\varepsilon > 0$. A similar result holds when B is given by equation (3).

8. What are the eigenvalues of a nilpotent matrix N?

9. Show that if all of the eigenvalues of the matrix A have negative real parts, then for all $x_0 \in \mathbf{R}^n$

$$\lim_{t \to \infty} x(t) = 0$$

where $x(t)$ is the solution of the initial value problem (4).

10. Suppose that the elementary blocks B in the Jordan form of the matrix A, given by (2) or (3), have no ones or I_2 blocks off the diagonal. (The matrix A is called *semisimple* in this case.) Show that if all of the eigenvalues of A have nonpositive real parts, then for each $x_0 \in \mathbf{R}^n$ there is a positive constant M such that $|x(t)| \leq M$ for all $t \geq 0$ where $x(t)$ is the solution of the initial value problem (4).

11. Show by example that if A is not semisimple, then even if all of the eigenvalues of A have nonpositive real parts, there is an $x_0 \in \mathbf{R}^n$ such that

$$\lim_{t \to \infty} |x(t)| = \infty.$$

Hint: Cf. Example 4 in Section 1.7.

12. For any solution $x(t)$ of the initial value problem (4) with $\det A \neq 0$ and $x_0 \neq 0$ show that exactly one of the following alternatives holds.

   (a) $\lim_{t \to \infty} x(t) = 0$ and $\lim_{t \to -\infty} |x(t)| = \infty$;

   (b) $\lim_{t \to \infty} |x(t)| = \infty$ and $\lim_{t \to -\infty} x(t) = 0$;

   (c) There are positive constants m and M such that for all $t \in \mathbf{R}$

   $$m \leq |x(t)| \leq M;$$

   (d) $\lim_{t \to \pm\infty} |x(t)| = \infty$;

   (e) $\lim_{t \to \infty} |x(t)| = \infty$, $\lim_{t \to -\infty} x(t)$ does not exist;

   (f) $\lim_{t \to -\infty} |x(t)| = \infty$, $\lim_{t \to \infty} x(t)$ does not exist.

Hint: See Problem 5 in Problem Set 9.

1.9 Stability Theory

In this section we define the *stable, unstable and center subspace*, E^s, E^u and E^c respectively, of a linear system

$$\dot{\mathbf{x}} = A\mathbf{x}. \tag{1}$$

Recall that E^s and E^u were defined in Section 1.2 in the case when A had distinct eigenvalues. We also establish some important properties of these subspaces in this section.

Let $\mathbf{w}_j = \mathbf{u}_j + i\mathbf{v}_j$; be a generalized eigenvector of the (real) matrix A corresponding to an eigenvalue $\lambda_j = a_j + ib_j$. Note that if $b_j = 0$ then $\mathbf{v}_j = 0$. And let

$$B = \{\mathbf{u}_1, \ldots, \mathbf{u}_k, \mathbf{u}_{k+1}, \mathbf{v}_{k+1}, \ldots, \mathbf{u}_m, \mathbf{v}_m\}$$

be a basis of $\mathbf{R}^n$ (with $n = 2m - k$) as established by Theorems 1 and 2 and the Remark in Section 1.7.

Definition 1. Let $\lambda_j = a_j + ib_j, \mathbf{w}_j = \mathbf{u}_j + i\mathbf{v}_j$ and B be as described above. Then

$$E^s = \text{Span}\{\mathbf{u}_j, \mathbf{v}_j \mid a_j < 0\}$$
$$E^c = \text{Span}\{\mathbf{u}_j, \mathbf{v}_j \mid a_j = 0\}$$

and

$$E^u = \text{Span}\{\mathbf{u}_j, \mathbf{v}_j \mid a_j > 0\};$$

i.e., E^s, E^c and E^u are the subspaces of $\mathbf{R}^n$ spanned by the real and imaginary parts of the generalized eigenvectors $\mathbf{w}_j$ corresponding to eigenvalues λ_j with negative, zero and positive real parts respectively.

Example 1. The matrix

$$A = \begin{bmatrix} -2 & -1 & 0 \\ 1 & -2 & 0 \\ 0 & 0 & 3 \end{bmatrix}$$

has eigenvectors

$$\mathbf{w}_1 = \mathbf{u}_1 + i\mathbf{v}_1 = \begin{bmatrix} 0 \\ 1 \\ 0 \end{bmatrix} + i \begin{bmatrix} 1 \\ 0 \\ 0 \end{bmatrix} \quad \text{corresponding to } \lambda_1 = -2 + i$$

and

$$\mathbf{u}_2 = \begin{bmatrix} 0 \\ 0 \\ 1 \end{bmatrix} \quad \text{corresponding to } \lambda_2 = 3.$$

The stable subspace E^s of (1) is the x_1, x_2 plane and the unstable subspace E^u of (1) is the x_3-axis. The phase portrait for the system (1) is shown in Figure 1 for this example.

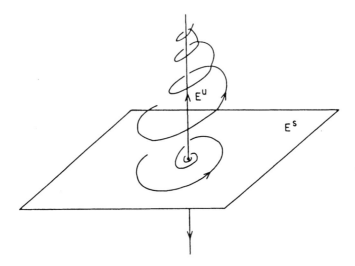

Figure 1. The stable and unstable subspaces E^s and E^u of the linear system (1).

Example 2. The matrix

$$A = \begin{bmatrix} 0 & -1 & 0 \\ 1 & 0 & 0 \\ 0 & 0 & 2 \end{bmatrix}$$

has $\lambda_1 = i, \mathbf{u}_1 = (0, 1, 0)^T, \mathbf{v}_1 = (1, 0, 0)^T, \lambda_2 = 2$ and $\mathbf{u}_2 = (0, 0, 1)^T$. The center subspace of (1) is the x_1, x_2 plane and the unstable subspace of (1) is the x_3-axis. The phase portrait for the system (1) is shown in Figure 2 for this example. Note that all solutions lie on the cylinders $x_1^2 + x_2^2 = c^2$.

In these examples we see that all solutions in E^s approach the equilibrium point $\mathbf{x} = \mathbf{0}$ as $t \to \infty$ and that all solutions in E^u approach the equilibrium point $\mathbf{x} = \mathbf{0}$ as $t \to -\infty$. Also, in the above example the solutions in E^c are bounded and if $\mathbf{x}(0) \neq \mathbf{0}$, then they are bounded away from $\mathbf{x} = \mathbf{0}$ for all $t \in \mathbf{R}$. We shall see that these statements about E^s and E^u are true in general; however, solutions in E^c need not be bounded as the next example shows.

1.9. Stability Theory

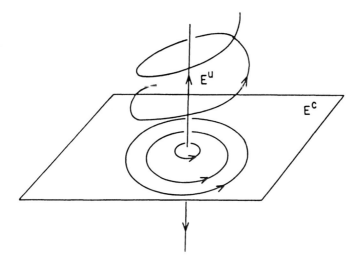

Figure 2. The center and unstable subspaces E^c and E^u of the linear system (1).

Example 3. Consider the linear system (1) with
$$A = \begin{bmatrix} 0 & 0 \\ 1 & 0 \end{bmatrix}; \quad \text{i.e.,} \quad \begin{matrix} \dot{x}_1 = 0 \\ \dot{x}_2 = x_1 \end{matrix}$$

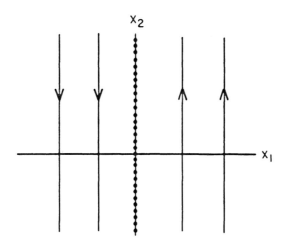

Figure 3. The center subspace E^c for (1).

We have $\lambda_1 = \lambda_2 = 0$, $\mathbf{u}_1 = (0,1)^T$ is an eigenvector and $\mathbf{u}_2 = (1,0)^T$ is a generalized eigenvector corresponding to $\lambda = 0$. Thus $E^c = \mathbf{R}^2$. The solution of (1) with $\mathbf{x}(0) = \mathbf{c} = (c_1, c_2)^T$ is easily found to be

$$x_1(t) = c_1$$
$$x_2(t) = c_1 t + c_2.$$

The phase portrait for (1) in this case is given in Figure 3. Some solutions (those with $c_1 = 0$) remain bounded while others do not.

We next describe the notion of the flow of a system of differential equations and show that the stable, unstable and center subspaces of (1) are invariant under the flow of (1).

By the fundamental theorem in Section 1.4, the solution to the initial value problem associated with (1) is given by

$$\mathbf{x}(t) = e^{At}\mathbf{x}_0.$$

The set of mappings $e^{At}\colon \mathbf{R}^n \to \mathbf{R}^n$ may be regarded as describing the motion of points $\mathbf{x}_0 \in \mathbf{R}^n$ along trajectories of (1). This set of mappings is called *the flow of the linear system* (1). We next define the important concept of a hyperbolic flow:

Definition 2. If all eigenvalues of the $n \times n$ matrix A have nonzero real part, then the flow $e^{At}\colon \mathbf{R}^n \to \mathbf{R}^n$ is called a *hyperbolic flow* and (1) is called a *hyperbolic linear system*.

Definition 3. A subspace $E \subset \mathbf{R}^n$ is said to be *invariant with respect to the flow* $e^{At}\colon \mathbf{R}^n \to \mathbf{R}^n$ if $e^{At}E \subset E$ for all $t \in \mathbf{R}$.

We next show that the stable, unstable and center subspaces, E^s, E^u and E^c of (1) are invariant under the flow e^{At} of the linear system (1); i.e., any solution starting in E^s, E^u or E^c at time $t = 0$ remains in E^s, E^u or E^c respectively for all $t \in \mathbf{R}$.

Lemma. *Let E be the generalized eigenspace of A corresponding to an eigenvalue λ. Then $AE \subset E$.*

Proof. Let $\{\mathbf{v}_1, \ldots, \mathbf{v}_k\}$ be a basis of generalized eigenvectors for E. Then given $\mathbf{v} \in E$,

$$\mathbf{v} = \sum_{j=1}^{k} c_j \mathbf{v}_j$$

and by linearity

$$A\mathbf{v} = \sum_{j=1}^{k} c_j A\mathbf{v}_j.$$

1.9. Stability Theory

Now since each v_j satisfies

$$(A - \lambda I)^{k_j} v_j = 0$$

for some minimal k_j, we have

$$(A - \lambda I) v_j = V_j$$

where $V_j \in \text{Ker}(A - \lambda I)^{k_j - 1} \subset E$. Thus, it follows by induction that $Av_j = \lambda v_j + V_j \in E$ and since E is a subspace of $\mathbf{R}^n$, it follows that

$$\sum_{j=1}^{k} c_j A v_j \in E;$$

i.e., $Av \in E$ and therefore $AE \subset E$.

Theorem 1. *Let A be a real $n \times n$ matrix. Then*

$$\mathbf{R}^n = E^s \oplus E^u \oplus E^c$$

where E^s, E^u and E^c are the stable, unstable and center subspaces of (1) respectively; furthermore, E^s, E^u and E^c are invariant with respect to the flow e^{At} of (1) respectively.

Proof. Since $B = \{u_1, \ldots, u_k, u_{k+1}, v_{k+1}, \ldots, u_m, v_m\}$ described at the beginning of this section is a basis for $\mathbf{R}^n$, it follows from the definition of E^s, E^u and E^c that

$$\mathbf{R}^n = E^s \oplus E^u \oplus E^c.$$

If $x_0 \in E^s$ then

$$x_0 = \sum_{j=1}^{n_s} c_j V_j$$

where $V_j = v_j$ or u_j and $\{V_j\}_{j=1}^{n_s} \subset B$ is a basis for the stable subspace E^s as described in Definition 1. Then by the linearity of e^{At}, it follows that

$$e^{At} x_0 = \sum_{j=1}^{n_s} c_j e^{At} V_j.$$

But

$$e^{At} V_j = \lim_{k \to \infty} \left[I + At + \cdots + \frac{A^k t^k}{k!} \right] V_j \in E^s$$

since for $j = 1, \ldots, n_s$ by the above lemma $A^k V_j \in E^s$ and since E^s is complete. Thus, for all $t \in \mathbf{R}$, $e^{At} x_0 \in E^s$ and therefore $e^{At} E^s \subset E^s$; i.e., E^s is invariant under the flow e^{At}. It can similarly be shown that E^u and E^c are invariant under the flow e^{At}.

We next generalize the definition of sinks and sources of two-dimensional systems given in Section 1.5.

Definition 4. If all of the eigenvalues of A have negative (positive) real parts, the origin is called *a sink (source)* for the linear system (1).

Example 4. Consider the linear system (1) with

$$A = \begin{bmatrix} -2 & -1 & 0 \\ 1 & -2 & 0 \\ 0 & 0 & -3 \end{bmatrix}.$$

We have eigenvalues $\lambda_1 = -2 + i$ and $\lambda_2 = -3$ and the same eigenvectors as in Example 1. $E^s = \mathbf{R}^3$ and the origin is a sink for this example. The phase portrait is shown in Figure 4.

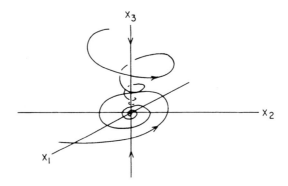

Figure 4. A linear system with a sink at the origin.

Theorem 2. *The following statements are equivalent:*

(a) *For all* $\mathbf{x}_0 \in \mathbf{R}^n$, $\lim_{t \to \infty} e^{At}\mathbf{x}_0 = 0$ *and for* $\mathbf{x}_0 \neq 0$, $\lim_{t \to -\infty} |e^{At}\mathbf{x}_0| = \infty$.

(b) *All eigenvalues of A have negative real part.*

(c) *There are positive constants a, c, m and M such that for all $\mathbf{x}_0 \in \mathbf{R}^n$*

$$|e^{At}\mathbf{x}_0| \leq Me^{-ct}|\mathbf{x}_0|$$

for $t \geq 0$ and

$$|e^{At}\mathbf{x}_0| \geq me^{-at}|\mathbf{x}_0|$$

for $t \leq 0$.

Proof (a $\Rightarrow$ b): If one of the eigenvalues $\lambda = a + ib$ has positive real part, $a > 0$, then by the theorem and corollary in Section 1.8, there exists an

1.9. Stability Theory

$\mathbf{x}_0 \in \mathbf{R}^n, \mathbf{x}_0 \neq \mathbf{0}$, such that $|e^{At}\mathbf{x}_0| \geq e^{at}|\mathbf{x}_0|$. Therefore $|e^{At}\mathbf{x}_0| \to \infty$ as $t \to \infty$; i.e.,

$$\lim_{t \to \infty} e^{At}\mathbf{x}_0 \neq \mathbf{0}.$$

And if one of the eigenvalues of A has zero real part, say $\lambda = ib$, then by the corollary in Section 1.8, there exists $\mathbf{x}_0 \in \mathbf{R}^n, \mathbf{x}_0 \neq \mathbf{0}$ such that at least one component of the solution is of the form $ct^k \cos bt$ or $ct^k \sin bt$ with $k \geq 0$. And once again

$$\lim_{t \to \infty} e^{At}\mathbf{x}_0 \neq \mathbf{0}.$$

Thus, if not all of the eigenvalues of A have negative real part, there exists $\mathbf{x}_0 \in \mathbf{R}^n$ such that $e^{At}\mathbf{x}_0 \not\to \mathbf{0}$ as $t \to \infty$; i.e., a $\Rightarrow$ b.

(b $\Rightarrow$ c): If all of the eigenvalues of A have negative real part, then it follows from the Jordan canonical form theorem and its corollary in Section 1.8 that there exist positive constants a, c, m and M such that for all $\mathbf{x}_0 \in \mathbf{R}^n |e^{At}\mathbf{x}_0| \leq Me^{-ct}|\mathbf{x}_0|$ for $t \geq 0$ and $|e^{At}\mathbf{x}_0| \geq me^{-at}|\mathbf{x}_0|$ for $t \leq 0$.

(c $\Rightarrow$ a): If this last pair of inequalities is satisfied for all $\mathbf{x}_0 \in \mathbf{R}^n$, it follows by taking the limit as $t \to \pm\infty$ on each side of the above inequalities that

$$\lim_{t \to \infty} |e^{At}\mathbf{x}_0| = 0 \quad \text{and that} \quad \lim_{t \to -\infty} |e^{At}\mathbf{x}_0| = \infty$$

for $\mathbf{x}_0 \neq \mathbf{0}$. This completes the proof of Theorem 2.

The next theorem is proved in exactly the same manner as Theorem 2 above using the theorem and its corollary in Section 1.8.

Theorem 3. *The following statements are equivalent:*

(a) *For all* $\mathbf{x}_0 \in \mathbf{R}^n, \lim_{t \to -\infty} e^{At}\mathbf{x}_0 = \mathbf{0}$ *and for* $\mathbf{x}_0 \neq \mathbf{0}, \lim_{t \to \infty} |e^{At}\mathbf{x}_0| = \infty$.

(b) *All eigenvalues of A have positive real part.*

(c) *There are positive constants a, c, m and M such that for all $\mathbf{x}_0 \in \mathbf{R}^n$*

$$|e^{At}\mathbf{x}_0| \leq M e^{ct}|\mathbf{x}_0|$$

for $t \leq 0$ and

$$|e^{At}\mathbf{x}_0| \geq m e^{at}|\mathbf{x}_0|$$

for $t \geq 0$.

Corollary. If $x_0 \in E^s$, then $e^{At}x_0 \in E^s$ for all $t \in \mathbf{R}$ and
$$\lim_{t \to \infty} e^{At}x_0 = 0.$$

And if $x_0 \in E^u$, then $e^{At}x_0 \in E^u$ for all $t \in \mathbf{R}$ and
$$\lim_{t \to -\infty} e^{At}x_0 = 0.$$

Thus, we see that all solutions of (1) which start in the stable manifold E^s of (1) remain in E^s for all t and approach the origin exponentially fast as $t \to \infty$; and all solutions of (1) which start in the unstable manifold E^u of (1) remain in E^u for all t and approach the origin exponentially fast as $t \to -\infty$. As we shall see in Chapter 2 there is an analogous result for nonlinear systems called the Stable Manifold Theorem; cf. Section 2.7 in Chapter 2.

PROBLEM SET 9

1. Find the stable, unstable and center subspaces E^s, E^u and E^c of the linear system (1) with the matrix

   (a) $A = \begin{bmatrix} 1 & 0 \\ 0 & -1 \end{bmatrix}$

   (b) $A = \begin{bmatrix} 0 & 1 \\ -1 & 0 \end{bmatrix}$

   (c) $A = \begin{bmatrix} 1 & 0 \\ 0 & 1 \end{bmatrix}$

   (d) $A = \begin{bmatrix} -1 & -3 \\ 0 & 2 \end{bmatrix}$

   (e) $A = \begin{bmatrix} 2 & 3 \\ 0 & -1 \end{bmatrix}$

   (f) $A = \begin{bmatrix} 2 & 4 \\ 0 & -2 \end{bmatrix}$

   (g) $A = \begin{bmatrix} 0 & 0 \\ 0 & -1 \end{bmatrix}$

   (h) $A = \begin{bmatrix} 0 & 0 \\ 1 & 0 \end{bmatrix}$

   (i) $A = \begin{bmatrix} -1 & -1 \\ 1 & -1 \end{bmatrix}.$

   Also, sketch the phase portrait in each of these cases. Which of these matrices define a hyperbolic flow, e^{At}?

1.9. Stability Theory

2. Same as Problem 1 for the matrices

(a) $A = \begin{bmatrix} -1 & 0 & 0 \\ 0 & -2 & 0 \\ 0 & 0 & 3 \end{bmatrix}$

(b) $A = \begin{bmatrix} 0 & -1 & 0 \\ 1 & 0 & 0 \\ 0 & 0 & -1 \end{bmatrix}$

(c) $A = \begin{bmatrix} -1 & -3 & 0 \\ 0 & 2 & 0 \\ 0 & 0 & -1 \end{bmatrix}$

(d) $A = \begin{bmatrix} 2 & 3 & 0 \\ 0 & -1 & 0 \\ 0 & 0 & -1 \end{bmatrix}$.

3. Solve the system
$$\dot{\mathbf{x}} = \begin{bmatrix} 0 & 2 & 0 \\ -2 & 0 & 0 \\ 2 & 0 & 6 \end{bmatrix} \mathbf{x}.$$

Find the stable, unstable and center subspaces E^s, E^u and E^c for this system and sketch the phase portrait. For $\mathbf{x}_0 \in E^c$, show that the sequence of points $\mathbf{x}_n = e^{An}\mathbf{x}_0 \in E^c$; similarly, for $\mathbf{x}_0 \in E^s$ or E^u, show that $\mathbf{x}_n \in E^s$ or E^u respectively.

4. Find the stable, unstable and center subspaces E^s, E^u and E^c for the linear system (1) with the matrix A given by

   (a) Problem 2(b) in Problem Set 7.
   (b) Problem 2(d) in Problem Set 7.

5. Let A be an $n \times n$ nonsingular matrix and let $\mathbf{x}(t)$ be the solution of the initial value problem (1) with $\mathbf{x}(0) = \mathbf{x}_0$. Show that

   (a) if $\mathbf{x}_0 \in E^s \sim \{\mathbf{0}\}$ then $\lim_{t \to \infty} \mathbf{x}(t) = \mathbf{0}$ and $\lim_{t \to -\infty} |\mathbf{x}(t)| = \infty$;

   (b) if $\mathbf{x}_0 \in E^u \sim \{\mathbf{0}\}$ then $\lim_{t \to \infty} |\mathbf{x}(t)| = \infty$ and $\lim_{t \to -\infty} \mathbf{x}(t) = \mathbf{0}$;

   (c) if $\mathbf{x}_0 \in E^c \sim \{\mathbf{0}\}$ and A is semisimple (cf. Problem 10 in Section 1.8), then there are positive constants m and M such that for all $t \in \mathbf{R}, m \leq |\mathbf{x}(t)| \leq M$;

   (d_1) if $\mathbf{x}_0 \in E^c \sim \{\mathbf{0}\}$ and A is not semisimple, then there is an $\mathbf{x}_0 \in \mathbf{R}^n$ such that $\lim_{t \to \pm\infty} |\mathbf{x}(t)| = \infty$;

   (d_2) if $E^s \neq \{\mathbf{0}\}$, $E^u \neq \{\mathbf{0}\}$, and $\mathbf{x}_0 \in E^s \oplus E^u \sim (E^s \cup E^u)$, then $\lim_{t \to \pm\infty} |\mathbf{x}(t)| = \infty$;

(e) if $E^u \neq \{\mathbf{0}\}$, $E^c \neq \{\mathbf{0}\}$ and $\mathbf{x}_0 \in E^u \oplus E^c \sim (E^u \cup E^c)$, then $\lim_{t \to \infty} |\mathbf{x}(t)| = \infty$; $\lim_{t \to -\infty} \mathbf{x}(t)$ does not exist;

(f) if $E^s \neq \{\mathbf{0}\}$, $E^c \neq \{\mathbf{0}\}$, and $\mathbf{x}_0 \in E^s \oplus E^c \sim (E^s \cup E^c)$, then $\lim_{t \to -\infty} |\mathbf{x}(t)| = \infty$, $\lim_{t \to \infty} \mathbf{x}(t)$ does not exist. Cf. Problem 12 in Problem Set 8.

6. Show that the only invariant lines for the linear system (1) with $\mathbf{x} \in \mathbf{R}^2$ are the lines $ax_1 + bx_2 = 0$ where $\mathbf{v} = (-b, a)^T$ is an eigenvector of A.

1.10 Nonhomogeneous Linear Systems

In this section we solve the nonhomogeneous linear system

$$\dot{\mathbf{x}} = A\mathbf{x} + \mathbf{b}(t) \tag{1}$$

where A is an $n \times n$ matrix and $\mathbf{b}(t)$ is a continuous vector valued function.

Definition. A *fundamental matrix solution* of

$$\dot{\mathbf{x}} = A\mathbf{x} \tag{2}$$

is any nonsingular $n \times n$ matrix function $\Phi(t)$ that satisfies

$$\Phi'(t) = A\Phi(t) \quad \text{for all} \quad t \in \mathbf{R}.$$

Note that according to the lemma in Section 1.4, $\Phi(t) = e^{At}$ is a fundamental matrix solution which satisfies $\Phi(0) = I$, the $n \times n$ identity matrix. Furthermore, any fundamental matrix solution $\Phi(t)$ of (2) is given by $\Phi(t) = e^{At}C$ for some nonsingular matrix C. Once we have found a fundamental matrix solution of (2), it is easy to solve the nonhomogeneous system (1). The result is given in the following theorem.

Theorem 1. *If $\Phi(t)$ is any fundamental matrix solution of (2), then the solution of the nonhomogeneous linear system (1) and the initial condition $\mathbf{x}(0) = \mathbf{x}_0$ is unique and is given by*

$$\mathbf{x}(t) = \Phi(t)\Phi^{-1}(0)\mathbf{x}_0 + \int_0^t \Phi(t)\Phi^{-1}(\tau)\mathbf{b}(\tau)d\tau. \tag{3}$$

Proof. For the function $\mathbf{x}(t)$ defined above,

$$\mathbf{x}'(t) = \Phi'(t)\Phi^{-1}(0)\mathbf{x}_0 + \Phi(t)\Phi^{-1}(t)\mathbf{b}(t)$$
$$+ \int_0^t \Phi'(t)\Phi^{-1}(\tau)\mathbf{b}(\tau)d\tau.$$

1.10. Nonhomogeneous Linear Systems

And since $\Phi(t)$ is a fundamental matrix solution of (2), it follows that

$$\mathbf{x}'(t) = A\left[\Phi(t)\Phi^{-1}(0)\mathbf{x}_0 + \int_0^t \Phi(t)\Phi^{-1}(\tau)\mathbf{b}(\tau)d\tau\right] + \mathbf{b}(t)$$
$$= A\mathbf{x}(t) + \mathbf{b}(t)$$

for all $t \in \mathbf{R}$. And this completes the proof of the theorem.

Remark 1. If the matrix A in (1) is time dependent, $A = A(t)$, then exactly the same proof shows that the solution of the nonhomogenous linear system (1) and the initial condition $\mathbf{x}(0) = \mathbf{x}_0$ is given by (3) provided that $\Phi(t)$ is a fundamental matrix solution of (2) with a variable coefficient matrix $A = A(t)$. For the most part, we do not consider solutions of (2) with $A = A(t)$ in this book. The reader should consult [C/L], [H] or [W] for a discussion of this topic which requires series methods and the theory of special functions.

Remark 2. With $\Phi(t) = e^{At}$, the solution of the nonhomogeneous linear system (1), as given in the above theorem, has the form

$$\mathbf{x}(t) = e^{At}\mathbf{x}_0 + e^{At}\int_0^t e^{-A\tau}\mathbf{b}(\tau)d\tau.$$

Example. Solve the forced harmonic oscillator problem

$$\ddot{x} + x = f(t).$$

This can be written as the nonhomogeneous system

$$\dot{x}_1 = -x_2$$
$$\dot{x}_2 = x_1 + f(t)$$

or equivalently in the form (1) with

$$A = \begin{bmatrix} 0 & -1 \\ 1 & 0 \end{bmatrix} \quad \text{and} \quad \mathbf{b}(t) = \begin{bmatrix} 0 \\ f(t) \end{bmatrix}.$$

In this case

$$e^{At} = \begin{bmatrix} \cos t & -\sin t \\ \sin t & \cos t \end{bmatrix} = R(t),$$

a rotation matrix; and

$$e^{-At} = \begin{bmatrix} \cos t & \sin t \\ -\sin t & \cos t \end{bmatrix} = R(-t).$$

The solution of the above system with initial condition $\mathbf{x}(0) = \mathbf{x}_0$ is thus given by

$$\mathbf{x}(t) = e^{At}\mathbf{x}_0 + e^{At}\int_0^t e^{-A\tau}\mathbf{b}(\tau)d\tau$$

$$= R(t)\mathbf{x}_0 + R(t)\int_0^t \begin{bmatrix} f(\tau)\sin\tau \\ f(\tau)\cos\tau \end{bmatrix} d\tau.$$

It follows that the solution $x(t) = x_1(t)$ of the original forced harmonic oscillator problem is given by

$$x(t) = x(0)\cos t - \dot{x}(0)\sin t + \int_0^t f(\tau)\sin(\tau - t)d\tau.$$

Problem Set 10

1. Just as the method of variation of parameters can be used to solve a nonhomogeneous linear differential equation, it can also be used to solve the nonhomogeneous linear system (1). To see how this method can be used to obtain the solution in the form (3), assume that the solution $\mathbf{x}(t)$ of (1) can be written in the form

$$\mathbf{x}(t) = \Phi(t)\mathbf{c}(t)$$

where $\Phi(t)$ is a fundamental matrix solution of (2). Differentiate this equation for $\mathbf{x}(t)$ and substitute it into (1) to obtain

$$\mathbf{c}'(t) = \Phi^{-1}(t)\mathbf{b}(t).$$

Integrate this equation and use the fact that $\mathbf{c}(0) = \Phi^{-1}(0)\mathbf{x}_0$ to obtain

$$\mathbf{c}(t) = \Phi^{-1}(0)\mathbf{x}_0 + \int_0^t \Phi^{-1}(\tau)\mathbf{b}(\tau)d\tau.$$

Finally, substitute the function $\mathbf{c}(t)$ into $\mathbf{x}(t) = \Phi(t)\mathbf{c}(t)$ to obtain (3).

2. Use Theorem 1 to solve the nonhomogeneous linear system

$$\dot{\mathbf{x}} = \begin{bmatrix} 1 & 1 \\ 0 & -1 \end{bmatrix}\mathbf{x} + \begin{pmatrix} t \\ 1 \end{pmatrix}$$

with the initial condition

$$\mathbf{x}(0) = \begin{pmatrix} 1 \\ 0 \end{pmatrix}.$$

1.10. Nonhomogeneous Linear Systems

3. Show that
$$\Phi(t) = \begin{bmatrix} e^{-2t}\cos t & -\sin t \\ e^{-2t}\sin t & \cos t \end{bmatrix}$$
is a fundamental matrix solution of the nonautonomous linear system
$$\dot{x} = A(t)x$$
with
$$A(t) = \begin{bmatrix} -2\cos^2 t & -1 - \sin 2t \\ 1 - \sin 2t & -2\sin^2 t \end{bmatrix}.$$
Find the inverse of $\Phi(t)$ and use Theorem 1 and Remark 1 to solve the nonhomogenous linear system
$$\dot{x} = A(t)x + b(t)$$
with $A(t)$ given above and $b(t) = (1, e^{-2t})^T$. Note that, in general, if $A(t)$ is a periodic matrix of period T, then corresponding to any fundamental matrix $\Phi(t)$, there exists a periodic matrix $P(t)$ of period $2T$ and a constant matrix B such that
$$\Phi(t) = P(t)e^{Bt}.$$
Cf. [C/L], p. 81. Show that $P(t)$ is a rotation matrix and $B = \mathrm{diag}[-2, 0]$ in this problem.

2

Nonlinear Systems: Local Theory

In Chapter 1 we saw that any linear system

$$\dot{\mathbf{x}} = A\mathbf{x} \tag{1}$$

has a unique solution through each point $\mathbf{x}_0$ in the phase space $\mathbf{R}^n$; the solution is given by $\mathbf{x}(t) = e^{At}\mathbf{x}_0$ and it is defined for all $t \in \mathbf{R}$. In this chapter we begin our study of nonlinear systems of differential equations

$$\dot{\mathbf{x}} = \mathbf{f}(\mathbf{x}) \tag{2}$$

where $\mathbf{f} \colon E \to \mathbf{R}^n$ and E is an open subset of $\mathbf{R}^n$. We show that under certain conditions on the function $\mathbf{f}$, the nonlinear system (2) has a unique solution through each point $\mathbf{x}_0 \in E$ defined on a maximal interval of existence $(\alpha, \beta) \subset \mathbf{R}$. In general, it is not possible to solve the nonlinear system (2); however, a great deal of qualitative information about the local behavior of the solution is determined in this chapter. In particular, we establish the Hartman–Grobman Theorem and the Stable Manifold Theorem which show that topologically the local behavior of the nonlinear system (2) near an equilibrium point $\mathbf{x}_0$ where $\mathbf{f}(\mathbf{x}_0) = \mathbf{0}$ is typically determined by the behavior of the linear system (1) near the origin when the matrix $A = D\mathbf{f}(\mathbf{x}_0)$, the derivative of $\mathbf{f}$ at $\mathbf{x}_0$. We also discuss some of the ramifications of these theorems for two-dimensional systems when $\det D\mathbf{f}(\mathbf{x}_0) \neq 0$ and cite some of the local results of Andronov et al. [A–I] for planar systems (2) with $\det D\mathbf{f}(\mathbf{x}_0) = 0$.

2.1 Some Preliminary Concepts and Definitions

Before beginning our discussion of the fundamental theory of nonlinear systems of differential equations, we present some preliminary concepts and definitions. First of all, in this book we shall only consider *autonomous* systems of ordinary differential equations

$$\dot{\mathbf{x}} = \mathbf{f}(\mathbf{x}) \tag{1}$$

as opposed to nonautonomous systems

$$\dot{\mathbf{x}} = \mathbf{f}(\mathbf{x}, t) \qquad (2)$$

where the function **f** can depend on the independent variable t; however, any nonautonomous system (2) with $\mathbf{x} \in \mathbf{R}^n$ can be written as an autonomous system (1) with $\mathbf{x} \in \mathbf{R}^{n+1}$ simply by letting $x_{n+1} = t$ and $\dot{x}_{n+1} = 1$. The fundamental theory for (1) and (2) does not differ significantly although it is possible to obtain the existence and uniqueness of solutions of (2) under slightly weaker hypotheses on **f** as a function of t; cf. for example Coddington and Levinson [C/L]. Also, see problem 3 in Problem Set 2.

Notice that the existence of the solution of the elementary differential equation

$$\dot{x} = f(t)$$

is given by

$$x(t) = x(0) + \int_0^t f(s)\, ds$$

if $f(t)$ is integrable. And in general, the differential equations (1) or (2) will have a solution if the function **f** is continuous; cf. [C/L], p. 6. However, continuity of the function **f** in (1) is not sufficient to guarantee uniqueness of the solution as the next example shows.

Example 1. The initial value problem

$$\dot{x} = 3x^{2/3}$$
$$x(0) = 0$$

has two different solutions through the point $(0, 0)$, namely

$$u(t) = t^3$$

and

$$v(t) \equiv 0$$

for all $t \in \mathbf{R}$. Clearly, each of these functions satisfies the differential equation for all $t \in \mathbf{R}$ as well as the initial condition $x(0) = 0$. (The first solution $u(t) = t^3$ can be obtained by the method of separation of variables.) Notice that the function $f(x) = 3x^{2/3}$ is continuous at $x = 0$ but that it is not differentiable there.

Another feature of nonlinear systems that differs from linear systems is that even when the function **f** in (1) is defined and continuous for all $\mathbf{x} \in \mathbf{R}^n$, the solution $\mathbf{x}(t)$ may become unbounded at some finite time $t = \beta$; i.e., the solution may only exist on some proper subinterval $(\alpha, \beta) \subset \mathbf{R}$. This is illustrated by the next example.

2.1. Some Preliminary Concepts and Definitions

Example 2. Consider the initial value problem

$$\dot{x} = x^2$$
$$x(0) = 1.$$

The solution, which can be found by the method of separation of variables, is given by

$$x(t) = \frac{1}{1-t}.$$

This solution is only defined for $t \in (-\infty, 1)$ and

$$\lim_{t \to 1^-} x(t) = \infty.$$

The interval $(-\infty, 1)$ is called the maximal interval of existence of the solution of this initial value problem. Notice that the function $x(t) = (1-t)^{-1}$ has another branch defined on the interval $(1, \infty)$; however, this branch is not considered as part of the solution of the initial value problem since the initial time $t = 0 \notin (1, \infty)$. This is made clear in the definition of a solution in Section 2.2.

Before stating and proving the fundamental existence-uniqueness theorem for the nonlinear system (1), it is first necessary to define some terminology and notation concerning the derivative $D\mathbf{f}$ of a function $\mathbf{f} \colon \mathbf{R}^n \to \mathbf{R}^n$.

Definition 1. The function $\mathbf{f} \colon \mathbf{R}^n \to \mathbf{R}^n$ is *differentiable* at $\mathbf{x}_0 \in \mathbf{R}^n$ if there is a linear transformation $D\mathbf{f}(\mathbf{x}_0) \in L(\mathbf{R}^n)$ that satisfies

$$\lim_{|\mathbf{h}| \to 0} \frac{|\mathbf{f}(\mathbf{x}_0 + \mathbf{h}) - \mathbf{f}(\mathbf{x}_0) - D\mathbf{f}(\mathbf{x}_0)\mathbf{h}|}{|\mathbf{h}|} = 0$$

The linear transformation $D\mathbf{f}(\mathbf{x}_0)$ is called the *derivative of* $\mathbf{f}$ *at* $\mathbf{x}_0$.

The following theorem, established for example on p. 215 in Rudin [R], gives us a method for computing the derivative in coordinates.

Theorem 1. *If* $\mathbf{f} \colon \mathbf{R}^n \to \mathbf{R}^n$ *is differentiable at* $\mathbf{x}_0$, *then the partial derivatives* $\frac{\partial f_i}{\partial x_j}$, $i, j = 1, \ldots, n$, *all exist at* $\mathbf{x}_0$ *and for all* $\mathbf{x} \in \mathbf{R}^n$,

$$D\mathbf{f}(\mathbf{x}_0)\mathbf{x} = \sum_{j=1}^n \frac{\partial \mathbf{f}}{\partial x_j}(\mathbf{x}_0) x_j.$$

Thus, if $\mathbf{f}$ is a differentiable function, the derivative $D\mathbf{f}$ is given by the $n \times n$ Jacobian matrix

$$D\mathbf{f} = \left[\frac{\partial f_i}{\partial x_j} \right].$$

Example 3. Find the derivative of the function

$$f(x) = \begin{bmatrix} x_1 - x_2^2 \\ -x_2 + x_1 x_2 \end{bmatrix}$$

and evaluate it at the point $x_0 = (1, -1)^T$. We first compute the Jacobian matrix of partial derivatives,

$$Df = \begin{bmatrix} \dfrac{\partial f_1}{\partial x_1} & \dfrac{\partial f_1}{\partial x_2} \\ \dfrac{\partial f_2}{\partial x_1} & \dfrac{\partial f_2}{\partial x_2} \end{bmatrix} = \begin{bmatrix} 1 & -2x_2 \\ x_2 & -1 + x_1 \end{bmatrix}$$

and then

$$Df(1, -1) = \begin{bmatrix} 1 & 2 \\ -1 & 0 \end{bmatrix}.$$

In most of the theorems in the remainder of this book, it is assumed that the function $f(x)$ is continuously differentiable; i.e., that the derivative $Df(x)$ considered as a mapping $Df \colon \mathbf{R}^n \to L(\mathbf{R}^n)$ is a continuous function of x in some open set $E \subset \mathbf{R}^n$. The linear spaces $\mathbf{R}^n$ and $L(\mathbf{R}^n)$ are endowed with the Euclidean norm $|\cdot|$ and the operator norm $\|\cdot\|$, defined in Section 1.3 of Chapter 1, respectively. Continuity is then defined as usual:

Definition 2. Suppose that V_1 and V_2 are two normed linear spaces with respective norms $\|\cdot\|_1$ and $\|\cdot\|_2$; i.e., V_1 and V_2 are linear spaces with norms $\|\cdot\|_1$ and $\|\cdot\|_2$ satisfying a–c in Section 1.3 of Chapter 1. Then

$$F \colon V_1 \to V_2$$

is *continuous* at $x_0 \in V_1$ if for all $\varepsilon > 0$ there exists a $\delta > 0$ such that $x \in V_1$ and $\|x - x_0\|_1 < \delta$ implies that

$$\|F(x) - F(x_0)\|_2 < \varepsilon.$$

And F is said to be continuous on the set $E \subset V_1$ if it is continuous at each point $x \in E$. If F is continuous on $E \subset V_1$, we write $F \in C(E)$.

Definition 3. Suppose that $f \colon E \to \mathbf{R}^n$ is differentiable on E. Then $f \in C^1(E)$ if the derivative $Df \colon E \to L(\mathbf{R}^n)$ is continuous on E.

The next theorem, established on p. 219 in Rudin [R], gives a simple test for deciding whether or not a function $f \colon E \to \mathbf{R}^n$ belongs to $C^1(E)$.

Theorem 2. *Suppose that E is an open subset of $\mathbf{R}^n$ and that $f \colon E \to \mathbf{R}^n$. Then $f \in C^1(E)$ iff the partial derivatives $\dfrac{\partial f_i}{\partial x_j}$, $i, j = 1, \ldots, n$, exist and are continuous on E.*

2.1. Some Preliminary Concepts and Definitions

Remark 1. For E an open subset of $\mathbf{R}^n$, the higher order derivatives $D^k\mathbf{f}(\mathbf{x}_0)$ of a function $\mathbf{f}\colon E \to \mathbf{R}^n$ are defined in a similar way and it can be shown that $\mathbf{f} \in C^k(E)$ if and only if the partial derivatives

$$\frac{\partial^k f_i}{\partial x_{j_1} \cdots \partial x_{j_k}}$$

with $i, j_1, \ldots, j_k = 1, \ldots, n$, exist and are continuous on E. Furthermore, $D^2\mathbf{f}(\mathbf{x}_0)\colon E \times E \to \mathbf{R}^n$ and for $(\mathbf{x}, \mathbf{y}) \in E \times E$ we have

$$D^2\mathbf{f}(\mathbf{x}_0)(\mathbf{x}, \mathbf{y}) = \sum_{j_1, j_2 = 1}^{n} \frac{\partial^2 \mathbf{f}(\mathbf{x}_0)}{\partial x_{j_1} \partial x_{j_2}} x_{j_1} y_{j_2}.$$

Similar formulas hold for $D^k\mathbf{f}(\mathbf{x}_0)\colon (E \times \cdots \times E) \to \mathbf{R}^n$; cf. [R], p. 235.

A function $\mathbf{f}\colon E \to \mathbf{R}^n$ is said to be *analytic* in the open set $E \subset \mathbf{R}^n$ if each component $f_j(\mathbf{x})$, $j = 1, \ldots, n$, is analytic in E, i.e., if for $j = 1, \ldots, n$ and $\mathbf{x}_0 \in E$, $f_j(\mathbf{x})$ has a Taylor series which converges to $f_j(\mathbf{x})$ in some neighborhood of $\mathbf{x}_0$ in E.

Problem Set 1

1. (a) Compute the derivative of the following functions

$$\mathbf{f}(\mathbf{x}) = \begin{bmatrix} x_1 + x_1 x_2^2 \\ -x_2 + x_2^2 + x_1^2 \end{bmatrix}, \quad \mathbf{f}(\mathbf{x}) = \begin{bmatrix} x_1 + x_1 x_2^2 + x_1 x_3^2 \\ -x_1 + x_2 - x_2 x_3 + x_1 x_2 x_3 \\ x_2 + x_3 - x_1^2 \end{bmatrix}.$$

   (b) Find the zeros of the above functions, i.e., the points $\mathbf{x}_0 \in \mathbf{R}^n$ where $\mathbf{f}(\mathbf{x}_0) = 0$, and evaluate $D\mathbf{f}(\mathbf{x})$ at these points.

   (c) For the first function $\mathbf{f}\colon \mathbf{R}^2 \to \mathbf{R}^2$ defined in part (a) above, compute $D^2\mathbf{f}(\mathbf{x}_0)(\mathbf{x}, \mathbf{y})$ where $\mathbf{x}_0 = (0, 1)$ is a zero of $\mathbf{f}$.

2. Find the largest open subset $E \subset \mathbf{R}^2$ for which

   (a) $\mathbf{f}(\mathbf{x}) = \begin{bmatrix} \frac{-x_1}{|\mathbf{x}|^3} \\ \frac{-x_2}{|\mathbf{x}|^3} \end{bmatrix}$ is continuously differentiable.

   (b) $\mathbf{f}(\mathbf{x}) = \begin{bmatrix} |\mathbf{x}| + \frac{1}{|x_1 - 1|} \\ \sqrt{x_1 + 1} - \sqrt{x_2 + 2} \end{bmatrix}$ is continuously differentiable.

3. Show that the initial value problem

$$\dot{x} = |x|^{1/2}$$
$$x(0) = 0$$

has four different solutions through the point $(0, 0)$. Sketch these solutions in the (t, x)-plane.

4. Show that the initial value problem

$$\dot{x} = x^3$$
$$x(0) = 2$$

has a solution on an interval $(-\infty, b)$ for some $b \in \mathbf{R}$. Sketch the solution in the (t, x)-plane and note the behavior of $x(t)$ as $t \to b^-$.

5. Show that the initial value problem

$$\dot{x} = \frac{1}{2x}$$
$$x(1) = 1$$

has a solution $x(t)$ on the interval $(0, \infty)$, that $x(t)$ is defined and continuous on $[0, \infty)$, but that $x'(0)$ does not exist.

6. Show that the function $\mathbf{F} \colon \mathbf{R}^2 \to L(\mathbf{R}^2)$ defined by

$$\mathbf{F}(\mathbf{x}) = \begin{bmatrix} x_1 & x_2 \\ -x_2 & x_1 \end{bmatrix}$$

is continuous for all $\mathbf{x} \in \mathbf{R}^2$ according to Definition 2.

2.2 The Fundamental Existence-Uniqueness Theorem

In this section, we establish the fundamental existence-uniqueness theorem for a nonlinear autonomous system of ordinary differential equations

$$\dot{\mathbf{x}} = \mathbf{f}(\mathbf{x}) \tag{1}$$

under the hypothesis that $\mathbf{f} \in C^1(E)$ where E is an open subset of $\mathbf{R}^n$. Picard's classical method of successive approximations is used to prove this theorem. The more modern approach based on the contraction mapping principle is relegated to the problems at the end of this section. The method of successive approximations not only allows us to establish the existence and uniqueness of the solution of the initial value problem associated with (1), but it also allows us to establish the continuity and differentiability of the solution with respect to initial conditions and parameters. This is done in the next section. The method is also used in the proof of the Stable Manifold Theorem in Section 2.7 and in the proof of the Hartman–Grobman Theorem in Section 2.8. The method of successive approximations is one of the basic tools used in the qualitative theory of ordinary differential equations.

2.2. The Fundamental Existence-Uniqueness Theorem

Definition 1. Suppose that $f \in C(E)$ where E is an open subset of $\mathbf{R}^n$. Then $\mathbf{x}(t)$ *is a solution of the differential equation* (1) *on an interval* I if $\mathbf{x}(t)$ is differentiable on I and if for all $t \in I$, $\mathbf{x}(t) \in E$ and

$$\mathbf{x}'(t) = \mathbf{f}(\mathbf{x}(t)).$$

And given $\mathbf{x}_0 \in E$, $\mathbf{x}(t)$ *is a solution of the initial value problem*

$$\dot{\mathbf{x}} = \mathbf{f}(\mathbf{x})$$
$$\mathbf{x}(t_0) = \mathbf{x}_0$$

on an interval I if $t_0 \in I$, $\mathbf{x}(t_0) = \mathbf{x}_0$ and $\mathbf{x}(t)$ is a solution of the differential equation (1) on the interval I.

In order to apply the method of successive approximations to establish the existence of a solution of (1), we need to define the concept of a Lipschitz condition and show that C^1 functions are locally Lipschitz.

Definition 2. Let E be an open subset of $\mathbf{R}^n$. A function $\mathbf{f}: E \to \mathbf{R}^n$ is said to *satisfy a Lipschitz condition on* E if there is a positive constant K such that for all $\mathbf{x}, \mathbf{y} \in E$

$$|\mathbf{f}(\mathbf{x}) - \mathbf{f}(\mathbf{y})| \leq K|\mathbf{x} - \mathbf{y}|.$$

The function $\mathbf{f}$ is said to be *locally Lipschitz* on E if for each point $\mathbf{x}_0 \in E$ there is an ε-neighborhood of $\mathbf{x}_0$, $N_\varepsilon(\mathbf{x}_0) \subset E$ and a constant $K_0 > 0$ such that for all $\mathbf{x}, \mathbf{y} \in N_\varepsilon(\mathbf{x}_0)$

$$|\mathbf{f}(\mathbf{x}) - \mathbf{f}(\mathbf{y})| \leq K_0|\mathbf{x} - \mathbf{y}|.$$

By an ε-neighborhood of a point $\mathbf{x}_0 \in \mathbf{R}^n$, we mean an open ball of positive radius ε; i.e.,

$$N_\varepsilon(\mathbf{x}_0) = \{\mathbf{x} \in \mathbf{R}^n \mid |\mathbf{x} - \mathbf{x}_0| < \varepsilon\}.$$

Lemma. *Let E be an open subset of $\mathbf{R}^n$ and let* $\mathbf{f}: E \to \mathbf{R}^n$. *Then, if* $\mathbf{f} \in C^1(E)$, $\mathbf{f}$ *is locally Lipschitz on* E.

Proof. Since E is an open subset of $\mathbf{R}^n$, given $\mathbf{x}_0 \in E$, there is an $\varepsilon > 0$ such that $N_\varepsilon(\mathbf{x}_0) \subset E$. Let

$$K = \max_{|\mathbf{x} - \mathbf{x}_0| \leq \varepsilon/2} \|D\mathbf{f}(\mathbf{x})\|,$$

the maximum of the continuous function $D\mathbf{f}(\mathbf{x})$ on the compact set $|\mathbf{x} - \mathbf{x}_0| \leq \varepsilon/2$. Let N_0 denote the $\varepsilon/2$-neighborhood of $\mathbf{x}_0$, $N_{\varepsilon/2}(\mathbf{x}_0)$. Then for $\mathbf{x}, \mathbf{y} \in N_0$, set $\mathbf{u} = \mathbf{y} - \mathbf{x}$. It follows that $\mathbf{x} + s\mathbf{u} \in N_0$ for $0 \leq s \leq 1$ since N_0 is a convex set. Define the function $\mathbf{F}: [0,1] \to \mathbf{R}^n$ by

$$\mathbf{F}(s) = \mathbf{f}(\mathbf{x} + s\mathbf{u}).$$

Then by the chain rule,

$$\mathbf{F}'(s) = D\mathbf{f}(\mathbf{x} + s\mathbf{u})\mathbf{u}$$

and therefore

$$\mathbf{f}(\mathbf{y}) - \mathbf{f}(\mathbf{x}) = \mathbf{F}(1) - \mathbf{F}(0)$$
$$= \int_0^1 \mathbf{F}'(s)\,ds = \int_0^1 D\mathbf{f}(\mathbf{x} + s\mathbf{u})\mathbf{u}\,ds.$$

It then follows from the lemma in Section 1.3 of Chapter 1 that

$$|\mathbf{f}(\mathbf{y}) - \mathbf{f}(\mathbf{x})| \le \int_0^1 |D\mathbf{f}(\mathbf{x} + s\mathbf{u})\mathbf{u}|\,ds$$
$$\le \int_0^1 \|D\mathbf{f}(\mathbf{x} + s\mathbf{u})\|\,|\mathbf{u}|\,ds$$
$$\le K|\mathbf{u}| = K|\mathbf{y} - \mathbf{x}|.$$

And this proves the lemma.

Picard's method of successive approximations is based on the fact that $\mathbf{x}(t)$ is a solution of the initial value problem

$$\dot{\mathbf{x}} = \mathbf{f}(\mathbf{x}) \qquad \qquad (2)$$
$$\mathbf{x}(0) = \mathbf{x}_0$$

if and only if $\mathbf{x}(t)$ is a continuous function that satisfies the integral equation

$$\mathbf{x}(t) = \mathbf{x}_0 + \int_0^t \mathbf{f}(\mathbf{x}(s))\,ds.$$

The successive approximations to the solution of this integral equation are defined by the sequence of functions

$$\mathbf{u}_0(t) = \mathbf{x}_0$$
$$\mathbf{u}_{k+1}(t) = \mathbf{x}_0 + \int_0^t \mathbf{f}(\mathbf{u}_k(s))\,ds \qquad (3)$$

for $k = 0, 1, 2, \ldots$. In order to illustrate the mechanics involved in the method of successive approximations, we use the method to solve an elementary linear differential equation

Example 1. Solve the initial value problem

$$\dot{x} = ax$$
$$x(0) = x_0$$

2.2. The Fundamental Existence-Uniqueness Theorem

by the method of successive approximations. Let

$$u_0(t) = x_0$$

and compute

$$u_1(t) = x_0 + \int_0^t a x_0 \, ds = x_0(1 + at)$$

$$u_2(t) = x_0 + \int_0^t a x_0 (1 + as) \, ds = x_0 \left(1 + at + a^2 \frac{t^2}{2}\right)$$

$$u_3(t) = x_0 + \int_0^t a x_0 \left(1 + as + a^2 \frac{s^2}{2}\right) ds$$

$$= x_0 \left(1 + at + a^2 \frac{t^2}{2!} + a^3 \frac{t^3}{3!}\right).$$

It follows by induction that

$$u_k(t) = x_0 \left(1 + at + \cdots + a^k \frac{t^k}{k!}\right)$$

and we see that

$$\lim_{k \to \infty} u_k(t) = x_0 e^{at}.$$

That is, the successive approximations converge to the solution $x(t) = x_0 e^{at}$ of the initial value problem.

In order to show that the successive approximations (3) converge to a solution of the initial value problem (2) on an interval $I = [-a, a]$, it is first necessary to review some material concerning the completeness of the linear space $C(I)$ of continuous functions on an interval $I = [-a, a]$. The norm on $C(I)$ is defined as

$$\|\mathbf{u}\| = \sup_I |\mathbf{u}(t)|.$$

Convergence in this norm is equivalent to uniform convergence.

Definition 3. Let V be a normed linear space. Then a sequence $\{\mathbf{u}_k\} \subset V$ is called a *Cauchy sequence* if for all $\varepsilon > 0$ there is an N such that $k, m \geq N$ implies that

$$\|\mathbf{u}_k - \mathbf{u}_m\| < \varepsilon.$$

The space V is called *complete* if every Cauchy sequence in V converges to an element in V.

The following theorem, proved for example in Rudin [R] on p. 151, establishes the completeness of the normed linear space $C(I)$ with $I = [-a, a]$.

Theorem. *For $I = [-a, a]$, $C(I)$ is a complete normed linear space.*

2. Nonlinear Systems: Local Theory

We can now prove the fundamental existence-uniqueness theorem for nonlinear systems.

Theorem (The Fundamental Existence-Uniqueness Theorem). *Let E be an open subset of $\mathbf{R}^n$ containing $\mathbf{x}_0$ and assume that $\mathbf{f} \in C^1(E)$. Then there exists an $a > 0$ such that the initial value problem*

$$\dot{\mathbf{x}} = \mathbf{f}(\mathbf{x})$$
$$\mathbf{x}(0) = \mathbf{x}_0$$

has a unique solution $\mathbf{x}(t)$ on the interval $[-a, a]$.

Proof. Since $\mathbf{f} \in C^1(E)$, it follows from the lemma that there is an ε-neighborhood $N_\varepsilon(\mathbf{x}_0) \subset E$ and a constant $K > 0$ such that for all $\mathbf{x}, \mathbf{y} \in N_\varepsilon(\mathbf{x}_0)$,

$$|\mathbf{f}(\mathbf{x}) - \mathbf{f}(\mathbf{y})| \leq K|\mathbf{x} - \mathbf{y}|.$$

Let $b = \varepsilon/2$. Then the continuous function $\mathbf{f}(\mathbf{x})$ is bounded on the compact set

$$N_0 = \{\mathbf{x} \in \mathbf{R}^n \mid |\mathbf{x} - \mathbf{x}_0| \leq b\}.$$

Let

$$M = \max_{\mathbf{x} \in N_0} |\mathbf{f}(\mathbf{x})|.$$

Let the successive approximations $\mathbf{u}_k(t)$ be defined by (3). Then assuming that there exists an $a > 0$ such that $\mathbf{u}_k(t)$ is defined and continuous on $[-a, a]$ and satisfies

$$\max_{[-a,a]} |\mathbf{u}_k(t) - \mathbf{x}_0| \leq b, \qquad (4)$$

it follows that $\mathbf{f}(\mathbf{u}_k(t))$ is defined and continuous on $[-a, a]$ and therefore that

$$\mathbf{u}_{k+1}(t) = \mathbf{x}_0 + \int_0^t \mathbf{f}(\mathbf{u}_k(s))\, ds$$

is defined and continuous on $[-a, a]$ and satisfies

$$|\mathbf{u}_{k+1}(t) - \mathbf{x}_0| \leq \int_0^t |\mathbf{f}(\mathbf{u}_k(s))|\, ds \leq Ma$$

for all $t \in [-a, a]$. Thus, choosing $0 < a \leq b/M$, it follows by induction that $\mathbf{u}_k(t)$ is defined and continuous and satisfies (4) for all $t \in [-a, a]$ and $k = 1, 2, 3, \ldots$.

Next, since for all $t \in [-a, a]$ and $k = 0, 1, 2, 3, \ldots$, $\mathbf{u}_k(t) \in N_0$, it follows from the Lipschitz condition satisfied by $\mathbf{f}$ that for all $t \in [-a, a]$

$$|\mathbf{u}_2(t) - \mathbf{u}_1(t)| \leq \int_0^t |\mathbf{f}(\mathbf{u}_1(s)) - \mathbf{f}(\mathbf{u}_0(s))|\, ds$$

2.2. The Fundamental Existence-Uniqueness Theorem

$$\leq K \int_0^t |u_1(s) - u_0(s)|\, ds$$
$$\leq Ka \max_{[-a,a]} |u_1(t) - x_0|$$
$$\leq Kab.$$

And then assuming that

$$\max_{[-a,a]} |u_j(t) - u_{j-1}(t)| \leq (Ka)^{j-1} b \tag{5}$$

for some integer $j \geq 2$, it follows that for all $t \in [-a, a]$

$$|u_{j+1}(t) - u_j(t)| \leq \int_0^t |f(u_j(s)) - f(u_{j-1}(s))|\, ds$$
$$\leq K \int_0^t |u_j(s) - u_{j-1}(s)|\, ds$$
$$\leq Ka \max_{[-a,a]} |u_j(t) - u_{j-1}(t)|$$
$$\leq (Ka)^j b.$$

Thus, it follows by induction that (5) holds for $j = 2, 3, \ldots$. Setting $\alpha = Ka$ and choosing $0 < a < 1/K$, we see that for $m > k \geq N$ and $t \in [-a, a]$

$$|u_m(t) - u_k(t)| \leq \sum_{j=k}^{m-1} |u_{j+1}(t) - u_j(t)|$$
$$\leq \sum_{j=N}^{\infty} |u_{j+1}(t) - u_j(t)|$$
$$\leq \sum_{j=N}^{\infty} \alpha^j b = \frac{\alpha^N}{1-\alpha} b.$$

This last quantity approaches zero as $N \to \infty$. Therefore, for all $\varepsilon > 0$ there exists an N such that $m, k \geq N$ implies that

$$\|u_m - u_k\| = \max_{[-a,a]} |u_m(t) - u_k(t)| < \varepsilon;$$

i.e., $\{u_k\}$ is a Cauchy sequence of continuous functions in $C([-a, a])$. It follows from the above theorem that $u_k(t)$ converges to a continuous function $u(t)$ uniformly for all $t \in [-a, a]$ as $k \to \infty$. And then taking the limit of both sides of equation (3) defining the successive approximations, we see that the continuous function

$$u(t) = \lim_{k \to \infty} u_k(t) \tag{6}$$

satisfies the integral equation

$$\mathbf{u}(t) = \mathbf{x}_0 + \int_0^t \mathbf{f}(\mathbf{u}(s))\, ds \qquad (7)$$

for all $t \in [-a, a]$. We have used the fact that the integral and the limit can be interchanged since the limit in (6) is uniform for all $t \in [-a, a]$. Then since $\mathbf{u}(t)$ is continuous, $\mathbf{f}(\mathbf{u}(t))$ is continuous and by the fundamental theorem of calculus, the right-hand side of the integral equation (7) is differentiable and

$$\mathbf{u}'(t) = \mathbf{f}(\mathbf{u}(t))$$

for all $t \in [-a, a]$. Furthermore, $\mathbf{u}(0) = \mathbf{x}_0$ and from (4) it follows that $\mathbf{u}(t) \in N_\varepsilon(\mathbf{x}_0) \subset E$ for all $t \in [-a, a]$. Thus $\mathbf{u}(t)$ is a solution of the initial value problem (2) on $[-a, a]$. It remains to show that it is the only solution.

Let $\mathbf{u}(t)$ and $\mathbf{v}(t)$ be two solutions of the initial value problem (2) on $[-a, a]$. Then the continuous function $|\mathbf{u}(t) - \mathbf{v}(t)|$ achieves its maximum at some point $t_1 \in [-a, a]$. It follows that

$$\|\mathbf{u} - \mathbf{v}\| = \max_{[-a,a]} |\mathbf{u}(t) - \mathbf{v}(t)|$$

$$= \left| \int_0^{t_1} \mathbf{f}(\mathbf{u}(s)) - \mathbf{f}(\mathbf{v}(s))\, ds \right|$$

$$\leq \int_0^{|t_1|} |\mathbf{f}(\mathbf{u}(s)) - \mathbf{f}(\mathbf{v}(s))|\, ds$$

$$\leq K \int_0^{|t_1|} |\mathbf{u}(s) - \mathbf{v}(s)|\, ds$$

$$\leq K a \max_{[-a,a]} |\mathbf{u}(t) - \mathbf{v}(t)|$$

$$\leq K a \|\mathbf{u} - \mathbf{v}\|.$$

But $Ka < 1$ and this last inequality can only be satisfied if $\|\mathbf{u} - \mathbf{v}\| = 0$. Thus, $\mathbf{u}(t) = \mathbf{v}(t)$ on $[-a, a]$. We have shown that the successive approximations (3) converge uniformly to a unique solution of the initial value problem (2) on the interval $[-a, a]$ where a is any number satisfying $0 < a < \min(\frac{b}{M}, \frac{1}{K})$.

Remark. Exactly the same method of proof shows that the initial value problem

$$\dot{\mathbf{x}} = \mathbf{f}(\mathbf{x})$$
$$\mathbf{x}(t_0) = \mathbf{x}_0$$

has a unique solution on some interval $[t_0 - a, t_0 + a]$.

2.2. The Fundamental Existence-Uniqueness Theorem

PROBLEM SET 2

1. (a) Find the first three successive approximations $u_1(t)$, $u_2(t)$ and $u_3(t)$ for the initial value problem

$$\dot{x} = x^2$$
$$x(0) = 1.$$

Also, use mathematical induction to show that for all $n \geq 1$,
$u_n(t) = 1 + t + \cdots + t^n + O(t^{n+1})$ as $t \to 0$.

(b) Solve the initial value problem in part (a) and show that the function $x(t) = 1/(1-t)$ is a solution of that initial value problem on the interval $(-\infty, 1)$ according to Definition 1. Also, show that the first $(n+1)$-terms in $u_n(t)$ agree with the first $(n+1)$-terms in the Taylor series for the function $x(t) = 1/(1-t)$ about $x = 0$.

(c) Show that the function $x(t) = (3t)^{1/3}$, which is defined and continuous for all $t \in \mathbf{R}$, is a solution of the differential equation

$$\dot{x} = \frac{1}{x^2}$$

for all $t \neq 0$ and that it is a solution of the corresponding initial value problem with $x(1/3) = 1$ on the interval $(0, \infty)$ according to Definition 1.

2. Let A be an $n \times n$ matrix. Show that the successive approximations (3) converge to the solution $\mathbf{x}(t) = e^{At}\mathbf{x}_0$ of the initial value problem

$$\dot{\mathbf{x}} = A\mathbf{x}$$
$$\mathbf{x}(0) = \mathbf{x}_0.$$

3. Use the method of successive approximations to show that if $\mathbf{f}(\mathbf{x}, t)$ is continuous in t for all t in some interval containing $t = 0$ and continuously differentiable in $\mathbf{x}$ for all $\mathbf{x}$ in some open set $E \subset \mathbf{R}^n$ containing $\mathbf{x}_0$, then there exists an $a > 0$ such that the initial value problem

$$\dot{\mathbf{x}} = \mathbf{f}(\mathbf{x}, t)$$
$$\mathbf{x}(0) = \mathbf{x}_0$$

has a unique solution $\mathbf{x}(t)$ on the interval $[-a, a]$. **Hint:** Define $\mathbf{u}_0(t) = \mathbf{x}_0$ and

$$\mathbf{u}_{k+1}(t) = \mathbf{x}_0 + \int_0^t \mathbf{f}(\mathbf{u}_k(s), s)\, ds$$

and show that the successive approximations $\mathbf{u}_k(t)$ converge uniformly to $\mathbf{x}(t)$ on $[-a, a]$ as in the proof of the fundamental existence-uniqueness theorem.

4. Use the method of successive approximations to show that if the matrix valued function $A(t)$ is continuous on $[-a_0, a_0]$ then there exists an $a > 0$ such that the initial value problem

$$\dot{\Phi} = A(t)\Phi$$
$$\Phi(0) = I$$

(where I is the $n \times n$ identity matrix) has a unique fundamental matrix solution $\Phi(t)$ on $[-a, a]$. **Hint:** Define $\Phi_0(t) = I$ and

$$\Phi_{k+1}(t) = I + \int_0^t A(s)\Phi_k(s)\,ds,$$

and use the fact that the continuous matrix valued function $A(t)$ satisfies $\|A(t)\| \leq M_0$ for all t in the compact set $[-a_0, a_0]$ to show that the successive approximations $\Phi_k(t)$ converge uniformly to $\Phi(t)$ on some interval $[-a, a]$ with $a < 1/M_0$ and $a \leq a_0$.

5. Let V be a normed linear space. If $T \colon V \to V$ satisfies

$$\|T(\mathbf{u}) - T(\mathbf{v})\| \leq c\|\mathbf{u} - \mathbf{v}\|$$

for all $\mathbf{u}$ and $\mathbf{v} \in V$ with $0 < c < 1$ then T is called a *contraction mapping*. The following theorem is proved for example in Rudin [R]:

Theorem (The Contraction Mapping Principle). *Let V be a complete normed linear space and $T \colon V \to V$ a contraction mapping. Then there exists a unique $\mathbf{u} \in V$ such that $T(\mathbf{u}) = \mathbf{u}$.*

Let $\mathbf{f} \in C^1(E)$ and $\mathbf{x}_0 \in E$. For $I = [-a, a]$ and $\mathbf{u} \in C(I)$, let

$$T(\mathbf{u})(t) = \mathbf{x}_0 + \int_0^t \mathbf{f}(\mathbf{u}(s))\,ds.$$

Define a closed subset V of $C(I)$ and apply the Contraction Mapping Principle to show that the integral equation (7) has a unique continuous solution $\mathbf{u}(t)$ for all $t \in [-a, a]$ provided the constant $a > 0$ is sufficiently small.

Hint: Since $\mathbf{f}$ is locally Lipschitz on E and $\mathbf{x}_0 \in E$, there are positive constants ε and K_0 such that the condition in Definition 2 is satisfied on $N_\varepsilon(\mathbf{x}_0) \subset E$. Let $V = \{\mathbf{u} \in C(I) \mid \|\mathbf{u} - \mathbf{x}_0\| \leq \varepsilon\}$. Then V is complete since it is a closed subset of $C(I)$. Show that (i) for all $\mathbf{u}, \mathbf{v} \in V$, $\|T(\mathbf{u}) - T(\mathbf{v})\| \leq aK_0\|\mathbf{u} - \mathbf{v}\|$ and that (ii) the positive constant a can be chosen sufficiently small that for $t \in [-a, a]$, $T \circ \mathbf{u}(t) \in N_\varepsilon(\mathbf{x}_0)$, i.e., $T \colon V \to V$.

6. Prove that $\mathbf{x}(t)$ is a solution of the initial value problem (2) for all $t \in I$ if and only if $\mathbf{x}(t)$ is a continuous function that satisfies the integral equation

$$\mathbf{x}(t) = \mathbf{x}_0 + \int_0^t \mathbf{f}(\mathbf{x}(s))\,ds$$

for all $t \in I$.

2.3. Dependence on Initial Conditions and Parameters

7. Under the hypothesis of the Fundamental Existence-Uniqueness Theorem, if $\mathbf{x}(t)$ is the solution of the initial value problem (2) on an interval I, prove that the second derivative $\ddot{\mathbf{x}}(t)$ is continuous on I.

8. Prove that if $\mathbf{f} \in C^1(E)$ where E is a compact convex subset of $\mathbf{R}^n$ then $\mathbf{f}$ satisfies a Lipschitz condition on E. **Hint:** Cf. Theorem 9.19 in [R].

9. Prove that if $\mathbf{f}$ satisfies a Lipschitz condition on E then $\mathbf{f}$ is uniformly continuous on E.

10. (a) Show that the function $f(x) = 1/x$ is not uniformly continuous on $E = (0,1)$. **Hint:** f is uniformly continuous on E if for all $\varepsilon > 0$ there exists a $\delta > 0$ such that for all $x, y \in E$ with $|x - y| < \delta$ we have $|f(x) - f(y)| < \varepsilon$. Thus, f is *not* uniformly continuous on E if there exists an $\varepsilon > 0$ such that for all $\delta > 0$ there exist $x, y \in E$ with $|x - y| < \delta$ such that $|f(x) - f(y)| \geq \varepsilon$. Choose $\varepsilon = 1$ and show that for all $\delta > 0$ with $\delta < 1$, $x = \delta/2$ and $y = \delta$ implies that $x, y \in (0,1)$, $|x - y| < \delta$ and $|f(x) - f(y)| > 1$.

    (b) Show that $f(x) = 1/x$ does not satisfy a Lipschitz condition on $(0,1)$.

11. Prove that if $\mathbf{f}$ is differentiable at $\mathbf{x}_0$ then there exists a $\delta > 0$ and a $K_0 > 0$ such that for all $\mathbf{x} \in N_\delta(\mathbf{x}_0)$
$$|\mathbf{f}(\mathbf{x}) - \mathbf{f}(\mathbf{x}_0)| \leq K_0 |\mathbf{x} - \mathbf{x}_0|.$$

2.3 Dependence on Initial Conditions and Parameters

In this section we investigate the dependence of the solution of the initial value problem
$$\begin{aligned} \dot{\mathbf{x}} &= \mathbf{f}(\mathbf{x}) \\ \mathbf{x}(0) &= \mathbf{y} \end{aligned} \qquad (1)$$
on the initial condition $\mathbf{y}$. If the differential equation depends on a parameter $\mu \in \mathbf{R}^m$, i.e., if the function $\mathbf{f}(\mathbf{x})$ in (1) is replaced by $\mathbf{f}(\mathbf{x}, \mu)$, then the solution $\mathbf{u}(t, \mathbf{y}, \mu)$ will also depend on the parameter μ. Roughly speaking, the dependence of the solution $\mathbf{u}(t, \mathbf{y}, \mu)$ on the initial condition $\mathbf{y}$ and the parameter μ is as continuous as the function $\mathbf{f}$. In order to establish this type of continuous dependence of the solution on initial conditions and parameters, we first establish a result due to T.H. Gronwall.

Lemma (Gronwall). *Suppose that $g(t)$ is a continuous real valued function that satisfies $g(t) \geq 0$ and*
$$g(t) \leq C + K \int_0^t g(s)\, ds$$

for all $t \in [0, a]$ where C and K are positive constants. It then follows that for all $t \in [0, a]$,

$$g(t) \leq Ce^{Kt}.$$

Proof. Let $G(t) = C + K \int_0^t g(s)\, ds$ for $t \in [0, a]$. Then $G(t) \geq g(t)$ and $G(t) > 0$ for all $t \in [0, a]$. It follows from the fundamental theorem of calculus that

$$G'(t) = Kg(t)$$

and therefore that

$$\frac{G'(t)}{G(t)} = \frac{Kg(t)}{G(t)} \leq \frac{KG(t)}{G(t)} = K$$

for all $t \in [0, a]$. And this is equivalent to saying that

$$\frac{d}{dt}(\log G(t)) \leq K$$

or

$$\log G(t) \leq Kt + \log G(0)$$

or

$$G(t) \leq G(0)e^{Kt} = Ce^{Kt}$$

for all $t \in [0, a]$, which implies that $g(t) \leq Ce^{Kt}$ for all $t \in [0, a]$.

Theorem 1 (Dependence on Initial Conditions). *Let E be an open subset of $\mathbf{R}^n$ containing $\mathbf{x}_0$ and assume that $\mathbf{f} \in C^1(E)$. Then there exists an $a > 0$ and a $\delta > 0$ such that for all $\mathbf{y} \in N_\delta(\mathbf{x}_0)$ the initial value problem*

$$\dot{\mathbf{x}} = \mathbf{f}(\mathbf{x})$$
$$\mathbf{x}(0) = \mathbf{y}$$

has a unique solution $\mathbf{u}(t, \mathbf{y})$ with $\mathbf{u} \in C^1(G)$ where $G = [-a, a] \times N_\delta(\mathbf{x}_0) \subset \mathbf{R}^{n+1}$; furthermore, for each $\mathbf{y} \in N_\delta(\mathbf{x}_0)$, $\mathbf{u}(t, \mathbf{y})$ is a twice continuously differentiable function of t for $t \in [-a, a]$.

Proof. Since $\mathbf{f} \in C^1(E)$, it follows from the lemma in Section 2.2 that there is an ε-neighborhood $N_\varepsilon(\mathbf{x}_0) \subset E$ and a constant $K > 0$ such that for all $\mathbf{x}$ and $\mathbf{y} \in N_\varepsilon(\mathbf{x}_0)$,

$$|\mathbf{f}(\mathbf{x}) - \mathbf{f}(\mathbf{y})| \leq K|\mathbf{x} - \mathbf{y}|.$$

As in the proof of the fundamental existence theorem, let $N_0 = \{\mathbf{x} \in \mathbf{R}^n \mid |\mathbf{x} - \mathbf{x}_0| \leq \varepsilon/2\}$, let M_0 be the maximum of $|\mathbf{f}(\mathbf{x})|$ on N_0 and let M_1 be the maximum of $\|D\mathbf{f}(\mathbf{x})\|$ on N_0. Let $\delta = \varepsilon/4$, and for $\mathbf{y} \in N_\delta(\mathbf{x}_0)$ define the successive approximations $\mathbf{u}_k(t, \mathbf{y})$ as

$$\mathbf{u}_0(t, \mathbf{y}) = \mathbf{y}$$
$$\mathbf{u}_{k+1}(t, \mathbf{y}) = \mathbf{y} + \int_0^t \mathbf{f}(\mathbf{u}_k(s, \mathbf{y}))\, ds. \tag{2}$$

2.3. Dependence on Initial Conditions and Parameters

Assume that $\mathbf{u}_k(t, \mathbf{y})$ is defined and continuous for all $(t, \mathbf{y}) \in G = [-a, a] \times N_\delta(\mathbf{x}_0)$ and that for all $\mathbf{y} \in N_\delta(\mathbf{x}_0)$

$$\|\mathbf{u}_k(t, \mathbf{y}) - \mathbf{x}_0\| < \varepsilon/2 \tag{3}$$

where $\|\cdot\|$ denotes the maximum over all $t \in [-a, a]$. This is clearly satisfied for $k = 0$. And assuming this is true for k, it follows that $\mathbf{u}_{k+1}(t, \mathbf{y})$, defined by the above successive approximations, is continuous on G. This follows since a continuous function of a continuous function is continuous and since the above integral of the continuous function $\mathbf{f}(\mathbf{u}_k(s, \mathbf{y}))$ is continuous in t by the fundamental theorem of calculus and also in $\mathbf{y}$; cf. Rudin [R] or Carslaw [C]. We also have

$$\|\mathbf{u}_{k+1}(t, \mathbf{y}) - \mathbf{y}\| \leq \int_0^t |\mathbf{f}(\mathbf{u}_k(s, \mathbf{y}))|\, ds \leq M_0 a$$

for $t \in [-a, a]$ and $\mathbf{y} \in N_\delta(\mathbf{x}_0) \subset N_0$. Thus, for $t \in [-a, a]$ and $\mathbf{y} \in N_\delta(\mathbf{x}_0)$ with $\delta = \varepsilon/4$, we have

$$\|\mathbf{u}_{k+1}(t, \mathbf{y}) - \mathbf{x}_0\| \leq \|\mathbf{u}_{k+1}(t, \mathbf{y}) - \mathbf{y}\| + \|\mathbf{y} - \mathbf{x}_0\|$$
$$\leq M_0 a + \varepsilon/4 < \varepsilon/2$$

provided $M_0 a < \varepsilon/4$, i.e., provided $a < \varepsilon/(4M_0)$. Thus, the above induction hypothesis holds for all $k = 1, 2, 3, \ldots$ and $(t, \mathbf{y}) \in G$ provided $a < \varepsilon/(4M_0)$.

We next show that the successive approximations $\mathbf{u}_k(t, \mathbf{y})$ converge uniformly to a continuous function $\mathbf{u}(t, \mathbf{y})$ for all $(t, \mathbf{y}) \in G$ as $k \to \infty$. As in the proof of the fundamental existence theorem,

$$\|\mathbf{u}_2(t, \mathbf{y}) - \mathbf{u}_1(t, \mathbf{y})\| \leq Ka\|\mathbf{u}_1(t, \mathbf{y}) - \mathbf{y}\|$$
$$\leq Ka\|\mathbf{u}_1(t, \mathbf{y}) - \mathbf{x}_0\| + Ka\|\mathbf{y} - \mathbf{x}_0\|$$
$$\leq Ka(\varepsilon/2 + \varepsilon/4) \leq Ka\varepsilon$$

for $(t, \mathbf{y}) \in G$. And then it follows exactly as in the proof of the fundamental existence theorem in Section 2.2 that

$$\|\mathbf{u}_{k+1}(t, \mathbf{y}) - \mathbf{u}_k(t, \mathbf{y})\| \leq (Ka)^k \varepsilon$$

for $(t, \mathbf{y}) \in G$ and consequently that the successive approximations converge uniformly to a continuous function $\mathbf{u}(t, \mathbf{y})$ for $(t, \mathbf{y}) \in G$ as $k \to \infty$ provided $a < 1/K$. Furthermore, the function $\mathbf{u}(t, \mathbf{y})$ satisfies

$$\mathbf{u}(t, \mathbf{y}) = \mathbf{y} + \int_0^t \mathbf{f}(\mathbf{u}(s, \mathbf{y}))\, ds$$

for $(t, \mathbf{y}) \in G$ and also $\mathbf{u}(0, \mathbf{y}) = \mathbf{y}$. And it follows from the inequality (3) that $\mathbf{u}(t, \mathbf{y}) \in N_{\varepsilon/2}(\mathbf{x}_0)$ for all $(t, \mathbf{y}) \in G$. Thus, by the fundamental theorem of calculus and the chain rule, it follows that

$$\dot{\mathbf{u}}(t, \mathbf{y}) = \mathbf{f}(\mathbf{u}(t, \mathbf{y}))$$

and that
$$\ddot{u}(t, y) = Df(u(t, y))\dot{u}(t, y)$$
for all $(t, y) \in G$; i.e., $u(t, y)$ is a twice continuously differentiable function of t which satisfies the initial value problem (1) for all $(t, y) \in G$. The uniqueness of the solution $u(t, y)$ follows from the fundamental theorem in Section 2.2.

We now show that $u(t, y)$ is a continuously differentiable function of y for all $(t, y) \in [-a, a] \times N_{\delta/2}(x_0)$. In order to do this, fix $y_0 \in N_{\delta/2}(x_0)$ and choose $h \in R^n$ such that $|h| < \delta/2$. Then $y_0 + h \in N_\delta(x_0)$. Let $u(t, y_0)$ and $u(t, y_0 + h)$ be the solutions of the initial value problem (1) with $y = y_0$ and with $y = y_0 + h$ respectively. It then follows that

$$|u(t, y_0 + h) - u(t, y_0)| \leq |h| + \int_0^t |f(u(s, y_0 + h)) - f(u(s, y_0))|\, ds$$

$$\leq |h| + K \int_0^t |u(s, y_0 + h) - u(s, y_0)|\, ds$$

for all $t \in [-a, a]$. Thus, it follows from Gronwall's Lemma that

$$|u(t, y_0 + h) - u(t, y_0)| \leq |h| e^{K|t|} \tag{4}$$

for all $t \in [-a, a]$. We next define $\Phi(t, y_0)$ to be the fundamental matrix solution of the initial value problem

$$\begin{aligned} \dot{\Phi} &= A(t, y_0)\Phi \\ \Phi(0, y_0) &= I \end{aligned} \tag{5}$$

with $A(t, y_0) = Df(u(t, y_0))$ and I the $n \times n$ identity matrix. The existence and continuity of $\Phi(t, y_0)$ on some interval $[-a, a]$ follow from the method of successive approximations as in problem 4 of Problem Set 2 and problem 4 in Problem Set 3. It then follows from the initial value problems for $u(t, y_0)$, $u(t, y_0 + h)$ and $\Phi(t, y_0)$ and Taylor's Theorem,

$$f(u) - f(u_0) = Df(u_0)(u - u_0) + R(u, u_0)$$

where $|R(u, u_0)|/|u - u_0| \to 0$ as $|u - u_0| \to 0$, that

$$|u(t, y_0) - u(t, y_0 + h) + \Phi(t, y_0)h| \leq \int_0^t |f(u(s, y_0))$$
$$- f(u(s, y_0 + h)) + Df(u(s, y_0))\Phi(s, y_0)h|\, ds$$
$$\leq \int_0^t \|Df(u(s, y_0))\| |u(s, y_0) - u(s, y_0 + h) + \Phi(s, y_0)h|\, ds$$
$$+ \int_0^t |R(u(s, y_0 + h), u(s, y_0))|\, ds \tag{6}$$

Since $|R(u, u_0)|/|u - u_0| \to 0$ as $|u - u_0| \to 0$ and since $u(s, y)$ is continuous on G, it follows that given any $\varepsilon_0 > 0$, there exists a $\delta_0 > 0$ such that if

2.3. Dependence on Initial Conditions and Parameters   83

$|h| < \delta_0$ then $|R(u(s, y_0), u(s, y_0 + h))| < \varepsilon_0 |u(s, y_0) - u(s, y_0 + h)|$ for all $s \in [-a, a]$. Thus, if we let

$$g(t) = |u(t, y_0) - u(t, y_0 + h) + \Phi(t, y_0)h|$$

it then follows from (4) and (6) that for all $t \in [-a, a]$, $y_0 \in N_{\delta/2}(x_0)$ and $|h| < \min(\delta_0, \delta/2)$ we have

$$g(t) \leq M_1 \int_0^t g(s)\, ds + \varepsilon_0 |h| a e^{Ka}.$$

Hence, it follows from Gronwall's Lemma that for any given $\varepsilon_0 > 0$

$$g(t) \leq \varepsilon_0 |h| a e^{Ka} e^{M_1 a}$$

for all $t \in [-a, a]$ provided $|h| < \min(\delta_0, \delta/2)$. Thus,

$$\lim_{|h| \to 0} \frac{|u(t, y_0) - u(t, y_0 + h) + \Phi(t, y_0)h|}{|h|} = 0$$

uniformly for all $t \in [-a, a]$. Therefore, according to Definition 1 in Section 2.1,

$$\frac{\partial u}{\partial y}(t, y_0) = \Phi(t, y_0)$$

for all $t \in [-a, a]$ where $\Phi(t, y_0)$ is the fundamental matrix solution of the initial value problem (5) which is continuous in t and in y_0 for all $t \in [-a, a]$ and $y_0 \in N_{\delta/2}(x_0)$. This completes the proof of the theorem.

Corollary. *Under the hypothesis of the above theorem,*

$$\Phi(t, y) = \frac{\partial u}{\partial y}(t, y)$$

for $t \in [-a, a]$ and $y \in N_\delta(x_0)$ if and only if $\Phi(t, y)$ is the fundamental matrix solution of

$$\dot{\Phi} = Df[u(t, y)]\Phi$$
$$\Phi(0, y) = I$$

for $t \in [-a, a]$ and $y \in N_\delta(x_0)$.

Remark 1. A similar proof shows that if $f \in C^r(E)$ then the solution $u(t, y)$ of the initial value problem (1) is in $C^r(G)$ where G is defined as in the above theorem. And if $f(x)$ is a (real) analytic function for $x \in E$ then $u(t, y)$ is analytic in the interior of G; cf. [C/L].

Remark 2. If $\mathbf{x}_0$ is an equilibrium point of (1), i.e., if $\mathbf{f}(\mathbf{x}_0) = 0$ so that $\mathbf{u}(t, \mathbf{x}_0) = \mathbf{x}_0$ for all $t \in \mathbf{R}$, then

$$\Phi(t, \mathbf{x}_0) = \frac{\partial \mathbf{u}}{\partial \mathbf{x}_0}(t, \mathbf{x}_0)$$

satisfies

$$\dot{\Phi} = D\mathbf{f}(\mathbf{x}_0)\Phi$$
$$\Phi(0, \mathbf{x}_0) = I.$$

And according to the Fundamental Theorem for Linear Systems

$$\Phi(t, \mathbf{x}_0) = e^{D\mathbf{f}(\mathbf{x}_0)t}.$$

Remark 3. It follows from the continuity of the solution $\mathbf{u}(t, \mathbf{y})$ of the initial value problem (1) that for each $t \in [-a, a]$

$$\lim_{\mathbf{y} \to \mathbf{x}_0} \mathbf{u}(t, \mathbf{y}) = \mathbf{u}(t, \mathbf{x}_0).$$

It follows from the inequality (4) that this limit is uniform for all $t \in [-a, a]$. We prove a slightly stronger version of this result in Theorem 4 of the next section.

Theorem 2 (Dependence on Parameters). *Let E be an open subset of $\mathbf{R}^{n+m}$ containing the point $(\mathbf{x}_0, \boldsymbol{\mu}_0)$ where $\mathbf{x}_0 \in \mathbf{R}^n$ and $\boldsymbol{\mu}_0 \in \mathbf{R}^m$ and assume that $\mathbf{f} \in C^1(E)$. It then follows that there exists an $a > 0$ and a $\delta > 0$ such that for all $\mathbf{y} \in N_\delta(\mathbf{x}_0)$ and $\boldsymbol{\mu} \in N_\delta(\boldsymbol{\mu}_0)$, the initial value problem*

$$\dot{\mathbf{x}} = \mathbf{f}(\mathbf{x}, \boldsymbol{\mu})$$
$$\mathbf{x}(0) = \mathbf{y}$$

has a unique solution $\mathbf{u}(t, \mathbf{y}, \boldsymbol{\mu})$ with $\mathbf{u} \in C^1(G)$ where $G = [-a, a] \times N_\delta(\mathbf{x}_0) \times N_\delta(\boldsymbol{\mu}_0)$.

This theorem follows immediately from the previous theorem by replacing the vectors $\mathbf{x}_0$, $\mathbf{x}$, $\dot{\mathbf{x}}$ and $\mathbf{y}$ by the vectors $(\mathbf{x}_0, \boldsymbol{\mu}_0)$, $(\mathbf{x}, \boldsymbol{\mu})$, $(\dot{\mathbf{x}}, \mathbf{0})$ and $(\mathbf{y}, \boldsymbol{\mu})$ or it can be proved directly using Gronwall's Lemma and the method of successive approximations; cf. problem 3 below.

PROBLEM SET 3

1. Use the fundamental theorem for linear systems in Chapter 1 to solve the initial value problem

$$\dot{\mathbf{x}} = A\mathbf{x}$$
$$\mathbf{x}(0) = \mathbf{y}.$$

2.3. Dependence on Initial Conditions and Parameters

Let $\mathbf{u}(t, \mathbf{y})$ denote the solution and compute

$$\Phi(t) = \frac{\partial \mathbf{u}}{\partial \mathbf{y}}(t, \mathbf{y}).$$

Show that $\Phi(t)$ is the fundamental matrix solution of

$$\dot{\Phi} = A\Phi$$
$$\Phi(0) = I.$$

2. (a) Solve the initial value problem

$$\dot{\mathbf{x}} = \mathbf{f}(\mathbf{x})$$
$$\mathbf{x}(0) = \mathbf{y}$$

for $\mathbf{f}(\mathbf{x}) = (-x_1, -x_2 + x_1^2, x_3 + x_1^2)^T$. Denote the solution by $\mathbf{u}(t, \mathbf{y})$ and compute

$$\Phi(t, \mathbf{y}) = \frac{\partial \mathbf{u}}{\partial \mathbf{y}}(t, \mathbf{y}).$$

Compute the derivative $D\mathbf{f}(\mathbf{x})$ for the given function $\mathbf{f}(\mathbf{x})$ and show that for all $t \in \mathbf{R}$ and $\mathbf{y} \in \mathbf{R}^3$, $\Phi(t, \mathbf{y})$ satisfies

$$\dot{\Phi} = A(t, \mathbf{y})\Phi$$
$$\Phi(0, \mathbf{y}) = I$$

where $A(t, \mathbf{y}) = D\mathbf{f}[\mathbf{u}(t, \mathbf{y})]$.

(b) Carry out the same steps for the above initial value problem with $\mathbf{f}(\mathbf{x}) = (x_1^2, x_2 + x_1^{-1})^T$.

3. Consider the initial value problem

$$\dot{\mathbf{x}} = \mathbf{f}(t, \mathbf{x}, \boldsymbol{\mu})$$
$$\mathbf{x}(0) = \mathbf{x}_0. \qquad (*)$$

Given that E is an open subset of $\mathbf{R}^{n+m+1}$ containing the point $(0, \mathbf{x}_0, \boldsymbol{\mu}_0)$ where $\mathbf{x}_0 \in \mathbf{R}^n$ and $\boldsymbol{\mu}_0 \in \mathbf{R}^m$ and that $\mathbf{f}$ and $\partial \mathbf{f}/\partial \mathbf{x}$ are continuous on E, use Gronwall's Lemma and the method of successive approximations to show that there is an $a > 0$ and a $\delta > 0$ such that the initial value problem $(*)$ has a unique solution $\mathbf{u}(t, \boldsymbol{\mu})$ continuous on $[-a, a] \times N_\delta(\boldsymbol{\mu}_0)$.

4. Let E be an open subset of $\mathbf{R}^n$ containing $\mathbf{y}_0$. Use the method of successive approximations and Gronwall's Lemma to show that if

$A(t, \mathbf{y})$ is continuous on $[-a_0, a_0] \times E$ then there exist an $a > 0$ and a $\delta > 0$ such that for all $\mathbf{y} \in N_\delta(\mathbf{y}_0)$ the initial value problem

$$\dot{\Phi} = A(t, \mathbf{y})\Phi$$
$$\Phi(0, \mathbf{y}) = I$$

has a unique solution $\Phi(t, \mathbf{y})$ continuous on $[-a, a] \times N_\delta(\mathbf{y}_0)$. **Hint:** Cf. problem 4 in Problem Set 2 and regard $\mathbf{y}$ as a parameter as in problem 3 above.

5. Let $\Phi(t)$ be the fundamental matrix solution of

$$\dot{\Phi} = A(t)\Phi$$

and the initial condition

$$\Phi(0) = I$$

for $t \in [0, a]$; cf. Problem 4 in Section 2. Use Liouville's Theorem (cf. [H], p. 46), which states that

$$\det \Phi(t) = \exp \int_0^t \operatorname{trace} A(s)\, ds,$$

to show that for all $t \in [0, a]$

$$\det \frac{\partial \mathbf{u}}{\partial \mathbf{y}}(t, \mathbf{x}_0) = \exp \int_0^t \nabla \cdot \mathbf{f}(\mathbf{u}(s, \mathbf{x}_0))\, ds$$

where $\mathbf{u}(t, \mathbf{y})$ is the solution of the initial value problem (1) as in the first theorem in this section. **Hint:** Use the Corollary in this section.

6. (Cf. Hartman [H], p. 96.) Let $\mathbf{f} \in C^1(E)$ where E is an open set in $\mathbf{R}^n$ containing the point $\mathbf{x}_0$. Let $\mathbf{u}(t, \mathbf{y}_0)$ be the unique solution of the initial value problem (1) for $t \in [0, a]$ with $\mathbf{y} = \mathbf{y}_0$. Show that the set of maps of $\mathbf{y}_0 \to \mathbf{y}$ defined by $\mathbf{y} = \mathbf{u}(t, \mathbf{y}_0)$ for each fixed $t \in [0, a]$ are volume preserving in E if and only if $\nabla \cdot \mathbf{f}(\mathbf{x}) = 0$ for all $\mathbf{x} \in E$. **Hint:** Recall that under a transformation of coordinates $\mathbf{y} = \mathbf{u}(\mathbf{x})$ which maps a region R_0 one-to-one and onto a region R_1, the volume of the region R_1 is given by

$$V = \int_{R_0} \cdots \int J(\mathbf{x})\, dx_1 \ldots dx_n$$

where the Jacobian determinant

$$J(\mathbf{x}) = \det \frac{\partial \mathbf{u}}{\partial \mathbf{x}}(\mathbf{x}).$$

2.4 The Maximal Interval of Existence

The fundamental existence-uniqueness theorem of Section 2.2 established that if $\mathbf{f} \in C^1(E)$ then the initial value problem

$$\dot{\mathbf{x}} = \mathbf{f}(\mathbf{x}) \qquad\qquad (1)$$
$$\mathbf{x}(0) = \mathbf{x}_0$$

has a unique solution defined on some interval $(-a, a)$. In this section we show that (1) has a unique solution $\mathbf{x}(t)$ defined on a maximal interval of existence (α, β). Furthermore, if $\beta < \infty$ and if the limit

$$\mathbf{x}_1 = \lim_{t \to \beta^-} \mathbf{x}(t)$$

exists then $\mathbf{x}_1 \in \dot{E}$, the boundary of E. The boundary of the open set E, $\dot{E} = \bar{E} \sim E$ where $\bar{E}$ denotes the closure of E. On the other hand, if the above limit exists and $\mathbf{x}_1 \in E$, then $\beta = \infty$, $\mathbf{f}(\mathbf{x}_1) = \mathbf{0}$ and $\mathbf{x}_1$ is an equilibrium point of (1) according to Definition 1 in Section 2.6 below; cf. Problem 5. The following examples illustrate these ideas.

Example 1 (Cf. Example 2 in Section 2.1). The initial value problem

$$\dot{x} = x^2 \qquad x(0) = 1$$

has the solution $x(t) = (1-t)^{-1}$ defined on its maximal interval of existence $(\alpha, \beta) = (-\infty, 1)$. Furthermore, $\lim_{t \to 1^-} x(t) = \infty$.

Example 2. The initial value problem

$$\dot{x} = -\frac{1}{2x}$$
$$x(0) = 1$$

has the solution $x(t) = \sqrt{1-t}$ defined on its maximal interval of existence $(\alpha, \beta) = (-\infty, 1)$. The function $f(x) = -1/(2x) \in C^1(E)$ where $E = (0, \infty)$ and $\dot{E} = \{0\}$. Note that

$$\lim_{t \to 1^-} x(t) = 0 \in \dot{E}.$$

Example 3. Consider the initial value problem

$$\dot{x}_1 = -\frac{x_2}{x_3^2}$$
$$\dot{x}_2 = \frac{x_1}{x_3^2}$$
$$\dot{x}_3 = 1$$

with $\mathbf{x}(1/\pi) = (0, -1, 1/\pi)^T$. The solution is

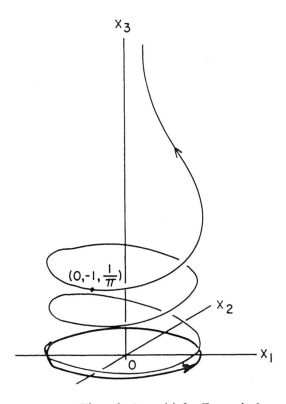

Figure 1. The solution $\mathbf{x}(t)$ for Example 3.

$$\mathbf{x}(t) = \begin{bmatrix} \sin 1/t \\ \cos 1/t \\ t \end{bmatrix}$$

on the maximal interval $(\alpha, \beta) = (0, \infty)$. Cf. Figure 1. At the finite endpoint $\alpha = 0$, $\lim_{t \to 0^+} \mathbf{x}(t)$ does not exist. Note, however, that the arc length

$$\int_{1/\pi}^t |\dot{\mathbf{x}}(\tau)| \, d\tau \geq \int_{1/\pi}^t \frac{\sqrt{x_1^2(\tau) + x_2^2(\tau)}}{x_3^2(\tau)} \, d\tau = \int_{1/\pi}^t \frac{d\tau}{\tau^2} = \frac{1}{t} - \pi \to \infty$$

as $t \to 0^+$. Cf. Problem 3.

We next establish the existence and some basic properties of the maximal interval of existence (α, β) of the solution $\mathbf{x}(t)$ of the initial value problem (1).

2.4. The Maximal Interval of Existence

Lemma 1. *Let E be an open subset of $\mathbf{R}^n$ containing $\mathbf{x}_0$ and suppose $\mathbf{f} \in C^1(E)$. Let $\mathbf{u}_1(t)$ and $\mathbf{u}_2(t)$ be solutions of the initial value problem (1) on the intervals I_1 and I_2. Then $0 \in I_1 \cap I_2$ and if I is any open interval containing 0 and contained in $I_1 \cap I_2$, it follows that $\mathbf{u}_1(t) = \mathbf{u}_2(t)$ for all $t \in I$.*

Proof. Since $\mathbf{u}_1(t)$ and $\mathbf{u}_2(t)$ are solutions of the initial value problem (1) on I_1 and I_2 respectively, it follows from Definition 1 in Section 2.2 that $0 \in I_1 \cap I_2$. And if I is an open interval containing 0 and contained in $I_1 \cap I_2$, then the fundamental existence-uniqueness theorem in Section 2.2 implies that $\mathbf{u}_1(t) = \mathbf{u}_2(t)$ on some open interval $(-a, a) \subset I$. Let I^* be the union of all such open intervals contained in I. Then I^* is the largest open interval contained in I on which $\mathbf{u}_1(t) = \mathbf{u}_2(t)$. Clearly, $I^* \subset I$ and if I^* is a proper subset of I, then one of the endpoints t_0 of I^* is contained in $I \subset I_1 \cap I_2$. It follows from the continuity of $\mathbf{u}_1(t)$ and $\mathbf{u}_2(t)$ on I that

$$\lim_{t \to t_0} \mathbf{u}_1(t) = \lim_{t \to t_0} \mathbf{u}_2(t).$$

Call this common limit $\mathbf{u}_0$. It then follows from the uniqueness of solutions that $\mathbf{u}_1(t) = \mathbf{u}_2(t)$ on some interval $I_0 = (t_0 - a, t_0 + a) \subset I$. Thus, $\mathbf{u}_1(t) = \mathbf{u}_2(t)$ on the interval $I^* \cup I_0 \subset I$ and I^* is a proper subset of $I^* \cup I_0$. But this contradicts the fact that I^* is the largest open interval contained in I on which $\mathbf{u}_1(t) = \mathbf{u}_2(t)$. Therefore, $I^* = I$ and we have $\mathbf{u}_1(t) = \mathbf{u}_2(t)$ for all $t \in I$.

Theorem 1. *Let E be an open subset of $\mathbf{R}^n$ and assume that $\mathbf{f} \in C^1(E)$. Then for each point $\mathbf{x}_0 \in E$, there is a maximal interval J on which the initial value problem (1) has a unique solution, $\mathbf{x}(t)$; i.e., if the initial value problem has a solution $\mathbf{y}(t)$ on an interval I then $I \subset J$ and $\mathbf{y}(t) = \mathbf{x}(t)$ for all $t \in I$. Furthermore, the maximal interval J is open; i.e., $J = (\alpha, \beta)$.*

Proof. By the fundamental existence-uniqueness theorem in Section 2.2, the initial value problem (1) has a unique solution on some open interval $(-a, a)$. Let (α, β) be the union of all open intervals I such that (1) has a solution on I. We define a function $\mathbf{x}(t)$ on (α, β) as follows: Given $t \in (\alpha, \beta)$, t belongs to some open interval I such that (1) has a solution $\mathbf{u}(t)$ on I; for this given $t \in (\alpha, \beta)$, define $\mathbf{x}(t) = \mathbf{u}(t)$. Then $\mathbf{x}(t)$ is a well-defined function of t since if $t \in I_1 \cap I_2$ where I_1 and I_2 are any two open intervals such that (1) has solutions $\mathbf{u}_1(t)$ and $\mathbf{u}_2(t)$ on I_1 and I_2 respectively, then by the lemma $\mathbf{u}_1(t) = \mathbf{u}_2(t)$ on the open interval $I_1 \cap I_2$. Also, $\mathbf{x}(t)$ is a solution of (1) on (α, β) since each point $t \in (\alpha, \beta)$ is contained in some open interval I on which the initial value problem (1) has a unique solution $\mathbf{u}(t)$ and since $\mathbf{x}(t)$ agrees with $\mathbf{u}(t)$ on I. The fact that J is open follows from the fact that any solution of (1) on an interval $(\alpha, \beta]$ can be uniquely continued to a solution on an interval $(\alpha, \beta + a)$ with $a > 0$ as in the proof of Theorem 2 below.

Definition. The interval (α, β) in Theorem 1 is called the *maximal interval of existence* of the solution $\mathbf{x}(t)$ of the initial value problem (1) or simply the maximal interval of existence of the initial value problem (1).

Theorem 2. *Let E be an open subset of $\mathbf{R}^n$ containing $\mathbf{x}_0$, let $\mathbf{f} \in C^1(E)$, and let (α, β) be the maximal interval of existence of the solution $\mathbf{x}(t)$ of the initial value problem (1). Assume that $\beta < \infty$. Then given any compact set $K \subset E$, there exists a $t \in (\alpha, \beta)$ such that $\mathbf{x}(t) \notin K$.*

Proof. Since $\mathbf{f}$ is continuous on the compact set K, there is a positive number M such that $|\mathbf{f}(\mathbf{x})| \leq M$ for all $\mathbf{x} \in K$. Let $\mathbf{x}(t)$ be the solution of the initial value problem (1) on its maximal interval of existence (α, β) and assume that $\beta < \infty$ and that $\mathbf{x}(t) \in K$ for all $t \in (\alpha, \beta)$. We first show that $\lim_{t \to \beta^-} \mathbf{x}(t)$ exists. If $\alpha < t_1 < t_2 < \beta$ then

$$|\mathbf{x}(t_1) - \mathbf{x}(t_2)| \leq \int_{t_1}^{t_2} |\mathbf{f}(\mathbf{x}(s))|\, ds \leq M|t_2 - t_1|.$$

Thus as t_1 and t_2 approach β from the left, $|\mathbf{x}(t_2) - \mathbf{x}(t_1)| \to 0$ which, by the Cauchy criterion for convergence in $\mathbf{R}^n$ (i.e., the completeness of $\mathbf{R}^n$) implies that $\lim_{t \to \beta^-} \mathbf{x}(t)$ exists. Let $\mathbf{x}_1 = \lim_{t \to \beta^-} \mathbf{x}(t)$. Then $\mathbf{x}_1 \in K \subset E$ since K is compact. Next define the function $\mathbf{u}(t)$ on $(\alpha, \beta]$ by

$$\mathbf{u}(t) = \begin{cases} \mathbf{x}(t) & \text{for } t \in (\alpha, \beta) \\ \mathbf{x}_1 & \text{for } t = \beta. \end{cases}$$

Then $\mathbf{u}(t)$ is differentiable on $(\alpha, \beta]$. Indeed,

$$\mathbf{u}(t) = \mathbf{x}_0 + \int_0^t \mathbf{f}(\mathbf{u}(s))\, ds$$

which implies that

$$\mathbf{u}'(\beta) = \mathbf{f}(\mathbf{u}(\beta));$$

i.e., $\mathbf{u}(t)$ is a solution of the initial value problem (1) on $(\alpha, \beta]$. The function $\mathbf{u}(t)$ is called the continuation of the solution $\mathbf{x}(t)$ to $(\alpha, \beta]$. Since $\mathbf{x}_1 \in E$, it follows from the fundamental existence-uniqueness theorem in Section 2.2 that the initial value problem $\dot{\mathbf{x}} = \mathbf{f}(\mathbf{x})$ together with $\mathbf{x}(\beta) = \mathbf{x}_1$ has a unique solution $\mathbf{x}_1(t)$ on some interval $(\beta - a, \beta + a)$. By the above lemma, $\mathbf{x}_1(t) = \mathbf{u}(t)$ on $(\beta - a, \beta)$ and $\mathbf{x}_1(\beta) = \mathbf{u}(\beta) = \mathbf{x}_1$. So if we define

$$\mathbf{v}(t) = \begin{cases} \mathbf{u}(t) & \text{for } t \in (\alpha, \beta] \\ \mathbf{x}_1(t) & \text{for } t \in [\beta, \beta + a), \end{cases}$$

then $\mathbf{v}(t)$ is a solution of the initial value problem (1) on $(\alpha, \beta + a)$. But this contradicts the fact that (α, β) is the maximal interval of existence for

2.4. The Maximal Interval of Existence

the initial value problem (1). Hence, if $\beta < \infty$, it follows that there exists a $t \in (\alpha, \beta)$ such that $\mathbf{x}(t) \notin K$.

If (α, β) is the maximal interval of existence for the initial value problem (1) then $0 \in (\alpha, \beta)$ and the intervals $[0, \beta)$ and $(\alpha, 0]$ are called the *right and left maximal intervals of existence* respectively. Essentially the same proof yields the following result.

Theorem 3. *Let E be an open subset of $\mathbf{R}^n$ containing $\mathbf{x}_0$, let $\mathbf{f} \in C^1(E)$, and let $[0, \beta)$ be the right maximal interval of existence of the solution $\mathbf{x}(t)$ of the initial value problem (1). Assume that $\beta < \infty$. Then given any compact set $K \subset E$, there exists a $t \in (0, \beta)$ such that $\mathbf{x}(t) \notin K$.*

Corollary 1. *Under the hypothesis of the above theorem, if $\beta < \infty$ and if $\lim\limits_{t \to \beta^-} \mathbf{x}(t)$ exists then $\lim\limits_{t \to \beta^-} \mathbf{x}(t) \in \dot{E}$.*

Proof. If $\mathbf{x}_1 = \lim\limits_{t \to \beta^-} \mathbf{x}(t)$, then the function

$$\mathbf{u}(t) = \begin{cases} \mathbf{x}(t) & \text{for } t \in [0, \beta) \\ \mathbf{x}_1 & \text{for } t = \beta \end{cases}$$

is continuous on $[0, \beta]$. Let K be the image of the compact set $[0, \beta]$ under the continuous map $\mathbf{u}(t)$; i.e.,

$$K = \{\mathbf{x} \in \mathbf{R}^n \mid \mathbf{x} = \mathbf{u}(t) \text{ for some } t \in [0, \beta]\}.$$

Then K is compact. Assume that $\mathbf{x}_1 \in E$. Then $K \subset E$ and it follows from Theorem 3 that there exists a $t \in (0, \beta)$ such that $\mathbf{x}(t) \notin K$. This is a contradiction and therefore $\mathbf{x}_1 \notin E$. But since $\mathbf{x}(t) \in E$ for all $t \in [0, \beta)$, it follows that $\mathbf{x}_1 = \lim\limits_{t \to \beta^-} \mathbf{x}(t) \in \bar{E}$. Therefore $\mathbf{x}_1 \in \bar{E} \sim E$; i.e., $\mathbf{x}_1 \in \dot{E}$.

Corollary 2. *Let E be an open subset of $\mathbf{R}^n$ containing $\mathbf{x}_0$, let $\mathbf{f} \in C^1(E)$, and let $[0, \beta)$ be the right maximal interval of existence of the solution $\mathbf{x}(t)$ of the initial value problem (1). Assume that there exists a compact set $K \subset E$ such that*

$$\{\mathbf{y} \in \mathbf{R}^n \mid \mathbf{y} = \mathbf{x}(t) \text{ for some } t \in [0, \beta)\} \subset K.$$

It then follows that $\beta = \infty$; i.e. the initial value problem (1) has a solution $\mathbf{x}(t)$ on $[0, \infty)$.

Proof. This corollary is just the contrapositive of the statement in Theorem 3.

We next prove the following theorem which strengthens the result on uniform convergence with respect to initial conditions in Remark 3 of Section 2.3.

Theorem 4. *Let E be an open subset of $\mathbf{R}^n$ containing $\mathbf{x}_0$ and let $\mathbf{f} \in C^1(E)$. Suppose that the initial value problem (1) has a solution $\mathbf{x}(t, \mathbf{x}_0)$ defined on a closed interval $[a, b]$. Then there exists a $\delta > 0$ and a positive constant K such that for all $\mathbf{y} \in N_\delta(\mathbf{x}_0)$ the initial value problem*

$$\dot{\mathbf{x}} = \mathbf{f}(\mathbf{x})$$
$$\mathbf{x}(0) = \mathbf{y} \qquad (2)$$

has a unique solution $\mathbf{x}(t, \mathbf{y})$ defined on $[a, b]$ which satisfies

$$|\mathbf{x}(t, \mathbf{y}) - \mathbf{x}(t, \mathbf{x}_0)| \leq |\mathbf{y} - \mathbf{x}_0| e^{K|t|}$$

and

$$\lim_{\mathbf{y} \to \mathbf{x}_0} \mathbf{x}(t, \mathbf{y}) = \mathbf{x}(t, \mathbf{x}_0)$$

uniformly for all $t \in [a, b]$.

Remark 1. If in Theorem 4 we have a function $\mathbf{f}(\mathbf{x}, \mu)$ depending on a parameter $\mu \in \mathbf{R}^m$ which satisfies $\mathbf{f} \in C^1(E)$ where E is an open subset of $\mathbf{R}^{n+m}$ containing $(\mathbf{x}_0, \mu_0)$, it can be shown that if for $\mu = \mu_0$ the initial value problem (1) has a solution $\mathbf{x}(t, \mathbf{x}_0, \mu_0)$ defined on a closed interval $a \leq t \leq b$, then there is a $\delta > 0$ and a $K > 0$ such that for all $\mathbf{y} \in N_\delta(\mathbf{x}_0)$ and $\mu \in N_\delta(\mu_0)$ the initial value problem

$$\dot{\mathbf{x}} = \mathbf{f}(\mathbf{x}, \mu)$$
$$\mathbf{x}(0) = \mathbf{y}$$

has a unique solution $\mathbf{x}(t, \mathbf{y}, \mu)$ defined for $a \leq t \leq b$ which satisfies

$$|\mathbf{x}(t, \mathbf{y}, \mu) - \mathbf{x}(t, \mathbf{x}_0, \mu_0)| \leq [|\mathbf{y} - \mathbf{x}_0| + |\mu - \mu_0|] e^{K|t|}$$

and

$$\lim_{(\mathbf{y}, \mu) \to (\mathbf{x}_0, \mu_0)} \mathbf{x}(t, \mathbf{y}, \mu) = \mathbf{x}(t, \mathbf{x}_0, \mu_0)$$

uniformly for all $t \in [a, b]$. Cf. [C/L], p. 58.

In order to prove this theorem, we first establish the following lemma.

Lemma 2. *Let E be an open subset of $\mathbf{R}^n$ and let A be a compact subset of E. Then if $\mathbf{f}: E \to \mathbf{R}^n$ is locally Lipschitz on E, it follows that $\mathbf{f}$ satisfies a Lipschitz condition on A.*

Proof. Let M be the maximal value of the continuous function $\mathbf{f}$ on the compact set A. Suppose that $\mathbf{f}$ does not satisfy a Lipschitz condition on A. Then for every $K > 0$, we can find $\mathbf{x}, \mathbf{y} \in A$ such that

$$|\mathbf{f}(\mathbf{y}) - \mathbf{f}(\mathbf{x})| > K|\mathbf{y} - \mathbf{x}|.$$

2.4. The Maximal Interval of Existence

In particular, there exist sequences $\mathbf{x}_n$ and $\mathbf{y}_n$ in A such that

$$|\mathbf{f}(\mathbf{y}_n) - \mathbf{f}(\mathbf{x}_n)| > n|\mathbf{y}_n - \mathbf{x}_n| \qquad (*)$$

for $n = 1, 2, 3, \ldots$. Since A is compact, there are convergent subsequences, call them $\mathbf{x}_n$ and $\mathbf{y}_n$ for simplicity in notation, such that $\mathbf{x}_n \to \mathbf{x}^*$ and $\mathbf{y}_n \to \mathbf{y}^*$ with $\mathbf{x}^*$ and $\mathbf{y}^*$ in A. It follows that $\mathbf{x}^* = \mathbf{y}^*$ since for all $n = 1, 2, 3, \ldots$

$$|\mathbf{y}^* - \mathbf{x}^*| = \lim_{n \to \infty} |\mathbf{y}_n - \mathbf{x}_n| \leq \frac{1}{n}|\mathbf{f}(\mathbf{y}_n) - \mathbf{f}(\mathbf{x}_n)| \leq \frac{2M}{n}.$$

Now, by hypotheses, there exists a neighborhood N_0 of $\mathbf{x}^*$ and a constant K_0 such that $\mathbf{f}$ satisfies a Lipschitz condition with Lipschitz constant K_0 for all $\mathbf{x}$ and $\mathbf{y} \in N_0$. But since $\mathbf{x}_n$ and $\mathbf{y}_n$ approach $\mathbf{x}^*$ as $n \to \infty$, it follows that $\mathbf{x}_n$ and $\mathbf{y}_n$ are in N_0 for n sufficiently large; i.e., for n sufficiently large

$$|\mathbf{f}(\mathbf{y}_n) - \mathbf{f}(\mathbf{x}_n)| \leq K|\mathbf{y}_n - \mathbf{x}_n|.$$

But for $n \geq K$, this contradicts the above inequality $(*)$ and this establishes the lemma.

Proof (of Theorem 4). Since $[a, b]$ is compact and $\mathbf{x}(t, \mathbf{x}_0)$ is a continuous function of t, $\{\mathbf{x} \in \mathbf{R}^n \mid \mathbf{x} = \mathbf{x}(t, \mathbf{x}_0) \text{ and } a \leq t \leq b\}$ is a compact subset of E. And since E is open, there exists an $\varepsilon > 0$ such that the compact set

$$A = \{\mathbf{x} \in \mathbf{R}^n \mid |\mathbf{x} - \mathbf{x}(t, \mathbf{x}_0)| \leq \varepsilon \text{ and } a \leq t \leq b\}$$

is a subset of E. Since $\mathbf{f} \in C^1(E)$, it follows from the lemma in Section 2.2 that $\mathbf{f}$ is locally Lipschitz in E; and then by the above lemma, $\mathbf{f}$ satisfies a Lipschitz condition

$$|\mathbf{f}(\mathbf{y}) - \mathbf{f}(\mathbf{x})| \leq K|\mathbf{y} - \mathbf{x}|$$

for all $\mathbf{x}, \mathbf{y} \in A$. Choose $\delta > 0$ so small that $\delta \leq \varepsilon$ and $\delta \leq \varepsilon e^{-K(b-a)}$. Let $\mathbf{y} \in N_\delta(\mathbf{x}_0)$ and let $\mathbf{x}(t, \mathbf{y})$ be the solution of the initial value problem (2) on its maximal interval of existence (α, β). We shall show that $[a, b] \subset (\alpha, \beta)$. Suppose that $\beta \leq b$. It then follows that $\mathbf{x}(t, \mathbf{y}) \in A$ for all $t \in (\alpha, \beta)$ because if this were not true then there would exist a $t^* \in (\alpha, \beta)$ such that $\mathbf{x}(t, \mathbf{x}_0) \in A$ for $t \in (\alpha, t^*]$ and $\mathbf{x}(t^*, \mathbf{y}) \in \dot{A}$. But then

$$|\mathbf{x}(t, \mathbf{y}) - \mathbf{x}(t, \mathbf{x}_0)| \leq |\mathbf{y} - \mathbf{x}_0| + \int_0^t |\mathbf{f}(\mathbf{x}(s, \mathbf{y})) - f(\mathbf{x}(s, \mathbf{x}_0)|\, ds$$

$$\leq |\mathbf{y} - \mathbf{x}_0| + K \int_0^t |\mathbf{x}(s, \mathbf{y}) - \mathbf{x}(s, \mathbf{x}_0)|\, ds$$

for all $t \in (\alpha, t^*]$. And then by Gronwall's Lemma in Section 2.3, it follows that

$$|\mathbf{x}(t^*, \mathbf{y}) - \mathbf{x}(t^*, \mathbf{x}_0)| \leq |\mathbf{y} - \mathbf{x}_0|e^{K|t^*|} < \delta e^{K(b-a)} < \varepsilon$$

since $t^* < \beta \leq b$. Thus $\mathbf{x}(t^*, \mathbf{y})$ is an interior point of A, a contradiction. Thus, $\mathbf{x}(t, \mathbf{y}) \in A$ for all $t \in (\alpha, \beta)$. But then by Theorem 2, (α, β) is not the maximal interval of existence of $\mathbf{x}(t, \mathbf{y})$, a contradiction. Thus $b < \beta$. It is similarly proved that $\alpha < a$. Hence, for all $\mathbf{y} \in N_\delta(\mathbf{x}_0)$, the initial value problem (2) has a unique solution defined on $[a, b]$. Furthermore, if we assume that there is a $t^* \in [a, b]$ such that $\mathbf{x}(t, \mathbf{y}) \in A$ for all $t \in [a, t^*)$ and $\mathbf{x}(t^*, \mathbf{y}) \in \dot{A}$, a repeat of the above argument based on Gronwall's Lemma leads to a contradiction and shows that $\mathbf{x}(t, \mathbf{y}) \in A$ for all $t \in [a, b]$ and hence that

$$|\mathbf{x}(t, \mathbf{y}) - \mathbf{x}(t, \mathbf{x}_0)| \leq |\mathbf{y} - \mathbf{x}_0| e^{K|t|}$$

for all $t \in [a, b]$. It then follows that

$$\lim_{\mathbf{y} \to \mathbf{x}_0} \mathbf{x}(t, \mathbf{y}) = \mathbf{x}(t, \mathbf{x}_0)$$

uniformly for all $t \in [a, b]$.

PROBLEM SET 4

1. Find the maximal interval of existence (α, β) for the following initial value problems and if $\alpha > -\infty$ or $\beta < \infty$ discuss the limit of the solution as $t \to \alpha^+$ or as $t \to \beta^-$ respectively:

    (a) $\dot{x} = x^2$
    $x(0) = x_0$

    (b) $\dot{x} = \sec x$
    $x(0) = 0$

    (c) $\dot{x} = x^2 - 4$
    $x(0) = 0$

    (d) $\dot{x} = x^3$
    $x(0) = x_0 > 0$

    (e) Problem 2(b) in Problem Set 3 with $y_1 > 0$.

2. Find the maximal interval of existence (α, β) for the following initial value problems and if $\alpha > -\infty$ or $\beta < \infty$ discuss the limit of the solution as $t \to \alpha^+$ or as $t \to \beta^-$ respectively:

    (a) $\dot{x}_1 = x_1^2$ $\quad x_1(0) = 1$
    $\dot{x}_2 = x_2 + x_1^{-1}$ $\quad x_2(0) = 1$

    (b) $\dot{x}_1 = \dfrac{1}{2x_1}$ $\quad x_1(0) = 1$
    $\dot{x}_2 = x_2^2$ $\quad x_2(0) = 1$

(c) $\begin{aligned} \dot{x}_1 &= \dfrac{1}{2x_1} & x_1(0) &= 1 \\ \dot{x}_2 &= x_1 & x_2(0) &= 1 \end{aligned}$

3. Let $f \in C^1(E)$ where E is an open set in $\mathbf{R}^n$ containing the point $\mathbf{x}_0$ and let $\mathbf{x}(t)$ be the solution of the initial value problem (1) on its maximal interval of existence (α, β). Prove that if $\beta < \infty$ and if the arc-length of the half-trajectory $\Gamma_+ = \{\mathbf{x} \in \mathbf{R}^n \mid \mathbf{x} = \mathbf{x}(t), 0 \leq t < \beta\}$ is finite, then it follows that the limit

$$\mathbf{x}_1 = \lim_{t \to \beta^-} \mathbf{x}(t)$$

exists. Then by Corollary 1, $\mathbf{x}_1 \in \dot{E}$. **Hint:** Assume that the above limit does not exist. This implies that there is a sequence t_n converging to β from the left such that $\mathbf{x}(t_n)$ is not Cauchy. Use this fact to show that the arc-length of Γ_+ is then unbounded.

4. Convert the system in Example 3 to cylindrical coordinates, (r, θ, z) as in Problem 9 in Section 1.5 of Chapter 1, and solve the resulting system with the initial conditions $r = 1$, $\theta = -\pi$, $z = 1/\pi$ at $t = 0$. What is the maximal interval of existence in this case?

5. Use Corollary 2 to show that if $\mathbf{x}_1 = \lim_{t \to \beta^-} \mathbf{x}(t)$ exists and $\mathbf{x}_1 \in E$, then $\beta = \infty$; and then show that $\mathbf{f}(\mathbf{x}_1) = 0$ and note that $\mathbf{x}(t) \equiv \mathbf{x}_1$ is a solution of (1) and $\mathbf{x}(0) = \mathbf{x}_1$.

2.5 The Flow Defined by a Differential Equation

In Section 1.9 of Chapter 1, we defined the flow, $e^{At}: \mathbf{R}^n \to \mathbf{R}^n$, of the linear system

$$\dot{\mathbf{x}} = A\mathbf{x}.$$

The mapping $\phi_t = e^{At}$ satisfies the following basic properties for all $\mathbf{x} \in \mathbf{R}^n$:

(i) $\phi_0(\mathbf{x}) = \mathbf{x}$

(ii) $\phi_s(\phi_t(\mathbf{x})) = \phi_{s+t}(\mathbf{x})$ for all $s, t \in \mathbf{R}$

(iii) $\phi_{-t}(\phi_t(\mathbf{x})) = \phi_t(\phi_{-t}(\mathbf{x})) = \mathbf{x}$ for all $t \in \mathbf{R}$.

Property (i) follows from the definition of e^{At}, property (ii) follows from Proposition 2 in Section 1.3 of Chapter 1, and property (iii) follows from Corollary 2 in Section 1.3 of Chapter 1.

In this section, we define the flow, ϕ_t, of the nonlinear system

$$\dot{\mathbf{x}} = \mathbf{f}(\mathbf{x}) \tag{1}$$

and show that it satisfies these same basic properties. In the following definition, we denote the maximal interval of existence (α, β) of the solution $\phi(t, \mathbf{x}_0)$ of the initial value problem

$$\dot{\mathbf{x}} = \mathbf{f}(\mathbf{x}) \\ \mathbf{x}(0) = \mathbf{x}_0 \tag{2}$$

by $I(\mathbf{x}_0)$ since the endpoints α and β of the maximal interval generally depend on $\mathbf{x}_0$; cf. problems 1(a) and (d) in Section 2.4.

Definition 1. Let E be an open subset of $\mathbf{R}^n$ and let $\mathbf{f} \in C^1(E)$. For $\mathbf{x}_0 \in E$, let $\phi(t, \mathbf{x}_0)$ be the solution of the initial value problem (2) defined on its maximal interval of existence $I(\mathbf{x}_0)$. Then for $t \in I(\mathbf{x}_0)$, the set of mappings ϕ_t defined by

$$\phi_t(\mathbf{x}_0) = \phi(t, \mathbf{x}_0)$$

is called *the flow of the differential equation* (1) or the flow defined by the differential equation (1); ϕ_t is also referred to as *the flow of the vector field* $\mathbf{f}(\mathbf{x})$.

If we think of the initial point $\mathbf{x}_0$ as being fixed and let $I = I(\mathbf{x}_0)$, then the mapping $\phi(\cdot, \mathbf{x}_0) \colon I \to E$ defines a solution curve or *trajectory* of the system (1) through the point $\mathbf{x}_0 \in E$. As usual, the mapping $\phi(\cdot, \mathbf{x}_0)$ is identified with its graph in $I \times E$ and a trajectory is visualized as a motion along a curve Γ through the point $\mathbf{x}_0$ in the subset E of the phase space $\mathbf{R}^n$; cf. Figure 1. On the other hand, if we think of the point $\mathbf{x}_0$ as varying throughout $K \subset E$, then the flow of the differential equation (1), $\phi_t \colon K \to E$ can be viewed as the motion of all the points in the set K; cf. Figure 2.

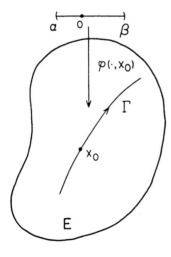
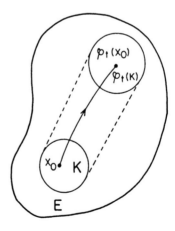

Figure 1. A trajectory Γ of the system (1).

Figure 2. The flow ϕ_t of the system (1).

2.5. The Flow Defined by a Differential Equation

If we think of the differential equation (1) as describing the motion of a fluid, then a trajectory of (1) describes the motion of an individual particle in the fluid while the flow of the differential equation (1) describes the motion of the entire fluid.

We now show that the basic properties (i)–(iii) of linear flows are also satisfied by nonlinear flows. But first we extend Theorem 1 of Section 2.3, establishing that $\phi(t, \mathbf{x}_0)$ is a locally smooth function, to a global result. Using the same notation as in Definition 1, let us define the set $\Omega \subset \mathbf{R} \times E$ as

$$\Omega = \{(t, \mathbf{x}_0) \in \mathbf{R} \times E \mid t \in I(\mathbf{x}_0)\}.$$

Example 1. Consider the differential equation

$$\dot{x} = \frac{1}{x}$$

with $f(x) = 1/x \in C^1(E)$ and $E = \{x \in \mathbf{R} \mid x > 0\}$. The solution of this differential equation and initial condition $x(0) = x_0$ is given by

$$\phi(t, x_0) = \sqrt{2t + x_0^2}$$

on its maximal interval of existence $I(x_0) = (-x_0^2/2, \infty)$. The region Ω for this problem is shown in Figure 3.

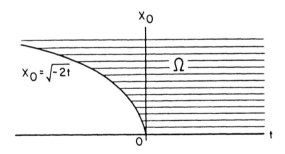

Figure 3. The region Ω.

Theorem 1. *Let E be an open subset of $\mathbf{R}^n$ and let $\mathbf{f} \in C^1(E)$. Then Ω is an open subset of $\mathbf{R} \times E$ and $\phi \in C^1(\Omega)$.*

Proof. If $(t_0, \mathbf{x}_0) \in \Omega$ and $t_0 > 0$, then according to the definition of the set Ω, the solution $\mathbf{x}(t) = \phi(t, \mathbf{x}_0)$ of the initial value problem (2) is defined on $[0, t_0]$. Thus, as in the proof of Theorem 2 in Section 2.4, the solution $\mathbf{x}(t)$ can be extended to an interval $[0, t_0 + \varepsilon]$ for some $\varepsilon > 0$; i.e., $\phi(t, \mathbf{x}_0)$ is defined on the closed interval $[t_0 - \varepsilon, t_0 + \varepsilon]$. It then follows from Theorem 4 in Section 2.4 that there exists a neighborhood of $\mathbf{x}_0$, $N_\delta(\mathbf{x}_0)$, such that

$\phi(t, \mathbf{y})$ is defined on $[t_0-\varepsilon, t_0+\varepsilon] \times N_\delta(\mathbf{x}_0)$; i.e., $(t_0-\varepsilon, t_0+\varepsilon) \times N_\delta(\mathbf{x}_0) \subset \Omega$. Therefore, Ω is open in $\mathbf{R} \times E$. It follows from Theorem 4 in Section 2.4 that $\phi \in C^1(G)$ where $G = (t_0 - \varepsilon, t_0 + \varepsilon) \times N_\delta(\mathbf{x}_0)$. A similar proof holds for $t_0 \leq 0$, and since $(t_0, \mathbf{x}_0)$ is an arbitrary point in Ω, it follows that $\phi \in C^1(\Omega)$.

Remark. Theorem 1 can be generalized to show that if $\mathbf{f} \in C^r(E)$ with $r \geq 1$, then $\phi \in C^r(\Omega)$ and that if $\mathbf{f}$ is analytic in E, then ϕ is analytic in Ω.

Theorem 2. *Let E be an open set of $\mathbf{R}^n$ and let $\mathbf{f} \in C^1(E)$. Then for all $\mathbf{x}_0 \in E$, if $t \in I(\mathbf{x}_0)$ and $s \in I(\phi_t(\mathbf{x}_0))$, it follows that $s + t \in I(\mathbf{x}_0)$ and*

$$\phi_{s+t}(\mathbf{x}_0) = \phi_s(\phi_t(\mathbf{x}_0)).$$

Proof. Suppose that $s > 0$, $t \in I(\mathbf{x}_0)$ and $s \in I(\phi_t(\mathbf{x}_0))$. Let the maximal interval $I(\mathbf{x}_0) = (\alpha, \beta)$ and define the function $\mathbf{x}: (\alpha, s+t] \to E$ by

$$\mathbf{x}(r) = \begin{cases} \phi(r, \mathbf{x}_0) & \text{if } \alpha < r \leq t \\ \phi(r - t, \phi_t(\mathbf{x}_0)) & \text{if } t \leq r \leq s + t. \end{cases}$$

Then $\mathbf{x}(r)$ is a solution of the initial value problem (2) on $(\alpha, s+t]$. Hence $s + t \in I(\mathbf{x}_0)$ and by uniqueness of solutions

$$\phi_{s+t}(\mathbf{x}_o) = \mathbf{x}(s+t) = \phi(s, \phi_t(\mathbf{x}_0)) = \phi_s(\phi_t(\mathbf{x}_0)).$$

If $s = 0$ the statement of the theorem follows immediately. And if $s < 0$, then we define the function $\mathbf{x}: [s+t, \beta) \to E$ by

$$\mathbf{x}(t) = \begin{cases} \phi(r, \mathbf{x}_0) & \text{if } t \leq r < \beta \\ \phi(r - t, \phi_t(\mathbf{x}_0)) & \text{if } s + t \leq r \leq t. \end{cases}$$

Then $\mathbf{x}(r)$ is a solution of the initial value problem (2) on $[s+t, \beta)$ and the last statement of the theorem follows from the uniqueness of solutions as above.

Theorem 3. *Under the hypotheses of Theorem 1, if $(t, \mathbf{x}_0) \in \Omega$ then there exists a neighborhood U of $\mathbf{x}_0$ such that $\{t\} \times U \subset \Omega$. It then follows that the set $V = \phi_t(U)$ is open in E and that*

$$\phi_{-t}(\phi_t(\mathbf{x})) = \mathbf{x} \ \text{for all } \mathbf{x} \in U$$

and

$$\phi_t(\phi_{-t}(\mathbf{y})) = \mathbf{y} \ \text{for all } \mathbf{y} \in V.$$

Proof. If $(t, \mathbf{x}_0) \in \Omega$ then if follows as in the proof of Theorem 1 that there exists a neighborhood of $\mathbf{x}_0$, $U = N_\delta(\mathbf{x}_0)$, such that $(t - \varepsilon, t + \varepsilon) \times U \subset \Omega$; thus, $\{t\} \times U \subset \Omega$. For $\mathbf{x} \in U$, let $\mathbf{y} = \phi_t(\mathbf{x})$ for all $t \in I(\mathbf{x})$. Then

2.5. The Flow Defined by a Differential Equation

$-t \in I(\mathbf{y})$ since the function $\mathbf{h}(s) = \phi(s+t, \mathbf{y})$ is a solution of (1) on $[-t, 0]$ that satisfies $\mathbf{h}(-t) = \mathbf{y}$; i.e., ϕ_{-t} is defined on the set $V = \phi_t(U)$. It then follows from Theorem 2 that $\phi_{-t}(\phi_t(\mathbf{x})) = \phi_0(\mathbf{x}) = \mathbf{x}$ for all $\mathbf{x} \in U$ and that $\phi_t(\phi_{-t}(\mathbf{y})) = \phi_0(\mathbf{y}) = \mathbf{y}$ for all $\mathbf{y} \in V$. It remains to prove that V is open. Let $V^* \supset V$ be the maximal subset of E on which ϕ_{-t} is defined. V^* is open because Ω is open and $\phi_{-t}: V^* \to E$ is continuous because by Theorem 1, ϕ is continuous. Therefore, the inverse image of the open set U under the continuous map ϕ_{-t}, i.e., $\phi_t(U)$, is open in E. Thus, V is open in E.

In Chapter 3 we show that the time along each trajectory of (1) can be rescaled, without affecting the phase portrait of (1), so that for all $\mathbf{x}_0 \in E$, the solution $\phi(t, \mathbf{x}_0)$ of the initial value problem (2) is defined for all $t \in \mathbf{R}$; i.e., for all $\mathbf{x}_0 \in E, I(\mathbf{x}_0) = (-\infty, \infty)$. This rescaling avoids some of the complications found in stating the above theorems. Once this rescaling has been made, it follows that $\Omega = \mathbf{R} \times E, \phi \in C^1(\mathbf{R} \times E), \phi_t \in C^1(E)$ for all $t \in \mathbf{R}$, and properties (i)–(iii) for the flow of the nonlinear system (1) hold for all $t \in \mathbf{R}$ and $\mathbf{x} \in E$ just as for the linear flow e^{At}. In the remainder of this chapter, and in particular in Sections 2.7 and 2.8 of this chapter, it will be assumed that this rescaling has been made so that for all $\mathbf{x}_0 \in E, \phi(t, \mathbf{x}_0)$ is defined for all $t \in \mathbf{R}$; i.e., we shall assume throughout the remainder of this chapter that the flow of the nonlinear system (1) $\phi_t \in C^1(E)$ for all $t \in \mathbf{R}$.

Definition 2. Let E be an open subset of $\mathbf{R}^n$, let $\mathbf{f} \in C^1(E)$, and let $\phi_t: E \to E$ be the flow of the differential equation (1) defined for all $t \in \mathbf{R}$. Then a set $S \subset E$ is called *invariant with respect to the flow* ϕ_t if $\phi_t(S) \subset S$ for all $t \in \mathbf{R}$ and S is called *positively (or negatively) invariant with respect to the flow* ϕ_t if $\phi_t(S) \subset S$ for all $t \geq 0$ (or $t \leq 0$).

In Section 1.9 of Chapter 1 we showed that the stable, unstable and center subspaces of the linear system $\dot{\mathbf{x}} = A\mathbf{x}$ are invariant under the linear flow $\phi_t = e^{At}$. A similar result is established in Section 2.7 for the nonlinear flow ϕ_t of (1).

Example 2. Consider the nonlinear system (1) with

$$\mathbf{f}(\mathbf{x}) = \begin{bmatrix} -x_1 \\ x_2 + x_1^2 \end{bmatrix}.$$

The solution of the initial value problem (1) together with the initial condition $\mathbf{x}(0) = \mathbf{c}$ is given by

$$\phi_t(\mathbf{c}) = \phi(t, \mathbf{c}) = \begin{bmatrix} c_1 e^{-t} \\ c_2 e^t + \frac{c_1^2}{3}(e^t - e^{-2t}) \end{bmatrix}.$$

We now show that the set

$$S = \{\mathbf{x} \in \mathbf{R}^2 \mid x_2 = -x_1^2/3\}$$

is invariant under the flow ϕ_t. This follows since if $\mathbf{c} \in S$ then $c_2 = -c_1^2/3$ and it follows that

$$\phi_t(\mathbf{c}) = \begin{bmatrix} c_1 e^{-t} \\ -\frac{c_1^2}{3} e^{-2t} \end{bmatrix} \in S.$$

Thus $\phi_t(S) \subset S$ for all $t \in \mathbf{R}$. The phase portrait for the nonlinear system (1) with $\mathbf{f}(\mathbf{x})$ given above is shown in Figure 4. The set S is called the stable manifold for this system. This is discussed in Section 2.7.

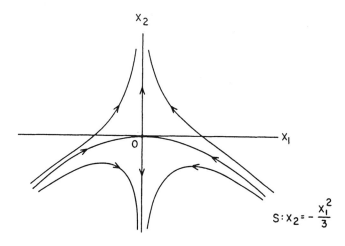

Figure 4. The invariant set S for the system (1).

PROBLEM SET 5

1. As in Example 1, sketch the region Ω in the (t, x_0) plane for the initial value problem

$$\dot{x} = x^2$$
$$x(0) = x_0.$$

2. Do problem 1 for the initial value problem

$$\dot{x} = -x^3$$
$$x(0) = x_0.$$

3. Sketch the flow for the linear system $\dot{\mathbf{x}} = A\mathbf{x}$ with

$$A = \begin{bmatrix} -1 & 0 \\ 0 & 2 \end{bmatrix}$$

2.6. Linearization

and, in particular, describe (with a rough sketch) what happens to a small neighborhood $N_\varepsilon(\mathbf{x}_0)$ of a point $\mathbf{x}_0$ on the negative x_1-axis, say $N_\varepsilon(-3, 0)$ with $\varepsilon = .2$. (Cf. Figure 1 in Section 1.1 of Chapter 1.)

4. Sketch the flow for the linear system $\dot{\mathbf{x}} = A\mathbf{x}$ with

$$A = \begin{bmatrix} -1 & -3 \\ 0 & 2 \end{bmatrix}$$

and describe $\phi_t(N_\varepsilon(\mathbf{x}_0))$ for $\mathbf{x}_0 = (-3, 0)$, $\varepsilon = .2$. (Cf. Figure 2 in Section 1.2 of Chapter 1.)

5. Determine the flow $\phi_t \colon \mathbf{R}^2 \to \mathbf{R}^2$ for the nonlinear system (1) with

$$\mathbf{f}(\mathbf{x}) = \begin{bmatrix} -x_1 \\ 2x_2 + x_1^2 \end{bmatrix}$$

and show that the set $S = \{\mathbf{x} \in \mathbf{R}^2 \mid x_2 = -x_1^2/4\}$ is invariant with respect to the flow ϕ_t.

6. Determine the flow $\phi_t \colon \mathbf{R}^3 \to \mathbf{R}^3$ for the nonlinear system (1) with

$$\mathbf{f}(\mathbf{x}) = \begin{bmatrix} -x_1 \\ -x_2 + x_1^2 \\ x_3 + x_1^2 \end{bmatrix}$$

and show that the set $S = \{\mathbf{x} \in \mathbf{R}^3 \mid x_3 = -x_1^2/3\}$ is invariant under the flow ϕ_t. Sketch the parabolic cylinder S. (Cf. Example 1 in Section 2.7.)

7. Determine the flow $\phi_t \colon \mathbf{R} \sim \{0\} \to \mathbf{R}$ for

$$\dot{x} = \frac{1}{x^2}$$

and for $x_0 \neq 0$ determine the maximal interval of existence $I(x_0) = (\alpha, \beta)$. If $\alpha > -\infty$ or $\beta < \infty$ show that

$$\lim_{t \to \alpha^+} \phi_t(x_0) \in \dot{E} \quad \text{or} \quad \lim_{t \to \beta^-} \phi_t(x_0) \in \dot{E}$$

where $E = \mathbf{R} \sim \{0\}$. Sketch the set $\Omega = \{(t, x_0) \in \mathbf{R}^2 \mid t \in I(x_0)\}$. Show that $\phi_t(\phi_s(x_0)) = \phi_{t+s}(x_0)$ for $s \in I(x_0)$ and $s + t \in I(x_0)$.

2.6 Linearization

A good place to start analyzing the nonlinear system

$$\dot{\mathbf{x}} = \mathbf{f}(\mathbf{x}) \tag{1}$$

is to determine the equilibrium points of (1) and to describe the behavior of (1) near its equilibrium points. In the next two sections it is shown that the local behavior of the nonlinear system (1) near a hyperbolic equilibrium point $\mathbf{x}_0$ is qualitatively determined by the behavior of the linear system

$$\dot{\mathbf{x}} = A\mathbf{x}, \qquad (2)$$

with the matrix $A = D\mathbf{f}(\mathbf{x}_0)$, near the origin. The linear function $A\mathbf{x} = D\mathbf{f}(\mathbf{x}_0)\mathbf{x}$ is called the *linear part of* $\mathbf{f}$ *at* $\mathbf{x}_0$.

Definition 1. A point $\mathbf{x}_0 \in \mathbf{R}^n$ is called an *equilibrium point* or *critical point* of (1) if $\mathbf{f}(\mathbf{x}_0) = 0$. An equilibrium point $\mathbf{x}_0$ is called a *hyperbolic equilibrium point* of (1) if none of the eigenvalues of the matrix $D\mathbf{f}(\mathbf{x}_0)$ have zero real part. The linear system (2) with the matrix $A = D\mathbf{f}(\mathbf{x}_0)$ is called *the linearization of* (1) *at* $\mathbf{x}_0$.

If $\mathbf{x}_0 = \mathbf{0}$ is an equilibrium point of (1), then $\mathbf{f}(\mathbf{0}) = \mathbf{0}$ and, by Taylor's Theorem,

$$\mathbf{f}(\mathbf{x}) = D\mathbf{f}(\mathbf{0})\mathbf{x} + \frac{1}{2}D^2\mathbf{f}(\mathbf{0})(\mathbf{x}, \mathbf{x}) + \cdots.$$

It follows that the linear function $D\mathbf{f}(\mathbf{0})\mathbf{x}$ is a good first approximation to the nonlinear function $\mathbf{f}(\mathbf{x})$ near $\mathbf{x} = \mathbf{0}$ and it is reasonable to expect that the behavior of the nonlinear system (1) near the point $\mathbf{x} = \mathbf{0}$ will be approximated by the behavior of its linearization at $\mathbf{x} = \mathbf{0}$. In Section 2.7 it is shown that this is indeed the case if the matrix $D\mathbf{f}(\mathbf{0})$ has no zero or pure imaginary eigenvalues.

Note that if $\mathbf{x}_0$ is an equilibrium point of (1) and $\phi_t \colon E \to \mathbf{R}^n$ is the flow of the differential equation (1), then $\phi_t(\mathbf{x}_0) = \mathbf{x}_0$ for all $t \in \mathbf{R}$. Thus, $\mathbf{x}_0$ is called *a fixed point of the flow* ϕ_t; it is also called a *zero*, a *critical point*, or a *singular point of the vector field* $\mathbf{f} \colon E \to \mathbf{R}^n$. We next give a rough classification of the equilibrium points of (1) according to the signs of the real parts of the eigenvalues of the matrix $D\mathbf{f}(\mathbf{x}_0)$. A finer classification is given in Section 2.10 for planar vector fields.

Definition 2. An equilibrium point $\mathbf{x}_0$ of (1) is called a *sink* if all of the eigenvalues of the matrix $D\mathbf{f}(\mathbf{x}_0)$ have negative real part; it is called a *source* if all of the eigenvalues of $D\mathbf{f}(\mathbf{x}_0)$ have positive real part; and it is called a *saddle* if it is a hyperbolic equilibrium point and $D\mathbf{f}(\mathbf{x}_0)$ has at least one eigenvalue with a positive real part and at least one with a negative real part.

Example 1. Let us classify all of the equilibrium points of the nonlinear system (1) with

$$\mathbf{f}(\mathbf{x}) = \begin{bmatrix} x_1^2 - x_2^2 - 1 \\ 2x_2 \end{bmatrix}.$$

2.6. Linearization

Clearly, $\mathbf{f}(\mathbf{x}) = \mathbf{0}$ at $\mathbf{x} = (1,0)^T$ and $\mathbf{x} = (-1,0)^T$ and these are the only equilibrium points of (1). The derivative

$$D\mathbf{f}(\mathbf{x}) = \begin{bmatrix} 2x_1 & -2x_2 \\ 0 & 2 \end{bmatrix}, D\mathbf{f}(1,0) = \begin{bmatrix} 2 & 0 \\ 0 & 2 \end{bmatrix},$$

and

$$D\mathbf{f}(-1,0) = \begin{bmatrix} -2 & 0 \\ 0 & 2 \end{bmatrix}.$$

Thus, $(1,0)$ is a source and $(-1,0)$ is a saddle.

In Section 2.8 we shall see that if $\mathbf{x}_0$ is a hyperbolic equilibrium point of (1) then the local behavior of the nonlinear system (1) is topologically equivalent to the local behavior of the linear system (2); i.e., there is a continuous one-to-one map of a neighborhood of $\mathbf{x}_0$ onto an open set U containing the origin, $H: N_\varepsilon(\mathbf{x}_0) \to U$, which transforms (1) into (2), maps trajectories of (1) in $N_\varepsilon(\mathbf{x}_0)$ onto trajectories of (2) in the open set U, and preserves the orientation of the trajectories by time, i.e., H preserves the direction of the flow along the trajectories.

Example 2. Consider the continuous map

$$H(\mathbf{x}) = \begin{bmatrix} x_1 \\ x_2 + \frac{x_1^2}{3} \end{bmatrix}$$

which maps $\mathbf{R}^2$ onto $\mathbf{R}^2$. It is not difficult to determine that the inverse of $\mathbf{y} = H(\mathbf{x})$ is given by

$$H^{-1}(\mathbf{y}) = \begin{bmatrix} y_1 \\ y_2 - \frac{y_1^2}{3} \end{bmatrix}$$

and that H^{-1} is a continuous mapping of $\mathbf{R}^2$ onto $\mathbf{R}^2$. Furthermore, the mapping H transforms the nonlinear system (1) with

$$\mathbf{f}(\mathbf{x}) = \begin{bmatrix} -x_1 \\ x_2 + x_1^2 \end{bmatrix}$$

into the linear system (2) with

$$A = D\mathbf{f}(0) = \begin{bmatrix} -1 & 0 \\ 0 & 1 \end{bmatrix}$$

in the sense that if $\mathbf{y} = H(\mathbf{x})$ then

$$\dot{\mathbf{y}} = \begin{bmatrix} \dot{x}_1 \\ \dot{x}_2 + \frac{2}{3}x_1\dot{x}_1 \end{bmatrix} = \begin{bmatrix} -x_1 \\ x_2 + x_1^2 + \frac{2}{3}x_1(-x_1) \end{bmatrix} = \begin{bmatrix} -y_1 \\ y_2 \end{bmatrix};$$

i.e.

$$\dot{\mathbf{y}} = \begin{bmatrix} -1 & 0 \\ 0 & 1 \end{bmatrix} \mathbf{y}.$$

We have used the fact that $\mathbf{x} = H^{-1}(\mathbf{y})$ implies that $x_1 = y_1$ and $x_2 = y_2 - y_1^2/3$ in obtaining the last step of the above equation for $\dot{\mathbf{y}}$. The phase portrait for the nonlinear system in this example is given in Figure 4 of Section 2.5 and the phase portrait for the linear system in this example is given in Figure 1 of Section 1.5 of Chapter 1. These two phase portraits are qualitatively the same.

PROBLEM SET 6

1. Classify the equilibrium points (as sinks, sources or saddles) of the nonlinear system (1) with $\mathbf{f}(\mathbf{x})$ given by

   (a) $\begin{bmatrix} x_1 - x_1 x_2 \\ x_2 - x_1^2 \end{bmatrix}$

   (b) $\begin{bmatrix} -4x_2 + 2x_1 x_2 - 8 \\ 4x_2^2 - x_1^2 \end{bmatrix}$

   (c) $\begin{bmatrix} 2x_1 - 2x_1 x_2 \\ 2x_2 - x_1^2 + x_2^2 \end{bmatrix}$

   (d) $\begin{bmatrix} -x_1 \\ -x_2 + x_1^2 \\ x_3 + x_1^2 \end{bmatrix}$

   (e) $\begin{bmatrix} x_2 - x_1 \\ kx_1 - x_2 - x_1 x_3 \\ x_1 x_2 - x_3 \end{bmatrix}$.

   Hint: In 1(e), the origin is a sink if $k < 1$ and a saddle if $k > 1$. It is a nonhyperbolic equilibrium point if $k = 1$.

2. Classify the equilibrium points of the Lorenz equation (1) with

$$\mathbf{f}(\mathbf{x}) = \begin{bmatrix} x_2 - x_1 \\ \mu x_1 - x_2 - x_1 x_3 \\ x_1 x_2 - x_3 \end{bmatrix}$$

   for $\mu > 0$. At what value of the parameter μ do two new equilibrium points "bifurcate" from the equilibrium point at the origin? **Hint:** For $\mu > 1$, the eigenvalues at the nonzero equilibrium points are $\lambda = -2$ and $\lambda = (-1 \pm \sqrt{5 - 4\mu})/2$.

3. Show that the continuous map $H\colon \mathbf{R}^3 \to \mathbf{R}^3$ defined by

$$H(\mathbf{x}) = \begin{bmatrix} x_1 \\ x_2 + x_1^2 \\ x_3 + \frac{x_1^2}{3} \end{bmatrix}$$

has a continuous inverse $H^{-1}: \mathbf{R}^3 \to \mathbf{R}^3$ and that the nonlinear system (1) with

$$\mathbf{f}(\mathbf{x}) = \begin{bmatrix} -x_1 \\ -x_2 + x_1^2 \\ x_3 + x_1^2 \end{bmatrix}$$

is transformed into the linear system (2) with $A = D\mathbf{f}(\mathbf{0})$ under this map; i.e., if $\mathbf{y} = H(\mathbf{x})$, show that $\dot{\mathbf{y}} = A\mathbf{y}$.

2.7 The Stable Manifold Theorem

The stable manifold theorem is one of the most important results in the local qualitative theory of ordinary differential equations. The theorem shows that near a hyperbolic equilibrium point $\mathbf{x}_0$, the nonlinear system

$$\dot{\mathbf{x}} = \mathbf{f}(\mathbf{x}) \tag{1}$$

has stable and unstable manifolds S and U tangent at $\mathbf{x}_0$ to the stable and unstable subspaces E^s and E^u of the linearized system

$$\dot{\mathbf{x}} = A\mathbf{x} \tag{2}$$

where $A = D\mathbf{f}(\mathbf{x}_0)$. Furthermore, S and U are of the same dimensions as E^s and E^u, and if ϕ_t is the flow of the nonlinear system (1), then S and U are positively and negatively invariant under ϕ_t respectively and satisfy

$$\lim_{t \to \infty} \phi_t(\mathbf{c}) = \mathbf{x}_0 \text{ for all } \mathbf{c} \in S$$

and

$$\lim_{t \to -\infty} \phi_t(\mathbf{c}) = \mathbf{x}_0 \text{ for all } \mathbf{c} \in U.$$

We first illustrate these ideas with an example and then make them more precise by proving the stable manifold theorem. It is assumed that the equilibrium point $\mathbf{x}_0$ is located at the origin throughout the remainder of this section. If this is not the case, then the equilibrium point $\mathbf{x}_0$ can be translated to the origin by the affine transformation of coordinates $\mathbf{x} \to \mathbf{x} - \mathbf{x}_0$.

Example 1. Consider the nonlinear system

$$\dot{x}_1 = -x_1$$
$$\dot{x}_2 = -x_2 + x_1^2$$
$$\dot{x}_3 = x_3 + x_1^2.$$

The only equilibrium point of this system is at the origin. The matrix
$$A = D\mathbf{f}(0) = \begin{bmatrix} -1 & 0 & 0 \\ 0 & -1 & 0 \\ 0 & 0 & 1 \end{bmatrix}.$$
Thus, the stable and unstable subspaces E^s and E^u of (2) are the x_1, x_2 plane and the x_3-axis respectively. After solving the first differential equation, $\dot{x}_1 = -x_1$, the nonlinear system reduces to two uncoupled first-order linear differential equations which are easily solved. The solution is given by
$$x_1(t) = c_1 e^{-t}$$
$$x_2(t) = c_2 e^{-t} + c_1^2(e^{-t} - e^{-2t})$$
$$x_3(t) = c_3 e^t + \frac{c_1^2}{3}(e^t - e^{-2t})$$
where $\mathbf{c} = \mathbf{x}(0)$. Clearly, $\lim_{t \to \infty} \phi_t(\mathbf{c}) = 0$ iff $c_3 + c_1^2/3 = 0$. Thus,
$$S = \{\mathbf{c} \in \mathbf{R}^3 \mid c_3 = -c_1^2/3\}.$$
Similarly, $\lim_{t \to -\infty} \phi_t(\mathbf{c}) = 0$ iff $c_1 = c_2 = 0$ and therefore
$$U = \{\mathbf{c} \in \mathbf{R}^3 \mid c_1 = c_2 = 0\}.$$
The stable and unstable manifolds for this system are shown in Figure 1. Note that the surface S is tangent to E^s, i.e., to the x_1, x_2 plane at the origin and that $U = E^u$.

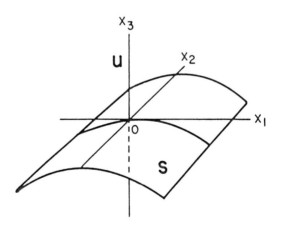

Figure 1

Before proving the stable manifold theorem, we first define the concept of a smooth surface or differentiable manifold.

2.7. The Stable Manifold Theorem

Definition 1. Let X be a metric space and let A and B be subsets of X. A *homeomorphism* of A onto B is a continuous one-to-one map of A onto B, $h: A \to B$, such that $h^{-1}: B \to A$ is continuous. The sets A and B are called *homeomorphic* or *topologically equivalent* if there is a homeomorphism of A onto B. If we wish to emphasize that h maps A onto B, we write $h: A \looparrowright B$.

Definition 2. An *n-dimensional differentiable manifold*, M (or a manifold of class C^k), is a connected metric space with an open covering $\{U_\alpha\}$, i.e., $M = \bigcup_\alpha U_\alpha$, such that

(1) for all α, U_α is homeomorphic to the open unit ball in $\mathbf{R}^n$, $B = \{\mathbf{x} \in \mathbf{R}^n \mid |\mathbf{x}| < 1\}$, i.e., for all α there exists a homeomorphism of U_α onto B, $\mathbf{h}_\alpha: U_\alpha \to B$, and

(2) if $U_\alpha \cap U_\beta \neq \emptyset$ and $\mathbf{h}_\alpha: U_\alpha \to B$, $\mathbf{h}_\beta: U_\beta \to B$ are homeomorphisms, then $\mathbf{h}_\alpha(U_\alpha \cap U_\beta)$ and $\mathbf{h}_\beta(U_\alpha \cap U_\beta)$ are subsets of $\mathbf{R}^n$ and the map

$$\mathbf{h} = \mathbf{h}_\alpha \circ \mathbf{h}_\beta^{-1}: \mathbf{h}_\beta(U_\alpha \cap U_\beta) \to \mathbf{h}_\alpha(U_\alpha \cap U_\beta)$$

is differentiable (or of class C^k) and for all $\mathbf{x} \in \mathbf{h}_\beta(U_\alpha \cap U_\beta)$, the Jacobian determinant $\det D\mathbf{h}(\mathbf{x}) \neq 0$. The manifold M is said to be *analytic* if the maps $\mathbf{h} = \mathbf{h}_\alpha \circ \mathbf{h}_\beta^{-1}$ are analytic.

The cylindrical surface S in the above example is a two-dimensional differentiable manifold. The projection of the x_1, x_2 plane onto S maps the unit disks centered at the points (m, n) in the $x_1 x_2$ plane onto homeomorphic images of the unit disk $B = \{\mathbf{x} \in \mathbf{R}^2 \mid x_1^2 + x_2^2 < 1\}$. These sets $U_{mn} \subset S$ then form a countable open cover of S in this case.

The pair $(U_\alpha, \mathbf{h}_\alpha)$ is called a *chart* for the manifold M and the set of all charts is called an *atlas* for M. The differentiable manifold M is called *orientable* if there is an atlas with $\det D\mathbf{h}_\alpha \circ \mathbf{h}_\beta^{-1}(\mathbf{x}) > 0$ for all α, β and $\mathbf{x} \in \mathbf{h}_\beta(U_\alpha \cap U_\beta)$. Cf. Problems 7 and 8.

Theorem (The Stable Manifold Theorem). *Let E be an open subset of $\mathbf{R}^n$ containing the origin, let $\mathbf{f} \in C^1(E)$, and let ϕ_t be the flow of the nonlinear system* (1). *Suppose that $\mathbf{f}(0) = 0$ and that $D\mathbf{f}(0)$ has k eigenvalues with negative real part and $n - k$ eigenvalues with positive real part. Then there exists a k-dimensional differentiable manifold S tangent to the stable subspace E^s of the linear system* (2) *at $\mathbf{0}$ such that for all $t \geq 0$, $\phi_t(S) \subset S$ and for all $\mathbf{x}_0 \in S$.*

$$\lim_{t \to \infty} \phi_t(\mathbf{x}_0) = 0;$$

and there exists an $n - k$ dimensional differentiable manifold U tangent to the unstable subspace E^u of (2) *at $\mathbf{0}$ such that for all $t \leq 0$, $\phi_t(U) \subset U$ and*

for all $\mathbf{x}_0 \in U$,
$$\lim_{t \to -\infty} \phi_t(\mathbf{x}_0) = 0.$$

Before proving this theorem, we remark that if $\mathbf{f} \in C^1(E)$ and $\mathbf{f}(0) = 0$, then the system (1) can be written as

$$\dot{\mathbf{x}} = A\mathbf{x} + \mathbf{F}(\mathbf{x}) \tag{3}$$

where $A = D\mathbf{f}(0)$, $\mathbf{F}(\mathbf{x}) = \mathbf{f}(\mathbf{x}) - A\mathbf{x}$, $\mathbf{F} \in C^1(E)$, $\mathbf{F}(0) = 0$ and $D\mathbf{F}(0) = 0$. This in turn implies that for all $\varepsilon > 0$ there is a $\delta > 0$ such that $|\mathbf{x}| \leq \delta$ and $|\mathbf{y}| \leq \delta$ imply that

$$|\mathbf{F}(\mathbf{x}) - \mathbf{F}(\mathbf{y})| \leq \varepsilon |\mathbf{x} - \mathbf{y}|. \tag{4}$$

Cf. problem 6.

Furthermore, as in Section 1.8 of Chapter 1, there is an $n \times n$ invertible matrix C such that

$$B = C^{-1}AC = \begin{bmatrix} P & 0 \\ 0 & Q \end{bmatrix}$$

where the eigenvalues $\lambda_1, \ldots, \lambda_k$ of the $k \times k$ matrix P have negative real part and the eigenvalues $\lambda_{k+1}, \ldots, \lambda_n$ of the $(n-k) \times (n-k)$ matrix Q have postive real part. We can choose $\alpha > 0$ sufficiently small that for $j = 1, \ldots, k$,

$$\mathrm{Re}(\lambda_j) < -\alpha < 0. \tag{5}$$

Letting $\mathbf{y} = C^{-1}\mathbf{x}$, the system (3) then has the form

$$\dot{\mathbf{y}} = B\mathbf{y} + \mathbf{G}(\mathbf{y}) \tag{6}$$

where $\mathbf{G}(\mathbf{y}) = C^{-1}\mathbf{F}(C\mathbf{y}) \in C^1(\tilde{E})$ where $\tilde{E} = C^{-1}(E)$ and $\mathbf{G}$ satisfies the Lipschitz-type condition (4) above.

It will be shown in the proof that there are $n - k$ differentiable functions $\psi_j(y_1, \ldots, y_k)$ such that the equations

$$y_j = \psi_j(y_1, \ldots, y_k), \quad j = k+1, \ldots, n$$

define a k-dimensional differentiable manifold $\tilde{S}$ in $\mathbf{y}$-space. The differentiable manifold S in $\mathbf{x}$-space is then obtained from $\tilde{S}$ under the linear transformation of coordinates $\mathbf{x} = C\mathbf{y}$.

Proof. Consider the system (6). Let

$$U(t) = \begin{bmatrix} e^{Pt} & 0 \\ 0 & 0 \end{bmatrix} \quad \text{and} \quad V(t) = \begin{bmatrix} 0 & 0 \\ 0 & e^{Qt} \end{bmatrix}.$$

Then $\dot{U} = BU$, $\dot{V} = BV$ and

$$e^{Bt} = U(t) + V(t).$$

2.7. The Stable Manifold Theorem

It is not difficult to see that with $\alpha > 0$ chosen as in (5), we can choose $K > 0$ sufficiently large and $\sigma > 0$ sufficiently small that

$$\|U(t)\| \leq Ke^{-(\alpha+\sigma)t} \text{ for all } t \geq 0$$

and

$$\|V(t)\| \leq Ke^{\sigma t} \text{ for all } t \leq 0.$$

Next consider the integral equation

$$\mathbf{u}(t,\mathbf{a}) = U(t)\mathbf{a} + \int_0^t U(t-s)\mathbf{G}(\mathbf{u}(s,\mathbf{a}))ds - \int_t^\infty V(t-s)\mathbf{G}(\mathbf{u}(s,\mathbf{a}))ds. \quad (7)$$

If $\mathbf{u}(t,\mathbf{a})$ is a continuous solution of this integral equation, then it is a solution of the differential equation (6). We now solve this integral equation by the method of successive approximations. Let

$$\mathbf{u}^{(0)}(t,\mathbf{a}) = 0$$

and

$$\mathbf{u}^{(j+1)}(t,\mathbf{a}) = U(t)\mathbf{a} + \int_0^t U(t-s)\mathbf{G}(\mathbf{u}^{(j)}(s,\mathbf{a}))ds$$
$$- \int_t^\infty V(t-s)\mathbf{G}(\mathbf{u}^{(j)}(s,\mathbf{a}))ds. \quad (8)$$

Assume that the induction hypothesis

$$|\mathbf{u}^{(j)}(t,\mathbf{a}) - \mathbf{u}^{(j-1)}(t,\mathbf{a})| \leq \frac{K|\mathbf{a}|e^{-\alpha t}}{2^{j-1}} \quad (9)$$

holds for $j = 1, 2, \ldots, m$ and $t \geq 0$. It clearly holds for $j = 1$ provided $t \geq 0$. Then using the Lipschitz-type condition (4) satisfied by the function $\mathbf{G}$ and the above estimates on $\|U(t)\|$ and $\|V(t)\|$, it follows from the induction hypothesis that for $t \geq 0$

$$|\mathbf{u}^{(m+1)}(t,\mathbf{a}) - \mathbf{u}^{(m)}(t,\mathbf{a})| \leq \int_0^t \|U(t-s)\|\varepsilon|\mathbf{u}^{(m)}(s,\mathbf{a}) - \mathbf{u}^{(m-1)}(s,\mathbf{a})|ds$$
$$+ \int_t^\infty \|V(t-s)\|\varepsilon|\mathbf{u}^{(m)}(s,\mathbf{a}) - \mathbf{u}^{(m-1)}(s,\mathbf{a})|ds$$
$$\leq \varepsilon \int_0^t Ke^{-(\alpha+\sigma)(t-s)} \frac{K|\mathbf{a}|e^{-\alpha s}}{2^{m-1}} ds$$
$$+ \varepsilon \int_0^\infty Ke^{\sigma(t-s)} \frac{K|\mathbf{a}|e^{-\alpha s}}{2^{m-1}} ds$$
$$\leq \frac{\varepsilon K^2 |\mathbf{a}|e^{-\alpha t}}{\sigma 2^{m-1}} + \frac{\varepsilon K^2 |\mathbf{a}|e^{-\alpha t}}{\sigma 2^{m-1}}$$
$$< \left(\frac{1}{4} + \frac{1}{4}\right) \frac{K|\mathbf{a}|e^{-\alpha t}}{2^{m-1}} = \frac{K|\mathbf{a}|e^{-\alpha t}}{2^m} \quad (10)$$

provided $\varepsilon K/\sigma < 1/4$; i.e., provided we choose $\varepsilon < \frac{\sigma}{4K}$. In order that the condition (4) hold for the function $\mathbf{G}$, it suffices to choose $K|\mathbf{a}| < \delta/2$; i.e., we choose $|\mathbf{a}| < \frac{\delta}{2K}$. It then follows by induction that (9) holds for all $j = 1, 2, 3, \ldots$ and $t \geq 0$. Thus, for $n > m > N$ and $t \geq 0$,

$$|\mathbf{u}^{(n)}(t, \mathbf{a}) - \mathbf{u}^{(m)}(t, \mathbf{a})| \leq \sum_{j=N}^{\infty} |\mathbf{u}^{(j+1)}(t, \mathbf{a}) - \mathbf{u}^{(j)}(t, \mathbf{a})|$$

$$\leq K|\mathbf{a}| \sum_{j=N}^{\infty} \frac{1}{2^j} = \frac{K|\mathbf{a}|}{2^{N-1}}.$$

This last quantity approaches zero as $N \to \infty$ and therefore $\{\mathbf{u}^{(j)}(t, \mathbf{a})\}$ is a Cauchy sequence of continuous functions. According to the theorem in Section 2.2,

$$\lim_{j \to \infty} \mathbf{u}^{(j)}(t, \mathbf{a}) = \mathbf{u}(t, \mathbf{a})$$

uniformly for all $t \geq 0$ and $|\mathbf{a}| < \delta/2K$. Taking the limit of both sides of (8), it follows from the uniform convergence that the continuous function $\mathbf{u}(t, \mathbf{a})$ satisfies the integral equation (7) and hence the differential equation (6). It follows by induction and the fact that $\mathbf{G} \in C^1(\tilde{E})$ that $\mathbf{u}^{(j)}(t, \mathbf{a})$ is a differentiable function of $\mathbf{a}$ for $t \geq 0$ and $|\mathbf{a}| < \delta/2K$. Thus, it follows from the uniform convergence that $\mathbf{u}(t, \mathbf{a})$ is a differentiable function of $\mathbf{a}$ for $t \geq 0$ and $|\mathbf{a}| < \delta/2K$. The estimate (10) implies that

$$|\mathbf{u}(t, \mathbf{a})| \leq 2K|\mathbf{a}|e^{-\alpha t} \tag{11}$$

for $t \geq 0$ and $|\mathbf{a}| < \delta/2K$.

It is clear from the integral equation (7) that the last $n - k$ components of the vector $\mathbf{a}$ do not enter the computation and hence they may be taken as zero. Thus, the components $u_j(t, \mathbf{a})$ of the solution $\mathbf{u}(t, \mathbf{a})$ satisfy the initial conditions

$$u_j(0, \mathbf{a}) = a_j \text{ for } j = 1, \ldots, k$$

and

$$u_j(0, \mathbf{a}) = -\left(\int_0^\infty V(-s)\mathbf{G}(\mathbf{u}(s, a_1, \ldots, a_k, 0))ds\right)_j \text{ for } j = k+1, \ldots, n.$$

For $j = k+1, \ldots, n$ we define the functions

$$\psi_j(a_1, \ldots, a_k) = u_j(0, a_1, \ldots, a_k, 0, \ldots, 0). \tag{12}$$

Then the initial values $y_j = u_j(0, a_1, \ldots, a_k, 0, \ldots, 0)$ satisfy

$$y_j = \psi_j(y_1, \ldots, y_k) \text{ for } j = k+1, \ldots, n$$

according to the definition (12). These equations then define a differentiable manifold $\tilde{S}$ for $\sqrt{y_1^2 + \cdots + y_k^2} < \delta/2K$. Furthermore, if $\mathbf{y}(t)$ is a solution

2.7. The Stable Manifold Theorem

of the differential equation (6) with $\mathbf{y}(0) \in \tilde{S}$, i.e., with $\mathbf{y}(0) = \mathbf{u}(0, \mathbf{a})$, then
$$\mathbf{y}(t) = \mathbf{u}(t, \mathbf{a}).$$

It follows from the estimate (11) that if $\mathbf{y}(t)$ is a solution of (6) with $\mathbf{y}(0) \in \tilde{S}$, then $\mathbf{y}(t) \to 0$ as $t \to \infty$. It can also be shown that if $\mathbf{y}(t)$ is a solution of (6) with $\mathbf{y}(0) \notin \tilde{S}$ then $\mathbf{y}(t) \not\to 0$ as $t \to \infty$; cf. Coddington and Levinson [C/L], p. 332. It therefore follows from Theorem 2 in Section 2.5 that if $\mathbf{y}(0) \in \tilde{S}$, then $\mathbf{y}(t) \in \tilde{S}$ for all $t \geq 0$. And it can be shown as in [C/L], p. 333 that

$$\frac{\partial \psi_j}{\partial y_i}(\mathbf{0}) = 0$$

for $i = 1, \ldots, k$ and $j = k+1, \ldots, n$; i.e., the differentiable manifold $\tilde{S}$ is tangent to the stable subspace $E^s = \{\mathbf{y} \in \mathbf{R}^n \mid y_1 = \cdots y_k = 0\}$ of the linear system $\dot{\mathbf{y}} = B\mathbf{y}$ at 0.

The existence of the unstable manifold $\tilde{U}$ of (6) is established in exactly the same way by considering the system (6) with $t \to -t$, i.e.,

$$\dot{\mathbf{y}} = -B\mathbf{y} - \mathbf{G}(\mathbf{y}).$$

The stable manifold for this system will then be the unstable manifold $\tilde{U}$ for (6). Note that it is also necessary to replace the vector $\mathbf{y}$ by the vector $(y_{k+1}, \ldots, y_n, y_1, \ldots, y_k)$ in order to determine the $n-k$ dimensional manifold $\tilde{U}$ by the above process. This completes the proof of the Stable Manifold Theorem.

Remark 1. The first rigorous results concerning invariant manifolds were due to Hadamard [10] in 1901, Liapunov [17] in 1907 and Perron [26] in 1928. They proved the existence of stable and unstable manifolds of systems of differential equations and of maps. (Cf. Theorem 3 in Section 4.8 of Chapter 4.) The proof presented in this section is due to Liapunov and Perron. Several recent results generalizing the results of the Stable Manifold Theorem have been given by Hale, Hirsch, Pugh, Shub and Smale to mention a few. Cf. [11, 14, 30]. We note that if the function $\mathbf{f} \in C^r(E)$ and $r \geq 1$, then the stable and unstable differentiable manifolds S and U of (1) are of class C^r. And if $\mathbf{f}$ is analytic in E then S and U are analytic manifolds.

We illustrate the construction of the successive approximations $\mathbf{u}^{(j)}(t, \mathbf{a})$ in the proof with an example.

Example 2. Consider the nonlinear system

$$\dot{x}_1 = -x_1 - x_2^2 \qquad \dot{x}_2 = x_2 + x_1^2.$$

We shall find the first three successive approximations $\mathbf{u}^{(1)}(t,\mathbf{a}), \mathbf{u}^{(2)}(t,\mathbf{a})$ and $\mathbf{u}^{(3)}(t,\mathbf{a})$ defined by (8) and use $\mathbf{u}^{(3)}(t,\mathbf{a})$ to approximate the function ψ_2 describing the stable manifold

$$S: x_2 = \psi_2(x_1).$$

For this problem, we have

$$A = B = \begin{bmatrix} -1 & 0 \\ 0 & 1 \end{bmatrix}, \quad F(\mathbf{x}) = G(\mathbf{x}) = \begin{bmatrix} -x_2^2 \\ x_1^2 \end{bmatrix},$$

$$U(t) = \begin{bmatrix} e^{-t} & 0 \\ 0 & 0 \end{bmatrix}, \quad V(t) = \begin{bmatrix} 0 & 0 \\ 0 & e^t \end{bmatrix} \quad \text{and} \quad \mathbf{a} = \begin{bmatrix} a_1 \\ 0 \end{bmatrix}.$$

We approximate the solution of the integral equation

$$\mathbf{u}(t,\mathbf{a}) = \begin{bmatrix} e^{-t}a_1 \\ 0 \end{bmatrix} + \int_0^t \begin{bmatrix} -e^{-(t-s)}u_2^2(s) \\ 0 \end{bmatrix} ds - \int_t^\infty \begin{bmatrix} 0 \\ e^{t-s}u_1^2(s) \end{bmatrix} ds$$

by the successive approximations

$$\mathbf{u}^{(0)}(t,\mathbf{a}) = 0$$

$$\mathbf{u}^{(1)}(t,\mathbf{a}) = \begin{bmatrix} e^{-t}a_1 \\ 0 \end{bmatrix}$$

$$\mathbf{u}^{(2)}(t,\mathbf{a}) = \begin{bmatrix} e^{-t}a_1 \\ 0 \end{bmatrix} - \int_t^\infty \begin{bmatrix} 0 \\ e^{t-s}e^{-2s}a_1^2 \end{bmatrix} ds = \begin{bmatrix} e^{-t}a_1 \\ -\frac{e^{-2t}}{3}a_1^2 \end{bmatrix}$$

$$\mathbf{u}^{(3)}(t,\mathbf{a}) = \begin{bmatrix} e^{-t}a_1 \\ 0 \end{bmatrix} - \frac{1}{9}\int_0^t \begin{bmatrix} e^{-(t-s)}e^{-4s}a_1^4 \\ 0 \end{bmatrix} ds - \int_t^\infty \begin{bmatrix} 0 \\ e^{t-s}e^{-2s}a_1^2 \end{bmatrix} ds$$

$$= \begin{bmatrix} e^{-t}a_1 + \frac{1}{27}(e^{-4t} - e^{-t})a_1^4 \\ -\frac{1}{3}e^{-2t}a_1^2 \end{bmatrix}.$$

It can be shown that $\mathbf{u}^{(4)}(t,\mathbf{a}) - \mathbf{u}^{(3)}(t,\mathbf{a}) = 0(a_1^5)$ and therefore the function $\psi_2(a_1) = u_2(0, a_1, 0)$ is approximated by

$$\psi_2(a_1) = -\frac{1}{3}a_1^2 + 0(a_1^5)$$

as $a_1 \to 0$. Hence, the stable manifold S is approximated by

$$S: x_2 = -\frac{x_1^2}{3} + 0(x_1^5)$$

as $x_1 \to 0$. The matrix $C = I$, the identity, for this example and hence the **x** and **y** spaces are the same. The unstable manifold U can be approximated by applying exactly the same procedure to the above system with $t \to -t$ and x_1 and x_2 interchanged. The stable manifold for the resulting system will then be the unstable manifold for the original system. We find

$$U: x_1 = -\frac{x_2^2}{3} + 0(x_2^5)$$

2.7. The Stable Manifold Theorem

as $x_2 \to 0$. The student should verify this calculation. The approximations for the stable and unstable manifolds S and U in a neighborhood of the origin and the stable and unstable subspaces E^s and E^u for $\dot{\mathbf{x}} = A\mathbf{x}$ are shown in Figure 2.

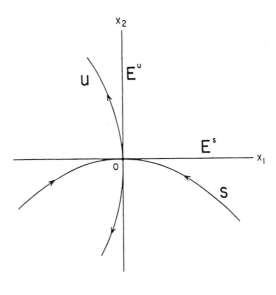

Figure 2

The stable and unstable manifolds S and U are only defined in a small neighborhood of the origin in the proof of the stable manifold theorem. S and U are therefore referred to as the *local* stable and unstable manifolds of (1) at the origin or simply as the *local* stable and unstable manifolds of the origin. We define the *global* stable and unstable manifolds of (1) at $\mathbf{0}$ by letting points in S flow backward in time and those in U flow forward in time.

Definition 3. Let ϕ_t be the flow of the nonlinear system (1). *The global stable and unstable manifolds of* (1) *at* $\mathbf{0}$ *are defined by*

$$W^s(\mathbf{0}) = \bigcup_{t \leq 0} \phi_t(S)$$

and

$$W^u(\mathbf{0}) = \bigcup_{t \geq 0} \phi_t(U)$$

respectively; $W^s(\mathbf{0})$ and $W^u(\mathbf{0})$ are also referred to as the *global stable and unstable manifolds of the origin* respectively. It can be shown that the global stable and unstable manifolds $W^s(\mathbf{0})$ and $W^u(\mathbf{0})$ are unique and

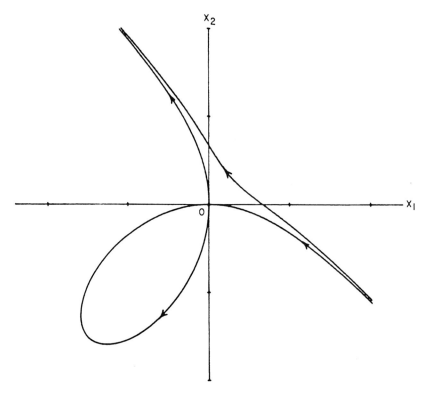

Figure 3

that they are invariant with respect to the flow ϕ_t; furthermore, for all $\mathbf{x} \in W^s(0)$, $\lim_{t\to\infty} \phi_t(\mathbf{x}) = 0$ and for all $\mathbf{x} \in W^u(0)$, $\lim_{t\to-\infty} \phi_t(\mathbf{x}) = 0$.

As in the proof of the Stable Manifold Theorem, it can be shown that in a small neighborhood, N, of a hyperbolic critical point at the origin, the local stable and unstable manifolds, S and U, of (1) at the origin are given by

$$S = \{\mathbf{x} \in N \mid \phi_t(\mathbf{x}) \to 0 \text{ as } t \to \infty \text{ and } \phi_t(\mathbf{x}) \in N \text{ for } t \geq 0\}$$

and

$$U = \{\mathbf{x} \in N \mid \phi_t(\mathbf{x}) \to 0 \text{ as } t \to -\infty \text{ and } \phi_t(\mathbf{x}) \in N \text{ for } t \leq 0\}$$

respectively. Cf. [G/H], p. 13.

Figure 3 shows some numerically computed solution curves for the system in Example 2. The global stable and unstable manifolds for this example are shown in Figure 4. Note that $W^s(0)$ and $W^u(0)$ intersect in a "homoclinic loop" at the origin. $W^s(0)$ and $W^u(0)$ are more properly called "branched manifolds" in this example.

2.7. The Stable Manifold Theorem

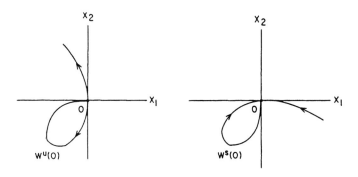

Figure 4

It follows from equation (11) in the proof of the stable manifold theorem that if $\mathbf{x}(t)$ is a solution of the differential equation (6) with $\mathbf{x}(0) \in S$, i.e., if $\mathbf{x}(t) = C\mathbf{y}(t)$ with $\mathbf{y}(0) = \mathbf{u}(0,\mathbf{a}) \in \tilde{S}$, then for any $\varepsilon > 0$ there exists a $\delta > 0$ such that if $|\mathbf{x}(0)| < \delta$ then

$$|\mathbf{x}(t)| \leq \varepsilon e^{-\alpha t}$$

for all $t \geq 0$. Just as in the proof of the stable manifold theorem, α is any positive number that satisfies $\text{Re}(\lambda_j) < -\alpha$ for $j = 1, \ldots, k$ where λ_j, $j = 1, \ldots, k$ are the eigenvalues of $D\mathbf{f}(0)$ with negative real part. This result shows that solutions starting in S, sufficiently near the origin, approach the origin exponentially fast as $t \to \infty$.

Corollary. *Under the hypotheses of the Stable Manifold Theorem, if S and U are the stable and unstable manifolds of (1) at the origin and if $\text{Re}(\lambda_j) < -\alpha < 0 < \beta < \text{Re}(\lambda_m)$ for $j = 1, \ldots, k$ and $m = k+1, \ldots, n$, then given $\varepsilon > 0$ there exists a $\delta > 0$ such that if $\mathbf{x}_0 \in N_\delta(0) \cap S$ then*

$$|\phi_t(\mathbf{x}_0)| \leq \varepsilon e^{-\alpha t}$$

for all $t \geq 0$ and if $\mathbf{x}_0 \in N_\delta(0) \cap U$ then

$$|\phi_t(\mathbf{x}_0)| \leq \varepsilon e^{\beta t}$$

for all $t \leq 0$.

We add one final result to this section which establishes the existence of an invariant *center manifold* $W^c(0)$ tangent to E^c at 0; cf., e.g., [G/H], p. 127 or [Ru], p. 32. The next theorem follows from the local center manifold theorem, Theorem 2 in Section 2.12, and the stable manifold theorem in this section.

Theorem (The Center Manifold Theorem). *Let* $\mathbf{f} \in C^r(E)$ *where* E *is an open subset of* $\mathbf{R}^n$ *containing the origin and* $r \geq 1$. *Suppose that* $\mathbf{f}(\mathbf{0}) = \mathbf{0}$ *and that* $D\mathbf{f}(\mathbf{0})$ *has* k *eigenvalues with negative real part,* j *eigenvalues with positive real part, and* $m = n - k - j$ *eigenvalues with zero real part. Then there exists an* m-*dimensional center manifold* $W^c(\mathbf{0})$ *of class* C^r *tangent to the center subspace* E^c *of* (2) *at* $\mathbf{0}$, *there exists a* k-*dimensional stable manifold* $W^s(\mathbf{0})$ *of class* C^r *tangent to the stable subspace* E^s *of* (2) *at* $\mathbf{0}$ *and there exists a* j-*dimensional unstable manifold* $W^u(\mathbf{0})$ *of class* C^r *tangent to the unstable subspace* E^u *of* (2) *at* $\mathbf{0}$; *furthermore,* $W^c(\mathbf{0})$, $W^s(\mathbf{0})$ *and* $W^u(\mathbf{0})$ *are invariant under the flow* ϕ_t *of* (1).

Example 3. Consider the system

$$\dot{x}_1 = x_1^2$$
$$\dot{x}_2 = -x_2.$$

The stable subspace E^s of the linearized system at the origin is the x_2-axis and the center subspace E^c is the x_1-axis. This system is easily solved and the phase portrait is shown in Figure 5.

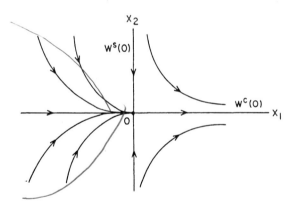

Figure 5. The phase portrait for the system in Example 3.

Any solution curve of the system in Example 3 to the left of the origin patched together with the positive x_1-axis at the origin gives a one-dimensional center manifold of class C^∞ which is tangent to E^c at the origin. This shows that, in general, the center manifold $W^c(\mathbf{0})$ is not unique; however, in Example 3 there is only one analytic center manifold, namely, the x_1-axis.

In Section 2.12, we give a method for approximating the shape of the analytic center manifold near a nonhyperbolic critical point and show that approximating the shape of the center manifold $W^c(\mathbf{0})$ near the nonhyperbolic critical point at the origin is essential in determining the flow on the center manifold; cf. Example 1 in Section 2.12. In Section 2.13, we present

2.7. The Stable Manifold Theorem

a method for simplifying the system of differential equations determining the flow on the center manifold.

PROBLEM SET 7

1. Write the system
$$\dot{x}_1 = x_1 + 6x_2 + x_1 x_2$$
$$\dot{x}_2 = 4x_1 + 3x_2 - x_1^2$$
in the form
$$\dot{\mathbf{y}} = B\mathbf{y} + \mathbf{G}(\mathbf{y})$$
where
$$B = \begin{bmatrix} \lambda_1 & 0 \\ 0 & \lambda_2 \end{bmatrix},$$
$\lambda_1 < 0$, $\lambda_2 > 0$ and $\mathbf{G}(\mathbf{y})$ is quadratic in y_1 and y_2.

2. Find the first three successive approximations $\mathbf{u}^{(1)}(t,\mathbf{a})$, $\mathbf{u}^{(2)}(t,\mathbf{a})$ and $\mathbf{u}^{(3)}(t,\mathbf{a})$ for
$$\dot{x}_1 = -x_1$$
$$\dot{x}_2 = x_2 + x_1^2$$
and use $\mathbf{u}^3(t,\mathbf{a})$ to approximate S near the origin. Also approximate the unstable manifold U near the origin for this system. Note that $\mathbf{u}^{(2)}(t,\mathbf{a}) = \mathbf{u}^{(3)}(t,\mathbf{a})$ and therefore $\mathbf{u}^{(j+1)}(t,\mathbf{a}) = \mathbf{u}^{(j)}(t,\mathbf{a})$ for $j \geq 2$. Thus $\mathbf{u}^{(j)}(t,\mathbf{a}) \to \mathbf{u}(t,\mathbf{a}) = \mathbf{u}^2(t,\mathbf{a})$ which gives the exact function defining S.

3. Solve the system in Problem 2 and show that S and U are given by
$$S: x_2 = -\frac{x_1^2}{3}$$
and
$$U: x_1 = 0.$$
Sketch S, U, E^s and E^u.

4. Find the first four successive approximations $\mathbf{u}^{(1)}(t,\mathbf{a})$, $\mathbf{u}^{(2)}(t,\mathbf{a})$, $\mathbf{u}^3(t,\mathbf{a})$, and $\mathbf{u}^{(4)}(t,\mathbf{a})$ for the system
$$\dot{x}_1 = -x_1$$
$$\dot{x}_2 = -x_2 + x_1^2$$
$$\dot{x}_3 = x_3 + x_2^2.$$
Show that $\mathbf{u}^{(3)}(t,\mathbf{a}) = \mathbf{u}^{(4)}(t,\mathbf{a}) = \cdots$ and hence $\mathbf{u}(t,\mathbf{a}) = \mathbf{u}^{(3)}(t,\mathbf{a})$. Find S and U for this problem.

5. Solve the above system and show that

$$S: x_3 = -\frac{1}{3}x_2^2 - \frac{1}{6}x_1^2 x_2 - \frac{1}{30}x_1^4$$

and

$$U: x_1 = x_2 = 0.$$

Note that these formulas actually determine the global stable and unstable manifolds $W^s(0)$ and $W^u(0)$ respectively.

6. Let E be an open subset of $\mathbf{R}^n$ containing the origin. Use the fact that if $\mathbf{F} \in C^1(E)$ then for all $\mathbf{x}, \mathbf{y} \in N_\delta(0) \subset E$ there exists a $\xi \in N_\delta(0)$ such that

$$|\mathbf{F}(\mathbf{x}) - \mathbf{F}(\mathbf{y})| \leq \|D\mathbf{F}(\xi)\| \, |\mathbf{x} - \mathbf{y}|$$

(cf. Theorem 5.19 and the proof of Theorem 9.19 in [R]) to prove that if $\mathbf{F} \in C^1(E)$ and $\mathbf{F}(0) = D\mathbf{F}(0) = 0$ then given any $\varepsilon > 0$ there exists a $\delta > 0$ such that for all $\mathbf{x}, \mathbf{y} \in N_\delta(0)$ we have

$$|\mathbf{F}(\mathbf{x}) - \mathbf{F}(\mathbf{y})| < \varepsilon |\mathbf{x} - \mathbf{y}|.$$

7. Show that the unit circle

$$C = S^1 = \{\mathbf{x} \in \mathbf{R}^2 \mid x^2 + y^2 = 1\}$$

is an orientable, one-dimensional, differentiable manifold; i.e., find an orientation-preserving atlas $(U_1, \mathbf{h}_1), \ldots, (U_4, \mathbf{h}_4)$ for C.

8. Show that the unit two-dimensional sphere

$$S^2 = \{\mathbf{x} \in \mathbf{R}^3 \mid x^2 + y^2 + z^2 = 1\}$$

is an orientable, two-dimensional, differentiable manifold. Do this using the following orientation-preserving atlas:

$U_1 = \{\mathbf{x} \in S^2 \mid z > 0\}, U_2 = \{\mathbf{x} \in S^2 \mid z < 0\}, U_3 = \{\mathbf{x} \in S^2 \mid y > 0\},$
$U_4 = \{\mathbf{x} \in S^2 \mid y < 0\}, U_5 = \{\mathbf{x} \in S^2 \mid x > 0\}, U_6 = \{\mathbf{x} \in S^2 \mid x < 0\},$

$\mathbf{h}_1(x, y, z) = (x, y), \mathbf{h}_1^{-1}(x, y) = (x, y, \sqrt{1 - x^2 - y^2})$
$\mathbf{h}_2(x, y, z) = (y, x), \mathbf{h}_2^{-1}(y, x) = (x, y, -\sqrt{1 - x^2 - y^2})$
$\mathbf{h}_3(x, y, z) = (z, x), \mathbf{h}_3^{-1}(z, x) = (x, \sqrt{1 - x^2 - z^2}, z)$
$\mathbf{h}_4(x, y, z) = (x, z), \mathbf{h}_4^{-1}(x, z) = (x, -\sqrt{1 - x^2 - z^2}, z)$
$\mathbf{h}_5(x, y, z) = (y, z), \mathbf{h}_5^{-1}(y, z) = (\sqrt{1 - y^2 - z^2}, y, z)$
$\mathbf{h}_6(x, y, z) = (z, y), \mathbf{h}_6^{-1}(z, y) = (-\sqrt{1 - y^2 - z^2}, y, z).$

To show this, compute $h_i \circ h_j^{-1}(x)$, $Dh_i \circ h_j^{-1}(x)$, and show that $\det Dh_i \circ h_j^{-1}(x) > 0$ for all $x \in h_j(U_i \cap U_j)$ where $U_i \cap U_j \neq \emptyset$. Note that you only need to do this for $j > i$, $U_1 \cap U_2 = U_3 \cap U_4 = U_5 \cap U_6 = \emptyset$, and that the atlas has been chosen to preserve outward normals on S^2. **Hint:** Start by showing that

$$h_1 \circ h_3^{-1}(z, x) = (x, \sqrt{1 - x^2 - z^2}),$$

$$Dh_1 \circ h_3^{-1}(z, x) = \begin{bmatrix} 0 & 1 \\ \frac{-z}{\sqrt{1-x^2-z^2}} & \frac{-x}{\sqrt{1-x^2-z^2}} \end{bmatrix}$$

and that

$$\det Dh_1 \circ h_3^{-1}(z, x) = \frac{z}{\sqrt{1 - x^2 - z^2}} > 0$$

for all $(z, x) \in h_3(U_1 \cap U_3) = \{(z, x) \in \mathbf{R}^2 \mid x^2 + z^2 < 1, z > 0\}$.

2.8 The Hartman–Grobman Theorem

The Hartman–Grobman Theorem is another very important result in the local qualitative theory of ordinary differential equations. The theorem shows that near a hyperbolic equilibrium point $\mathbf{x}_0$, the nonlinear system

$$\dot{\mathbf{x}} = \mathbf{f}(\mathbf{x}) \tag{1}$$

has the same qualitative structure as the linear system

$$\dot{\mathbf{x}} = A\mathbf{x} \tag{2}$$

with $A = D\mathbf{f}(\mathbf{x}_0)$. Throughout this section we shall assume that the equilibrium point $\mathbf{x}_0$ has been translated to the origin.

Definition 1. Two autonomous systems of differential equations such as (1) and (2) are said to be *topologically equivalent* in a neighborhood of the origin or to have the *same qualitative structure near the origin* if there is a homeomorphism H mapping an open set U containing the origin onto an open set V containing the origin which maps trajectories of (1) in U onto trajectories of (2) in V and preserves their orientation by time in the sense that if a trajectory is directed from $\mathbf{x}_1$ to $\mathbf{x}_2$ in U, then its image is directed from $H(\mathbf{x}_1)$ to $H(\mathbf{x}_2)$ in V. If the homeomorphism H preserves the parameterization by time, then the systems (1) and (2) are said to be *topologically conjugate* in a neighborhood of the origin.

Before stating the Hartman–Grobman Theorem, we consider a simple example of two topologically conjugate linear systems.

Example 1. Consider the linear systems $\dot{\mathbf{x}} = A\mathbf{x}$ and $\dot{\mathbf{y}} = B\mathbf{y}$ with

$$A = \begin{bmatrix} -1 & -3 \\ -3 & -1 \end{bmatrix} \quad \text{and} \quad B = \begin{bmatrix} 2 & 0 \\ 0 & -4 \end{bmatrix}.$$

Let $H(\mathbf{x}) = R\mathbf{x}$ where the matrix

$$R = \frac{1}{\sqrt{2}} \begin{bmatrix} 1 & -1 \\ 1 & 1 \end{bmatrix} \quad \text{and} \quad R^{-1} = \frac{1}{\sqrt{2}} \begin{bmatrix} 1 & 1 \\ -1 & 1 \end{bmatrix}.$$

Then $B = RAR^{-1}$ and letting $\mathbf{y} = H(\mathbf{x}) = R\mathbf{x}$ or $\mathbf{x} = R^{-1}\mathbf{y}$ gives us

$$\dot{\mathbf{y}} = RAR^{-1}\mathbf{y} = B\mathbf{y}.$$

Thus, if $\mathbf{x}(t) = e^{At}\mathbf{x}_0$ is the solution of the first system through $\mathbf{x}_0$, then $\mathbf{y}(t) = H(\mathbf{x}(t)) = R\mathbf{x}(t) = Re^{At}\mathbf{x}_0 = e^{Bt}R\mathbf{x}_0$ is the solution of the second system through $R\mathbf{x}_0$; i.e., H maps trajectories of the first system onto trajectories of the second system and it preserves the parameterization since

$$He^{At} = e^{Bt}H.$$

The mapping $H(\mathbf{x}) = R\mathbf{x}$ is simply a rotation through 45° and it is clearly a homeomorphism. The phase portraits of these two systems are shown in Figure 1.

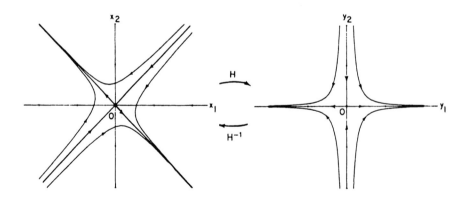

Figure 1

Theorem (The Hartman–Grobman Theorem). *Let E be an open subset of $\mathbf{R}^n$ containing the origin, let $\mathbf{f} \in C^1(E)$, and let ϕ_t be the flow of the nonlinear system (1). Suppose that $\mathbf{f}(0) = 0$ and that the matrix $A = D\mathbf{f}(0)$ has no eigenvalue with zero real part. Then there exists a homeomorphism H of an open set U containing the origin onto an open set V containing*

2.8. The Hartman–Grobman Theorem

the origin such that for each $\mathbf{x}_0 \in U$, there is an open interval $I_0 \subset \mathbf{R}$ containing zero such that for all $\mathbf{x}_0 \in U$ and $t \in I_0$

$$H \circ \phi_t(\mathbf{x}_0) = e^{At} H(\mathbf{x}_0);$$

i.e., H maps trajectories of (1) near the origin onto trajectories of (2) near the origin and preserves the parameterization by time.

Outline of the Proof. Consider the nonlinear system (1) with $\mathbf{f} \in C^1(E)$, $\mathbf{f}(0) = 0$ and $A = D\mathbf{f}(0)$.

1. Suppose that the matrix A is written in the form

$$A = \begin{bmatrix} P & 0 \\ 0 & Q \end{bmatrix}$$

where the eigenvalues of P have negative real part and the eigenvalues of Q have positive real part.

2. Let ϕ_t be the flow of the nonlinear system (1) and write the solution

$$\mathbf{x}(t, \mathbf{x}_0) = \phi_t(\mathbf{x}_0) = \begin{bmatrix} \mathbf{y}(t, \mathbf{y}_0, \mathbf{z}_0) \\ \mathbf{z}(t, \mathbf{y}_0, \mathbf{z}_0) \end{bmatrix}$$

where

$$\mathbf{x}_0 = \begin{bmatrix} \mathbf{y}_0 \\ \mathbf{z}_0 \end{bmatrix} \in \mathbf{R}^n,$$

$\mathbf{y}_0 \in E^s$, the stable subspace of A and $\mathbf{z}_0 \in E^u$, the unstable subspace of A.

3. Define the functions

$$\tilde{\mathbf{Y}}(\mathbf{y}_0, \mathbf{z}_0) = \mathbf{y}(1, \mathbf{y}_0, \mathbf{z}_0) - e^P \mathbf{y}_0$$

and

$$\tilde{\mathbf{Z}}(\mathbf{y}_0, \mathbf{z}_0) = \mathbf{z}(1, \mathbf{y}_0, \mathbf{z}_0) - e^Q \mathbf{z}_0.$$

Then $\tilde{\mathbf{Y}}(0) = \tilde{\mathbf{Z}}(0) = D\tilde{\mathbf{Y}}(0) = D\tilde{\mathbf{Z}}(0) = 0$. And since $\mathbf{f} \in C^1(E)$, $\tilde{\mathbf{Y}}(\mathbf{y}_0, \mathbf{z}_0)$ and $\tilde{\mathbf{Z}}(\mathbf{y}_0, \mathbf{z}_0)$ are continuously differentiable. Thus,

$$\|D\tilde{\mathbf{Y}}(\mathbf{y}_0, \mathbf{z}_0)\| \leq a$$

and

$$\|D\tilde{\mathbf{Z}}(\mathbf{y}_0, \mathbf{z}_0)\| \leq a$$

on the compact set $|\mathbf{y}_0|^2 + |\mathbf{z}_0|^2 \leq s_0^2$. The constant a can be taken as small as we like by choosing s_0 sufficiently small. We let $\mathbf{Y}(\mathbf{y}_0, \mathbf{z}_0)$ and $\mathbf{Z}(\mathbf{y}_0, \mathbf{z}_0)$ be smooth functions which are equal to $\tilde{\mathbf{Y}}(\mathbf{y}_0, \mathbf{z}_0)$ and $\tilde{\mathbf{Z}}(\mathbf{y}_0, \mathbf{z}_0)$ for $|\mathbf{y}_0|^2 + |\mathbf{z}_0|^2 \leq (s_0/2)^2$ and zero for $|\mathbf{y}_0|^2 + |\mathbf{z}_0|^2 \geq s_0^2$. Then by the mean value theorem

$$|\mathbf{Y}(\mathbf{y}_0, \mathbf{z}_0)| \leq a\sqrt{|\mathbf{y}_0|^2 + |\mathbf{z}_0|^2} \leq a(|\mathbf{y}_0| + |\mathbf{z}_0|)$$

and
$$|Z(y_0, z_0)| \leq a\sqrt{|y_0|^2 + |z_0|^2} \leq a(|y_0| + |z_0|)$$
for all $(y_0, z_0) \in \mathbf{R}^n$. We next let $B = e^P$ and $C = e^Q$. Then assuming that we have carried out the normalization in Problem 7 in Section 1.8 of Chapter 1, cf. Hartman [H], p. 233, we have
$$b = \|B\| < 1 \quad \text{and} \quad c = \|C^{-1}\| < 1.$$

4. For
$$\mathbf{x} = \begin{bmatrix} \mathbf{y} \\ \mathbf{z} \end{bmatrix} \in \mathbf{R}^n$$
define the transformations
$$L(\mathbf{y}, \mathbf{z}) = \begin{bmatrix} B\mathbf{y} \\ C\mathbf{z} \end{bmatrix}$$
and
$$T(\mathbf{y}, \mathbf{z}) = \begin{bmatrix} B\mathbf{y} + \mathbf{Y}(\mathbf{y}, \mathbf{z}) \\ C\mathbf{z} + \mathbf{Z}(\mathbf{y}, \mathbf{z}) \end{bmatrix};$$
i.e., $L(\mathbf{x}) = e^A \mathbf{x}$ and locally $T(\mathbf{x}) = \phi_1(\mathbf{x})$.

Lemma. *There exists a homeomorphism H of an open set U containing the origin onto an open set V containing the origin such that*
$$H \circ T = L \circ H.$$

We establish this lemma using the method of successive approximations. For $\mathbf{x} \in \mathbf{R}^n$, let
$$H(\mathbf{x}) = \begin{bmatrix} \Phi(\mathbf{y}, \mathbf{z}) \\ \Psi(\mathbf{y}, \mathbf{z}) \end{bmatrix}.$$
Then $H \circ T = L \circ H$ is equivalent to the pair of equations
$$\begin{aligned} B\Phi(\mathbf{y}, \mathbf{z}) &= \Phi(B\mathbf{y} + \mathbf{Y}(\mathbf{y}, \mathbf{z}), C\mathbf{z} + \mathbf{Z}(\mathbf{y}, \mathbf{z})) \\ C\Psi(\mathbf{y}, \mathbf{z}) &= \Psi(B\mathbf{y} + \mathbf{Y}(\mathbf{y}, \mathbf{z}), C\mathbf{z} + \mathbf{Z}(\mathbf{y}, \mathbf{z})). \end{aligned} \quad (3)$$

First of all, define the successive approximations for the second equation by
$$\begin{aligned} \Psi_0(\mathbf{y}, \mathbf{z}) &= \mathbf{z} \\ \Psi_{k+1}(\mathbf{y}, \mathbf{z}) &= C^{-1} \Psi_k(B\mathbf{y} + \mathbf{Y}(\mathbf{y}, \mathbf{z}), C\mathbf{z} + \mathbf{Z}(\mathbf{y}, \mathbf{z})). \end{aligned} \quad (4)$$
It then follows by an easy induction argument that for $k = 0, 1, 2, \ldots$, the $\Psi_k(\mathbf{y}, \mathbf{z})$ are continuous and satisfy $\Psi_k(\mathbf{y}, \mathbf{z}) = \mathbf{z}$ for $|\mathbf{y}| + |\mathbf{z}| \geq 2s_0$. We next prove by induction that for $j = 1, 2, \ldots$
$$|\Psi_j(\mathbf{y}, \mathbf{z}) - \Psi_{j-1}(\mathbf{y}, \mathbf{z})| \leq Mr^j(|\mathbf{y}| + |\mathbf{z}|)^\delta \quad (5)$$

2.8. The Hartman–Grobman Theorem

where $r = c[2\max(a,b,c)]^\delta$ with $\delta \in (0,1)$ chosen sufficiently small so that $r < 1$ (which is possible since $c < 1$) and $M = ac(2s_0)^{1-\delta}/r$. First of all for $j = 1$

$$\begin{aligned}
|\Psi_1(\mathbf{y},\mathbf{z}) - \Psi_0(\mathbf{y},\mathbf{z})| &= |C^{-1}\Psi_0(B\mathbf{y}+\mathbf{Y}(\mathbf{y},\mathbf{z}), C\mathbf{z}+\mathbf{Z}(\mathbf{y},\mathbf{z})) - \mathbf{z}| \\
&= |C^{-1}(C\mathbf{z}+\mathbf{Z}(\mathbf{y},\mathbf{z})) - \mathbf{z}| \\
&= |C^{-1}\mathbf{Z}(\mathbf{y},\mathbf{z})| \leq \|C^{-1}\||\mathbf{Z}(\mathbf{y},\mathbf{z})| \\
&\leq ca(|\mathbf{y}|+|\mathbf{z}|) \leq Mr(|\mathbf{y}|+|\mathbf{z}|)^\delta
\end{aligned}$$

since $\mathbf{Z}(\mathbf{y},\mathbf{z}) = 0$ for $|\mathbf{y}|+|\mathbf{z}| \geq 2s_0$. And then assuming that the induction hypothesis holds for $j = 1,\ldots,k$ we have

$$\begin{aligned}
|\Psi_{k+1}(\mathbf{y},\mathbf{z}) - \Psi_k(\mathbf{y},\mathbf{z})| &= |C^{-1}\Psi_k(B\mathbf{y}+\mathbf{Y}(\mathbf{y},\mathbf{z}), C\mathbf{z}+\mathbf{Z}(\mathbf{y},\mathbf{z})) \\
&\quad - C^{-1}\Psi_{k-1}(B\mathbf{y}+\mathbf{Y}(\mathbf{y},\mathbf{z}), C\mathbf{z}+\mathbf{Z}(\mathbf{y},\mathbf{z}))| \\
&\leq \|C^{-1}\||\Psi_k('') - \Psi_{k-1}('')| \\
&\leq cMr^k[|B\mathbf{y}+\mathbf{Y}(\mathbf{y},\mathbf{z})| + |C\mathbf{z}+\mathbf{Z}(\mathbf{y},\mathbf{z})|]^\delta \\
&\leq cMr^k[b|\mathbf{y}| + 2a(|\mathbf{y}|+|\mathbf{z}|) + c|\mathbf{z}|]^\delta \\
&\leq cMr^k[2\max(a,b,c)]^\delta(|\mathbf{y}|+|\mathbf{z}|)^\delta \\
&= Mr^{k+1}(|\mathbf{y}|+|\mathbf{z}|)^\delta
\end{aligned}$$

Thus, just as in the proof of the fundamental theorem in Section 2.2 and the stable manifold theorem in Section 2.7, $\Psi_k(\mathbf{y},\mathbf{z})$ is a Cauchy sequence of continuous functions which converges uniformly as $k \to \infty$ to a continuous function $\Psi(\mathbf{y},\mathbf{z})$. Also, $\Psi(\mathbf{y},\mathbf{z}) = \mathbf{z}$ for $|\mathbf{y}|+|\mathbf{z}| \geq 2s_0$. Taking limits in (4) shows that $\Psi(\mathbf{y},\mathbf{z})$ is a solution of the second equation in (3).

The first equation in (3) can be written as

$$B^{-1}\Phi(\mathbf{y},\mathbf{z}) = \Phi(B^{-1}\mathbf{y} + \mathbf{Y}_1(\mathbf{y},\mathbf{z}), C^{-1}\mathbf{z} + \mathbf{Z}_1(\mathbf{y},\mathbf{z})) \tag{6}$$

where the functions $\mathbf{Y}_1$ and $\mathbf{Z}_1$ are defined by the inverse of T (which exists if the constant a is sufficiently small, i.e., if s_0 is sufficiently small) as follows:

$$T^{-1}(\mathbf{y},\mathbf{z}) = \begin{bmatrix} B^{-1}\mathbf{y} + \mathbf{Y}_1(\mathbf{y},\mathbf{z}) \\ C^{-1}\mathbf{z} + \mathbf{Z}_1(\mathbf{y},\mathbf{z}) \end{bmatrix}.$$

Then equation (6) can be solved for $\Phi(\mathbf{y},\mathbf{z})$ by the method of successive approximations exactly as above with $\Phi_0(\mathbf{y},\mathbf{z}) = \mathbf{y}$ since $b = \|B\| < 1$. We therefore obtain the continuous map

$$H(\mathbf{y},\mathbf{z}) = \begin{bmatrix} \Phi(\mathbf{y},\mathbf{z}) \\ \Psi(\mathbf{y},\mathbf{z}) \end{bmatrix}.$$

And it follows as on pp. 248–249 in Hartman [H] that H is a homeomorphism of $\mathbf{R}^n$ onto $\mathbf{R}^n$.

5. We now let H_0 be the homeomorphism defined above and let L^t and T^t be the one-parameter families of transformations defined by

$$L^t(\mathbf{x}_0) = e^{At}\mathbf{x}_0 \quad \text{and} \quad T^t(\mathbf{x}_0) = \phi_t(\mathbf{x}_0).$$

Define

$$H = \int_0^1 L^{-s} H_0 T^s \, ds.$$

It then follows using the above lemma that there exists a neighborhood of the origin for which

$$L^t H = \int_0^1 L^{t-s} H_0 T^{s-t} \, ds \, T^t$$

$$= \int_{-t}^{1-t} L^{-s} H_0 T^s \, ds \, T^t$$

$$= \left[\int_{-t}^0 L^{-s} H_0 T^s \, ds + \int_0^{1-t} L^{-s} H_0 T^s \, ds \right] T^t$$

$$= \int_0^1 L^{-s} H_0 T^s \, ds \, T^t = H T^t$$

since by the above lemma $H_0 = L^{-1} H_0 T$ which implies that

$$\int_{-t}^0 L^{-s} H_0 T^s \, ds = \int_{-t}^0 L^{-s-1} H_0 T^{s+1} \, ds$$

$$= \int_{1-t}^1 L^{-s} H_0 T^s \, ds.$$

Thus, $H \circ T^t = L^t H$ or equivalently

$$H \circ \phi_t(\mathbf{x}_0) = e^{At} H(\mathbf{x}_0)$$

and it can be shown as on pp. 250–251 of Hartman [H] that H is a homeomorphism on $\mathbf{R}^n$. This completes the outline of the proof of the Hartman–Grobman Theorem.

We now illustrate how the successive approximations, defined by (4), can be used to obtain the homeomorphism $H(\mathbf{x}) = [\Phi(\mathbf{y},\mathbf{z}), \Psi(\mathbf{y},\mathbf{z})]^T$. In the following example, the successive approximations actually converge to a global homeomorphism which maps solutions of (1) onto solutions of (2) for all $\mathbf{x}_0 \in \mathbf{R}^2$ and for all $t \in \mathbf{R}$. Of course, this does not happen in general; cf. Problem 4.

Example 2. Consider the system

$$\dot{y} = -y$$
$$\dot{z} = z + y^2.$$

2.8. The Hartman–Grobman Theorem

The solution with $y(0) = y_0$ and $z(0) = z_0$ is given by

$$y(t) = y_0 e^{-t}$$
$$z(t) = z_0 e^t + \frac{y_0^2}{3}(e^t - e^{-2t}).$$

Thus in the context of the above proof $B = e^{-1}$, $C = e$, $Y(y_0, z_0) = 0$ and

$$Z(y_0, z_0) = k y_0^2 \quad \text{with} \quad k = \frac{e^3 - 1}{3e^2}.$$

Therefore, solving

$$C\Psi(y, z) = \Psi(By + Y(y, z), Cz + Z(y, z))$$

is equivalent to solving

$$e\Psi(y, z) = \Psi(e^{-1}y, ez + ky^2). \tag{7}$$

The successive approximations defined by (4) are given by

$$\Psi_0(y, z) = z$$
$$\Psi_1(y, z) = z + k e^{-1} y^2$$
$$\Psi_2(y, z) = z + k e^{-1}(1 + e^{-3}) y^2$$
$$\Psi_3(y, z) = z + k e^{-1}(1 + e^{-3} + e^{-6}) y^2$$
$$\vdots$$
$$\Psi_m(y, z) = z + k e^{-1}[1 + e^{-3} + \cdots + (e^{-3})^{m-1}] y^2.$$

The serious student should verify these calculations. Thus, as $m \to \infty$

$$\Psi_m(y, z) \to \Psi(y, z) = z + \frac{y^2}{3}$$

uniformly for $(y, z) \in \mathbf{R}^2$. The function $\Psi(y, z)$ satisfies equation (7). Similarly, the function $\Phi(y, z)$ is found by solving

$$B^{-1}\Phi(y, z) = \Phi(B^{-1}y + Y_1(y, z), C^{-1}z + Z_1(y, z))$$

where $Y_1(y, z) = 0$ and $Z_1(y, z) = -eky^2$. This leads to

$$\Phi(y, z) = y$$

for all $(y, z) \in \mathbf{R}^2$.

Thus, the homeomorphism H_0 is given by

$$H_0(y, z) = \begin{bmatrix} y \\ z + \frac{1}{3}y^2 \end{bmatrix}$$

and

$$H_0^{-1}(y, z) = \begin{bmatrix} y \\ z - \frac{1}{3}y^2 \end{bmatrix}.$$

For $L(y,z) = e^A(y,z)^T = (e^{-1}y, ez)^T$ and $T(y,z) = \phi_1(y,z) = (e^{-1}y, ez + ky^2)^T$, it follows that $H_0 \circ T = L \circ H_0$ as in the above lemma. The homeomorphism H is then given by

$$H = \int_0^1 L^{-s} H_0 T^s \, ds$$

where

$$L^t(y,z) = \begin{bmatrix} e^{-t}y \\ e^t z \end{bmatrix}$$

and

$$T^t(y,z) = \begin{bmatrix} e^{-t}y \\ e^t z + \frac{y^2}{3}(e^t - e^{-2t}) \end{bmatrix}.$$

Thus,

$$H(y,z) = \begin{bmatrix} y \\ z + \frac{y^2}{3} \end{bmatrix}$$

and

$$L^t H(y,z) = \begin{bmatrix} e^{-t}y \\ e^t z + e^t \frac{y^2}{3} \end{bmatrix} = H \circ T^t(y,z).$$

The student should verify these computations. The phase portraits for the nonlinear system (1) and the linear system (2) of this example are shown in Figure 2. The stable subspace $E^s = \{(y,z) \in \mathbf{R}^2 \mid z = 0\}$ gets mapped onto the stable manifold $W^s(0) = \{(y,z) \in \mathbf{R}^2 \mid z = -y^2/3\}$ by the homeomorphism H^{-1}; and the unstable subspace $E^u = \{(y,z) \in \mathbf{R}^2 \mid y = 0\}$ gets mapped onto the unstable manifold $W^u(0) = \{(y,z) \in \mathbf{R}^2 \mid y = 0\}$ by H^{-1}. Trajectories, such as $z = 1/y$, of the linear system get mapped onto trajectories, such as $z = \frac{1}{y} - \frac{y^2}{3}$, by H^{-1} and H preserves the parameterization by time.

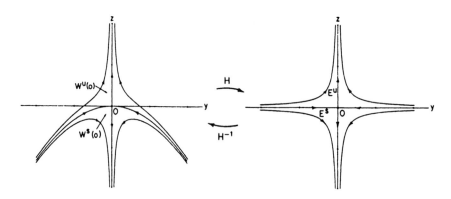

Figure 2

2.8. The Hartman–Grobman Theorem

The above proof of the Hartman–Grobman Theorem is due to P. Hartman; cf. [H], pp. 244–251. It was proved independently by P. Hartman [12] and the Russian mathematician D.M. Grobman [9] in 1959. This theorem with $\mathbf{f}$, H and H^{-1} analytic was proved by H. Poincaré in 1879, cf. [28], under the assumptions that the elementary divisors of A (cf. [Cu], p. 219) are simple and that the eigenvalues $\lambda_1, \ldots, \lambda_n$ of A lie in a half plane in C and satisfy

$$\lambda_j \neq m_1 \lambda_1 + \cdots + m_n \lambda_n \tag{8}$$

for all sets of non-negative integers $(m_1, \ldots, m_n)$ satisfying $m_1 + \cdots + m_n > 1$. An analogous result for smooth $\mathbf{f}$, H and H^{-1} was established by S. Sternberg [33] in 1957. And for $\mathbf{f}$ of class C^2, Hartman [13] in 1960 proved that there exists a continuously differentiable map H with a continuously differentiable inverse H^{-1} (i.e., a C^1-*diffeomorphism*) satisfying the conclusions of the above theorem even without the Diophantine conditions (8) on λ_j:

Theorem (Hartman). *Let E be an open subset of $\mathbf{R}^n$ containing the point $\mathbf{x}_0$, let $\mathbf{f} \in C^2(E)$, and let ϕ_t be the flow of the nonlinear system (1). Suppose that $\mathbf{f}(\mathbf{x}_0) = 0$ and that all of the eigenvalues $\lambda_1, \ldots, \lambda_n$ of the matrix $A = D\mathbf{f}(\mathbf{x}_0)$ have negative (or positive) real part. Then there exists a C^1-diffeomorphism H of a neighborhood U of $\mathbf{x}_0$ onto an open set V containing the origin such that for each $\mathbf{x} \in U$ there is an open interval $I(\mathbf{x}) \subset \mathbf{R}$ containing zero such that for all $\mathbf{x} \in U$ and $t \in I(\mathbf{x})$*

$$H \circ \phi_t(\mathbf{x}) = e^{At} H(\mathbf{x}). \tag{9}$$

The student should compare this theorem and the Hartman–Grobman Theorem in the context of Example 5 in Section 2.10 of this chapter. Also, it should be noted that assuming the existence of higher derivatives of $\mathbf{f}$ does not imply the existence of higher derivatives of H without Diophantine-type conditions on λ_j. In fact, Sternberg showed that in general even if $\mathbf{f}$ is analytic, there does not exist a mapping H with H and H^{-1} of class C^2 satisfying (9); cf. Problem 5.

Problem Set 8

1. Solve the system

$$\dot{y}_1 = -y_1$$
$$\dot{y}_2 = -y_2 + z^2$$
$$\dot{z} = z$$

   and show that the successive approximations $\Phi_k \to \Phi$ and $\Psi_k \to \Psi$ as $k \to \infty$ for all $\mathbf{x} = (y_1, y_2, z) \in \mathbf{R}^3$. Define $H_0 = (\Phi, \Psi)^T$ and use this homeomorphism to find

$$H = \int_0^1 L^{-s} H_0 T^s \, ds.$$

Use the homeomorphism H to find the stable and unstable manifolds

$$W^s(0) = H^{-1}(E^s) \quad \text{and} \quad W^u(0) = H^{-1}(E^u)$$

for this system.
Hint: You should find

$$H(y_1, y_2, z) = (y_1, y_2 - z^2/3, z)^T$$
$$W^s(0) = \{\mathbf{x} \in \mathbf{R}^3 \mid z = 0\}$$

and

$$W^u(0) = \{\mathbf{x} \in \mathbf{R}^3 \mid y_1 = 0, y_2 = z^2/3\}.$$

2. Same thing for

$$\dot{y} = -y$$
$$\dot{z}_1 = z_1$$
$$\dot{z}_2 = z_2 + y^2 + y z_1.$$

3. Same thing for

$$\dot{y}_1 = -y_1$$
$$\dot{y}_2 = -y_2 + y_1^2$$
$$\dot{z} = z + y_1^2$$

Hint: Here the successive approximations for Φ as given by (3) converge globally.

4. Show that the successive approximations for Φ as given by (3) or (6) do not converge globally for the system

$$\dot{y}_1 = -y_1$$
$$\dot{y}_2 = -y_2 + y_1^2 z$$
$$\dot{z} = z.$$

5. Show that the successive approximations (4) for $H(\mathbf{z}) = \Psi(\mathbf{z})$ do not converge globally for the analytic system

$$\dot{z}_1 = 2z_1$$
$$\dot{z}_2 = 4z_2 + z_1^2.$$

Furthermore, show that if $H(\mathbf{z})$ is any C^2-function which satisfies (9), then the Jacobian $J(\mathbf{z}) = \det DH(\mathbf{z})$ vanishes at $\mathbf{z} = \mathbf{0}$. This in turn implies that the inverse H^{-1}, if it exists, is not differentiable at $\mathbf{z} = \mathbf{0}$. (Why?)

Hint: First solve this system to obtain $\mathbf{z}(t) = \phi_t(\mathbf{z}_0)$. Then show that if $H = (H_1, H_2)^T$ satisfies the above equation, (9), for $t = 1$ we get

$$CH(\mathbf{z}) = H(C\mathbf{z} + \mathbf{e}_2 e^4 z_1^2)$$

where $C = e^A$, $A = \mathrm{diag}(2,4)$, $\mathbf{e}_2 = (0,1)^T$ and $ln\, e = 1$. Differentiate the second component of this equation partially with respect to z_1 to get

$$\frac{\partial H_2}{\partial z_1}(0,0) = 0.$$

A second partial differentiation with respect to z_1 then shows that

$$\frac{\partial H_2}{\partial z_2}(0,0) = 0;$$

i.e. the Jacobian $J(\mathbf{0}) = \det DH(\mathbf{0}) = 0$.

2.9 Stability and Liapunov Functions

In this section we discuss the stability of the equilibrium points of the nonlinear system

$$\dot{\mathbf{x}} = \mathbf{f}(\mathbf{x}). \tag{1}$$

The stability of any hyperbolic equilibrium point $\mathbf{x}_0$ of (1) is determined by the signs of the real parts of the eigenvalues λ_j of the matrix $D\mathbf{f}(\mathbf{x}_0)$. A hyperbolic equilibrium point $\mathbf{x}_0$ is asymptotically stable iff $\mathrm{Re}(\lambda_j) < 0$ for $j = 1, \ldots, n$; i.e., iff $\mathbf{x}_0$ is a sink. And a hyperbolic equilibrium point $\mathbf{x}_0$ is unstable iff it is either a source or a saddle. The stability of nonhyperbolic equilibrium points is typically more difficult to determine. A method, due to Liapunov, that is very useful for deciding the stability of nonhyperbolic equilibrium points is presented in this section. Additional methods are presented in Sections 2.11–2.13.

Definition 1. Let ϕ_t denote the flow of the differential equation (1) defined for all $t \in \mathbf{R}$. An equilibrium point $\mathbf{x}_0$ of (1) is *stable* if for all $\varepsilon > 0$ there exists a $\delta > 0$ such that for all $\mathbf{x} \in N_\delta(\mathbf{x}_0)$ and $t \geq 0$ we have

$$\phi_t(\mathbf{x}) \in N_\varepsilon(\mathbf{x}_0);$$

The equilibrium point $\mathbf{x}_0$ is *unstable* if it is not stable. And $\mathbf{x}_0$ is *asymptotically stable* if it is stable and if there exists a $\delta > 0$ such that for all $\mathbf{x} \in N_\delta(\mathbf{x}_0)$ we have

$$\lim_{t \to \infty} \phi_t(\mathbf{x}) = \mathbf{x}_0.$$

Note that the above limit being satisfied for all x in some neighborhood of x_0 does not imply that x_0 is stable; cf. Problem 7 at the end of Section 3.7.

It can be seen from the phase portraits in Section 1.5 of Chapter 1 that a stable node or focus of a linear system in $\mathbf{R}^2$ is an asymptotically stable equilibrium point; an unstable node or focus or a saddle of a linear system in $\mathbf{R}^2$ is an unstable equilibrium point; and a center of a linear system in $\mathbf{R}^2$ is a stable equilibrium point which is not asymptotically stable.

It follows from the Stable Manifold Theorem and the Hartman–Grobman Theorem that any sink of (1) is asymptotically stable and any source or saddle of (1) is unstable. Hence, *any hyperbolic equilibrium point of* (1) *is either asymptotically stable or unstable.* The corollary in Section 2.7 provides even more information concerning the local behavior of solutions near a sink:

Theorem 1. *If x_0 is a sink of the nonlinear system* (1) *and* $\mathrm{Re}(\lambda_j) < -\alpha < 0$ *for all of the eigenvalues λ_j of the matrix $Df(x_0)$, then given $\varepsilon > 0$ there exists a $\delta > 0$ such that for all $x \in N_\delta(x_0)$, the flow $\phi_t(x)$ of* (1) *satisfies*

$$|\phi_t(x) - x_0| \leq \varepsilon e^{-\alpha t}$$

for all $t \geq 0$.

Since hyperbolic equilibrium points are either asymptotically stable or unstable, the only time that an equilibrium point x_0 of (1) can be stable but not asymptotically stable is when $Df(x_0)$ has a zero eigenvalue or a pair of complex-conjugate, pure-imaginary eigenvalues $\lambda = \pm ib$. It follows from the next theorem, proved in [H/S], that all other eigenvalues λ_j of $Df(x_0)$ must satisfy $\mathrm{Re}(\lambda_j) \leq 0$ if x_0 is stable.

Theorem 2. *If x_0 is a stable equilibrium point of* (1), *no eigenvalue of $Df(x_0)$ has positive real part.*

We see that stable equilibrium points which are not asymptotically stable can only occur at nonhyperbolic equilibrium points. But the question as to whether a nonhyperbolic equilibrium point is stable, asymptotically stable or unstable is a delicate question.

The following method, due to Liapunov (in his 1892 doctoral thesis), is very useful in answering this question.

Definition 2. If $f \in C^1(E)$, $V \in C^1(E)$ and ϕ_t is the flow of the differential equation (1), then for $x \in E$ the derivative of the function $V(x)$ along the solution $\phi_t(x)$

$$\dot{V}(x) = \frac{d}{dt}V(\phi_t(x))|_{t=0} = DV(x)f(x).$$

2.9. Stability and Liapunov Functions

The last equality follows from the chain rule. If $\dot V(\mathbf{x})$ is negative in E then $V(\mathbf{x})$ decreases along the solution $\phi_t(\mathbf{x}_0)$ through $\mathbf{x}_0 \in E$ at $t = 0$. Furthermore, in $\mathbf{R}^2$, if $\dot V(\mathbf{x}) \leq 0$ with equality only at $\mathbf{x} = \mathbf{0}$, then for small positive C, the family of curves $V(\mathbf{x}) = C$ constitutes a family of closed curves enclosing the origin and the trajectories of (1) cross these curves from their exterior to their interior with increasing t; i.e., the origin of (1) is asymptotically stable. A function $V \colon \mathbf{R}^n \to \mathbf{R}$ satisfying the hypotheses of the next theorem is called a *Liapunov function*.

Theorem 3. *Let E be an open subset of $\mathbf{R}^n$ containing $\mathbf{x}_0$. Suppose that $\mathbf{f} \in C^1(E)$ and that $\mathbf{f}(\mathbf{x}_0) = \mathbf{0}$. Suppose further that there exists a real valued function $V \in C^1(E)$ satisfying $V(\mathbf{x}_0) = 0$ and $V(\mathbf{x}) > 0$ if $\mathbf{x} \neq \mathbf{x}_0$. Then (a) if $\dot V(\mathbf{x}) \leq 0$ for all $\mathbf{x} \in E$, $\mathbf{x}_0$ is stable; (b) if $\dot V(\mathbf{x}) < 0$ for all $\mathbf{x} \in E \sim \{\mathbf{x}_0\}$, $\mathbf{x}_0$ is asymptotically stable; (c) if $\dot V(\mathbf{x}) > 0$ for all $\mathbf{x} \in E \sim \{\mathbf{x}_0\}$, $\mathbf{x}_0$ is unstable.*

Proof. Without loss of generality, we shall assume that the equilibrium point $\mathbf{x}_0 = \mathbf{0}$. (a) Choose $\varepsilon > 0$ sufficiently small that $\overline{N_\varepsilon(\mathbf{0})} \subset E$ and let m_ε be the minimum of the continuous function $V(\mathbf{x})$ on the compact set

$$S_\varepsilon = \{\mathbf{x} \in \mathbf{R}^n \mid |\mathbf{x}| = \varepsilon\}.$$

Then since $V(\mathbf{x}) > 0$ for $\mathbf{x} \neq \mathbf{0}$, it follows that $m_\varepsilon > 0$. Since $V(\mathbf{x})$ is continuous and $V(\mathbf{0}) = 0$, it follows that there exists a $\delta > 0$ such that $|\mathbf{x}| < \delta$ implies that $V(\mathbf{x}) < m_\varepsilon$. Since $\dot V(\mathbf{x}) \leq 0$ for $\mathbf{x} \in E$, it follows that $V(\mathbf{x})$ is decreasing along trajectories of (1). Thus, if ϕ_t is the flow of the differential equation (1), it follows that for all $\mathbf{x}_0 \in N_\delta(\mathbf{0})$ and $t \geq 0$ we have

$$V(\phi_t(\mathbf{x}_0)) \leq V(\mathbf{x}_0) < m_\varepsilon.$$

Now suppose that for $|\mathbf{x}_0| < \delta$ there is a $t_1 > 0$ such that $|\phi_{t_1}(\mathbf{x}_0)| = \varepsilon$; i.e., such that $\phi_{t_1}(\mathbf{x}_0) \in S_\varepsilon$. Then since m_ε is the minimum of $V(\mathbf{x})$ on S_ε, this would imply that

$$V(\phi_{t_1}(\mathbf{x}_0)) \geq m_\varepsilon$$

which contradicts the above inequality. Thus for $|\mathbf{x}_0| < \delta$ and $t \geq 0$ it follows that $|\phi_t(\mathbf{x}_0)| < \varepsilon$; i.e., $\mathbf{0}$ is a stable equilibrium point.

(b) Suppose that $\dot V(\mathbf{x}) < 0$ for all $\mathbf{x} \in E$. Then $V(\mathbf{x})$ is strictly decreasing along trajectories of (1). Let ϕ_t be the flow of (1) and let $\mathbf{x}_0 \in N_\delta(\mathbf{0})$, the neighborhood defined in part (a). Then, by part (a), if $|\mathbf{x}_0| < \delta$, $\phi_t(\mathbf{x}_0) \subset \overline{N_\varepsilon(\mathbf{0})}$ for all $t \geq 0$. Let $\{t_k\}$ be any sequence with $t_k \to \infty$. Then since $\overline{N_\varepsilon(\mathbf{0})}$ is compact, there is a subsequence of $\{\phi_{t_k}(\mathbf{x}_0)\}$ that converges to a point in $\overline{N_\varepsilon(\mathbf{0})}$. But for any subsequence $\{t_n\}$ of $\{t_k\}$ such that $\{\phi_{t_n}(\mathbf{x}_0)\}$ converges, we show below that the limit is zero. It then follows that $\phi_{t_k}(\mathbf{x}_0) \to \mathbf{0}$ for any sequence $t_k \to \infty$ and therefore that $\phi_t(\mathbf{x}_0) \to \mathbf{0}$ as $t \to \infty$; i.e., that $\mathbf{0}$ is asymptotically stable. It remains to show that if $\phi_{t_n}(\mathbf{x}_0) \to \mathbf{y}_0$, then $\mathbf{y}_0 = \mathbf{0}$. Since $V(\mathbf{x})$ is strictly decreasing along tra-

jectories of (1) and since $V(\phi_{t_n}(\mathbf{x}_0)) \to V(\mathbf{y}_0)$ by the continuity of V, it follows that
$$V(\phi_t(\mathbf{x}_0)) > V(\mathbf{y}_0)$$
for all $t > 0$. But if $\mathbf{y}_0 \neq \mathbf{0}$, then for $s > 0$ we have $V(\phi_s(\mathbf{y}_0)) < V(\mathbf{y}_0)$ and, by continuity, it follows that for all $\mathbf{y}$ sufficiently close to $\mathbf{y}_0$ we have $V(\phi_s(\mathbf{y})) < V(\mathbf{y}_0)$ for $s > 0$. But then for $\mathbf{y} = \phi_{t_n}(\mathbf{x}_0)$ and n sufficiently large, we have
$$V(\phi_{s+t_n}(\mathbf{x}_0)) < V(\mathbf{y}_0)$$
which contradicts the above inequality. Therefore $\mathbf{y}_0 = \mathbf{0}$ and it follows that $\mathbf{0}$ is asymptotically stable.

(c) Let M be the maximum of the continuous function $V(\mathbf{x})$ on the compact set $\overline{N_\varepsilon(\mathbf{0})}$. Since $\dot{V}(\mathbf{x}) > 0$, $V(\mathbf{x})$ is strictly increasing along trajectories of (1). Thus, if ϕ_t is the flow of (1), then for any $\delta > 0$ and $\mathbf{x}_0 \in N_\delta(\mathbf{0}) \sim \{\mathbf{0}\}$ we have
$$V(\phi_t(\mathbf{x}_0)) > V(\mathbf{x}_0) > 0$$
for all $t > 0$. And since $\dot{V}(\mathbf{x})$ is positive definite, this last statement implies that
$$\inf_{t \geq 0} \dot{V}(\phi_t(\mathbf{x}_0)) = m > 0.$$
Thus,
$$V(\phi_t(\mathbf{x}_0)) - V(\mathbf{x}_0) \geq mt$$
for all $t \geq 0$. Therefore,
$$V(\phi_t(\mathbf{x}_0)) > mt > M$$
for t sufficiently large; i.e., $\phi_t(\mathbf{x}_0)$ lies outside the closed set $\overline{N_\varepsilon(\mathbf{0})}$. Hence, $\mathbf{0}$ is unstable.

Remark. If $\dot{V}(\mathbf{x}) = 0$ for all $\mathbf{x} \in E$ then the trajectories of (1) lie on the surfaces in $\mathbf{R}^n$ (or curves in $\mathbf{R}^2$) defined by
$$V(\mathbf{x}) = c.$$

Example 1. Consider the system
$$\dot{x}_1 = -x_2^3$$
$$\dot{x}_2 = x_1^3.$$
The origin is a nonhyperbolic equilibrium point of this system and
$$V(\mathbf{x}) = x_1^4 + x_2^4$$
is a Liapunov function for this system. In fact
$$\dot{V}(\mathbf{x}) = 4x_1^3 \dot{x}_1 + 4x_2^3 \dot{x}_2 = 0.$$

2.9. Stability and Liapunov Functions

Hence the solution curves lie on the closed curves

$$x_1^4 + x_2^4 = c^2$$

which encircle the origin. The origin is thus a stable equilibrium point of this system which is not asymptotically stable. Note that $D\mathbf{f}(\mathbf{0}) = 0$ for this example; i.e., $D\mathbf{f}(\mathbf{0})$ has two zero eigenvalues.

Example 2. Consider the system

$$\dot{x}_1 = -2x_2 + x_2 x_3$$
$$\dot{x}_2 = x_1 - x_1 x_3$$
$$\dot{x}_3 = x_1 x_2.$$

The origin is an equilibrium point for this system and

$$D\mathbf{f}(\mathbf{0}) = \begin{bmatrix} 0 & -2 & 0 \\ 1 & 0 & 0 \\ 0 & 0 & 0 \end{bmatrix}.$$

Thus $D\mathbf{f}(\mathbf{0})$ has eigenvalues $\lambda_1 = 0$, $\lambda_{2,3} = \pm 2i$; i.e., $\mathbf{x} = \mathbf{0}$ is a nonhyperbolic equilibrium point. So we use Liapunov's method. But how do we find a suitable Liapunov function? A function of the form

$$V(\mathbf{x}) = c_1 x_1^2 + c_2 x_2^2 + c_3 x_3^2$$

with positive constants c_1, c_2 and c_3 is usually worth a try, at least when the system contains some linear terms. Computing $\dot{V}(\mathbf{x}) = DV(\mathbf{x})\mathbf{f}(\mathbf{x})$, we find

$$\frac{1}{2}\dot{V}(\mathbf{x}) = (c_1 - c_2 + c_3)x_1 x_2 x_3 + (-2c_1 + c_2)x_1 x_2.$$

Hence if $c_2 = 2c_1$ and $c_3 = c_1 > 0$ we have $V(\mathbf{x}) > 0$ for $\mathbf{x} \neq \mathbf{0}$ and $\dot{V}(\mathbf{x}) = 0$ for all $\mathbf{x} \in \mathbf{R}^3$ and therefore by Theorem 3, $\mathbf{x} = \mathbf{0}$ is stable. Furthermore, choosing $c_1 = c_3 = 1$ and $c_2 = 2$, we see that the trajectories of this system lie on the ellipsoids $x_1^2 + 2x_2^2 + x_3^2 = c^2$.

We commented earlier that all sinks are asymptotically stable. However, as the next example shows, not all asymptotically stable equilibrium points are sinks. (Of course, a *hyperbolic* equilibrium point is asymptotically stable iff it is a sink.)

Example 3. Consider the following modification of the system in Example 2:

$$\dot{x}_1 = -2x_2 + x_2 x_3 - x_1^3$$
$$\dot{x}_2 = x_1 - x_1 x_3 - x_2^3$$
$$\dot{x}_3 = x_1 x_2 - x_3^3.$$

The Liapunov function of Example 2,
$$V(\mathbf{x}) = x_1^2 + 2x_2^2 + x_3^2,$$
satisfies $V(\mathbf{x}) > 0$ and
$$\dot{V}(\mathbf{x}) = -2(x_1^4 + 2x_2^4 + x_3^4) < 0$$
for $\mathbf{x} \neq \mathbf{0}$. Therefore, by Theorem 3, the origin is asymptotically stable, but it is not a sink since the eigenvalues $\lambda_1 = 0$, $\lambda_{2,3} = \pm 2i$ do not have negative real part.

Example 4. Consider the second-order differential equation
$$\ddot{x} + q(x) = 0$$
where the continuous function $q(x)$ satisfies $xq(x) > 0$ for $x \neq 0$. This differential equation can be written as the system
$$\dot{x}_1 = x_2$$
$$\dot{x}_2 = -q(x_1)$$
where $x_1 = x$. The total energy of the system
$$V(\mathbf{x}) = \frac{x_2^2}{2} + \int_0^{x_1} q(s)\,ds$$
(which is the sum of the kinetic energy $\frac{1}{2}\dot{x}_1^2$ and the potential energy) serves as a Liapunov function for this system.
$$\dot{V}(\mathbf{x}) = q(x_1)x_2 + x_2[-q(x_1)] = 0.$$
The solution curves are given by $V(\mathbf{x}) = c$; i.e., the energy is constant on the solution curves or trajectories of this system; and the origin is a stable equilibrium point.

Problem Set 9

1. Discuss the stability of the equilibrium points of the systems in Problem 1 of Problem Set 6.

2. Determine the stability of the equilibrium points of the system (1) with $\mathbf{f}(\mathbf{x})$ given by

   (a) $\begin{bmatrix} x_1^2 - x_2^2 - 1 \\ 2x_2 \end{bmatrix}$

   (b) $\begin{bmatrix} x_2 - x_1^2 + 2 \\ 2x_2^2 - 2x_1 x_2 \end{bmatrix}$

2.9. Stability and Liapunov Functions

(c) $\begin{bmatrix} -4x_1 - 2x_2 + 4 \\ x_1 x_2 \end{bmatrix}$.

3. Use the Liapunov function $V(\mathbf{x}) = x_1^2 + x_2^2 + x_3^2$ to show that the origin is an asymptotically stable equilibrium point of the system

$$\dot{\mathbf{x}} = \begin{bmatrix} -x_2 - x_1 x_2^2 + x_3^2 - x_1^3 \\ x_1 + x_3^3 - x_2^3 \\ -x_1 x_3 - x_3 x_1^2 - x_2 x_3^2 - x_3^5 \end{bmatrix}.$$

Show that the trajectories of the linearized system $\dot{\mathbf{x}} = D\mathbf{f}(\mathbf{0})\mathbf{x}$ for this problem lie on circles in planes parallel to the x_1, x_2 plane; hence, the origin is stable, but not asymptotically stable for the linearized system.

4. It was shown in Section 1.5 of Chapter 1 that the origin is a center for the linear system

$$\dot{\mathbf{x}} = \begin{bmatrix} 0 & -1 \\ 1 & 0 \end{bmatrix} \mathbf{x}.$$

The addition of nonlinear terms to the right-hand side of this linear system changes the stability of the origin. Use the Liapunov function $V(\mathbf{x}) = x_1^2 + x_2^2$ to establish the following results:

(a) The origin is an asymptotically stable equilibrium point of

$$\dot{\mathbf{x}} = \begin{bmatrix} 0 & -1 \\ 1 & 0 \end{bmatrix} \mathbf{x} + \begin{bmatrix} -x_1^3 - x_1 x_2^2 \\ -x_2^3 - x_2 x_1^2 \end{bmatrix}.$$

(b) The origin is an unstable equilibrium point of

$$\dot{\mathbf{x}} = \begin{bmatrix} 0 & -1 \\ 1 & 0 \end{bmatrix} \mathbf{x} + \begin{bmatrix} x_1^3 + x_1 x_2^2 \\ x_2^3 + x_2 x_1^2 \end{bmatrix}.$$

(c) The origin is a stable equilibrium point which is not asymptotically stable for

$$\dot{\mathbf{x}} = \begin{bmatrix} 0 & -1 \\ 1 & 0 \end{bmatrix} \mathbf{x} + \begin{bmatrix} -x_1 x_2 \\ x_1^2 \end{bmatrix}.$$

What are the solution curves in this case?

5. Use appropriate Liapunov functions to determine the stability of the equilibrium points of the following systems:

(a) $\begin{aligned} \dot{x}_1 &= -x_1 + x_2 + x_1 x_2 \\ \dot{x}_2 &= x_1 - x_2 - x_1^2 - x_2^3 \end{aligned}$

(b) $\begin{aligned} \dot{x}_1 &= x_1 - 3x_2 + x_1^3 \\ \dot{x}_2 &= -x_1 + x_2 - x_2^2 \end{aligned}$

(c) $\begin{aligned}\dot{x}_1 &= -x_1 - 2x_2 + x_1 x_2^2 \\ \dot{x}_2 &= 3x_1 - 3x_2 + x_2^3\end{aligned}$

(d) $\begin{aligned}\dot{x}_1 &= -4x_2 + x_1^2 \\ \dot{x}_2 &= 4x_1 + x_2^2\end{aligned}$

6. Let $A(t)$ be a continuous real-valued square matrix. Show that every solution of the nonautonomous linear system

$$\dot{\mathbf{x}} = A(t)\mathbf{x}$$

satisfies

$$|\mathbf{x}(t)| \leq |\mathbf{x}(0)| \exp \int_0^t \|A(s)\|\, ds.$$

And then show that if $\int_0^\infty \|A(s)\|\, ds < \infty$, then every solution of this system has a finite limit as t approaches infinity.

7. Show that the second-order differential equation

$$\ddot{x} + f(x)\dot{x} + g(x) = 0$$

can be written as the Lienard system

$$\begin{aligned}\dot{x}_1 &= x_2 - F(x_1) \\ \dot{x}_2 &= -g(x_1)\end{aligned}$$

where

$$F(x_1) = \int_0^{x_1} f(s)\, ds.$$

Let

$$G(x_1) = \int_0^{x_1} g(s)\, ds$$

and suppose that $G(x) > 0$ and $g(x)F(x) > 0$ (or $g(x)F(x) < 0$) in a deleted neighborhood of the origin. Show that the origin is an asymptotically stable equilibrium point (or an unstable equilibrium point) of this system.

8. Apply the previous results to the van der Pol equation

$$\ddot{x} + \varepsilon(x^2 - 1)\dot{x} + x = 0.$$

2.10 Saddles, Nodes, Foci and Centers

In Section 1.5 of Chapter 1, a linear system

$$\dot{\mathbf{x}} = A\mathbf{x} \tag{1}$$

2.10. Saddles, Nodes, Foci and Centers

where $\mathbf{x} \in \mathbf{R}^2$ was said to have a saddle, node, focus or center at the origin if its phase portrait was linearly equivalent to one of the phase portraits in Figures 1–4 in Section 1.5 of Chapter 1 respectively; i.e., if there exists a nonsingular linear transformation which reduces the matrix A to one of the canonical matrices B in Cases I–IV of Section 1.5 in Chapter 1 respectively. For example, the linear system (1) of the example in Section 2.8 of this chapter has a saddle at the origin.

In Section 2.6, a nonlinear system

$$\dot{\mathbf{x}} = \mathbf{f}(\mathbf{x}) \tag{2}$$

was said to have a saddle, a sink or a source at a hyperbolic equilibrium point $\mathbf{x}_0$ if the linear part of $\mathbf{f}$ at $\mathbf{x}_0$ had eigenvalues with both positive and negative real parts, only had eigenvalues with negative real parts, or only had eigenvalues with positive real parts, respectively.

In this section, we define the concept of a topological saddle for the nonlinear system (2) with $\mathbf{x} \in \mathbf{R}^2$ and show that if $\mathbf{x}_0$ is a hyperbolic equilibrium point of (2) then it is a topological saddle if and only if it is a saddl' of (2); i.e., a hyperbolic equilibrium point $\mathbf{x}_0$ is a topological saddle for (2) if and only if the origin is a saddle for (1) with $A = D\mathbf{f}(\mathbf{x}_0)$. We discuss topological saddles for nonhyperbolic equilibrium points of (2) with $\mathbf{x} \in \mathbf{R}^2$ in the next section. We also refine the classification of sinks of the nonlinear system (2) into stable nodes and foci and show that, under slightly stronger hypotheses on the function $\mathbf{f}$, i.e., stronger than $\mathbf{f} \in C^1(E)$, a hyperbolic critical point $\mathbf{x}_0$ is a stable node or focus for the nonlinear system (2) if and only if it is respectively a stable node or focus for the linear system (1) with $A = D\mathbf{f}(\mathbf{x}_0)$. Similarly, a source of (2) is either an unstable node or focus of (2) as defined below. Finally, we define centers and center-foci for the nonlinear system (2) and show that, under the addition of nonlinear terms, a center of the linear system (1) may become either a center, a center-focus, or a stable or unstable focus of (2).

Before defining these various types of equilibrium points for planar systems (2), it is convenient to introduce polar coordinates (r, θ) and to rewrite the system (2) in polar coordinates. In this section we let $\mathbf{x} = (x, y)^T$, $f_1(\mathbf{x}) = P(x, y)$ and $f_2(\mathbf{x}) = Q(x, y)$. The nonlinear system (2) can then be written as

$$\begin{aligned} \dot{x} &= P(x, y) \\ \dot{y} &= Q(x, y). \end{aligned} \tag{3}$$

If we let $r^2 = x^2 + y^2$ and $\theta = \tan^{-1}(y/x)$, then we have

$$r\dot{r} = x\dot{x} + y\dot{y}$$

and

$$r^2 \dot{\theta} = x\dot{y} - y\dot{x}.$$

It follows that for $r > 0$, the nonlinear system (3) can be written in terms of polar coordinates as

$$\dot{r} = P(r\cos\theta, r\sin\theta)\cos\theta + Q(r\cos\theta, r\sin\theta)\sin\theta$$
$$r\dot{\theta} = Q(r\cos\theta, r\sin\theta)\cos\theta - P(r\cos\theta, r\sin\theta)\sin\theta \quad (4)$$

or as

$$\frac{dr}{d\theta} = F(r, \theta) \equiv \frac{r[P(r\cos\theta, r\sin\theta)\cos\theta + Q(r\cos\theta, r\sin\theta)\sin\theta]}{Q(r\cos\theta, r\sin\theta)\cos\theta - P(r\cos\theta, r\sin\theta)\sin\theta}. \quad (5)$$

Writing the system of differential equations (3) in polar coordinates will often reveal the nature of the equilibrium point or critical point at the origin. This is illustrated by the next three examples; cf. Problem 4 in Problem Set 9.

Example 1. Write the system

$$\dot{x} = -y - xy$$
$$\dot{y} = x + x^2$$

in polar coordinates. For $r > 0$ we have

$$\dot{r} = \frac{x\dot{x} + y\dot{y}}{r} = \frac{-xy - x^2y + xy + x^2y}{r} = 0$$

and

$$\dot{\theta} = \frac{x\dot{y} - y\dot{x}}{r^2} = \frac{x^2 + x^3 + y^2 + xy^2}{r^2} = 1 + x > 0$$

for $x > -1$. Thus, along any trajectory of this system in the half plane $x > -1$, $r(t)$ is constant and $\theta(t)$ increases without bound as $t \to \infty$. That is, the phase portrait in a neighborhood of the origin is equivalent to the phase portrait in Figure 4 of Section 1.5 in Chapter 1 and the origin is called a center for this nonlinear system.

Example 2. Consider the system

$$\dot{x} = -y - x^3 - xy^2$$
$$\dot{y} = x - y^3 - x^2y.$$

In polar coordinates, for $r > 0$, we have

$$\dot{r} = -r^3$$

and

$$\dot{\theta} = 1.$$

Thus $r(t) = r_0(1 + 2r_0^2 t)^{-1/2}$ for $t > -1/(2r_0^2)$ and $\theta(t) = \theta_0 + t$. We see that $r(t) \to 0$ and $\theta(t) \to \infty$ as $t \to \infty$ and the phase portrait for this system in

2.10. Saddles, Nodes, Foci and Centers

a neighborhood of the origin is qualitatively equivalent to the first figure in Figure 3 in Section 1.5 of Chapter 1. The origin is called a stable focus for this nonlinear system.

Example 3. Consider the system

$$\dot{x} = -y + x^3 + xy^2$$
$$\dot{y} = x + y^3 + x^2 y.$$

In this case, we have for $r > 0$

$$\dot{r} = r^3$$

and

$$\dot{\theta} = 1.$$

Thus, $r(t) = r_0(1 - 2r_0^2 t)^{-1/2}$ for $t < 1/(2r_0^2)$ and $\theta(t) = \theta_0 + t$. We see that $r(t) \to 0$ and $|\theta(t)| \to \infty$ as $t \to -\infty$. The phase portrait in a neighborhood of the origin is qualitatively equivalent to the second figure in Figure 3 in Section 1.5 of Chapter 1 with the arrows reversed and the origin is called an unstable focus for this nonlinear system.

We now give precise geometrical definitions for a center, a center-focus, a stable and unstable focus, a stable and unstable node and a topological saddle of the nonlinear system (3). We assume that $\mathbf{x}_0 \in \mathbf{R}^2$ is an isolated equilibrium point of the nonlinear system (3) which has been translated to the origin; $r(t, r_0, \theta_0)$ and $\theta(t, r_0, \theta_0)$ will denote the solution of the nonlinear system (4) with $r(0) = r_0$ and $\theta(0) = \theta_0$.

Definition 1. The origin is called a *center* for the nonlinear system (2) if there exists a $\delta > 0$ such that every solution curve of (2) in the deleted neighborhood $N_\delta(0) \sim \{0\}$ is a closed curve with $\mathbf{0}$ in its interior.

Definition 2. The origin is called a *center-focus* for (2) if there exists a sequence of closed solution curves Γ_n with Γ_{n+1} in the interior of Γ_n such that $\Gamma_n \to 0$ as $n \to \infty$ and such that every trajectory between Γ_n and Γ_{n+1} spirals toward Γ_n or Γ_{n+1} as $t \to \pm\infty$.

Definition 3. The origin is called a *stable focus* for (2) if there exists a $\delta > 0$ such that for $0 < r_0 < \delta$ and $\theta_0 \in \mathbf{R}$, $r(t, r_0, \theta_0) \to 0$ and $|\theta(t, r_0, \theta_0)| \to \infty$ as $t \to \infty$. It is called an *unstable focus* if $r(t, r_0, \theta_0) \to 0$ and $|\theta(t, r_0, \theta_0)| \to \infty$ as $t \to -\infty$. Any trajectory of (2) which satisfies $r(t) \to 0$ and $|\theta(t)| \to \infty$ as $t \to \pm\infty$ is said to *spiral toward the origin* as $t \to \pm\infty$.

Definition 4. The origin is called a *stable node* for (2) if there exists a $\delta > 0$ such that for $0 < r_0 < \delta$ and $\theta_0 \in \mathbf{R}$, $r(t, r_0, \theta_0) \to 0$ as $t \to \infty$ and

$\lim_{t \to \infty} \theta(t, r_0, \theta_0)$ exists; i.e., each trajectory in a deleted neighborhood of the origin approaches the origin along a well-defined tangent line as $t \to \infty$. The origin is called an *unstable node* if $r(t, r_0, \theta_0) \to 0$ as $t \to -\infty$ and $\lim_{t \to -\infty} \theta(t, r_0, \theta_0)$ exists for all $r_0 \in (0, \delta)$ and $\theta_0 \in \mathbf{R}$. The origin is called a *proper node* for (2) if it is a node and if every ray through the origin is tangent to some trajectory of (2).

Definition 5. The origin is a *(topological) saddle* for (2) if there exist two trajectories Γ_1 and Γ_2 which approach $\mathbf{0}$ as $t \to \infty$ and two trajectories Γ_3 and Γ_4 which approach $\mathbf{0}$ as $t \to -\infty$ and if there exists a $\delta > 0$ such that all other trajectories which start in the deleted neighborhood of the origin $N_\delta(\mathbf{0}) \sim \{\mathbf{0}\}$ leave $N_\delta(\mathbf{0})$ as $t \to \pm\infty$. The special trajectories $\Gamma_1, \ldots, \Gamma_4$ are called *separatrices*.

For a (topological) saddle, the stable manifold at the origin $S = \Gamma_1 \cup \Gamma_2 \cup \{\mathbf{0}\}$ and the unstable manifold at the origin $U = \Gamma_3 \cup \Gamma_4 \cup \{\mathbf{0}\}$. If the trajectory Γ_i approaches the origin along a ray making an angle θ_i with the x-axis where $\theta_i \in (-\pi, \pi]$ for $i = 1, \ldots, 4$, then $\theta_2 = \theta_1 \pm \pi$ and $\theta_4 = \theta_3 \pm \pi$. This follows by considering the possible directions in which a trajectory of (2), written in polar form (4), can approach the origin; cf. equation (6) below. The following theorems, proved in [A–I], are useful in this regard. The first theorem is due to Bendixson [B].

Theorem 1 (Bendixson). *Let E be an open subset of $\mathbf{R}^2$ containing the origin and let $\mathbf{f} \in C^1(E)$. If the origin is an isolated critical point of (2), then either every neighborhood of the origin contains a closed solution curve with $\mathbf{0}$ in its interior or there exists a trajectory approaching $\mathbf{0}$ as $t \to \pm\infty$.*

Theorem 2. *Suppose that $P(x, y)$ and $Q(x, y)$ in (3) are analytic functions of x and y in some open subset E of $\mathbf{R}^2$ containing the origin and suppose that the Taylor expansions of P and Q about $(0, 0)$ begin with mth-degree terms $P_m(x, y)$ and $Q_m(x, y)$ with $m \geq 1$. Then any trajectory of (3) which approaches the origin as $t \to \infty$ either spirals toward the origin as $t \to \infty$ or it tends toward the origin in a definite direction $\theta = \theta_0$ as $t \to \infty$. If $xQ_m(x, y) - yP_m(x, y)$ is not identically zero, then all directions of approach, θ_0, satisfy the equation*

$$\cos\theta_0 Q_m(\cos\theta_0, \sin\theta_0) - \sin\theta_0 P_m(\cos\theta_0, \sin\theta_0) = 0.$$

Furthermore, if one trajectory of (3) spirals toward the origin as $t \to \infty$ then all trajectories of (3) in a deleted neighborhood of the origin spiral toward $\mathbf{0}$ as $t \to \infty$.

It follows from this theorem that if P and Q begin with first-degree terms, i.e., if

$$P_1(x, y) = ax + by$$

2.10. Saddles, Nodes, Foci and Centers

and
$$Q_1(x, y) = cx + dy$$
with a, b, c and d not all zero, then the only possible directions in which trajectories can approach the origin are given by directions θ which satisfy

$$b\sin^2\theta + (a - d)\sin\theta\cos\theta - c\cos^2\theta = 0. \tag{6}$$

For $\cos\theta \neq 0$ in this equation, i.e., if $b \neq 0$, this equation is equivalent to

$$b\tan^2\theta + (a - d)\tan\theta - c = 0. \tag{6'}$$

This quadratic has at most two solutions $\theta \in (-\pi/2, \pi/2]$ and if $\theta = \theta_1$ is a solution then $\theta = \theta_1 \pm \pi$ are also solutions. Finding the solutions of (6') is equivalent to finding the directions determined by the eigenvectors of the matrix
$$A = \begin{bmatrix} a & b \\ c & d \end{bmatrix}.$$

The next theorem follows immediately from the Stable Manifold Theorem and the Hartman–Grobman Theorem. It establishes that if the origin is a hyperbolic equilibrium point of the nonlinear system (2), then it is a (topological) saddle for (2) if and only if it is a saddle for its linearization at the origin. Furthermore, the directions θ_j along which the separatrices Γ_j approach the origin are solutions of (6).

Theorem 3. *Suppose that E is an open subset of $\mathbf{R}^2$ containing the origin and that $\mathbf{f} \in C^1(E)$. If the origin is a hyperbolic equilibrium point of the nonlinear system (2), then the origin is a (topological) saddle for (2) if and only if the origin is a saddle for the linear system (1) with $A = D\mathbf{f}(0)$.*

Example 4. According to the above theorem, the origin is a (topological) saddle or saddle for the nonlinear system

$$\dot{x} = x + 2y + x^2 - y^2$$
$$\dot{y} = 3x + 4y - 2xy$$

since the determinant of the linear part $\delta = -2$; cf. Theorem 1 in Section 1.5 of Chapter 1. Furthermore, the directions in which the separatrices approach the origin as $t \to \pm\infty$ are given by solutions of (6'):

$$2\tan^2\theta - 3\tan\theta - 3 = 0.$$

That is,
$$\theta = \tan^{-1}\left(\frac{3 \pm \sqrt{33}}{4}\right)$$

and we have $\theta_1 \simeq 65.42°$, $\theta_3 \simeq -34.46°$. At any point on the positive x-axis near the origin, the vector field defined by this system points upward since

$\dot{y} > 0$ there. This determines the direction of the flow defined by the above system. The local phase portraits for the linear part of this vector field as well as for the nonlinear system are shown in Figure 1. The qualitative behavior in a neighborhood of the origin is the same for either system.

The next example, due to Perron, shows that a node for a linear system may change to a focus with the addition of nonlinear terms. Note that the

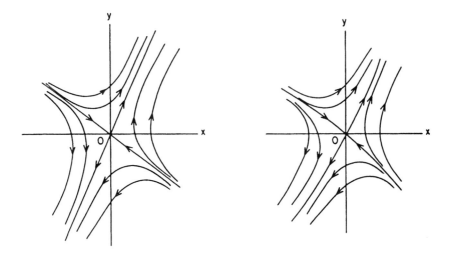

Figure 1. A saddle for the linear system and a topological saddle for the nonlinear system of Example 4.

vector field defined in this example $\mathbf{f} \in C^1(\mathbf{R}^2)$ but that $\mathbf{f} \notin C^2(\mathbf{R}^2)$; cf. Problem 2 at the end of this section. This example shows that the hypothesis $\mathbf{f} \in C^1(E)$ is not strong enough to imply that the phase portrait of a nonlinear system (2) is diffeomorphic to the phase portrait of its linearization. (Hartman's Theorem at the end of Section 2.8 shows that $\mathbf{f} \in C^2(E)$ is sufficient.)

Example 5. Consider the nonlinear system

$$\dot{x} = -x - \frac{y}{\ln\sqrt{x^2 + y^2}}$$

$$\dot{y} = -y + \frac{x}{\ln\sqrt{x^2 + y^2}}$$

for $x^2 + y^2 \neq 0$ and define $\mathbf{f}(0) = 0$. In polar coordinates we have

$$\dot{r} = -r$$

$$\dot{\theta} = \frac{1}{\ln r}.$$

2.10. Saddles, Nodes, Foci and Centers

Thus, $r(t) = r_0 e^{-t}$ and $\theta(t) = \theta_0 - \ln(1 - t/\ln r_0)$. We see that for $r_0 < 1$, $r(t) \to 0$ and $|\theta(t)| \to \infty$ as $t \to \infty$ and therefore, according to Definition 4, the origin is a stable focus for this nonlinear system; however, it is a stable proper node for the linearized system.

The next theorem, proved in [A–I], shows that under the stronger hypothesis that $\mathbf{f} \in C^2(E)$, i.e., under the hypothesis that $P(x,y)$ and $Q(x,y)$ have continuous second partials in a neighborhood of the origin, we find that nodes and foci of a linear system persist under the addition of nonlinear terms. Cf. Hartman's Theorem at the end of Section 2.8.

Theorem 4. *Let E be an open subset of $\mathbf{R}^2$ containing the origin and let $\mathbf{f} \in C^2(E)$. Suppose that the origin is a hyperbolic critical point of (2). Then the origin is a stable (or unstable) node for the nonlinear system (2) if and only if it is a stable (or unstble) node for the linear system (1) with $A = D\mathbf{f}(0)$. And the origin is a stable (or unstable) focus for the nonlinear system (2) if and only if it is a stable (or unstable) focus for the linear system (1) with $A = D\mathbf{f}(0)$.*

Remark. Under the hypotheses of Theorem 4, it follows that the origin is a proper node for the nonlinear system (2) if and only if it is a proper node for the linear system (1) with $A = D\mathbf{f}(0)$. And under the weaker hypothesis that $\mathbf{f} \in C^1(E)$, it still follows that if the origin is a focus for the linear system (1) with $A = D\mathbf{f}(0)$, then it is a focus for the nonlinear system (2); cf. [C/L].

Examples 1–3 above show that a center for a linear system can either remain a center or change to a stable or an unstable focus with the addition of nonlinear terms. The following example shows that a center for a linear system may also become a center-focus under the addition of nonlinear terms; and the following theorem shows that these are the only possibilities.

Example 6. Consider the nonlinear system

$$\dot{x} = -y + x\sqrt{x^2 + y^2}\sin(1/\sqrt{x^2 + y^2})$$
$$\dot{y} = x + y\sqrt{x^2 + y^2}\sin(1/\sqrt{x^2 + y^2})$$

for $x^2 + y^2 \neq 0$ where we define $\mathbf{f}(0) = 0$. In polar coordinates, we have

$$\dot{r} = r^2 \sin(1/r)$$
$$\dot{\theta} = 1$$

for $r > 0$ with $\dot{r} = 0$ at $r = 0$. Clearly, $\dot{r} = 0$ for $r = 1/(n\pi)$; i.e., each of the circles $r = 1/(n\pi)$ is a trajectory of this system. Furthermore, for $n\pi < 1/r < (n+1)\pi$, $\dot{r} < 0$ if n is odd and $\dot{r} > 0$ if n is even; i.e., the trajectories between the circles $r = 1/(n\pi)$ spiral inward or outward to one of these circles. Thus, we see that the origin is a center-focus for this nonlinear system according to Definition 2 above.

Theorem 5. *Let E be an open subset of $\mathbf{R}^2$ containing the origin and let $\mathbf{f} \in C^1(E)$ with $\mathbf{f}(0) = 0$. Suppose that the origin is a center for the linear system (1) with $A = D\mathbf{f}(0)$. Then the origin is either a center, a center-focus or a focus for the nonlinear system (2).*

Proof. We may assume that the matrix $A = D\mathbf{f}(0)$ has been transformed to its canonical form

$$A = \begin{bmatrix} 0 & -b \\ b & 0 \end{bmatrix}$$

with $b \neq 0$. Assume that $b > 0$; otherwise we can apply the linear transformation $t \to -t$. The nonlinear system (3) then has the form

$$\dot{x} = -by + p(x,y)$$
$$\dot{y} = bx + q(x,y).$$

Since $\mathbf{f} \in C^1(E)$, it follows that $|p(x,y)/r| \to 0$ and $|q(x,y)/r| \to 0$ as $r \to 0$; i.e., $p = o(r)$ and $q = o(r)$ as $r \to 0$. Cf. Problem 3. Thus, in polar coordinates we have $\dot{r} = o(r)$ and $\dot{\theta} = b + o(1)$ as $r \to 0$. Therefore, there exists a $\delta > 0$ such that $\dot{\theta} \geq b/2 > 0$ for $0 < r \leq \delta$. Thus for $0 < r_0 \leq \delta$ and $\theta_0 \in \mathbf{R}$, $\theta(t, r_0, \theta_0) \geq bt/2 + \theta_0 \to \infty$ as $t \to \infty$; and $\theta(t, r_0, \theta_0)$ is a monotone increasing function of t. Let $t = h(\theta)$ be the inverse of this monotone function. Define $\tilde{r}(\theta) = r(h(\theta), r_0, \theta_0)$ for $0 < r_0 \leq \delta$ and $\theta_0 \in \mathbf{R}$. Then $\tilde{r}(\theta)$ satisfies the differential equation (5) which has the form

$$\frac{d\tilde{r}}{d\theta} = \tilde{F}(\tilde{r}, \theta) = \frac{\cos\theta\, p(\tilde{r}\cos\theta, \tilde{r}\sin\theta) + \sin\theta\, q(\tilde{r}\cos\theta, \tilde{r}\sin\theta)}{b + (\cos\theta/\tilde{r})q(\tilde{r}\cos\theta, \tilde{r}\sin\theta) - (\sin\theta/\tilde{r})p(\tilde{r}\cos\theta, \tilde{r}\sin\theta)}.$$

Suppose that the origin is not a center or a center-focus for the nonlinear system (3). Then for $\delta > 0$ sufficiently small, there are no closed trajectories of (3) in the deleted neighborhood $N_\delta(0) \sim \{0\}$. Thus for $0 < r_0 < \delta$ and $\theta_0 \in \mathbf{R}$, either $\tilde{r}(\theta_0+2\pi) < \tilde{r}(\theta_0)$ or $\tilde{r}(\theta_0+2\pi) > \tilde{r}(\theta_0)$. Assume that the first case holds. The second case is treated in a similar manner. If $\tilde{r}(\theta_0 + 2\pi) < \tilde{r}(\theta_0)$ then $\tilde{r}(\theta_0 + 2k\pi) < \tilde{r}(\theta_0 + 2(k-1)\pi)$ for $k = 1, 2, 3, \ldots$. Otherwise we would have two trajectories of (3) through the same point which is impossible. The sequence $\tilde{r}(\theta_0 + 2k\pi)$ is monotone decreasing and bounded below by zero; therefore, the following limit exists and is nonnegative:

$$\tilde{r}_1 = \lim_{k \to \infty} \tilde{r}(\theta_0 + 2k\pi).$$

If $\tilde{r}_1 = 0$ then $\tilde{r}(\theta) \to 0$ as $\theta \to \infty$; i.e., $r(t, r_0, \theta_0) \to 0$ and $\theta(t, r_0, \theta_0) \to \infty$ as $t \to \infty$ and the origin is a stable focus of (3). If $\tilde{r}_1 > 0$ then since $|\tilde{F}(r,\theta)| \leq M$ for $0 \leq r \leq \delta$ and $0 \leq \theta \leq 2\pi$, the sequence $\tilde{r}(\theta_0+\theta+2k\pi)$ is equicontinuous on $[0, 2\pi]$. Therefore, by Ascoli's Lemma, cf. Theorem 7.25 in Rudin [R], there exists a uniformly convergent subsequence of $\tilde{r}(\theta_0 + \theta + 2k\pi)$ converging to a solution $\tilde{r}_1(\theta)$ which satisfies $\tilde{r}_1(\theta) = \tilde{r}_1(\theta + 2k\pi)$; i.e., $\tilde{r}_1(\theta)$ is a non-zero periodic solution of (5). This contradicts the fact that

2.10. Saddles, Nodes, Foci and Centers

there are no closed trajectories of (3) in $N_\delta(0) \sim \{0\}$ when the origin is not a center or a center focus of (3). Thus if the origin is not a center or a center focus of (3), $\tilde{r}_1 = 0$ and the origin is a focus of (3). This completes the proof of the theorem.

A center-focus cannot occur in an analytic system. This is a consequence of Dulac's Theorem discussed in Section 3.3 of Chapter 3. We therefore have the following corollary of Theorem 5 for analytic systems.

Corollary. *Let E be an open subset of $\mathbf{R}^2$ containing the origin and let $\mathbf{f}$ be analytic in E with $\mathbf{f}(0) = 0$. Suppose that the origin is a center for the linear system (1) with $A = D\mathbf{f}(0)$. Then the origin is either a center or a focus for the nonlinear system (2).*

As we noted in the previous section, Liapunov's method is one tool that can be used to distinguish a center from a focus for a nonlinear system; cf. Problem 4 in Problem Set 9. Another approach is to write the system in polar coordinates as in Examples 1–3 above. Yet another approach is to look for symmetries in the differential equations. The easiest symmetries to see are symmetries with respect to the x and y axes.

Definition 6. The system (3) is said to be *symmetric with respect to the x-axis* if it is invariant under the transformation $(t, y) \to (-t, -y)$; it is said to be *symmetric with respect to the y-axis* if it is invariant under the transformation $(t, x) \to (-t, -x)$.

Note that the system in Example 1 is symmetric with respect to the x-axis, but not with respect to the y-axis.

Theorem 6. *Let E be an open subset of $\mathbf{R}^2$ containing the origin and let $\mathbf{f} \in C^1(E)$ with $\mathbf{f}(0) = 0$. If the nonlinear system (2) is symmetric with respect to the x-axis or the y-axis, and if the origin is a center for the linear system (1) with $A = D\mathbf{f}(0)$, then the origin is a center for the nonlinear system (2).*

The idea of the proof of this theorem is that by Theorem 5, any trajectory of (3) in $N_\delta(0)$ which crosses the positive x-axis will also cross the negative x-axis. If the system (3) is symmetric with respect to the x-axis, then the trajectories of (3) in $N_\delta(0)$ will be symmetric with respect to the x-axis and hence all trajectories of (3) in $N_\delta(0)$ will be closed; i.e., the origin will be a center for (3).

PROBLEM SET 10

1. Write the following systems in polar coordinates and determine if the origin is a center, a stable focus or an unstable focus.

(a) $\begin{aligned}\dot{x} &= x - y \\ \dot{y} &= x + y\end{aligned}$

(b) $\begin{aligned}\dot{x} &= -y + xy^2 \\ \dot{y} &= x + y^3\end{aligned}$

(c) $\begin{aligned}\dot{x} &= -y + x^5 \\ \dot{y} &= x + y^5\end{aligned}$

2. Let
$$f(x) = \begin{cases} \dfrac{x}{\ln|x|} & \text{for } x \neq 0 \\ 0 & \text{for } x = 0 \end{cases}$$

Show that $f'(0) = \lim_{x \to 0} f'(x) = 0$; i.e., $f \in C^1(\mathbf{R})$, but that $f''(0)$ is undefined.

3. Show that if $x = 0$ is a zero of the function $f: \mathbf{R} \to \mathbf{R}$ and $f \in C^1(\mathbf{R})$ then
$$f(x) = Df(0)x + F(x)$$
where $|F(x)/x| \to 0$ as $x \to 0$. Show that this same result holds for $f: \mathbf{R}^2 \to \mathbf{R}^2$, i.e., show that $|\mathbf{F}(\mathbf{x})|/|\mathbf{x}| \to 0$ as $\mathbf{x} \to \mathbf{0}$.

Hint: Use Definition 1 in Section 2.1.

4. Determine the nature of the critical points of the following nonlinear systems (Cf. Problem 1 in Section 2.6); be as specific as possible.

(a) $\begin{aligned}\dot{x} &= x - xy \\ \dot{y} &= y - x^2\end{aligned}$

(b) $\begin{aligned}\dot{x} &= -4y + 2xy - 8 \\ \dot{y} &= 4y^2 - x^2\end{aligned}$

(c) $\begin{aligned}\dot{x} &= 2x - 2xy \\ \dot{y} &= 2y - x^2 + y^2\end{aligned}$

(d) $\begin{aligned}\dot{x} &= -x^2 - y^2 + 1 \\ \dot{y} &= 2x\end{aligned}$

(e) $\begin{aligned}\dot{x} &= -x^2 - y^2 + 1 \\ \dot{y} &= 2xy\end{aligned}$

(f) $\begin{aligned}\dot{x} &= x^2 - y^2 - 1 \\ \dot{y} &= 2y\end{aligned}$

2.11 Nonhyperbolic Critical Points in R^2

In this section we present some results on nonhyperbolic critical points of planar analytic systems. This work originated with Poincaré [P] and was extended by Bendixson [B] and more recently by Andronov et al. [A–I]. We assume that the origin is an isolated critical point of the planar system

$$\dot{x} = P(x,y)$$
$$\dot{y} = Q(x,y) \tag{1}$$

where P and Q are analytic in some neighborhood of the origin. In Sections 2.9 and 2.10 we have already presented some results for the case when the matrix of the linear part $A = D\mathbf{f}(0)$ has pure imaginary eigenvalues, i.e., when the origin is a center for the linearized system. In this section we give some results established in [A–I] for the case when the matrix A has one or two zero eigenvalues, but $A \neq 0$. And these results are extended to higher dimensions in Section 2.12.

First of all, note that if P and Q begin with mth-degree terms P_m and Q_m, then it follows from Theorem 2 in Section 2.10 that if the function

$$g(\theta) = \cos\theta\, Q_m(\cos\theta, \sin\theta) - \sin\theta\, P_m(\cos\theta, \sin\theta)$$

is not identically zero, then there are at most $2(m+1)$ directions $\theta = \theta_0$ along which a trajectory of (1) may approach the origin. These directions are given by solutions of the equation $g(\theta) = 0$. Suppose that $g(\theta)$ is not identically zero, then the solution curves of (1) which approach the origin along these tangent lines divide a neighborhood of the origin into a finite number of open regions called *sectors*. These sectors will be of one of three types as described in the following definitions; cf. [A–I] or [L]. The trajectories which lie on the boundary of a hyperbolic sector are called *separatrices*. Cf. Definition 1 in Section 3.11.

Definition 1. A sector which is topologically equivalent to the sector shown in Figure 1(a) is called *a hyperbolic sector*. A sector which is topologically equivalent to the sector shown in Figure 1(b) is called *a parabolic sector*. And a sector which is topologically equivalent to the sector shown in Figure 1(c) is called *an elliptic sector*.

Figure 1. (a) A hyperbolic sector. (b) A parabolic sector. (c) An elliptic sector.

In Definition 1, the homeomorphism establishing the topological equivalence of a sector to one of the sectors in Figure 1 need not preserve the direction of the flow; i.e., each of the sectors in Figure 1 with the arrows reversed are sectors of the same type. For example, a saddle has a deleted neighborhood consisting of four hyperbolic sectors and four separatrices. And a proper node has a deleted neighborhood consisting of one parabolic sector. According to Theorem 2 below, the system

$$\dot{x} = y$$
$$\dot{y} = -x^3 + 4xy$$

has an elliptic sector at the origin; cf. Problem 1 below. The phase portrait for this system is shown in Figure 2. Every trajectory which approaches the origin does so tangent to the x-axis.

A deleted neighborhood of the origin consists of one elliptic sector, one hyperbolic sector, two parabolic sectors, and four separatrices. Cf. Definition 1 and Problem 5 in Section 3.11. This type of critical point is called *a critical point with an elliptic domain*; cf. [A–I].

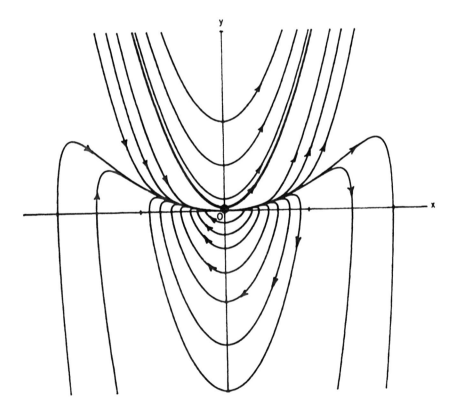

Figure 2. A critical point with an elliptic domain at the origin.

2.11. Nonhyperbolic Critical Points in R^2

Another type of nonhyperbolic critical point for a planar system is a saddle-node. A *saddle-node* consists of two hyperbolic sectors and one parabolic sector (as well as three separatrices and the critical point itself). According to Theorem 1 below, the system

$$\dot{x} = x^2$$
$$\dot{y} = y$$

has a saddle-node at the origin; cf. Problem 2. Even without Theorem 1, this system is easy to discuss since it can be solved explicitly for $x(t) = (1/x_0 - t)^{-1}$ and $y(t) = y_0 e^t$. The phase portrait for this system is shown in Figure 3.

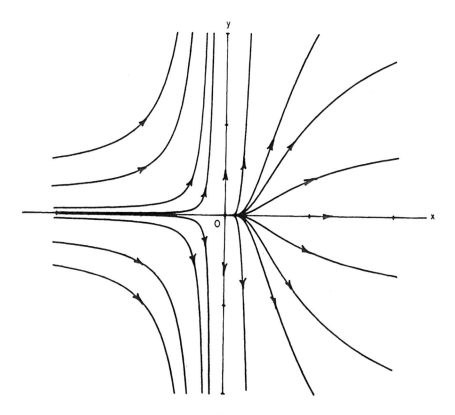

Figure 3. A saddle-node at the origin.

One other type of behavior that can occur at a nonhyperbolic critical point is illustrated by the following example:

$$\dot{x} = y$$
$$\dot{y} = x^2.$$

The phase portrait for this system is shown in Figure 4. We see that a deleted neighborhood of the origin consists of two hyperbolic sectors and two separatrices. This type of critical point is called a *cusp*.

As we shall see, besides the familiar types of critical points for planar analytic systems discussed in Section 2.10, i.e., nodes, foci, (topological) saddles and centers, the only other types of critical points that can occur for (1) when $A \neq 0$ are saddle-nodes, critical points with elliptic domains and cusps.

We first consider the case when the matrix A has one zero eigenvalue, i.e., when $\det A = 0$, but $\operatorname{tr} A \neq 0$. In this case, as in Chapter 1 and as is shown in [A–I] on p. 338, the system (1) can be put into the form

$$\begin{aligned} \dot{x} &= p_2(x, y) \\ \dot{y} &= y + q_2(x, y) \end{aligned} \qquad (2)$$

where p_2 and q_2 are analytic in a neighborhood of the origin and have expansions that begin with second-degree terms in x and y. The following theorem is proved on p. 340 in [A–I]. Cf. Section 2.12.

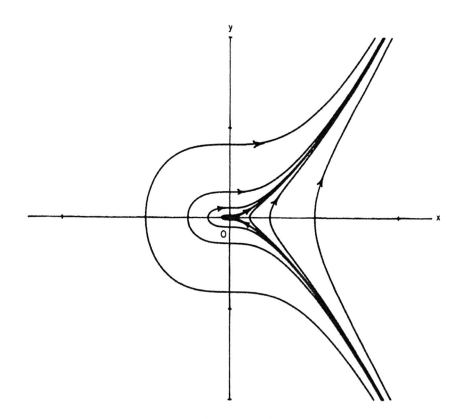

Figure 4. A cusp at the origin.

2.11. Nonhyperbolic Critical Points in R^2

Theorem 1. *Let the origin be an isolated critical point for the analytic system (2). Let $y = \phi(x)$ be the solution of the equation $y + q_2(x, y) = 0$ in a neighborhood of the origin and let the expansion of the function $\psi(x) = p_2(x, \phi(x))$ in a neighborhood of $x = 0$ have the form $\psi(x) = a_m x^m + \cdots$ where $m \geq 2$ and $a_m \neq 0$. Then (1) for m odd and $a_m > 0$, the origin is an unstable node, (2) for m odd and $a_m < 0$, the origin is a (topological) saddle and (3) for m even, the origin is a saddle-node.*

Next consider the case when A has two zero eigenvalues, i.e., $\det A = 0$, $\operatorname{tr} A = 0$, but $A \neq 0$. In this case it is shown in [A–I], p. 356, that the system (1) can be put in the "normal" form

$$\begin{aligned} \dot{x} &= y \\ \dot{y} &= a_k x^k [1 + h(x)] + b_n x^n y [1 + g(x)] + y^2 R(x, y) \end{aligned} \quad (3)$$

where $h(x)$, $g(x)$ and $R(x, y)$ are analytic in a neighborhood of the origin, $h(0) = g(0) = 0$, $k \geq 2$, $a_k \neq 0$ and $n \geq 1$. Cf. Section 2.13. The next two theorems are proved on pp. 357–362 in [A–I].

Theorem 2. *Let $k = 2m+1$ with $m \geq 1$ in (3) and let $\lambda = b_n^2 + 4(m+1)a_k$. Then if $a_k > 0$, the origin is a (topological) saddle. If $a_k < 0$, the origin is (1) a focus or a center if $b_n = 0$ and also if $b_n \neq 0$ and $n > m$ or if $n = m$ and $\lambda < 0$, (2) a node if $b_n \neq 0$, n is an even number and $n < m$ and also if $b_n \neq 0$, n is an even number, $n = m$ and $\lambda \geq 0$ and (3) a critical point with an elliptic domain if $b_n \neq 0$, n is an odd number and $n < m$ and also if $b_n \neq 0$, n is an odd number, $n = m$ and $\lambda \geq 0$.*

Theorem 3. *Let $k = 2m$ with $m \geq 1$ in (3). Then the origin is (1) a cusp if $b_n = 0$ and also if $b_n \neq 0$ and $n \geq m$ and (2) a saddle-node if $b_n \neq 0$ and $n < m$.*

We see that if $D\mathbf{f}(\mathbf{x}_0)$ has one zero eigenvalue, then the critical point $\mathbf{x}_0$ is either a node, a (topological) saddle, or a saddle-node; and if $D\mathbf{f}(\mathbf{x}_0)$ has two zero eigenvalues, then the critical point $\mathbf{x}_0$ is either a focus, a center, a node, a (topological) saddle, a saddle-node, a cusp, or a critical point with an elliptic domain.

Finally, what if the matrix $A = 0$? In this case, the behavior near the origin can be very complex. If P and Q begin with mth-degree terms, then the separatrices may divide a neighborhood of the origin into $2(m+1)$ sectors of various types. The number of elliptic sectors minus the number of hyperbolic sectors is always an even number and this number is related to the index of the critical point discussed in Section 3.12 of Chapter 3. For example, the homogenous quadratic system

$$\begin{aligned} \dot{x} &= x^2 + xy \\ \dot{y} &= \frac{1}{2}y^2 + xy \end{aligned}$$

has the phase portrait shown in Figure 5. There are two elliptic sectors and two parabolic sectors at the origin. All possible types of phase portraits for homogenous, quadratic systems have been classified by the Russian mathematician L.S. Lyagina [19]. For more information on the topic, cf. the book by Nemytskii and Stepanov [N/S].

Remark. A critical point, x_0, of (1) for which $Df(x_0)$ has a zero eigenvalue is often referred to as a multiple critical point. The reason for this is made clear in Section 4.2 of Chapter 4 where it is shown that a multiple critical point of (1) can be made to split into a number of hyperbolic critical points under a suitable perturbation of (1).

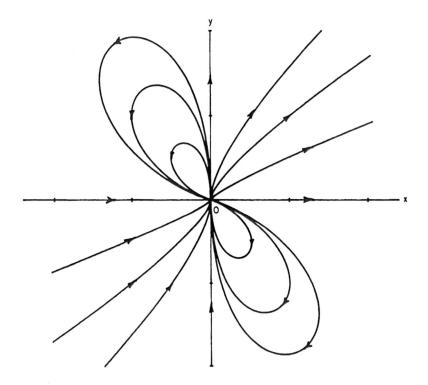

Figure 5. A nonhyperbolic critical point with two elliptic sectors and two parabolic sectors.

PROBLEM SET 11

1. Use Theorem 2 to show that the system

$$\dot{x} = y$$
$$\dot{y} = -x^3 + 4xy$$

2.11. Nonhyperbolic Critical Points in R^2

has a critical point with an elliptic domain at the origin. Note that $y = x^2/(2 \pm \sqrt{2})$ are two invariant curves of this system which bound two parabolic sectors.

2. Use Theorem 1 to determine the nature of the critical point at the origin for the following systems:

(a) $\dot{x} = x^2$
$\dot{y} = y$

(b) $\dot{x} = x^2 + y^2$
$\dot{y} = y - x^2$

(c) $\dot{x} = y^2 + x^3$
$\dot{y} = y - x^2$

(d) $\dot{x} = y^2 - x^3$
$\dot{y} = y - x^2$

(e) $\dot{x} = y^2 + x^4$
$\dot{y} = y - x^2 + y^2$

(f) $\dot{x} = y^2 - x^4$
$\dot{y} = y - x^2 + y^2$

3. Use Theorem 2 or Theorem 3 to determine the nature of the critical point at the origin for the following systems:

(a) $\dot{x} = y$
$\dot{y} = x^2$

(b) $\dot{x} = y$
$\dot{y} = x^2 + 2x^2 y + xy^2$

(c) $\dot{x} = y$
$\dot{y} = x^4 + xy$

(d) $\dot{x} = y$
$\dot{y} = -x^3 - x^2 y$

(e) $\dot{x} = y$
$\dot{y} = x^3 - x^2 y$

(f) $\dot{x} = x + y$
$\dot{y} = -x - y - x^3 + y^4$

Hint: For part (f), let $\xi = x$ and $\eta = x + y$ to put the system in the form (3).

2.12 Center Manifold Theory

In Section 2.8 we presented the Hartman–Grobman Theorem, which showed that, in a neighborhood of a hyperbolic critical point $\mathbf{x}_0 \in E$, the nonlinear system

$$\dot{\mathbf{x}} = \mathbf{f}(\mathbf{x}) \tag{1}$$

is topologically conjugate to the linear system

$$\dot{\mathbf{x}} = A\mathbf{x}, \tag{2}$$

with $A = D\mathbf{f}(\mathbf{x}_0)$, in a neighborhood of the origin. The Hartman–Grobman Theorem therefore completely solves the problem of determining the stability and qualitative behavior in a neighborhood of a hyperbolic critical point of a nonlinear system. In the last section, we gave some results for determining the stability and qualitative behavior in a neighborhood of a nonhyperbolic critical point of the nonlinear system (1) with $\mathbf{x} \in \mathbf{R}^2$ where $\det A = 0$ but $A \neq 0$. In this section, we present the Local Center Manifold Theorem, which generalizes Theorem 1 of the previous section to higher dimensions and shows that the qualitative behavior in a neighborhood of a nonhyperbolic critical point $\mathbf{x}_0$ of the nonlinear system (1) with $\mathbf{x} \in \mathbf{R}^n$ is determined by its behavior on the center manifold near $\mathbf{x}_0$. Since the center manifold is generally of smaller dimension than the system (1), this simplifies the problem of determining the stability and qualitative behavior of the flow near a nonhyperbolic critical point of (1). Of course, we still must determine the qualitative behavior of the flow on the center manifold near the hyperbolic critical point. If the dimension of the center manifold $W^c(\mathbf{x}_0)$ is one, this is trivial; and if the dimension of $W^c(\mathbf{x}_0)$ is two and a linear term is present in the differential equation determining the flow on $W^c(\mathbf{x}_0)$, then Theorems 2 and 3 in the previous section or the method in Section 2.9 can be used to determine the flow on $W^c(\mathbf{x}_0)$. The remaining cases must be treated as they appear; however, in the next section we will present a method for simplifying the nonlinear part of the system of differential equations that determines the flow on the center manifold.

Let us begin as we did in the proof of the Stable Manifold Theorem in Section 2.7 by noting that if $\mathbf{f} \in C^1(E)$ and $\mathbf{f}(\mathbf{0}) = \mathbf{0}$, then the system (1) can be written in the form of equation (6) in Section 2.7 where, in this case, the matrix $A = D\mathbf{f}(\mathbf{0}) = \text{diag}[C, P, Q]$ and the square matrix C has c eigenvalues with zero real parts, the square matrix P has s eigenvalues with negative real parts, and the square matrix Q has u eigenvalues with positive real parts; i.e., the system (1) can be written in diagonal form

$$\begin{aligned} \dot{\mathbf{x}} &= C\mathbf{x} + \mathbf{F}(\mathbf{x},\mathbf{y},\mathbf{z}) \\ \dot{\mathbf{y}} &= P\mathbf{y} + \mathbf{G}(\mathbf{x},\mathbf{y},\mathbf{z}) \\ \dot{\mathbf{z}} &= Q\mathbf{z} + \mathbf{H}(\mathbf{x},\mathbf{y},\mathbf{z}), \end{aligned} \tag{3}$$

where $(\mathbf{x},\mathbf{y},\mathbf{z}) \in \mathbf{R}^c \times \mathbf{R}^s \times \mathbf{R}^u$, $\mathbf{F}(\mathbf{0}) = \mathbf{G}(\mathbf{0}) = \mathbf{H}(\mathbf{0}) = \mathbf{0}$, and $D\mathbf{F}(\mathbf{0}) = D\mathbf{G}(\mathbf{0}) = D\mathbf{H}(\mathbf{0}) = \mathbf{O}$.

2.12. Center Manifold Theory

We first shall present the theory for the case when $u = 0$ and treat the general case at the end of this section. In the case when $u = 0$, it follows from the center manifold theorem in Section 2.7 that for $(\mathbf{F}, \mathbf{G}) \in C^r(E)$ with $r \geq 1$, there exists an s-dimensional invariant stable manifold $W^s(0)$ tangent to the stable subspace E^s of (1) at 0 and there exists a c-dimensional invariant center manifold $W^c(0)$ tangent to the center subspace E^c of (1) at 0. It follows that the local center manifold of (3) at 0,

$$W^c_{\text{loc}}(0) = \{(\mathbf{x}, \mathbf{y}) \in \mathbf{R}^c \times \mathbf{R}^s \mid \mathbf{y} = \mathbf{h}(\mathbf{x}) \text{ for } |\mathbf{x}| < \delta\} \qquad (4)$$

for some $\delta > 0$, where $\mathbf{h} \in C^r(N_\delta(0))$, $\mathbf{h}(0) = 0$, and $D\mathbf{h}(0) = O$ since $W^c(0)$ is tangent to the center subspace $E^c = \{(\mathbf{x}, \mathbf{y}) \in \mathbf{R}^c \times \mathbf{R}^s \mid \mathbf{y} = 0\}$ at the origin. This result is part of the Local Center Manifold Theorem, stated below, which is proved by Carr in [Ca].

Theorem 1 (The Local Center Manifold Theorem). *Let $\mathbf{f} \in C^r(E)$, where E is an open subset of $\mathbf{R}^n$ containing the origin and $r \geq 1$. Suppose that $\mathbf{f}(0) = 0$ and that $D\mathbf{f}(0)$ has c eigenvalues with zero real parts and s eigenvalues with negative real parts, where $c + s = n$. The system (1) then can be written in diagonal form*

$$\dot{\mathbf{x}} = C\mathbf{x} + \mathbf{F}(\mathbf{x}, \mathbf{y})$$
$$\dot{\mathbf{y}} = P\mathbf{y} + \mathbf{G}(\mathbf{x}, \mathbf{y}),$$

where $(\mathbf{x}, \mathbf{y}) \in \mathbf{R}^c \times \mathbf{R}^s$, C is a square matrix with c eigenvalues having zero real parts, P is a square matrix with s eigenvalues with negative real parts, and $\mathbf{F}(0) = \mathbf{G}(0) = 0$, $D\mathbf{F}(0) = D\mathbf{G}(0) = O$; furthermore, there exists a $\delta > 0$ and a function $\mathbf{h} \in C^r(N_\delta(0))$ that defines the local center manifold (4) and satisfies

$$D\mathbf{h}(\mathbf{x})[C\mathbf{x} + \mathbf{F}(\mathbf{x}, \mathbf{h}(\mathbf{x}))] - P\mathbf{h}(\mathbf{x}) - \mathbf{G}(\mathbf{x}, \mathbf{h}(\mathbf{x})) = 0 \qquad (5)$$

for $|\mathbf{x}| < \delta$; and the flow on the center manifold $W^c(0)$ is defined by the system of differential equations

$$\dot{\mathbf{x}} = C\mathbf{x} + \mathbf{F}(\mathbf{x}, \mathbf{h}(\mathbf{x})) \qquad (6)$$

for all $\mathbf{x} \in \mathbf{R}^c$ with $|\mathbf{x}| < \delta$.

Equation (5) for the function $\mathbf{h}(\mathbf{x})$ follows from the fact that the center manifold $W^c(0)$ is invariant under the flow defined by the system (1) by substituting $\dot{\mathbf{x}}$ and $\dot{\mathbf{y}}$ from the above differential equations in Theorem 1 into the equation

$$\dot{\mathbf{y}} = D\mathbf{h}(\mathbf{x})\dot{\mathbf{x}},$$

which follows from the chain rule applied to the equation $\mathbf{y} = \mathbf{h}(\mathbf{x})$ defining the center manifold. Even though equation (5) is a quasilinear partial differential equation for the components of $\mathbf{h}(\mathbf{x})$, which is difficult if not impossible to solve for $\mathbf{h}(\mathbf{x})$, it gives us a method for approximating the function $\mathbf{h}(\mathbf{x})$ to any degree of accuracy that we wish, provided that the integer r in Theorem 1 is sufficiently large. This is accomplished by substituting the series expansions for the components of $\mathbf{h}(\mathbf{x})$ into equation (5); cf. Theorem 2.1.3 in [Wi-II]. This is illustrated in the following examples, which also show that it is necessary to approximate the shape of the local center manifold $W^c_{loc}(0)$ in order to correctly determine the flow on $W^c(0)$ near the origin. Before presenting these examples, we note that for $c = 1$ and $s = 1$, the Local Center Manifold Theorem given above is the same as Theorem 1 in the previous section (if we let $t \to -t$ in equation (2) in Section 2.11). Thus, Theorem 1 above is a generalization of Theorem 1 in Section 2.11 to higher dimensions. Also, in the case when $c = s = 1$, as in Theorem 1 in Section 2.11, it is only necessary to solve the algebraic equation determined by setting the last two terms in equation (5) equal to zero in order to determine the correct flow on $W^c(0)$.

It should be noted that while there may be many different functions $\mathbf{h}(\mathbf{x})$ which determine different center manifolds for (3), the flows on the various center manifolds are determined by (6) and they are all topologically equivalent in a neighborhood of the origin. Furthermore, for analytic systems, if the Taylor series for the function $\mathbf{h}(\mathbf{x})$ converges in a neighborhood of the origin, then the analytic center manifold $\mathbf{y} = \mathbf{h}(\mathbf{x})$ is unique; however, not all analytic (or polynomial) systems have an analytic center manifold. Cf. Problem 4.

Example 1. Consider the following system with $c = s = 1$:

$$\dot{x} = x^2 y - x^5$$
$$\dot{y} = -y + x^2.$$

In this case, we have $C = O$, $P = [-1]$, $F(x, y) = x^2 y - x^5$, and $G(x, y) = x^2$. We substitute the expansions

$$h(x) = ax^2 + bx^3 + 0(x^4) \quad \text{and} \quad Dh(x) = 2ax + 3bx^2 + 0(x^3)$$

into equation (5) to obtain

$$(2ax + 3bx^2 + \cdots)(ax^4 + bx^5 + \cdots - x^5) + ax^2 + bx^3 + \cdots - x^2 = 0.$$

Setting the coefficients of like powers of x equal to zero yields $a - 1 = 0$, $b = 0$, $c = 0, \ldots$. Thus,

$$h(x) = x^2 + 0(x^5).$$

Substituting this result into equation (6) then yields

$$\dot{x} = x^4 + 0(x^5)$$

2.12. Center Manifold Theory

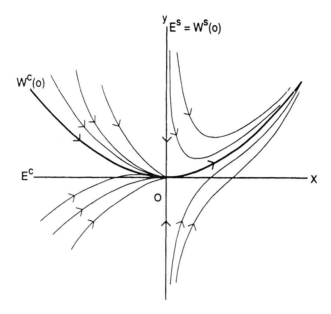

Figure 1. The phase portrait for the system in Example 1.

on the center manifold $W^c(0)$ near the origin. This implies that the local phase portrait is given by Figure 1. We see that the origin is a saddle-node and that it is unstable. However, if we were to use the center subspace approximation for the local center manifold, i.e., if we were to set $y = 0$ in the first differential equation in this example, we would obtain

$$\dot{x} = -x^5$$

and arrive at the incorrect conclusion that the origin is a stable node for the system in this example.

The idea in Theorem 1 in the previous section now becomes apparent in light of the Local Center Manifold Theorem; i.e., when the flow on the center manifold has the form

$$\dot{x} = a_m x^m + \cdots$$

near the origin, then for m even (as in Example 1 above) equation (2) in Section 2.11 has a saddle-node at the origin, and for m odd we get a topological saddle or a node at the origin, depending on whether the sign of a_m is the same as or the opposite of the sign of $\dot{y}$ near the origin.

Example 2. Consider the following system with $c = 2$ and $s = 1$:

$$\dot{x}_1 = x_1 y - x_1 x_2^2$$
$$\dot{x}_2 = x_2 y - x_2 x_1^2$$
$$\dot{y} = -y + x_1^2 + x_2^2.$$

In this example, we have $C = O$, $P = [-1]$,

$$\mathbf{F}(\mathbf{x}, y) = \begin{pmatrix} x_1 y - x_1 x_2^2 \\ x_2 y - x_2 x_1^2 \end{pmatrix} \quad \text{and} \quad G(\mathbf{x}, y) = x_1^2 + x_2^2.$$

We substitute the expansions

$$h(\mathbf{x}) = ax_1^2 + bx_1 x_2 + cx_2^2 + 0(|\mathbf{x}|^3)$$

and

$$Dh(\mathbf{x}) = [2ax_1 + bx_2, bx_1 + 2cx_2] + 0(|\mathbf{x}|^2)$$

into equation (5) to obtain

$$(2ax_1 + bx_2)[x_1(ax_1^2 + bx_1 x_2 + cx_2^2) - x_1 x_2^2]$$
$$+ (bx_1 + 2cx_2)[x_2(ax_1^2 + bx_1 x_2 + cx_2^2) - x_2 x_1^2]$$
$$+ (ax_1^2 + bx_1 x_2 + cx_2^2) - (x_1^2 + x_2^2) + 0(|\mathbf{x}|^3) = 0.$$

Since this is an identity for all x_1, x_2 with $|\mathbf{x}| < \delta$, we obtain $a = 1$, $b = 0$, $c = 1, \ldots$. Thus,

$$h(x_1, x_2) = x_1^2 + x_2^2 + 0(|\mathbf{x}|^3).$$

Substituting this result into equation (6) then yields

$$\dot{x}_1 = x_1^3 + 0(|\mathbf{x}|^4)$$
$$\dot{x}_2 = x_2^3 + 0(|\mathbf{x}|^4)$$

on the center manifold $W^c(0)$ near the origin. Since $r\dot{r} = x_1^4 + x_2^4 + O(|\mathbf{x}|^5) > 0$ for $0 < r < \delta$, this implies that the local phase portrait near the origin is given as in Figure 2. We see that the origin is a type of topological saddle that is unstable. However, if we were to use the center subspace approximation for the local center manifold, i.e., if we were to set $y = 0$ in the first two differential equations in this example, we would obtain

$$\dot{x}_1 = -x_1 x_2^2$$
$$\dot{x}_2 = -x_2 x_1^2$$

and arrive at the incorrect conclusion that the origin is a stable nonisolated critical point for the system in this example.

2.12. Center Manifold Theory

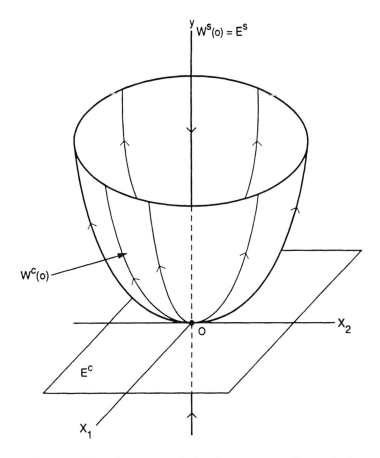

Figure 2. The phase portrait for the system in Example 2.

Example 3. For our last example, we consider the system

$$\dot{x}_1 = x_2 + y$$
$$\dot{x}_2 = y + x_1^2$$
$$\dot{y} = -y + x_2^2 + x_1 y.$$

The linear part of this system can be reduced to Jordan form by the matrix of (generalized) eigenvectors

$$P = \begin{bmatrix} 1 & 0 & 0 \\ 0 & 1 & -1 \\ 0 & 0 & 1 \end{bmatrix} \quad \text{with} \quad P^{-1} = \begin{bmatrix} 1 & 0 & 0 \\ 0 & 1 & 1 \\ 0 & 0 & 1 \end{bmatrix}.$$

This yields the following system in diagonal form

$$\dot{x}_1 = x_2$$

$$\dot{x}_2 = x_1^2 + (x_2 - y)^2 + x_1 y$$
$$\dot{y} = -y + (x_2 - y)^2 + x_1 y$$

with $c = 2$, $s = 1$, $P = [-1]$,

$$C = \begin{bmatrix} 0 & 1 \\ 0 & 0 \end{bmatrix}, \quad \mathbf{F}(\mathbf{x}, y) = \begin{pmatrix} 0 \\ x_1^2 + (x_2 - y)^2 + x_1 y \end{pmatrix},$$

and

$$G(\mathbf{x}, y) = (x_2 - y)^2 + x_1 y.$$

Let us substitute the expansions for $h(\mathbf{x})$ and $Dh(\mathbf{x})$ in Example 2 into equation (5) to obtain

$$(2ax_1 + bx_2)x_2 + (bx_1 + 2cx_2)[x_1^2 + (x_2 - y)^2 + x_1 y] + (ax_1^2 + bx_1 x_2 + cx_2^2)$$
$$- (x_2 - ax_1^2 - bx_1 x_2 - cx_2^2)^2 - x_1(ax_1^2 + bx_1 x_2 + cx_2^2) + 0(|\mathbf{x}|^3) = 0.$$

Since this is an identity for all x_1, x_2 with $|\mathbf{x}| < \delta$, we obtain $a = 0$, $b = 0$, $c = 1, \ldots$, i.e.,

$$h(\mathbf{x}) = x_2^2 + 0(|\mathbf{x}|^3).$$

Substituting this result into equation (6) then yields

$$\dot{x}_1 = x_2$$
$$\dot{x}_2 = x_1^2 + x_2^2 + O(|\mathbf{x}|^3)$$

on the center manifold $W^c(\mathbf{0})$ near the origin.

Theorem 3 in Section 2.11 then implies that the origin is a cusp for this system. The phase portrait for the system in this example is therefore topologically equivalent to the phase portrait in Figure 3.

As on pp. 203–204 in [Wi-II], the above results can be generalized to the case when the dimension of the unstable manifold $u \neq 0$ in the system (3). In that case, the local center manifold is given by

$$W^c_{\mathrm{loc}}(\mathbf{0}) = \{(\mathbf{x}, \mathbf{y}, \mathbf{z}) \in \mathbf{R}^c \times \mathbf{R}^s \times \mathbf{R}^u \mid \mathbf{y} = \mathbf{h}_1(\mathbf{x}) \text{ and } \mathbf{z} = \mathbf{h}_2(\mathbf{x}) \text{ for } |\mathbf{x}| < \delta\}$$

for some $\delta > 0$, where $\mathbf{h}_1 \in C^r(N_\delta(\mathbf{0}))$, $\mathbf{h}_2 \in C^r(N_\delta(\mathbf{0}))$, $\mathbf{h}_1(\mathbf{0}) = \mathbf{0}$, $\mathbf{h}_2(\mathbf{0}) = \mathbf{0}$, $D\mathbf{h}_1(\mathbf{0}) = 0$, and $D\mathbf{h}_2(\mathbf{0}) = 0$ since $W^c(\mathbf{0})$ is tangent to the center subspace $E^c = \{(\mathbf{x}, \mathbf{y}, \mathbf{z}) \in \mathbf{R}^c \times \mathbf{R}^s \times \mathbf{R}^u \mid \mathbf{y} = \mathbf{z} = \mathbf{0}\}$ at the origin. The functions $\mathbf{h}_1(\mathbf{x})$ and $\mathbf{h}_2(\mathbf{x})$ can be approximated to any desired degree of accuracy (provided that r is sufficiently large) by substituting their power series expansions into the following equations:

$$D\mathbf{h}_1(\mathbf{x})[C\mathbf{x} + \mathbf{F}(\mathbf{x}, \mathbf{h}_1(\mathbf{x}), \mathbf{h}_2(\mathbf{x}))] - P\mathbf{h}_1(\mathbf{x}) - \mathbf{G}(\mathbf{x}, \mathbf{h}_1(\mathbf{x}), \mathbf{h}_2(\mathbf{x})) = 0$$
$$D\mathbf{h}_2(\mathbf{x})[C\mathbf{x} + \mathbf{F}(\mathbf{x}, \mathbf{h}_1(\mathbf{x}), \mathbf{h}_2(\mathbf{x}))] - Q\mathbf{h}_2(\mathbf{x}) - \mathbf{H}(\mathbf{x}, \mathbf{h}_1(\mathbf{x}), \mathbf{h}_2(\mathbf{x})) = 0. \quad (7)$$

2.12. Center Manifold Theory

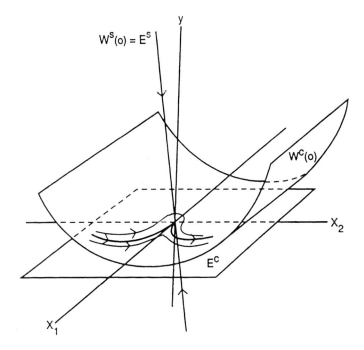

Figure 3. The phase portrait for the system in Example 3.

The next theorem, proved by Carr in [Ca], is analogous to the Hartman–Grobman Theorem except that, in order to determine completely the qualitative behavior of the flow near a nonhyperbolic critical point, one must be able to determine the qualitative behavior of the flow on the center manifold, which is determined by the first system of differential equations in the following theorem. The nonlinear part of that system, i.e., the function $\mathbf{F}(\mathbf{x}, \mathbf{h}_1(\mathbf{x}), \mathbf{h}_2(\mathbf{x}))$, can be simplified using the normal form theory in the next section.

Theorem 2. *Let E be an open subset of $\mathbf{R}^n$ containing the origin, and let $\mathbf{f} \in C^1(E)$; suppose that $\mathbf{f}(0) = 0$ and that the $n \times n$ matrix $D\mathbf{f}(0) = \text{diag}[C, P, Q]$, where the square matrix C has c eigenvalues with zero real parts, the square matrix P has s eigenvalues with negative real parts, and the square matrix Q has u eigenvalues with positive real parts. Then there exists C^1 functions $\mathbf{h}_1(\mathbf{x})$ and $\mathbf{h}_2(\mathbf{x})$ satisfying (7) in a neighborhood of the origin such that the nonlinear system (1), which can be written in the form (3), is topologically conjugate to the C^1 system*

$$\dot{\mathbf{x}} = C\mathbf{x} + \mathbf{F}(\mathbf{x}, \mathbf{h}_1(\mathbf{x}), \mathbf{h}_2(\mathbf{x}))$$
$$\dot{\mathbf{y}} = P\mathbf{y}$$
$$\dot{\mathbf{z}} = Q\mathbf{z}$$

for $(\mathbf{x}, \mathbf{y}, \mathbf{z}) \in \mathbf{R}^c \times \mathbf{R}^s \times \mathbf{R}^u$ in a neighborhood of the origin.

Problem Set 12

1. Consider the system in Example 3 in Section 2.7:
$$\dot{x} = x^2$$
$$\dot{y} = -y.$$
By substituting the expansion
$$h(x) = ax^2 + bx^3 + \cdots$$
into equation (5) show that the analytic center manifold for this system is defined by the function $h(x) \equiv 0$ for all $x \in \mathbf{R}$; i.e., show that the analytic center manifold $W^c(0) = E^c$ for this system. Also, show that for each $c \in \mathbf{R}$, the function
$$h(x, c) = \begin{cases} 0 & \text{for } x \geq 0 \\ ce^{1/x} & \text{for } x < 0 \end{cases}$$
defines a C^∞ center manifold for this system, i.e., show that it satisfies equation (5). Also, graph $h(x, c)$ for various $c \in \mathbf{R}$.

2. Use Theorem 1 to determine the qualitative behavior near the non-hyperbolic critical point at the origin for the system
$$\dot{x} = y$$
$$\dot{y} = -y + \alpha x^2 + xy$$
for $\alpha \neq 0$ and for $\alpha = 0$; i.e., follow the procedure in Example 1 after diagonalizing the system as in Example 3.

3. Same thing as in Problem 2 for the system
$$\dot{x} = xy$$
$$\dot{y} = -y - x^2.$$

4. Same thing as in Problem 2 for the system
$$\dot{x} = -x^3$$
$$\dot{y} = -y + x^2.$$
Also, show that this system has no analytic center manifold, i.e., show that if $h(x) = a_2 x^2 + a_3 x^3 + \cdots$, then it follows from (5) that $a_2 = 1, a_{2k+1} = 0$ and $a_{n+2} = na_n$ for n even; i.e., the Taylor series for $h(x)$ diverges for $x \neq 0$.

5. (a) Use Theorem 1 to find the approximation (6) for the flow on the local center manifold for the system
$$\dot{x}_1 = -x_2 + x_1 y$$
$$\dot{x}_2 = x_1 + x_2 y$$
$$\dot{y} = -y - x_1^2 - x_2^2 + y^2$$

and sketch the phase portrait for this system near the origin.

Hint: Convert the approximation (6) to polar coordinates and show that the origin is a stable focus of (6) in the x_1, x_2 plane.

(b) Same thing for the system

$$\dot{x}_1 = x_1 y - x_1 x_2^2$$
$$\dot{x}_2 = -2x_1^2 x_2 - x_1^4 + y^2$$
$$\dot{y} = -y + x_1^2 + x_2^2.$$

Hint: Show that the approximation (6) has a saddle-node in the x_1, x_2 plane.

(c) Same thing for the system

$$\dot{x}_1 = x_1^2 y - x_1^2 x_2^2$$
$$\dot{x}_2 = -2x_1^2 x_2^2 + y^2$$
$$\dot{y} = -y + x_1^2 + x_2^2.$$

Hint: Show that the approximation (6) has a nonhyperbolic critical point at the origin with two hyperbolic sectors in the x_1, x_2 plane.

6. Use Theorem 1 to find the approximation (6) for the flow on the local center manifold for the system

$$\dot{x} = ax^2 + bxy + cy^2$$
$$\dot{y} = -y + dx^2 + exy + fy^2,$$

and show that if $a \neq 0$, then the origin is a saddle-node. What type of critical point is at the origin if $a = 0$ and $bd \neq 0$? What if $a = b = 0$ and $cd \neq 0$? What if $a = d = 0$?

7. Use Theorem 1 to find the approximation (6) for the flow on the local center manifold for the system

$$\dot{x}_1 = -x_2^3 - x_1^3$$
$$\dot{x}_2 = x_1 y - x_2^3$$
$$\dot{y} = -y + x_1^2.$$

What type of critical point is at the origin of this system?

2.13 Normal Form Theory

The Hartman–Grobman Theorem shows us that in a neighborhood of a hyperbolic critical point, the qualitative behavior of a nonlinear system

$$\dot{\mathbf{x}} = \mathbf{f}(\mathbf{x}) \tag{1}$$

where $\mathbf{x} \in \mathbf{R}^n$ is determined by its linear part. In Section 1.8 we saw that the linear part of (1) can be put into Jordan canonical form $\dot{\mathbf{x}} = J\mathbf{x}$, which makes it easy to solve the linear system. The Local Center Manifold Theorem in the previous section showed us that, in a neighborhood of a nonhyperbolic critical point, determining the qualitative behavior of (1) could be reduced to the problem of determining the qualitative behavior of the nonlinear system

$$\dot{\mathbf{x}} = J\mathbf{x} + \mathbf{F}(\mathbf{x}) \tag{2}$$

on the center manifold. Since the dimension of the center manifold is typically less than n, this simplifies the problem of determining the qualitative behavior of the system (1) near a nonhyperbolic critical point. However, analyzing this system still may be a difficult task. The normal form theory allows us to simplify the nonlinear part, $\mathbf{F}(\mathbf{x})$, of (2) in order to make this task as easy as possible. This is accomplished by making a nonlinear, analytic transformation of coordinates of the form

$$\mathbf{x} = \mathbf{y} + \mathbf{h}(\mathbf{y}), \tag{3}$$

where $\mathbf{h}(\mathbf{y}) = 0(|\mathbf{y}|^2)$ as $|\mathbf{y}| \to 0$. Let us illustrate this idea with an example.

Example 1. The linear part of the system

$$\dot{x}_1 = x_2 + ax_1^2$$
$$\dot{x}_2 = x_1^2 + ex_1 x_2 + x_1^3$$

is already in Jordan form, and the 2×2 matrix

$$J = \begin{bmatrix} 0 & 1 \\ 0 & 0 \end{bmatrix}$$

has two zero eigenvalues. In order to reduce this system to the normal form used in Theorems 2 and 3 in Section 2.11, we let

$$\mathbf{x} = \mathbf{y} + \begin{pmatrix} 0 \\ -ay_1^2 \end{pmatrix},$$

i.e., we let $x_1 = y_1$ and $x_2 = y_2 - ay_1^2$ or, equivalently, $y_2 = x_2 + ax_1^2$. Under this nonlinear transformation of coordinates, it is easy to show that the above system of differential equations is transformed into the system

$$\dot{y}_1 = y_2$$
$$\dot{y}_2 = y_1^2 + (e + 2a)y_1 y_2 + (1 - ae)y_1^3.$$

The serious student should verify this computation. It then follows from Theorem 3 in Section 2.11 that the origin is a cusp for the nonlinear system in this example. Note that the qualitative behavior of the system (2) is invariant under the nonlinear change of coordinates (3), which has an inverse

2.13. Normal Form Theory

in a small neighborhood of the origin since it is a near-identity transformation; i.e., the two systems in this example are topologically conjugate, and therefore they have the same qualitative behavior in a neighborhood of the origin.

The method of reducing the system (2) to its "normal form" by means of a near-identity transformation of coordinates of the form (3) originated in the Ph.D. thesis of Poincaré; cf. [28]. The method was used by Andronov et al. in their study of planar systems [A-I], and an excellent modern-day description of the method can be found in [G/H] or [Wi-II]. If we rewrite the system (2) as

$$\dot{\mathbf{x}} = J\mathbf{x} + \mathbf{F}_2(\mathbf{x}) + 0(|\mathbf{x}|^3)$$

as $|\mathbf{x}| \to 0$, where $\mathbf{F}_2$ consists entirely of second-degree terms, then substituting the transformation (3) into the above equation and using the Chain Rule leads to

$$[I + Dh(\mathbf{y})]\dot{\mathbf{y}} = J\mathbf{y} + J\mathbf{h}(\mathbf{y}) + \mathbf{F}_2(\mathbf{y}) + 0(|\mathbf{y}^3|)$$

as $|\mathbf{y}| \to 0$. And then, since

$$[I + Dh(\mathbf{y})]^{-1} = I - Dh(\mathbf{y}) + 0(|\mathbf{y}|^2)$$

as $|\mathbf{y}| \to 0$, it follows that

$$\dot{\mathbf{y}} = J\mathbf{y} + J\mathbf{h}(\mathbf{y}) - Dh(\mathbf{y})J\mathbf{y} + \mathbf{F}_2(\mathbf{y}) + 0(|\mathbf{y}|^3)$$
$$= J\mathbf{y} + \tilde{\mathbf{F}}_2(\mathbf{y}) + 0(|\mathbf{y}|^3)$$

as $|\mathbf{y}| \to 0$, where

$$\tilde{\mathbf{F}}_2(\mathbf{y}) = J\mathbf{h}_2(\mathbf{y}) - Dh_2(\mathbf{y})J\mathbf{y} + \mathbf{F}_2(\mathbf{y}) \qquad (4)$$

and where we have written $\mathbf{h}(\mathbf{y}) = \mathbf{h}_2(\mathbf{y}) + 0(|\mathbf{y}|^3)$ as $|\mathbf{y}| \to 0$. We now want to choose $\mathbf{h}(\mathbf{y})$ in order to make $\tilde{\mathbf{F}}_2(\mathbf{y})$ as simple as possible. Ideally, this would mean choosing $\mathbf{h}(\mathbf{y})$ such that $\tilde{\mathbf{F}}_2(\mathbf{y}) = 0$; however, this is not always possible. For example, in the system in Example 1 above, we have

$$J = \begin{bmatrix} 0 & 1 \\ 0 & 0 \end{bmatrix} \quad \text{and} \quad \mathbf{F}_2(\mathbf{y}) = \begin{pmatrix} ay_1^2 \\ y_1^2 + ey_1 y_2 \end{pmatrix}.$$

If we let $\mathbf{h}(\mathbf{y}) = \mathbf{h}_2(\mathbf{y}) + 0(|\mathbf{y}|^3)$ and substitute the function

$$\mathbf{h}_2(\mathbf{y}) = \begin{pmatrix} a_{20}y_1^2 + a_{11}y_1 y_2 + a_{02}y_2^2 \\ b_{20}y_1^2 + b_{11}y_1 y_2 + b_{02}y_2^2 \end{pmatrix} \qquad (5)$$

into (4), we obtain

$$\tilde{\mathbf{F}}_2(\mathbf{y}) = \begin{pmatrix} b_{20}y_1^2 + (b_{11} - 2a_{20})y_1 y_2 + (b_{02} - a_{11})y_2^2 \\ -2b_{20}y_1 y_2 - b_{11}y_2^2 \end{pmatrix} + \begin{pmatrix} ay_1^2 \\ y_1^2 + ey_1 y_2 \end{pmatrix}$$
$$= \begin{pmatrix} 0 \\ y_1^2 + (e + 2a)y_1 y_2 \end{pmatrix}$$

for the choice of coefficients $b_{20} = -a$ and $a_{ij} = b_{ij} = 0$ otherwise in (5). As we shall see, this is as "simple" as we can make the quadratic terms in the system in Example 1 above. In other words, if we want to reduce the first component of $\tilde{\mathbf{F}}_2$ to zero, then the y_1^2 and the $y_1 y_2$ terms in the second component of $\tilde{\mathbf{F}}(\mathbf{y})$ cannot be eliminated by a nonlinear transformation of coordinates of the form (3). These terms are referred to as "resonance" terms; cf. [Wi-II].

In order to get a clearer understanding of what is going on in the above example, let us view the function

$$L_J[\mathbf{h}(\mathbf{y})] \equiv J\mathbf{h}(\mathbf{y}) - D\mathbf{h}(\mathbf{y}) J\mathbf{y} \tag{6}$$

as a linear transformation on the space H_2 of all second-degree polynomials of the form (5); i.e., in the case of systems in $\mathbf{R}^2$ we consider L_J as a linear operator on the six-dimensional vector space

$$H_2 = \operatorname{Span}\left\{\begin{pmatrix} x^2 \\ 0 \end{pmatrix}, \begin{pmatrix} xy \\ 0 \end{pmatrix}, \begin{pmatrix} y^2 \\ 0 \end{pmatrix}, \begin{pmatrix} 0 \\ x^2 \end{pmatrix}, \begin{pmatrix} 0 \\ xy \end{pmatrix}, \begin{pmatrix} 0 \\ y^2 \end{pmatrix}\right\}.$$

We now write $\mathbf{x} = (x, y)^T \in \mathbf{R}^2$ for ease in notation. A typical element $\mathbf{h}_2 \in H_2$ then is given by

$$\mathbf{h}_2(x) = \begin{pmatrix} a_{20} x^2 + a_{11} xy + a_{02} y^2 \\ b_{20} x^2 + b_{11} xy + b_{02} y^2 \end{pmatrix}, \tag{7}$$

and it is not difficult to compute

$$L_J[\mathbf{h}_2(\mathbf{x})] = \begin{pmatrix} b_{20} x^2 + (b_{11} - 2a_{20})xy + (b_{02} - a_{11})y^2 \\ -2b_{20} xy - b_{11} y^2 \end{pmatrix}.$$

Thus,

$$L_J(H_2) = \operatorname{Span}\left\{\begin{pmatrix} x^2 \\ -2xy \end{pmatrix}, \begin{pmatrix} xy \\ -y^2 \end{pmatrix}, \begin{pmatrix} xy \\ 0 \end{pmatrix}, \begin{pmatrix} y^2 \\ 0 \end{pmatrix}\right\}$$

$$= \operatorname{Span}\left\{\begin{pmatrix} x^2 \\ -2xy \end{pmatrix}, \begin{pmatrix} 0 \\ y^2 \end{pmatrix}, \begin{pmatrix} xy \\ 0 \end{pmatrix}, \begin{pmatrix} y^2 \\ 0 \end{pmatrix}\right\}.$$

Thus we see that

$$H_2 = L_J(H_2) \oplus G_2,$$

where

$$G_2 = \operatorname{Span}\left\{\begin{pmatrix} 0 \\ x^2 \end{pmatrix}, \begin{pmatrix} 0 \\ xy \end{pmatrix}\right\}.$$

This shows that any system of the form

$$\dot{x} = y + ax^2 + bxy + cy^2 + 0(|\mathbf{x}|^3)$$
$$\dot{y} = dx^2 + exy + fy^2 + 0(|\mathbf{x}|^3)$$

2.13. Normal Form Theory

with a general quadratic term $\mathbf{F}_2 \in H_2$ can be reduced, by a nonlinear transformation of coordinates $\mathbf{x} \to \mathbf{x} + \mathbf{h}_2(\mathbf{x})$ with $\mathbf{h}_2 \in H_2$ given by (7), to a system of the form

$$\dot{x} = y + 0(|\mathbf{x}|^3)$$
$$\dot{y} = dx^2 + (e+2a)xy + 0(|\mathbf{x}|^3)$$

with $\tilde{\mathbf{F}}_2 \in G_2$. In fact, in Problem 1 you will be asked to show that the function $\mathbf{h}_2(\mathbf{x})$, given by (7) with $a_{02} = a_{11} = 0$, $a_{20} = (b+f)/2$, $b_{02} = -c$, $b_{11} = f$, and $b_{20} = -a$, accomplishes this task.

This process can be generalized as follows: Writing the system (2) in the form

$$\dot{\mathbf{x}} = J\mathbf{x} + \mathbf{F}_2(\mathbf{x}) + \mathbf{F}_3(\mathbf{x}) + 0(|\mathbf{x}|^4)$$

as $|\mathbf{x}| \to 0$, where $\mathbf{F}_2 \in H_2$ and $\mathbf{F}_3 \in H_3$, we first can reduce this to a system of the form

$$\dot{\mathbf{x}} = J\mathbf{x} + \tilde{\mathbf{F}}_2(\mathbf{x}) + \mathbf{F}_3^*(\mathbf{x}) + 0(|\mathbf{x}|^4)$$

as $|\mathbf{x}| \to 0$ with $\tilde{\mathbf{F}}_2 \in G_2$ and $\mathbf{F}_3^* \in H_3$ by a nonlinear transformation of coordinates (3) with $\mathbf{h}(\mathbf{x}) = \mathbf{h}_2(\mathbf{x}) \in H_2$ as was done above for the case when (2) is a system with $\mathbf{x} \in \mathbf{R}^2$. We then can apply the nonlinear transformation of coordinates (3) with $\mathbf{h}(\mathbf{x}) = \mathbf{h}_3(\mathbf{x}) \in H_3$ to this latter system in order to reduce it to a system of the form

$$\dot{\mathbf{x}} = J\mathbf{x} + \tilde{\mathbf{F}}_2(\mathbf{x}) + \tilde{\mathbf{F}}_3(\mathbf{x}) + 0(|\mathbf{x}|^4)$$

with $\tilde{\mathbf{F}}_3 \in G_3$. This is illustrated in Problems 3 and 5. This process then can be continued to obtain the so-called "normal form" of the system (2) to any desired degree. This is the content of the Normal Form Theorem on p. 216 in [Wi-II]. It should be noted that the normal form of the system (2) is not unique. This can be seen since, for example, in $\mathbf{R}^2$, the linear space of polynomials G_2 complementary to $L_J(H_2)$ with the matrix J given in Example 1 is spanned by $\{(x^2, 0)^T, (0, x^2)^T\}$ as well as by $\{(0, xy)^T, (0, x^2)^T\}$.

Remark 1. As was shown above (and in Problem 1), any system of the form

$$\dot{\mathbf{x}} = J\mathbf{x} + \mathbf{F}(\mathbf{x})$$

with

$$J = \begin{bmatrix} 0 & 1 \\ 0 & 0 \end{bmatrix} \quad \text{and} \quad \mathbf{F}(\mathbf{x}) = \begin{pmatrix} ax^2 + bxy + cy^2 \\ dx^2 + exy + fy^2 \end{pmatrix} + 0(|\mathbf{x}|^3)$$

as $|\mathbf{x}| \to 0$ can be put into the normal form

$$\dot{x} = y + 0(|\mathbf{x}|^3)$$
$$\dot{y} = dx^2 + (e+2a)xy + 0(|\mathbf{x}|^3)$$

as $|\mathbf{x}| \to 0$; furthermore, by letting $y \to y + 0(|\mathbf{x}|^3)$, the quantity in the first of the above differential equations, this can be reduced to the normal form

$$\dot{x} = y$$
$$\dot{y} = dx^2 + (e + 2a)xy + 0(|\mathbf{x}|^3)$$

used in Theorems 2 and 3 in Section 2.11. Then for $d \neq 0$ and $(e+2a) \neq 0$, it follows from Theorems 2 and 3 in Section 2.11 that the $0(|\mathbf{x}|^3)$ terms do not affect the qualitative nature of the nonhyperbolic critical point at the origin. Also they do not affect the types of qualitative dynamical behavior that can occur in "unfolding" this nonhyperbolic critical point, as is discussed in Chapter 4. Thus, we can delete these terms in studying the bifurcations that take place in a neighborhood of this nonhyperbolic critical point and, after an appropriate normalization of coordinates, we can use one of the following two normal forms for studying these bifurcations:

$$\dot{x} = y$$
$$\dot{y} = x^2 \pm xy.$$

Cf. Section 4.13, where we also see that all possible types of dynamical behavior that can occur in systems "close" to this system are determined by the "universal unfoldings" of these two normal forms given by

$$\dot{x} = y$$
$$\dot{y} = \mu_1 + \mu_2 y + x^2 \pm xy.$$

We have almost entirely focused on the case when the linear part J of the system (1) is non-zero and has two zero eigenvalues. The case when J has two pure imaginary eigenvalues is more complicated, but it is treated in Section 3.3 of [G/H] where it is shown that any C^3 system of the form

$$\dot{x} = -y + O(|\mathbf{x}|^2)$$
$$\dot{y} = x + O(|\mathbf{x}|^2)$$

can be put into the normal form

$$\dot{x} = -y + (ax - by)(x^2 + y^2) + O(|\mathbf{x}|^4)$$
$$\dot{y} = x + (ay + bx)(x^2 + y^2) + O(|\mathbf{x}|^4)$$

for suitable constants a and b. Cf. Theorem 2 in Section 4.4. And, as we shall see in Problem 1(b) in Section 4.4, the origin is then a stable (or an unstable) weak focus of multiplicity 1 if $a < 0$ (or if $a > 0$).

Problem Set 13

1. Consider the quadratic system

$$\dot{x} = y + ax^2 + bxy + cy^2$$
$$\dot{y} = dx^2 + exy + fy^2$$

2.13. Normal Form Theory

with

$$J = \begin{bmatrix} 0 & 1 \\ 0 & 0 \end{bmatrix} \quad \text{and} \quad \mathbf{F}_2(\mathbf{x}) = \begin{pmatrix} ax^2 + bxy + cy^2 \\ dx^2 + exy + fy^2 \end{pmatrix}.$$

For $\mathbf{h}_2(\mathbf{x})$ given by (7), compute $L_J[\mathbf{h}_2(\mathbf{x})]$, defined by (6), and show that for $a_{02} = a_{11} = 0, a_{20} = (b+f)/2, b_{02} = -c, b_{11} = f$, and $b_{20} = -a$

$$\tilde{\mathbf{F}}_2(\mathbf{x}) = L_J[\mathbf{h}_2(\mathbf{x})] + \mathbf{F}_2(\mathbf{x}) = \begin{pmatrix} 0 \\ dx^2 + (e+2a)xy \end{pmatrix}.$$

2. Let

$$H_3 = \text{Span}\left\{\begin{pmatrix} x^3 \\ 0 \end{pmatrix}, \begin{pmatrix} x^2y \\ 0 \end{pmatrix}, \begin{pmatrix} xy^2 \\ 0 \end{pmatrix}, \begin{pmatrix} y^3 \\ 0 \end{pmatrix},\right.$$
$$\left.\begin{pmatrix} 0 \\ x^3 \end{pmatrix}, \begin{pmatrix} 0 \\ x^2y \end{pmatrix}, \begin{pmatrix} 0 \\ xy^2 \end{pmatrix}, \begin{pmatrix} 0 \\ y^3 \end{pmatrix}\right\}$$

$$J = \begin{bmatrix} 0 & 1 \\ 0 & 0 \end{bmatrix}$$

and

$$\mathbf{h}_3(\mathbf{x}) = \begin{pmatrix} a_{30}x^3 + a_{21}x^2y + a_{12}xy^2 + a_{03}y^3 \\ b_{30}x^3 + b_{21}x^2y + b_{12}xy^2 + b_{03}y^3 \end{pmatrix} \in H_3.$$

Show that

$$L_J(H_3) = \text{Span}\left\{\begin{pmatrix} x^3 \\ -3x^2y \end{pmatrix}, \begin{pmatrix} x^2y \\ 0 \end{pmatrix}, \begin{pmatrix} xy^2 \\ 0 \end{pmatrix}, \begin{pmatrix} y^3 \\ 0 \end{pmatrix},\right.$$
$$\left.\begin{pmatrix} 0 \\ xy^2 \end{pmatrix}, \begin{pmatrix} 0 \\ y^3 \end{pmatrix}\right\},$$

and that $H_3 = L_J(H_3) \oplus G_3$, where

$$G_3 = \text{Span}\left\{\begin{pmatrix} 0 \\ x^3 \end{pmatrix}, \begin{pmatrix} 0 \\ x^2y \end{pmatrix}\right\}.$$

3. Use the results in Problem 2 to show that a planar system of the form

$$\dot{\mathbf{x}} = J\mathbf{x} + \mathbf{F}_3(\mathbf{x}) + 0(|\mathbf{x}|^4)$$

with J given in Problem 2 and $\mathbf{F}_3 \in H_3$ can be reduced to the normal form

$$\dot{x} = y + 0(|\mathbf{x}|^4)$$
$$\dot{y} = ax^3 + bx^2y + 0(|\mathbf{x}|^4)$$

for some $a, b \in \mathbf{R}$. For $a \neq 0$, what type of critical point is at the origin of this system according to Theorem 2 in Section 2.11? (Consider the two cases $a > 0$ and $a < 0$ separately.)

4. For the 2 × 2 matrix J in Problem 1, show that

$$H_4 = L_J(H_4) \oplus G_4,$$

where

$$G_4 = \text{Span}\left\{\begin{pmatrix} 0 \\ x^4 \end{pmatrix}, \begin{pmatrix} 0 \\ x^3 y \end{pmatrix}\right\}.$$

What type of critical point is at the origin of the system

$$\dot{\mathbf{x}} = J\mathbf{x} + \mathbf{F}_4(\mathbf{x})$$

with $\mathbf{F}_4 \in H_4$ if the second component of $\mathbf{F}_4(\mathbf{x})$ contains an x^4 term?
Hint: See Theorem 3 in Section 2.11.

5. Show that the quadratic part of the cubic system

$$\dot{x} = y + x^2 - x^3 + xy^2 - y^3$$
$$\dot{y} = x^2 - 2xy + x^3 - x^2 y$$

can be reduced to normal form using the transformation defined in Example 1, and show that this yields the system

$$\dot{x} = y - x^3 + xy^2 - y^3 + 0(|\mathbf{x}|^4)$$
$$\dot{y} = x^2 + 3x^3 + x^2 y + 0(|\mathbf{x}|^4)$$

as $|\mathbf{x}| \to 0$. Then determine a nonlinear transformation of coordinates of the form (3) with $\mathbf{h}(\mathbf{x}) = \mathbf{h}_3(\mathbf{x}) \in H_3$ that reduces this system to the system

$$\dot{x} = y + 0(|\mathbf{x}|^4)$$
$$\dot{y} = x^2 + 3x^3 - 2x^2 y + 0(|\mathbf{x}|^4)$$

as $|\mathbf{x}| \to 0$, which is said to be in normal form (to degree three).

Remark 2. As in Remark 1 above, it follows from Theorems 2 and 3 in Section 2.11 that the x^3 and $0(|\mathbf{x}|^4)$ terms in the above system of differential equations do not affect the nature of the nonhyperbolic critical point at the origin. Thus, we might expect that

$$\dot{x} = y$$
$$\dot{y} = x^2 - 2x^2 y$$

is an appropriate normal form for studying the bifurcations that take place in a neighborhood of this nonhyperbolic critical point; however, we see at the end of Section 4.13 that the $x^2 y$ term can also be eliminated and that the appropriate normal form for studying these bifurcations is given by

$$\dot{x} = y$$
$$\dot{y} = x^2 \pm x^3 y$$

in which case all possible types of dynamical behavior that can occur in systems "close" to this system are given by the "universal unfolding" of this normal form given by

$$\dot{x} = y$$
$$\dot{y} = \mu_1 + \mu_2 y + \mu_3 xy + x^2 \pm x^3 y.$$

2.14 Gradient and Hamiltonian Systems

In this section we study two interesting types of systems which arise in physical problems and from which we draw a wealth of examples of the general theory.

Definition 1. Let E be an open subset of $\mathbf{R}^{2n}$ and let $H \in C^2(E)$ where $H = H(\mathbf{x}, \mathbf{y})$ with $\mathbf{x}, \mathbf{y} \in \mathbf{R}^n$. A system of the form

$$\dot{\mathbf{x}} = \frac{\partial H}{\partial \mathbf{y}}$$
$$\dot{\mathbf{y}} = -\frac{\partial H}{\partial \mathbf{x}}, \tag{1}$$

where

$$\frac{\partial H}{\partial \mathbf{x}} = \left(\frac{\partial H}{\partial x_1}, \ldots, \frac{\partial H}{\partial x_n}\right)^T \quad \text{and} \quad \frac{\partial H}{\partial \mathbf{y}} = \left(\frac{\partial H}{\partial y_1}, \ldots, \frac{\partial H}{\partial y_n}\right)^T,$$

is called a *Hamiltonian system* with n degrees of freedom on E.

For example, the Hamiltonian function

$$H(\mathbf{x}, \mathbf{y}) = (x_1^2 + x_2^2 + y_1^2 + y_2^2)/2$$

is the energy function for the spherical pendulum

$$\dot{x}_1 = y_1$$
$$\dot{x}_2 = y_2$$
$$\dot{y}_1 = -x_1$$
$$\dot{y}_2 = -x_2$$

which is discussed in Section 3.6 of Chapter 3. This system is equivalent to the pair of uncoupled harmonic oscillators

$$\ddot{x}_1 + x_1 = 0$$
$$\ddot{x}_2 + x_2 = 0.$$

All Hamiltonian systems are conservative in the sense that the Hamiltonian function or the total energy $H(\mathbf{x}, \mathbf{y})$ remains constant along trajectories of the system.

Theorem 1 (Conservation of Energy). *The total energy $H(\mathbf{x},\mathbf{y})$ of the Hamiltonian system (1) remains constant along trajectories of (1).*

Proof. The total derivative of the Hamiltonian function $H(\mathbf{x},\mathbf{y})$ along a trajectory $\mathbf{x}(t), \mathbf{y}(t)$ of (1)

$$\frac{dH}{dt} = \frac{\partial H}{\partial \mathbf{x}} \cdot \dot{\mathbf{x}} + \frac{\partial H}{\partial \mathbf{y}} \cdot \dot{\mathbf{y}} = \frac{\partial H}{\partial \mathbf{x}} \cdot \frac{\partial H}{\partial \mathbf{y}} - \frac{\partial H}{\partial \mathbf{y}} \cdot \frac{\partial H}{\partial \mathbf{x}} = 0.$$

Thus, $H(\mathbf{x},\mathbf{y})$ is constant along any solution curve of (1) and the trajectories of (1) lie on the surfaces $H(\mathbf{x}, \mathbf{y}) = $ constant.

We next establish some very specific results about the nature of the critical points of Hamiltonian systems with one degree of freedom. Note that the equilibrium points or critical points of the system (1) correspond to the critical points of the Hamiltonian function $H(\mathbf{x},\mathbf{y})$ where $\frac{\partial H}{\partial \mathbf{x}} = \frac{\partial H}{\partial \mathbf{y}} = 0$. We may, without loss of generality, assume that the critical point in question has been translated to the origin.

Lemma. *If the origin is a focus of the Hamiltonian system*

$$\dot{x} = H_y(x,y)$$
$$\dot{y} = -H_x(x,y), \qquad (1')$$

then the origin is not a strict local maximum or minimum of the Hamiltonian function $H(x,y)$.

Proof. Suppose that the origin is a stable focus for $(1')$. Then according to Definition 3 in Section 2.10, there is an $\varepsilon > 0$ such that for $0 < r_0 < \varepsilon$ and $\theta_0 \in \mathbf{R}$, the polar coordinates of the solution of $(1')$ with $r(0) = r_0$ and $\theta(0) = \theta_0$ satisfy $r(t, r_0, \theta_0) \to 0$ and $|\theta(t, r_0, \theta_0)| \to \infty$ as $t \to \infty$; i.e., for $(x_0, y_0) \in N_\varepsilon(0) \sim \{0\}$, the solution $(x(t, x_0, y_0), y(t, x_0, y_0)) \to 0$ as $t \to \infty$. Thus, by Theorem 1 and the continuity of $H(x,y)$ and the solution, it follows that

$$H(x_0, y_0) = \lim_{t \to \infty} H(x(t, x_0, y_0), y(t, x_0, y_0)) = H(0,0)$$

for all $(x_0, y_0) \in N_\varepsilon(0)$. Thus, the origin is not a strict local maximum or minimum of the function $H(x,y)$; i.e., it is not true that $H(x,y) > H(0,0)$ or $H(x,y) < H(0,0)$ for all points (x,y) in a deleted neighborhood of the origin. A similar argument applies when the origin is an unstable focus of $(1')$.

2.14. Gradient and Hamiltonian Systems

Definition 2. A critical point of the system

$$\dot{x} = f(x)$$

at which $Df(x_0)$ has no zero eigenvalues is called a *nondegenerate critical point* of the system, otherwise, it is called a *degenerate critical point* of the system.

Note that any nondegenerate critical point of a planar system is either a hyperbolic critical point of the system or a center of the linearized system.

Theorem 2. *Any nondegenerate critical point of an analytic Hamiltonian system (1') is either a (topological) saddle or a center; furthermore, (x_0, y_0) is a (topological) saddle for (1') iff it is a saddle of the Hamiltonian function $H(x,y)$ and a strict local maximum or minimum of the function $H(x,y)$ is a center for (1').*

Proof. We assume that the critical point is at the origin. Thus, $H_x(0,0) = H_y(0,0) = 0$ and the linearization of (1') at the origin is

$$\dot{x} = Ax \qquad (2)$$

where

$$A = \begin{bmatrix} H_{yx}(0,0) & H_{yy}(0,0) \\ -H_{xx}(0,0) & -H_{xy}(0,0) \end{bmatrix}.$$

We see that $\operatorname{tr} A = 0$ and that $\det A = H_{xx}(0)H_{yy}(0) - H_{xy}^2(0)$. Thus, the critical point at the origin is a saddle of the function $H(x,y)$ iff $\det A < 0$ iff it is a saddle for the linear system (2) iff it is a (topological) saddle for the Hamiltonian system (1') according to Theorem 3 in Section 2.10. Also, according to Theorem 1 in Section 1.5 of Chapter 1, if $\operatorname{tr} A = 0$ and $\det A > 0$, the origin is a center for the linear system (2). And then according to the Corollary in Section 2.10, the origin is either a center or a focus for (1'). Thus, if the nondegenerate critical point $(0,0)$ is a strict local maximum or minimum of the function $H(x,y)$, then $\det A > 0$ and, according to the above lemma, the origin is not a focus for (1'); i.e., the origin is a center for the Hamiltonian system (1').

One particular type of Hamiltonian system with one degree of freedom is the *Newtonian system* with one degree of freedom,

$$\ddot{x} = f(x)$$

where $f \in C^1(a,b)$. This differential equation can be written as a system in $\mathbf{R}^2$:

$$\begin{aligned} \dot{x} &= y \\ \dot{y} &= f(x). \end{aligned} \qquad (3)$$

The total energy for this system $H(x,y) = T(y) + U(x)$ where $T(y) = y^2/2$ is the kinetic energy and

$$U(x) = -\int_{x_0}^x f(s)\,ds$$

is the potential energy. With this definition of $H(x,y)$ we see that the Newtonian system (3) can be written as a Hamiltonian system. It is not difficult to establish the following facts for the Newtonian system (2); cf. Problem 9 at the end of this section.

Theorem 3. *The critical points of the Newtonian system (3) all lie on the x-axis. The point $(x_0, 0)$ is a critical point of the Newtonian system (3) iff it is a critical point of the function $U(x)$, i.e., a zero of the function $f(x)$. If $(x_0, 0)$ is a strict local maximum of the analytic function $U(x)$, it is a saddle for (3). If $(x_0, 0)$ is a strict local minimum of the analytic function $U(x)$, it is a center for (3). If $(x_0, 0)$ is a horizontal inflection point of the function $U(x)$, it is a cusp for the system (3). And finally, the phase portrait of (3) is symmetric with respect to the x-axis.*

Example 1. Let us construct the phase portrait for the undamped pendulum

$$\ddot{x} + \sin x = 0.$$

This differential equation can be written as a Newtonian system

$$\dot{x} = y$$
$$\dot{y} = -\sin x$$

where the potential energy

$$U(x) = \int_0^x \sin t\, dt = 1 - \cos x.$$

The graph of the function $U(x)$ and the phase portrait for the undamped pendulum, which follows from Theorem 3, are shown in Figure 1 below.

Note that the origin in the phase portrait for the undamped pendulum shown in Figure 1 corresponds to the stable equilibrium position of the pendulum hanging straight down. The critical points at $(\pm\pi, 0)$ correspond to the unstable equilibrium position where the pendulum is straight up. Trajectories near the origin are nearly circles and are approximated by the solution curves of the linear pendulum

$$\ddot{x} + x = 0.$$

2.14. Gradient and Hamiltonian Systems

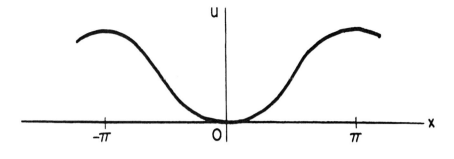

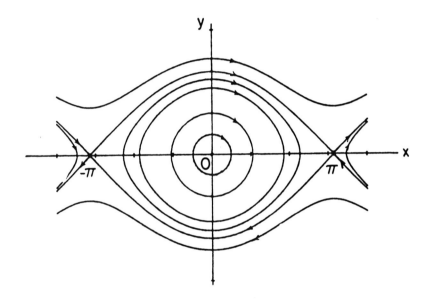

Figure 1. The phase portrait for the undamped pendulum.

The closed trajectories encircling the origin describe the usual periodic motions associated with a pendulum where the pendulum swings back and forth. The separatrices connecting the saddles at $(\pm\pi, 0)$ correspond to motions with total energy $H = 2$ in which case the pendulum approaches the unstable vertical position as $t \to \pm\infty$. And the trajectories outside the separatrix loops, where $H > 2$, correspond to motions where the pendulum goes over the top.

Definition 3. Let E be an open subset of $\mathbf{R}^n$ and let $V \in C^2(E)$. A system of the form
$$\dot{\mathbf{x}} = -\operatorname{grad} V(\mathbf{x}), \tag{4}$$
where
$$\operatorname{grad} V = \left(\frac{\partial V}{\partial x_1}, \ldots, \frac{\partial V}{\partial x_n} \right)^T,$$
is called a *gradient system* on E.

Note that the equilibrium points or critical points of the gradient system (4) correspond to the critical points of the function $V(\mathbf{x})$ where $\operatorname{grad} V(\mathbf{x}) = 0$. Points where $\operatorname{grad} V(\mathbf{x}) \neq 0$ are called regular points of the function $V(\mathbf{x})$. At regular points of $V(\mathbf{x})$, the gradient vector $\operatorname{grad} V(\mathbf{x})$ is perpendicular to the level surface $V(\mathbf{x}) = $ constant through the point. And it is easy to show that at a critical point $\mathbf{x}_0$ of $V(\mathbf{x})$, which is a strict local minimum of $V(\mathbf{x})$, the function $V(\mathbf{x}) - V(\mathbf{x}_0)$ is a strict Liapunov function for the system (4) in some neighborhood of $\mathbf{x}_0$; cf. Problem 7 at the end of this section. We therefore have the following theorem:

Theorem 4. *At regular points of the function $V(\mathbf{x})$, trajectories of the gradient system (4) cross the level surfaces $V(\mathbf{x}) = $ constant orthogonally. And strict local minima of the function $V(\mathbf{x})$ are asymptotically stable equilibrium points of (4).*

Since the linearization of (4) at any critical point $\mathbf{x}_0$ of (4) has a matrix
$$A = \left[\frac{\partial^2 V(\mathbf{x}_0)}{\partial x_i \partial x_j} \right]_{i,j=1,\ldots,n}$$
which is symmetric, the eigenvalues of A are all real and A is diagonalizable with respect to an orthonormal basis; cf. [H/S].

Once again, for planar gradient systems, we can be very specific about the nature of the critical points of the system:

Theorem 5. *Any nondegenerate critical point of an analytic gradient system (4) on $\mathbf{R}^2$ is either a saddle or a node; furthermore, if (x_0, y_0) is a saddle of the function $V(x,y)$, it is a saddle of (4) and if (x_0, y_0) is a strict local maximum or minimum of the function $V(x,y)$, it is respectively an unstable or a stable node for (4).*

Example 2. Let $V(x,y) = x^2(x-1)^2 + y^2$. The gradient system (4) then has the form
$$\dot{x} = -4x(x-1)(x-1/2)$$
$$\dot{y} = -2y.$$

There are critical points at $(0,0), (1/2, 0)$ and $(1,0)$. It follows from Theorem 5 that $(0,0)$ and $(1,0)$ are stable nodes and that $(1/2, 0)$ is a saddle

2.14. Gradient and Hamiltonian Systems

for this system; cf. Problem 8 at the end of this section. The level curves $V(x, y) = $ constant and the trajectories of this system are shown in Figure 2.

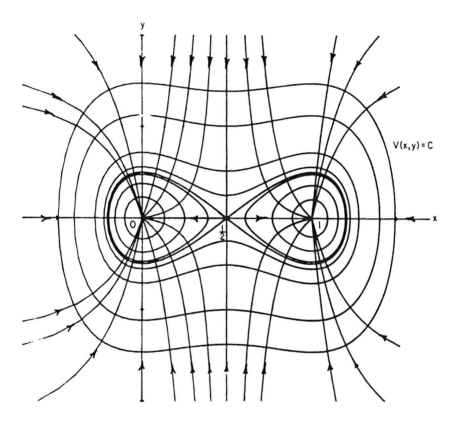

Figure 2. The level curves $V(x, y) = $ constant (closed curves) and the trajectories of the gradient system in Example 2.

One last topic, which shows that there is an interesting relationship between gradient and Hamiltonian systems, is considered in this section. We only give the details for planar systems.

Definition 4. Consider the planar system

$$\dot{x} = P(x, y)$$
$$\dot{y} = Q(x, y). \qquad (5)$$

The system orthogonal to (5) is defined as the system

$$\dot{x} = Q(x, y)$$
$$\dot{y} = -P(x, y). \qquad (6)$$

Clearly, (5) and (6) have the same critical points and at regular points, trajectories of (5) are orthogonal to the trajectories of (6). Furthermore, centers of (5) correspond to nodes of (6), saddles of (5) correspond to saddles of (6), and foci of (5) correspond to foci of (6). Also, if (5) is a Hamiltonian system with $P = H_y$ and $Q = -H_x$, then (6) is a gradient system and conversely.

Theorem 6. *The system (5) is a Hamiltonian system iff the system (6) orthogonal to (5) is a gradient system.*

In higher dimensions, we have that if (1) is a Hamiltonian system with n degrees of freedom then the system

$$\dot{\mathbf{x}} = -\frac{\partial H}{\partial \mathbf{x}}$$
$$\dot{\mathbf{y}} = -\frac{\partial H}{\partial \mathbf{y}} \quad (7)$$

orthogonal to (1) is a gradient system in $\mathbf{R}^{2n}$ and the trajectories of the gradient system (7) cross the surfaces $H(\mathbf{x}, \mathbf{y}) = $ constant orthogonally. In Example 2 if we take $H(x, y) = V(x, y)$, then Figure 2 illustrates the orthogonality of the trajectories of the Hamiltonian and gradient flows, the Hamiltonian flow swirling clockwise.

PROBLEM SET 14

1. (a) Show that the system

   $$\dot{x} = a_{11}x + a_{12}y + Ax^2 - 2Bxy + Cy^2$$
   $$\dot{y} = a_{21}x - a_{11}y + Dx^2 - 2Axy + By^2$$

   is a Hamiltonian system with one degree of freedom; i.e., find the Hamiltonian function $H(x, y)$ for this system.

   (b) Given $\mathbf{f} \in C^1(E)$, where E is an open, simply connected subset of $\mathbf{R}^2$, show that the system

   $$\dot{\mathbf{x}} = \mathbf{f}(\mathbf{x})$$

   is a Hamiltonian system on E iff $\nabla \cdot \mathbf{f}(\mathbf{x}) = 0$ for all $\mathbf{x} \in E$.

2. Find the Hamiltonian function for the system

   $$\dot{x} = y$$
   $$\dot{y} = -x + x^2$$

   and, using Theorem 3, sketch the phase portrait for this system.

3. Same as Problem 2 for the system

$$\dot{x} = y$$
$$\dot{y} = -x + x^3.$$

4. Given the function $U(x)$ pictured below:

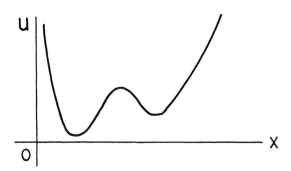

sketch the phase portrait for the Hamiltonian system with Hamiltonian $H(x, y) = y^2/2 + U(x)$.

5. For each of the following Hamiltonian functions, sketch the phase portraits for the Hamiltonian system (1) and the gradient system (7) orthogonal to (1). Draw both phase portraits on the same phase plane.

(a) $H(x, y) = x^2 + 2y^2$
(b) $H(x, y) = x^2 - y^2$
(c) $H(x, y) = y \sin x$
(d) $H(x, y) = x^2 - y^2 - 2x + 4y + 5$
(e) $H(x, y) = 2x^2 - 2xy + 5y^2 + 4x + 4y + 4$
(f) $H(x, y) = x^2 - 2xy - y^2 + 2x - 2y + 2.$

6. For the gradient system (4), with $V(x, y, z)$ given below, sketch some of the surfaces $V(x, y, z) =$ constant and some of the trajectories of the system.

(a) $V(x, y, z) = x^2 + y^2 - z$
(b) $V(x, y, z) = x^2 + 2y^2 + z^2$
(c) $V(x, y, z) = x^2(x - 1) + y^2(y - 2) + z^2.$

7. Show that if $\mathbf{x}_0$ is a strict local minimum of $V(\mathbf{x})$ then the function $V(\mathbf{x}) - V(\mathbf{x}_0)$ is a strict Liapunov function for the gradient system (4).

8. Show that the function $V(x, y) = x^2(x - 1)^2 + y^2$ has strict local minima at $(0, 0)$ and $(1, 0)$ and a saddle at $(1/2, 0)$, and therefore that the gradient system (4) with $V(x, y)$ given above has stable nodes at $(0, 0)$ and $(1, 0)$ and a saddle at $(1/2, 0)$.

9. Prove that the critical point $(x_0, 0)$ of the Newtonian system (3) is a saddle if it is a strict local maximum of the function $U(x)$ and that it is a center if it is a strict local minimum of the function $U(x)$. Also, show that if $(x_0, 0)$ is a horizontal inflection point of $U(x)$ then it is a cusp of the system (3); cf. Figure 4 in Section 2.11.

10. Prove that if the system (5) has a nondegenerate critical point at the origin which is a stable focus with the flow swirling counterclockwise around the origin, then the system (6) orthogonal to (5) has an unstable focus at the origin with the flow swirling counterclockwise around the origin. **Hint:** In this case, the system (5) is linearly equivalent to

$$\dot{x} = ax - by + \text{higher degree terms}$$
$$\dot{y} = bx + ay + \text{higher degree terms}$$

with $a < 0$ and $b > 0$. What does this tell you about the system (6) orthogonal to (5)? Consider other variations on this theme; e.g., what if the origin has an unstable clockwise focus?

11. Show that the planar two-body problem

$$\ddot{x} = -\frac{x}{(x^2 + y^2)^{3/2}}$$
$$\ddot{y} = -\frac{y}{(x^2 + y^2)^{3/2}}$$

can be written as a Hamiltonian system with two degrees of freedom on $E = \mathbf{R}^4 \sim \{\mathbf{0}\}$. What is the gradient system orthogonal to this system?

12. Show that the flow defined by a Hamiltonian system with one-degree of freedom is area preserving. **Hint:** Cf. Problem 6 in Section 2.3.

3

Nonlinear Systems: Global Theory

In Chapter 2 we saw that any nonlinear system

$$\dot{\mathbf{x}} = \mathbf{f}(\mathbf{x}), \tag{1}$$

with $\mathbf{f} \in C^1(E)$ and E an open subset of $\mathbf{R}^n$, has a unique solution $\phi_t(\mathbf{x}_0)$, passing through a point $\mathbf{x}_0 \in E$ at time $t = 0$ which is defined for all $t \in I(\mathbf{x}_0)$, the maximal interval of existence of the solution. Furthermore, the flow ϕ_t of the system satisfies (i) $\phi_0(\mathbf{x}) = \mathbf{x}$ and (ii) $\phi_{t+s}(\mathbf{x}) = \phi_t(\phi_s(\mathbf{x}))$ for all $\mathbf{x} \in E$ and the function $\phi(t, \mathbf{x}) = \phi_t(\mathbf{x})$ defines a C^1-map $\phi \colon \Omega \to E$ where $\Omega = \{(t, \mathbf{x}) \in \mathbf{R} \times E \mid t \in I(\mathbf{x})\}$.

In this chapter we define a dynamical system as a C^1-map $\phi \colon \mathbf{R} \times E \to E$ which satisfies (i) and (ii) above. We first show that we can rescale the time in any C^1-system (1) so that for all $\mathbf{x} \in E$, the maximal interval of existence $I(\mathbf{x}) = (-\infty, \infty)$. Thus any C^1-system (1), after an appropriate rescaling of the time, defines a dynamical system $\phi \colon \mathbf{R} \times E \to E$ where $\phi(t, \mathbf{x}) = \phi_t(\mathbf{x})$ is the solution of (1) with $\phi_0(\mathbf{x}) = \mathbf{x}$. We next consider limit sets and attractors of dynamical systems. Besides equilibrium points and periodic orbits, a dynamical system can have homoclinic loops or separatrix cycles as well as strange attractors as limit sets. We study periodic orbits in some detail and give the Stable Manifold Theorem for periodic orbits as well as several examples which illustrate the general theory in this chapter. Determining the nature of limit sets of nonlinear systems (1) with $n \geq 3$ is a challenging problem which is the subject of much mathematical research at this time.

The last part of this chapter is restricted to planar systems where the theory is more complete. The Poincaré–Bendixson Theorem, established in Section 3.7, implies that for planar systems any ω-limit set is either a critical point, a limit cycle or a union of separatrix cycles. Determining the number of limit cycles of polynomial systems in $\mathbf{R}^2$ is another difficult problem in the theory. This problem was posed in 1900 by David Hilbert as one of the problems in his famous list of outstanding mathematical problems at the turn of the century. This problem remains unsolved today even for quadratic systems in $\mathbf{R}^2$. Some results on the number of limit cycles of planar systems are established in Section 3.8 of this chapter. We conclude this chapter with a technique, based on the Poincaré–Bendixson Theorem and some projective geometry, for constructing the global phase portrait for a planar dynamical system. The global phase portrait determines the qualitative behavior of every solution $\phi_t(\mathbf{x})$ of the system (1) for all $t \in (-\infty, \infty)$ as well as for all $\mathbf{x} \in \mathbf{R}^2$. This qualitative information combined with the quantitative information about individual trajectories that can be obtained on a computer is generally as close as we can come to solving a nonlinear system of differential equations; but, in a sense, this information is better than obtaining a formula for the solution since it geometrically describes the behavior of every solution for all time.

3.1 Dynamical Systems and Global Existence Theorems

A dynamical system gives a functional description of the solution of a physical problem or of the mathematical model describing the physical problem. For example, the motion of the undamped pendulum discussed in Section 2.14 of Chapter 2 is a dynamical system in the sense that the motion of the pendulum is described by its position and velocity as functions of time and the initial conditions.

Mathematically speaking, a dynamical system is a function $\phi(t, \mathbf{x})$, defined for all $t \in \mathbf{R}$ and $\mathbf{x} \in E \subset \mathbf{R}^n$, which describes how points $\mathbf{x} \in E$ move with respect to time. We require that the family of maps $\phi_t(\mathbf{x}) = \phi(t, \mathbf{x})$ have the properties of a flow defined in Section 2.5 of Chapter 2.

Definition 1. A *dynamical system* on E is a C^1-map

$$\phi: \mathbf{R} \times E \to E$$

where E is an open subset of $\mathbf{R}^n$ and if $\phi_t(\mathbf{x}) = \phi(t, \mathbf{x})$, then ϕ_t satisfies

(i) $\phi_0(\mathbf{x}) = \mathbf{x}$ for all $\mathbf{x} \in E$ and

(ii) $\phi_t \circ \phi_s(\mathbf{x}) = \phi_{t+s}(\mathbf{x})$ for all $s, t \in \mathbf{R}$ and $\mathbf{x} \in E$.

Remark 1. It follows from Definition 1 that for each $t \in \mathbf{R}$, ϕ_t is a C^1-map of E into E which has a C^1-inverse, ϕ_{-t}; i.e., ϕ_t with $t \in \mathbf{R}$ is a one-parameter family of diffeomorphisms on E that forms a commutative group under composition.

It is easy to see that if A is an $n \times n$ matrix then the function $\phi(t, \mathbf{x}) = e^{At}\mathbf{x}$ defines a dynamical system on $\mathbf{R}^n$ and also, for each $\mathbf{x}_0 \in \mathbf{R}^n$, $\phi(t, \mathbf{x}_0)$ is the solution of the initial value problem

$$\dot{\mathbf{x}} = A\mathbf{x}$$
$$\mathbf{x}(0) = \mathbf{x}_0.$$

3.1. Dynamical Systems and Global Existence Theorems

In general, if $\phi(t, \mathbf{x})$ is a dynamical system on $E \subset \mathbf{R}^n$, then the function

$$\mathbf{f}(\mathbf{x}) = \frac{d}{dt}\phi(t,\mathbf{x})|_{t=0}$$

defines a C^1-vector field on E and for each $\mathbf{x}_0 \in E$, $\phi(t, \mathbf{x}_0)$ is the solution of the initial value problem

$$\dot{\mathbf{x}} = \mathbf{f}(\mathbf{x})$$
$$\mathbf{x}(0) = \mathbf{x}_0.$$

Furthermore, for each $\mathbf{x}_0 \in E$, the maximal interval of existence of $\phi(t, \mathbf{x}_0)$, $I(\mathbf{x}_0) = (-\infty, \infty)$. Thus, each dynamical system gives rise to a C^1-vector field $\mathbf{f}$ and the dynamical system describes the solution set of the differential equation defined by this vector field. Conversely, given a differential equation (1) with $\mathbf{f} \in C^1(E)$ and E an open subset of $\mathbf{R}^n$, the solution $\phi(t, \mathbf{x}_0)$ of the initial value problem (1) with $\mathbf{x}_0 \in E$ will be a dynamical system on E if and only if for all $\mathbf{x}_0 \in E$, $\phi(t, \mathbf{x}_0)$ is defined for all $t \in \mathbf{R}$; i.e., if and only if for all $\mathbf{x}_0 \in E$, the maximal interval of existence $I(\mathbf{x}_0)$ of $\phi(t, \mathbf{x}_0)$ is $(-\infty, \infty)$. In this case we say that $\phi(t, \mathbf{x}_0)$ is *the dynamical system on E defined by* (1).

The next theorem shows that any C^1-vector field $\mathbf{f}$ defined on all of $\mathbf{R}^n$ leads to a dynamical system on $\mathbf{R}^n$. While the solutions $\phi(t, \mathbf{x}_0)$ of the original system (1) may not be defined for all $t \in \mathbf{R}$, the time t can be rescaled along trajectories of (1) to obtain a topologically equivalent system for which the solutions are defined for all $t \in \mathbf{R}$.

Before stating this theorem, we generalize the notion of topological equivalent systems defined in Section 2.8 of Chapter 2 for a neighborhood of the origin.

Definition 2. Suppose that $\mathbf{f} \in C^1(E_1)$ and $\mathbf{g} \in C^1(E_2)$ where E_1 and E_2 are open subsets of $\mathbf{R}^n$. Then the two autonomous systems of differential equations

$$\dot{\mathbf{x}} = \mathbf{f}(\mathbf{x}) \quad (1)$$

and

$$\dot{\mathbf{x}} = \mathbf{g}(\mathbf{x}) \quad (2)$$

are said to be *topologically equivalent* if there is a homeomorphism $H: E_1 \to E_2$ which maps trajectories of (1) onto trajectories of (2) and preserves their orientation by time. In this case, the vector fields $\mathbf{f}$ and $\mathbf{g}$ are also said to be topologically equivalent. If $E = E_1 = E_2$ then the systems (1) and (2) are said to be *topologically equivalent on E* and the vector fields $\mathbf{f}$ and $\mathbf{g}$ are said to be topologically equivalent on E.

Remark 2. Note that while the homeomorphism H in this definition preserves the orientation of trajectory by time and gives us a continuous deformation of the phase portrait of (1) in the phase space E_1 onto the phase portrait of (2) in the phase space E_2, it need not preserve the parameterization by time along the trajectories. In fact, if ϕ_t is the flow on E_1 defined by (1) and we assume that (2) defines a dynamical system ψ_t on E_2, then the systems (1) and (2) are topologically equivalent if and only if there is a homeomorphism $H\colon E_1 \looparrowright E_2$ and for each $\mathbf{x} \in E_1$ there is a continuously differentiable function $t(\mathbf{x}, \tau)$ defined for all $\tau \in \mathbf{R}$ such that $\partial t/\partial \tau > 0$ and

$$H \circ \phi_{t(\mathbf{x},\tau)}(\mathbf{x}) = \psi_\tau \circ H(\mathbf{x})$$

for all $\mathbf{x} \in E_1$ and $\tau \in \mathbf{R}$. In general, two autonomous systems are topologically equivalent on E if and only if they are both topologically equivalent to some autonomous system of differential equations defining a dynamical system on E (cf. Theorem 2 below). As was noted in the above definitions, if the two systems (1) and (2) are topologically equivalent, then the vector fields $\mathbf{f}$ and $\mathbf{g}$ are said to be topologically equivalent; on the other hand, if the homeomorphism H does preserve the parameterization by time, then the vector fields $\mathbf{f}$ and $\mathbf{g}$ are said to be *topologically conjugate*. Clearly, if two vector fields $\mathbf{f}$ and $\mathbf{g}$ are topologically conjugate, then they are topologically equivalent.

Theorem 1 (Global Existence Theorem). *For* $\mathbf{f} \in C^1(\mathbf{R}^n)$ *and for each* $\mathbf{x}_0 \in \mathbf{R}^n$, *the initial value problem*

$$\dot{\mathbf{x}} = \frac{\mathbf{f}(\mathbf{x})}{1 + |\mathbf{f}(\mathbf{x})|}$$

$$\mathbf{x}(0) = \mathbf{x}_0 \tag{3}$$

has a unique solution $\mathbf{x}(t)$ *defined for all* $t \in \mathbf{R}$, *i.e.*, *(3) defines a dynamical system on* $\mathbf{R}^n$; *furthermore, (3) is topologically equivalent to (1) on* $\mathbf{R}^n$.

Remark 3. The systems (1) and (3) are topologically equivalent on $\mathbf{R}^n$ since the time t along the solutions $\mathbf{x}(t)$ of (1) has simply been rescaled according to the formula

$$\tau = \int_0^t [1 + |\mathbf{f}(\mathbf{x}(s))|]\, ds; \tag{4}$$

i.e., the homeomorphism H in Definition 2 is simply the identity on $\mathbf{R}^n$. The solution $\mathbf{x}(t)$ of (1), with respect to the new time τ, then satisfies

$$\frac{d\mathbf{x}}{d\tau} = \frac{d\mathbf{x}}{dt} \bigg/ \frac{d\tau}{dt} = \frac{\mathbf{f}(\mathbf{x})}{1 + |\mathbf{f}(\mathbf{x})|};$$

i.e., $\mathbf{x}(t(\tau))$ is the solution of (3) where $t(\tau)$ is the inverse of the strictly increasing function $\tau(t)$ defined by (4). The function $\tau(t)$ maps the maximal

3.1. Dynamical Systems and Global Existence Theorems

interval of existence (α, β) of the solution $\mathbf{x}(t)$ of (1) one-to-one and onto $(-\infty, \infty)$, the maximal interval of existence of (3).

Proof (of Theorem 1). It is not difficult to show that if $\mathbf{f} \in C^1(\mathbf{R}^n)$ then the function
$$\frac{\mathbf{f}}{1+|\mathbf{f}|} \in C^1(\mathbf{R}^n);$$

cf. Problem 3 at the end of this section. For $\mathbf{x}_0 \in \mathbf{R}^n$, let $\mathbf{x}(t)$ be the solution of the initial value problem (3) on its maximal interval of existence (α, β). Then by Problem 6 in Section 2.2, $\mathbf{x}(t)$ satisfies the integral equation

$$\mathbf{x}(t) = \mathbf{x}_0 + \int_0^t \frac{\mathbf{f}(\mathbf{x}(s))}{1+|\mathbf{f}(\mathbf{x}(s))|} ds$$

for all $t \in (\alpha, \beta)$ and since $|\mathbf{f}(\mathbf{x})|/(1+|\mathbf{f}(\mathbf{x})|) \leq 1$, it follows that

$$|\mathbf{x}(t)| \leq |\mathbf{x}_0| + \int_0^{|t|} ds = |\mathbf{x}_0| + |t|$$

for all $t \in (\alpha, \beta)$. Suppose that $\beta < \infty$. Then

$$|\mathbf{x}(t)| \leq |\mathbf{x}_0| + \beta$$

for all $t \in [0, \beta)$; i.e., for all $t \in [0, \beta)$, the solution of (3) through the point $\mathbf{x}_0$ at time $t = 0$ is contained in the compact set

$$K = \{\mathbf{x} \in \mathbf{R}^n \mid |\mathbf{x}| \leq |\mathbf{x}_0| + \beta\} \subset \mathbf{R}^n.$$

But then, by Corollary 2 in Section 2.4 of Chapter 2, $\beta = \infty$, a contradiction. Therefore, $\beta = \infty$. A similar proof shows that $\alpha = -\infty$. Thus, for all $\mathbf{x}_0 \in \mathbf{R}^n$, the maximal interval of existence of the solution $\mathbf{x}(t)$ of the initial value problem (3), $(\alpha, \beta) = (-\infty, \infty)$.

Example 1. As in Problem 1(a) in Problem Set 4 of Chapter 2, the maximal interval of existence of the solution

$$x(t) = \frac{x_0}{1 - x_0 t}$$

of the initial value problem

$$\dot{x} = x^2$$
$$x(0) = x_0$$

is $(-\infty, 1/x_0)$ for $x_0 > 0$, $(1/x_0, \infty)$ for $x_0 < 0$ and $(-\infty, \infty)$ for $x_0 = 0$. The phase portrait in $\mathbf{R}^1$ is simply

The related initial value problem

$$\dot{x} = \frac{x^2}{1+x^2}$$
$$x(0) = x_0$$

has a unique solution $x(t)$ defined on $(-\infty, \infty)$ which is given by

$$2x(t) = t + x_0 - \frac{1}{x_0} + \frac{x_0}{|x_0|}\sqrt{t^2 + 2\left(x_0 - \frac{1}{x_0}\right)t + \left(x_0 + \frac{1}{x_0}\right)^2}$$

for $x_0 \neq 0$ and $x(t) \equiv 0$ for $x_0 = 0$. It follows that for $x_0 > 0$, $x(t) \to \infty$ as $t \to \infty$ and $x(t) \to 0$ as $t \to -\infty$; and for $x_0 < 0$, $x(t) \to 0$ as $t \to \infty$ and $x(t) \to -\infty$ as $t \to -\infty$. The phase portrait for the second system above is exactly the same as the phase portrait for the first system. In this example, the function $\tau(t)$ defined by (4) is given by

$$\tau(t) = t + \frac{x_0^2 t}{1 - x_0 t}.$$

For $x_0 = 0$, $\tau(t) = t$; for $x_0 > 0$, $\tau(t)$ maps the maximal interval $(-\infty, 1/x_0)$ one-to-one and onto $(-\infty, \infty)$; and for $x_0 < 0$, $\tau(t)$ maps the maximal interval $(1/x_0, \infty)$ one-to-one and onto $(-\infty, \infty)$.

If $f \in C^1(E)$ with E a proper subset of $\mathbf{R}^n$, the above normalization will not, in general, lead to a dynamical system as the next example shows.

Example 2. For $x_0 > 0$ the initial value problem

$$\dot{x} = \frac{1}{2x}$$
$$x(0) = x_0$$

has the unique solution $x(t) = \sqrt{t + x_0^2}$ defined on its maximal interval of existence $I(x_0) = (-x_0^2, \infty)$. The function $f(x) = 1/(2x) \in C^1(E)$ where $E = (0, \infty)$. We have $x(t) \to 0 \in \dot{E}$ as $t \to -x_0^2$. The related initial value problem

$$\dot{x} = \frac{1/2x}{1 + (1/2x)} = \frac{1}{2x + 1}$$
$$x(0) = x_0$$

has the unique solution

$$x(t) = -\frac{1}{2} + \sqrt{t + (x_0 + 1/2)^2}$$

defined on its maximal interval of existence $I(x_0) = (-(x_0 + 1/2)^2, \infty)$. We see that in this case $I(x_0) \neq \mathbf{R}$.

3.1. Dynamical Systems and Global Existence Theorems

However, a slightly more subtle rescaling of the time along trajectories of (1) does lead to a dynamical system equivalent to (1) even when E is a proper subset of $\mathbf{R}^n$. This idea is due to Vinograd; cf. [N/S].

Theorem 2. *Suppose that E is an open subset of $\mathbf{R}^n$ and that $\mathbf{f} \in C^1(E)$. Then there is a function $\mathbf{F} \in C^1(E)$ such that*

$$\dot{\mathbf{x}} = \mathbf{F}(\mathbf{x}) \tag{5}$$

defines a dynamical system on E and such that (5) is topologically equivalent to (1) on E.

Proof. First of all, as in Theorem 1, the function

$$\mathbf{g}(\mathbf{x}) = \frac{\mathbf{f}(\mathbf{x})}{1 + |\mathbf{f}(\mathbf{x})|} \in C^1(E),$$

$|\mathbf{g}(\mathbf{x})| \leq 1$ and the systems (1) and (3) are topologically equivalent on E. Furthermore, solutions $\mathbf{x}(t)$ of (3) satisfy

$$\int_0^t |\dot{\mathbf{x}}(t')| \, dt' = \int_0^t |\mathbf{g}(\mathbf{x}(t'))| \, dt' \leq |t|;$$

i.e., for finite t, the trajectory defined by $\mathbf{x}(t)$ has finite arc length. Let (α, β) be the maximal interval of existence of $\mathbf{x}(t)$ and suppose that $\beta < \infty$. Then since the arc length of the half-trajectory defined by $\mathbf{x}(t)$ for $t \in (0, \beta)$ is finite, the half-trajectory defined by $\mathbf{x}(t)$ for $t \in [0, \beta)$ must have a limit point

$$\mathbf{x}_1 = \lim_{t \to \beta^-} \mathbf{x}(t) \in \dot{E}$$

(unlike Example 3 in Section 2.4 of Chapter 2). Cf. Corollary 1 and Problem 3 in Section 2.4 of Chapter 2. Now define the closed set $K = \mathbf{R}^n \sim E$ and let

$$G(\mathbf{x}) = \frac{d(\mathbf{x}, K)}{1 + d(\mathbf{x}, K)}$$

where $d(\mathbf{x}, \mathbf{y})$ denotes the distance between $\mathbf{x}$ and $\mathbf{y}$ in $\mathbf{R}^n$ and

$$d(\mathbf{x}, K) = \inf_{\in K} d(\mathbf{x}, \mathbf{y});$$

i.e., for $\mathbf{x} \in E$, $d(\mathbf{x}, K)$ is the distance of $\mathbf{x}$ from the boundary $\dot{E}$ of E. Then the function $G \in C^1(\mathbf{R}^n)$, $0 \leq G(\mathbf{x}) \leq 1$ and for $\mathbf{x} \in K$, $G(\mathbf{x}) = 0$. Let $\mathbf{F}(\mathbf{x}) = \mathbf{g}(\mathbf{x})G(\mathbf{x})$. Then $\mathbf{F} \in C^1(E)$ and the system (5), $\dot{\mathbf{x}} = \mathbf{F}(\mathbf{x})$, is topologically equivalent to (3) on E since we have simply rescaled the time along trajectories of (3); i.e., the homeomorphism H in Definition 2 is simply the identity on E. Furthermore, the system (5) has a bounded right-hand side and therefore its trajectories have finite arc-length for finite

t. To prove that (5) defines a dynamical system on E, it suffices to show that all half-trajectories of (5) which (a) start in E, (b) have finite arc length s_0, and (c) terminate at a limit point $\mathbf{x}_1 \in \dot{E}$ are defined for all $t \in [0, \infty)$. Along any solution $\mathbf{x}(t)$ of (5), $\frac{ds}{dt} = |\dot{\mathbf{x}}(t)|$ and hence

$$t = \int_0^s \frac{ds'}{|\mathbf{F}(\mathbf{x}(t(s')))|}$$

where $t(s)$ is the inverse of the strictly increasing function $s(t)$ defined by

$$s = \int_0^t |\mathbf{F}(\mathbf{x}(t'))| dt'$$

for $s > 0$. But for each point $\mathbf{x} = \mathbf{x}(t(s))$ on the half-trajectory we have

$$G(\mathbf{x}) = \frac{d(\mathbf{x}, K)}{1 + d(\mathbf{x}, K)} < d(\mathbf{x}, K) = \inf_{\mathbf{y} \in K} d(\mathbf{x}, \mathbf{y}) \leq d(\mathbf{x}, \mathbf{x}_1) \leq s_0 - s.$$

And therefore since $0 < |g(\mathbf{x})| \leq 1$, we have

$$t \geq \int_0^s \frac{ds'}{s_0 - s'} = \log \frac{s_0 - s}{s_0}$$

and hence $t \to \infty$ as $s \to s_0$; i.e., the half-trajectory defined by $\mathbf{x}(t)$ is defined for all $t \in [0, \infty)$; i.e., $\beta = \infty$. Similarly, it can be shown that $\alpha = -\infty$ and hence, the system (5) defines a dynamical system on E which is topologically equivalent to (1) on E.

For $\mathbf{f} \in C^1(E)$, E an open subset of $\mathbf{R}^n$, Theorem 2 implies that there is no loss in generality in assuming that the system (1) defines a dynamical system $\phi(t, \mathbf{x}_0)$ on E. Throughout the remainder of this book we therefore make this assumption; i.e., we assume that for all $\mathbf{x}_0 \in E$, the maximal interval of existence $I(\mathbf{x}_0) = (-\infty, \infty)$. In the next section, we then go on to discuss the limit sets of trajectories $\mathbf{x}(t)$ of (1) as $t \to \pm\infty$. However, we first present two more global existence theorems which are of some interest.

Theorem 3. *Suppose that* $\mathbf{f} \in C^1(\mathbf{R}^n)$ *and that* $\mathbf{f}(\mathbf{x})$ *satisfies the global Lipschitz condition*

$$|\mathbf{f}(\mathbf{x})| - \mathbf{f}(\mathbf{y})| \leq M|\mathbf{x} - \mathbf{y}|$$

for all $\mathbf{x}, \mathbf{y} \in \mathbf{R}^n$. *Then for* $\mathbf{x}_0 \in \mathbf{R}^n$, *the initial value problem* (1) *has a unique solution* $\mathbf{x}(t)$ *defined for all* $t \in \mathbf{R}$.

Proof. Let $\mathbf{x}(t)$ be the solution of the initial value problem (1) on its maximal interval of existence (α, β). Then using the fact that $d|\mathbf{x}(t)|/dt \leq |\dot{\mathbf{x}}(t)|$ and the triangle inequality,

$$\frac{d}{dt}|\mathbf{x}(t) - \mathbf{x}_0| \leq |\dot{\mathbf{x}}(t)| = |\mathbf{f}(\mathbf{x}(t))|$$
$$\leq |\mathbf{f}(\mathbf{x}(t)) - \mathbf{f}(\mathbf{x}_0)| + |\mathbf{f}(\mathbf{x}_0)|$$
$$\leq M|\mathbf{x}(t) - \mathbf{x}_0| + |\mathbf{f}(\mathbf{x}_0)|.$$

3.1. Dynamical Systems and Global Existence Theorems

Thus, if we assume that $\beta < \infty$, then the function $g(t) = |\mathbf{x}(t) - \mathbf{x}_0|$ satisfies

$$g(t) = \int_0^t \frac{dg(s)}{ds}\,ds \leq |\mathbf{f}(\mathbf{x}_0)|\beta + M\int_0^t g(s)\,ds$$

for all $t \in [0, \beta)$. It then follows from Gronwall's Lemma in Section 2.3 of Chapter 2 that

$$|\mathbf{x}(t) - \mathbf{x}_0| \leq \beta |\mathbf{f}(\mathbf{x}_0)| e^{M\beta}$$

for all $t \in [0, \beta)$; i.e., the trajectory of (1) through the point $\mathbf{x}_0$ at time $t = 0$ is contained in the compact set

$$K = \{\mathbf{x} \in \mathbf{R}^n \mid |\mathbf{x} - \mathbf{x}_0| \leq \beta|\mathbf{f}(\mathbf{x}_0)|e^{M\beta}\} \subset \mathbf{R}^n.$$

But then by Corollary 2 in Section 2.4 of Chapter 2, it follows that $\beta = \infty$, a contradiction. Therefore, $\beta = \infty$ and it can similarly be shown that $\alpha = -\infty$. Thus, for all $\mathbf{x}_0 \in \mathbf{R}^n$, the maximal interval of existence of the solution $\mathbf{x}(t)$ of the initial value problem (1), $I(\mathbf{x}_0) = (-\infty, \infty)$.

If $\mathbf{f} \in C^1(M)$ where M is a compact subset of $\mathbf{R}^n$, then $\mathbf{f}$ satisfies a global Lipschitz condition on M and we have a result similar to the above theorem for $\mathbf{x}_0 \in M$. This result has been extended to compact manifolds by Chillingworth; cf. [G/H], p. 7. A C^1-vector field on a manifold M is defined at the end of Section 3.10.

Theorem 4 (Chillingworth). *Let M be a compact manifold and let $\mathbf{f} \in C^1(M)$. Then for $\mathbf{x}_0 \in M$, the initial value problem (1) has a unique solution $\mathbf{x}(t)$ defined for all $t \in \mathbf{R}$.*

Problem Set 1

1. If $\mathbf{f}(\mathbf{x}) = A\mathbf{x}$ with
   $$A = \begin{bmatrix} -1 & -3 \\ 0 & 2 \end{bmatrix}$$
   find the dynamical system defined by the initial value problem (1).

2. If $f(x) = x^2$, find the dynamical system defined by the initial value problem (3), and show that it agrees with the result in Example 1.

3. (a) Show that if E is an open subset of $\mathbf{R}$ and $f \in C^1(E)$ then the function
   $$F(x) = \frac{f(x)}{1 + |f(x)|}$$
   satisfies $F \in C^1(E)$.
   Hint: Show that if $f(x) \neq 0$ at $x \in E$ then
   $$F'(x) = \frac{f'(x)}{(1 + |f(x)|)^2}$$

and that if for $x_0 \in E$, $f(x_0) = 0$ then $F'(x_0) = f'(x_0)$ and that $\lim_{x \to x_0} F'(x) = F'(x_0)$.

(b) Extend the results of part (a) to $\mathbf{f} \in C^1(E)$ for E an open subset of $\mathbf{R}^n$.

4. Show that the function

$$f(x) = \frac{1}{1+x^2}$$

satisfies a global Lipschitz condition on $\mathbf{R}$ and find the dynamical system defined by the initial value problem (1) for this function.

5. Another way to rescale the time along trajectories of (1) is to define

$$\tau = \int_0^t [1 + |\mathbf{f}(\mathbf{x}(s))|^2] \, ds$$

This leads to the initial value problem

$$\dot{\mathbf{x}} = \frac{\mathbf{f}(\mathbf{x})}{1 + |\mathbf{f}(\mathbf{x})|^2}$$

for the function $\mathbf{x}(t(\tau))$. Prove a result analogous to Theorem 1 for this system.

6. Two vector fields $\mathbf{f}, \mathbf{g} \in C^k(\mathbf{R}^n)$ are said to be C^k-*equivalent on* $\mathbf{R}^n$ if there is a homeomorphism H of $\mathbf{R}^n$ with $H, H^{-1} \in C^k(\mathbf{R}^n)$ which maps trajectories of (1) onto trajectories of (2) and preserves their orientation by time; H is called a C^k-diffeomorphism on $\mathbf{R}^n$. If ϕ_t and ψ_t are dynamical systems on $\mathbf{R}^n$ defined by (1) and (2) respectively, then $\mathbf{f}$ and $\mathbf{g}$ are C^k-equivalent on $\mathbf{R}^n$ if and only if there exists a C^k-diffeomorphism on $\mathbf{R}^n$ and for each $\mathbf{x} \in \mathbf{R}^n$ there exists a strictly increasing C^k-function $\tau(\mathbf{x}, t)$ such that $\partial \tau / \partial t > 0$ and

$$H \circ \phi_t(\mathbf{x}) = \psi_{\tau(\mathbf{x},t)} \circ H(\mathbf{x}) \qquad (*)$$

for all $x \in \mathbf{R}^n$ and $t \in \mathbf{R}$. If $\mathbf{f}$ and $\mathbf{g}$ are C^1-equivalent on $\mathbf{R}^n$,

(a) prove that equilibrium points of (1) are mapped to equilibrium points of (2) under H and

(b) prove that periodic orbits of (1) are mapped onto periodic orbits of (2) and that if t_0 is the period of a periodic orbit $\phi_t(\mathbf{x}_0)$ of (1) then $\tau_0 = \tau(\mathbf{x}_0, t_0)$ is the period of the periodic orbit $\psi_\tau(H(\mathbf{x}_0))$ of (2).

Hint: In order to prove (a), differentiate $(*)$ with respect to t, using the chain rule, and show that if $\mathbf{f}(\mathbf{x}_0) = 0$ then $\mathbf{g}(\psi_\tau(H(\mathbf{x}_0))) = 0$ for all $\tau \in \mathbf{R}$; i.e., $\psi_\tau(H(\mathbf{x}_0)) = H(\mathbf{x}_0)$ for all $\tau \in \mathbf{R}$.

3.2. Limit Sets and Attractors

7. If **f** and **g** are C^2 equivalent on $\mathbf{R}^n$, prove that at an equilibrium point $\mathbf{x}_0$ of (1), the eigenvalues of $D\mathbf{f}(\mathbf{x}_0)$ and the eigenvalues of $D\mathbf{g}(H(\mathbf{x}_0))$ differ by the positive multiplicative constant $k_0 = \frac{\partial \tau}{\partial t}(\mathbf{x}_0, 0)$.

   Hint: Differentiate (*) twice, first with respect to t and then, after setting $t = 0$, with respect to $\mathbf{x}$ and then show that $D\mathbf{f}(\mathbf{x}_0)$ and $k_0 D\mathbf{g}(H(\mathbf{x}_0))$ are similar.

8. Two vector fields, $\mathbf{f}, \mathbf{g} \in C^k(\mathbf{R}^n)$ are said to be C^k-*conjugate on* $\mathbf{R}^n$ if there is a C^k-diffeomorphism of $\mathbf{R}^n$ which maps trajectories of (1) onto trajectories of (2) and preserves the parameterization by time. If ϕ_t and ψ_t are the dynamical systems on $\mathbf{R}^n$ defined by (1) and (2) respectively, then **f** and **g** are C^k-conjugate on $\mathbf{R}^n$ if and only if there exists a C_k-diffeomorphism H such that
$$H \circ \phi_t = \psi_t \circ H$$
on $\mathbf{R}^n$. Prove that if **f** and **g** are C^2-conjugate on $\mathbf{R}^n$ then at an equilibrium point $\mathbf{x}_0$ of (1), the eigenvalues of $D\mathbf{f}(\mathbf{x}_0)$ and $D\mathbf{g}(H(\mathbf{x}_0))$ are equal.

9. (Discrete Dynamical Systems). If ϕ_t is a dynamical system on an open set $E \subset \mathbf{R}^n$, then the mapping $\mathbf{F} \colon E \to E$ defined by $\mathbf{F}(\mathbf{x}) = \phi_\tau(\mathbf{x})$, where τ is any fixed time in $\mathbf{R}$, is a diffeomorphism on E which defines a discrete dynamical system. The discrete dynamical system consists of the iterates of $\mathbf{F}$; i.e., for each $\mathbf{x} \in E$ we get a sequence of points $\mathbf{F}(\mathbf{x}), \mathbf{F}^2(\mathbf{x}), \mathbf{F}^3(\mathbf{x}), \ldots$, in E. Iterates of the Poincaré map $\mathbf{P}$, discussed in Section 3.4, also define a discrete dynamical system. In fact, any map $\mathbf{F} \colon E \to E$ which is a diffeomorphism on E defines a discrete dynamical system $\{\mathbf{F}^n\}$, $n = 0, \pm 1, \pm 2, \ldots$, on E. In this context, show that for $\mu \neq 0$ the mapping $\mathbf{F}(x, y) = (y, \mu x + y - y^3)$ is a diffeomorphism and find its inverse. For $\mu > 0$ show that $(\sqrt{\mu}, \sqrt{\mu})$ is a fixed point of $\mathbf{F}$; i.e., show that $\mathbf{F}(\sqrt{\mu}, \sqrt{\mu}) = (\sqrt{\mu}, \sqrt{\mu})$.

3.2 Limit Sets and Attractors

Consider the autonomous system
$$\dot{\mathbf{x}} = \mathbf{f}(\mathbf{x}) \tag{1}$$
with $\mathbf{f} \in C^1(E)$ where E is an open subset of $\mathbf{R}^n$. In Section 3.1 we saw that there is no loss in generality in assuming that the system (1) defines a dynamical system $\phi(t, \mathbf{x})$ on E. For $\mathbf{x} \in E$, the function $\phi(\cdot, \mathbf{x}) \colon \mathbf{R} \to E$ defines a *solution curve, trajectory,* or *orbit* of (1) through the point $\mathbf{x}_0$ in E. If we identify the function $\phi(\cdot, \mathbf{x})$ with its graph, we can think of a trajectory through the point $\mathbf{x}_0 \in E$ as a motion along the curve
$$\Gamma_{\mathbf{x}_0} = \{\mathbf{x} \in E \mid \mathbf{x} = \phi(t, \mathbf{x}_0), t \in \mathbf{R}\}$$

defined by (1). We shall also refer to $\Gamma_{\mathbf{x}_0}$ as the trajectory of (1) through the point $\mathbf{x}_0$ at time $t = 0$. If the point $\mathbf{x}_0$ plays no role in the discussion, we simply denote the trajectory by Γ and draw the curve Γ in the subset E of the phase space $\mathbf{R}^n$ with an arrow indicating the direction of the motion along Γ with increasing time. Cf. Figure 1. By the *positive half-trajectory* through the point $\mathbf{x}_0 \in E$, we mean the motion along the curve

$$\Gamma_{\mathbf{x}_0}^+ = \{\mathbf{x} \in E \mid \mathbf{x} = \phi(t, \mathbf{x}_0), t \geq 0\}$$

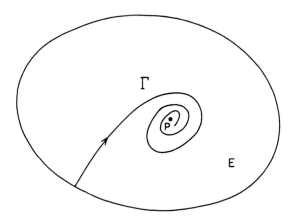

Figure 1. A trajectory Γ of (1) which approaches the ω-limit point $\mathbf{p} \in E$ as $t \to \infty$.

$\Gamma_{\mathbf{x}_0}^-$, is similarly defined. Any trajectory $\Gamma = \Gamma^+ \cup \Gamma^-$.

Definition 1. A point $\mathbf{p} \in E$ is an ω-*limit point* of the trajectory $\phi(\cdot, \mathbf{x})$ of the system (1) if there is a sequence $t_n \to \infty$ such that

$$\lim_{n \to \infty} \phi(t_n, \mathbf{x}) = \mathbf{p}.$$

Similarly, if there is a sequence $t_n \to -\infty$ such that

$$\lim_{n \to \infty} \phi(t_n, \mathbf{x}) = \mathbf{q},$$

and the point $\mathbf{q} \in E$, then the point $\mathbf{q}$ is called an α-*limit point* of the trajectory $\phi(\cdot, \mathbf{x})$ of (1). The set of all ω-limit points of a trajectory Γ is called the ω-*limit set of* Γ and it is denoted by $\omega(\Gamma)$. The set of all α-limit points of a trajectory Γ is called the α-*limit set of* Γ and it is denoted by $\alpha(\Gamma)$. The set of all limit points of Γ, $\alpha(\Gamma) \cup \omega(\Gamma)$ is called the *limit set of* Γ.

3.2. Limit Sets and Attractors

Theorem 1. *The α and ω-limit sets of a trajectory Γ of (1), $\alpha(\Gamma)$ and $\omega(\Gamma)$, are closed subsets of E and if Γ is contained in a compact subset of $\mathbf{R}^n$, then $\alpha(\Gamma)$ and $\omega(\Gamma)$, are non-empty, connected, compact subsets of E.*

Proof. It follows from Definition 1 that $\omega(\Gamma) \subset E$. In order to show that $\omega(\Gamma)$ is a closed subset of E, we let $\mathbf{p}_n$ be a sequence of points in $\omega(\Gamma)$ with $\mathbf{p}_n \to \mathbf{p} \in \mathbf{R}^n$ and show that $\mathbf{p} \in \omega(\Gamma)$. Let $\mathbf{x}_0 \in \Gamma$. Then since $\mathbf{p}_n \in \omega(\Gamma)$, it follows that for each $n = 1, 2, \ldots$, there is a sequence $t_k^{(n)} \to \infty$ as $k \to \infty$ such that

$$\lim_{k \to \infty} \phi(t_k^{(n)}, \mathbf{x}_0) = \mathbf{p}_n.$$

Furthermore, we may assume that $t_k^{(n+1)} > t_k^{(n)}$ since otherwise we can choose a subsequence of $t_k^{(n)}$ with this property. The above equation implies that for all $n \geq 2$, there is a sequence of integers $K(n) > K(n-1)$ such that for $k \geq K(n)$,

$$|\phi(t_k^{(n)}, \mathbf{x}_0) - \mathbf{p}_n| < \frac{1}{n}.$$

Let $t_n = t_{K(n)}^{(n)}$. Then $t_n \to \infty$ and by the triangle inequality,

$$|\phi(t_n, \mathbf{x}_0) - \mathbf{p}| \leq |\phi(t_n, \mathbf{x}_0) - \mathbf{p}_n| + |\mathbf{p}_n - \mathbf{p}| \leq \frac{1}{n} + |\mathbf{p}_n - \mathbf{p}| \to 0$$

as $n \to \infty$. Thus $\mathbf{p} \in \omega(\Gamma)$.

If $\Gamma \subset K$, a compact subset of $\mathbf{R}^n$, and $\phi(t_n, \mathbf{x}_0) \to \mathbf{p} \in \omega(\Gamma)$, then $\mathbf{p} \in K$ since $\phi(t_n, \mathbf{x}_0) \in \Gamma \subset K$ and K is compact. Thus, $\omega(\Gamma) \subset K$ and therefore $\omega(\Gamma)$ is compact since a closed subset of a compact set is compact. Furthermore, $\omega(\Gamma) \neq 0$ since the sequence of points $\phi(n, \mathbf{x}_0) \in K$ contains a convergent subsequence which converges to a point in $\omega(\Gamma) \subset K$. Finally, suppose that $\omega(\Gamma)$ is not connected. Then there exist two nonempty, disjoint, closed sets A and B such that $\omega(\Gamma) = A \cup B$. Since A and B are both bounded, they are a finite distance δ apart where the distance from A to B

$$d(A, B) = \inf_{\mathbf{x} \in A, \mathbf{y} \in B} |\mathbf{x} - \mathbf{y}|.$$

Since the points of A and B are ω-limit points of Γ, there exists arbitrarily large t such that $\phi(t, \mathbf{x}_0)$ are within $\delta/2$ of A and there exists arbitrarily large t such that the distance of $\phi(t, \mathbf{x}_0)$ from A is greater than $\delta/2$. Since the distance $d(\phi(t, \mathbf{x}_0), A)$ of $\phi(t, \mathbf{x}_0)$ from A is a continuous function of t, it follows that there must exist a sequence $t_n \to \infty$ such that $d(\phi(t_n, \mathbf{x}_0), A) = \delta/2$. Since $\{\phi(t_n, \mathbf{x}_0)\} \subset K$ there is a subsequence converging to a point $\mathbf{p} \in \omega(\Gamma)$ with $d(\mathbf{p}, A) = \delta/2$. But, then $d(\mathbf{p}, B) \geq d(A, B) - d(\mathbf{p}, A) = \delta/2$ which implies that $\mathbf{p} \notin A$ and $\mathbf{p} \notin B$; i.e., $\mathbf{p} \notin \omega(\Gamma)$, a contradiction. Thus, $\omega(\Gamma)$ is connected. A similar proof serves to establish these same results for $\alpha(\Gamma)$.

Theorem 2. *If* **p** *is an ω-limit point of a trajectory Γ of* (1), *then all other points of the trajectory $\phi(\cdot, \mathbf{p})$ of* (1) *through the point* **p** *are also ω-limit points of Γ; i.e., if* $\mathbf{p} \in \omega(\Gamma)$ *then* $\Gamma_{\mathbf{p}} \subset \omega(\Gamma)$ *and similarly if* $\mathbf{p} \in \alpha(\Gamma)$ *then* $\Gamma_{\mathbf{p}} \subset \alpha(\Gamma)$.

Proof. Let $\mathbf{p} \in \omega(\Gamma)$ where Γ is the trajectory $\phi(\cdot, \mathbf{x}_0)$ of (1) through the point $\mathbf{x}_0 \in E$. Let **q** be a point on the trajectory $\phi(\cdot, \mathbf{p})$ of (1) through the point **p**; i.e., $\mathbf{q} = \phi(\tilde{t}, \mathbf{p})$ for some $\tilde{t} \in \mathbf{R}$. Since **p** is an ω-limit point of the trajectory $\phi(\cdot, \mathbf{x}_0)$, there is a sequence $t_n \to \infty$ such that $\phi(t_n, \mathbf{x}_0) \to \mathbf{p}$. Thus by Theorem 1 in Section 2.3 of Chapter 2 (on continuity with respect to initial conditions) and property (ii) of dynamical systems,

$$\phi(t_n + \tilde{t}, \mathbf{x}_0) = \phi(\tilde{t}, \phi(t_n, \mathbf{x}_0)) \to \phi(\tilde{t}, \mathbf{p}) = \mathbf{q}.$$

And since $t_n + \tilde{t} \to \infty$, the point **q** is an ω-limit point of $\phi(\cdot, \mathbf{x}_0)$. A similar proof holds when **p** is an α-limit point of Γ and this completes the proof of the theorem.

It follows from this theorem that for all points $\mathbf{p} \in \omega(\Gamma)$, $\phi_t(\mathbf{p}) \in \omega(\Gamma)$ for all $t \in \mathbf{R}$; i.e., $\phi_t(\omega(\Gamma)) \subset \omega(\Gamma)$. Thus, according to Definition 2 in Section 2.5 of Chapter 2, we have the following result.

Corollary. $\alpha(\Gamma)$ *and* $\omega(\Gamma)$ *are invariant with respect to the flow ϕ_t of (1).*

The α- and ω-limit sets of a trajectory Γ of (1) are thus closed invariant subsets of E. In the next definition, a *neighborhood of a set A* is any open set U containing A and we say that $\mathbf{x}(t) \to A$ as $t \to \infty$ if the distance $d(\mathbf{x}(t), A) \to 0$ as $t \to \infty$.

Definition 2. A closed invariant set $A \subset E$ is called *an attracting set* of (1) if there is some neighborhood U of A such that for all $\mathbf{x} \in U$, $\phi_t(\mathbf{x}) \in U$ for all $t \geq 0$ and $\phi_t(\mathbf{x}) \to A$ as $t \to \infty$. An *attractor* of (1) is an attracting set which contains a dense orbit.

Note that any equilibrium point $\mathbf{x}_0$ of (1) is its own α and ω-limit set since $\phi(t, \mathbf{x}_0) = \mathbf{x}_0$ for all $t \in \mathbf{R}$. And if a trajectory Γ of (1) has a unique ω-limit point $\mathbf{x}_0$, then by the above Corollary, $\mathbf{x}_0$ is an equilibrium point of (1). A stable node or focus, defined in Section 2.10 of Chapter 2, is the ω-limit set of every trajectory in some neighborhood of the point; and a stable node or focus of (1) is an attractor of (1). However, not every ω-limit set of a trajectory of (1) is an attracting set of (1); for example, a saddle $\mathbf{x}_0$ of a planar system (1) is the ω-limit set of three trajectories in a neighborhood $N(\mathbf{x}_0)$, but no other trajectories through points in $N(\mathbf{x}_0)$ approach $\mathbf{x}_0$ as $t \to \infty$.

If **q** is any regular point in $\alpha(\Gamma)$ or $\omega(\Gamma)$ then the trajectory through **q** is called a *limit orbit* of Γ. Thus, by Theorem 2, we see that $\alpha(\Gamma)$ and $\omega(\Gamma)$ consist of equilibrium points and limit orbits of (1). We now consider some specific examples of limit sets and attractors.

3.2. Limit Sets and Attractors

Example 1. Consider the system

$$\dot{x} = -y + x(1 - x^2 - y^2)$$
$$\dot{y} = x + y(1 - x^2 - y^2).$$

In polar coordinates, we have

$$\dot{r} = r(1 - r^2)$$
$$\dot{\theta} = 1.$$

We see that the origin is an equilibrium point of this system; the flow spirals around the origin in the counter-clockwise direction; it spirals outward for $0 < r < 1$ since $\dot{r} > 0$ for $0 < r < 1$; and it spirals inward for $r > 1$ since $\dot{r} < 0$ for $r > 1$. The counter-clockwise flow on the unit circle describes a trajectory Γ_0 of (1) since $\dot{r} = 0$ on $r = 1$. The trajectory through the point $(\cos\theta_0, \sin\theta_0)$ on the unit circle at $t = 0$ is given by $\mathbf{x}(t) = (\cos(t + \theta_0), \sin(t + \theta_0))^T$. The phase portrait for this system is shown in Figure 2. The trajectory Γ_0 is called a stable limit cycle. A precise definition of a limit cycle is given in the next section.

The stable limit cycle Γ_0 of the system in Example 1, shown in Figure 2, is the ω-limit set of every trajectory of this system except the equilibrium point at the origin. Γ_0 is composed of one limit orbit and Γ_0 is its own α and ω-limit set. It is made clear by this example that what we really mean by a *trajectory* or *orbit* Γ of the system (1) is the equivalence class of solution curves $\phi(\cdot, \mathbf{x})$ with $\mathbf{x} \in \Gamma$; cf. Problem 3. We typically pick one representative $\phi(\cdot, \mathbf{x}_0)$ with $\mathbf{x}_0 \in \Gamma$, to describe the trajectory and refer to it as the trajectory $\Gamma_{\mathbf{x}_0}$ through the point $\mathbf{x}_0$ at time $t = 0$. In the next section we show that any stable limit cycle of (1) is an attractor of (1).

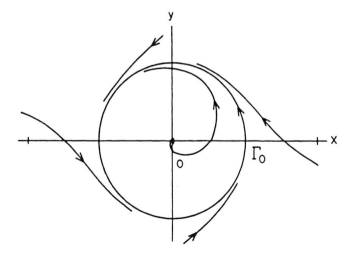

Figure 2. A stable limit cycle Γ_0 which is an attractor of (1).

We next present some examples of limit sets and attractors in $\mathbf{R}^3$.

Example 2. The system

$$\dot{x} = -y + x(1 - z^2 - x^2 - y^2)$$
$$\dot{y} = x + y(1 - z^2 - x^2 - y^2)$$
$$\dot{z} = 0$$

has the unit two-dimensional sphere S^2 together with that portion of the z-axis outside S^2 as an attracting set. Each plane $z = z_0$ is an invariant set and for $|z_0| < 1$ the ω-limit set of any trajectory not on the z-axis is a stable cycle (defined in the next section) on S^2. Cf. Figure 3.

Example 3. The system

$$\dot{x} = -y + x(1 - x^2 - y^2)$$
$$\dot{y} = x + y(1 - x^2 - y^2)$$
$$\dot{z} = \alpha$$

has the z-axis and the cylinder $x^2 + y^2 = 1$ as invariant sets. The cylinder is an attracting set; cf. Figure 4 where $\alpha > 0$.

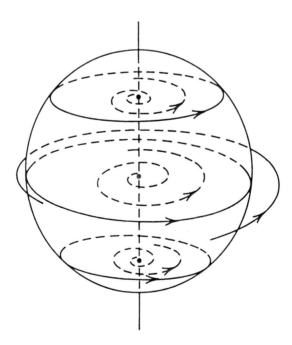

Figure 3. A dynamical system with S^2 as part of its attracting set.

3.2. Limit Sets and Attractors

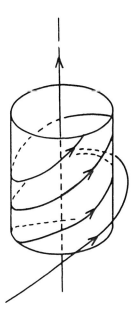

Figure 4. A dynamical system with a cylinder as its attracting set.

If in Example 3 we identify the points $(x, y, 0)$ and $(x, y, 2\pi)$ in the planes $z = 0$ and $z = 2\pi$, we get a flow in $\mathbf{R}^3$ with a two-dimensional invariant torus T^2 as an attracting set. The z-axis gets mapped onto an unstable cycle Γ (defined in the next section). And if α is an irrational multiple of π then the torus T^2 is an attractor (cf. problem 2) and it is the ω-limit set of every trajectory except the cycle Γ. Cf. Figure 5. Several other examples of flows with invariant tori are given in Section 3.6.

In Section 3.7 we establish the Poincaré–Bendixson Theorem which shows that the α and ω-limit sets of any two-dimensional system are fairly simple objects. In fact, it is shown in Section 3.7 that they are either equilibrium points, limit cycles or a union of separatrix cycles (defined in the next section). However, for higher dimensional systems, the α and ω-limit sets may be quite complicated as the next example indicates. A study of the strange types of limit sets that can occur in higher dimensional systems is one of the primary objectives of the book by Guckenheimer and Holmes [G/H]. An in-depth (numerical) study of the "strange attractor" for the Lorenz system in the next example has been carried out by Sparrow [S].

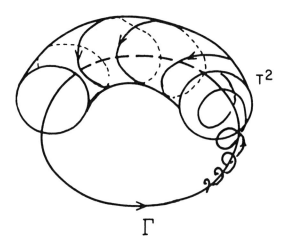

Figure 5. A dynamical system with an invariant torus as an attracting set.

Example 4. (The Lorenz System). The original work of Lorenz in 1963 as well as the more recent work of Sparrow [S] indicates that for certain values of the parameters σ, ρ and β, the system

$$\dot{x} = \sigma(y - x)$$
$$\dot{y} = \rho x - y - xz$$
$$\dot{z} = -\beta z + xy$$

has a strange attracting set. For example for $\sigma = 10$, $\rho = 28$ and $\beta = 8/3$, a single trajectory of this system is shown in Figure 6 along with a "branched surface" S. The attractor A of this system is made up of an infinite number of branched surfaces S which are interleaved and which intersect; however, the trajectories of this system in A do not intersect but move from one branched surface to another as they circulate through the apparent branch. The numerical results in [S] and the related theoretical work in [G/H] indicate that the closed invariant set A contains (i) a countable set of periodic orbits of arbitrarily large period, (ii) an uncountable set of nonperiodic motions and (iii) a dense orbit; cf. [G/H], p. 95. The attracting set A having these properties is referred to as a strange attractor. This example is discussed more fully in Section 4.5 of Chapter 4 in this book.

3.2. Limit Sets and Attractors

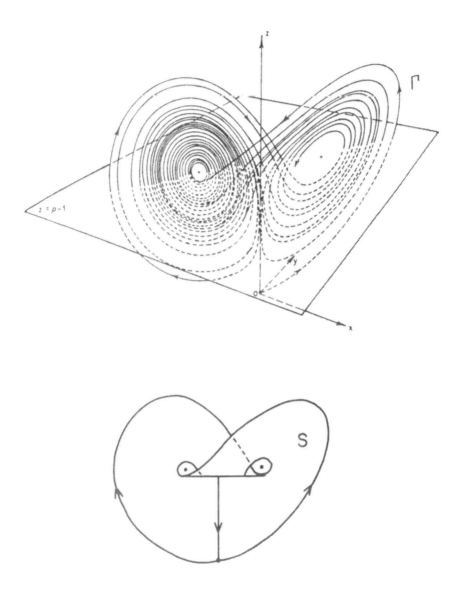

Figure 6. A trajectory Γ of the Lorenz system and the corresponding branched surface S. Reprinted with permission from Guckenheimer and Holmes [G/H].

PROBLEM SET 2

1. Sketch the phase portrait and show that the interval $[-1, 1]$ on the x-axis is an attracting set for the system

$$\dot{x} = x - x^3$$
$$\dot{y} = -y.$$

Is the interval $[-1, 1]$ an attractor? Are either of the intervals $(0, 1]$ or $[1, \infty)$ attractors? Are any of the infinite intervals $(0, \infty)$, $[0, \infty)$, $(-1, \infty)$, $[-1, \infty)$ or $(-\infty, \infty)$ on the x-axis attracting sets for this system?

2. (Flow on a torus; cf. [H/S], p. 241). Identify $\mathbf{R}^4$ with $\mathbf{C}^2$ having two complex coordinates (w, z) and consider the linear system

$$\dot{w} = 2\pi i w$$
$$\dot{z} = 2\pi \alpha i z$$

where α is an irrational real number.

(a) Set $a = e^{2\pi \alpha i}$ and show that the set $\{a^n \in \mathbf{C} \mid n = 1, 2, \ldots\}$ is dense in the unit circle $C = \{z \in \mathbf{C} \mid |z| = 1\}$.

(b) Let ϕ_t be the flow of this system. Show that for any integer n,

$$\phi_n(w, z) = (w, a^n z).$$

(c) Let $\mathbf{x}_0 = (w_0, z_0)$ belong to the torus $T^2 = C \times C \subset \mathbf{C}^2$. Use (a) and (b) to show that $\omega(\Gamma_{\mathbf{x}_0}) = T^2$.

(d) Find $\omega(\Gamma)$ and $\alpha(\Gamma)$ for any trajectory Γ of this system.

Hint: In part (a), use the following theorem to show that for $\epsilon > 0$ there are positive integers m and n such that

$$|a^n - 1| = |e^{2\pi \alpha n i} - e^{2\pi m i}| < \epsilon.$$

Theorem (Hurwitz). *Given any irrational number α and any $N > 0$, there exist positive integers m and n such that $n > N$ and*

$$\left|\alpha - \frac{m}{n}\right| < \frac{1}{n^2}.$$

It then follows that for $\theta = 2\pi n \alpha - 2\pi m$ and the integer K chosen such that $K|\theta| < 2\pi < (K+1)|\theta|$, any point on the circle C is within an ϵ distance of one of the points in $\{(a^n)^k \in \mathbf{C} \mid k = 1, \ldots, K\}$.

3. Let ϕ_t be the flow of the system (1). Define two solution curves

$$\Gamma_j = \{(t, \mathbf{x}) \in \mathbf{R} \times E \mid t \in \mathbf{R} \text{ and } \mathbf{x} = \phi(t, \mathbf{x}_j)\}$$

$j = 1, 2$ of (1) to be equivalent, $\Gamma_1 \sim \Gamma_2$, if there is a $t_0 \in \mathbf{R}$ such that for all $t \in \mathbf{R}$

$$\phi_t(\mathbf{x}_2) = \phi_{t+t_0}(\mathbf{x}_1).$$

Show that $\sim$ is an equivalence relation; i.e., show that $\sim$ is reflexive, symmetric and transitive.

3.2. Limit Sets and Attractors

4. Sketch the phase portrait of a planar system having

    (a) a trajectory Γ with $\alpha(\Gamma) = \omega(\Gamma) = \{\mathbf{x}_0\}$, but $\Gamma \neq \{\mathbf{x}_0\}$.

    (b) a trajectory Γ such that $\omega(\Gamma)$ consists of one limit orbit (cf. Example 1).

    (c) a trajectory Γ such that $\omega(\Gamma)$ consists of one limit orbit and one equilibrium point.

    (d) a trajectory Γ such that $\omega(\Gamma)$ consists of two limit orbits and one equilibrium point (cf. Example 2 in Section 2.14 of Chapter 2).

    (e) a trajectory Γ such that $\omega(\Gamma)$ consists of two limit orbits and two equilibrium points (cf. Example 1 in Section 2.14 of Chapter 2).

    (f) a trajectory Γ such that $\omega(\Gamma)$ consists of five limit orbits and three equilibrium points.

    Can $\omega(\Gamma)$ consist of one limit orbit and two equilibrium points? How many different topological types, in $\mathbf{R}^2$, are there in case (d) above?

5. (a) According to the corollary of Theorem 2, every ω-limit set is an invariant set of the flow ϕ_t of (1). Give an example to show that not every set invariant with respect to the flow ϕ_t of (1) is the α or ω-limit set of a trajectory of (1).

    (b) As we mentioned above, any stable limit cycle Γ is an attracting set; furthermore, Γ is the ω-limit set of every trajectory in a neighborhood of Γ. Give an example to show that not every attracting set A is the ω-limit set of a trajectory in a neighborhood of A.

    (c) Is the cylinder in Example 3 an attractor of the system in that example?

6. Consider the Lorenz system

$$\dot{x} = \sigma(y - x)$$
$$\dot{y} = \rho x - y - xz$$
$$\dot{z} = xy - \beta z$$

    with $\sigma > 0$, $\rho > 0$ and $\beta > 0$.

    (a) Show that this system is invariant under the transformation $(x, y, z, t) \to (-x, -y, z, t)$.

    (b) Show that the z-axis is invariant under the flow of this system and that it consists of three trajectories.

(c) Show that this system has equilibrium points at the origin and at $(\pm\sqrt{\beta(\rho-1)}, \pm\sqrt{\beta(\rho-1)}, \rho-1)$ for $\rho > 1$. For $\rho > 1$ show that there is a one-dimensional unstable manifold $W^u(0)$ at the origin.

(d) For $\rho \in (0,1)$ use the Liapunov function $V(x,y,z) = \rho x^2 + \sigma y^2 + \sigma z^2$ to show that the origin is globally stable; i.e., for $\rho \in (0,1)$, the origin is the ω-limit set of every trajectory of this system.

3.3 Periodic Orbits, Limit Cycles and Separatrix Cycles

In this section we discuss periodic orbits or cycles, limit cycles and separatrix cycles of a dynamical system $\phi(t, \mathbf{x})$ defined by

$$\dot{\mathbf{x}} = \mathbf{f}(\mathbf{x}) \tag{1}$$

Definition 1. A *cycle* or *periodic orbit* of (1) is any closed solution curve of (1) which is not an equilibrium point of (1). A periodic orbit Γ is called *stable* if for each $\epsilon > 0$ there is a neighborhood U of Γ such that for all $\mathbf{x} \in U$, $d(\Gamma_{\mathbf{x}}^+, \Gamma) < \epsilon$; i.e., if for all $\mathbf{x} \in U$ and $t \geq 0$, $d(\phi(t,\mathbf{x}), \Gamma) < \epsilon$. A periodic orbit Γ is called *unstable* if it is not stable; and Γ is called *asymptotically stable* if it is stable and if for all points $\mathbf{x}$ in some neighborhood U of Γ

$$\lim_{t \to \infty} d(\phi(t,\mathbf{x}), \Gamma) = 0.$$

Cycles of the system (1) correspond to periodic solutions of (1) since $\phi(\cdot, \mathbf{x}_0)$ defines a closed solution curve of (1) if and only if for all $t \in \mathbf{R}$ $\phi(t+T, \mathbf{x}_0) = \phi(t, \mathbf{x}_0)$ for some $T > 0$. The minimal T for which this equality holds is called the *period* of the periodic orbit $\phi(\cdot, \mathbf{x}_0)$. Note that a center, defined in Section 2.10 of Chapter 2, is an equilibrium point surrounded by a continuous band of cycles. In general, the period T will vary continuously as we move along a continuous curve intersecting this family of cycles; however, in the case of a center for a linear system, the period is the same for each periodic orbit in the family. Each periodic orbit in the family of cycles encircling a center is stable but not asymptotically stable. We shall see in Section 3.5 that a periodic orbit

$$\Gamma\colon \mathbf{x} = \gamma(t) \qquad 0 \leq t \leq T$$

of (1) if asymptotically stable only if

$$\int_0^T \nabla \cdot \mathbf{f}(\gamma(t))\,dt \leq 0.$$

3.3. Periodic Orbits, Limit Cycles and Separatrix Cycles

For planar systems we shall see in Section 3.4 that this condition with strict inequality is both necessary and sufficient for the asymptotic stability of a simple limit cycle Γ. An asymptotically stable cycle is referred to as an ω-limit cycle and any ω-limit cycle is an attractor of (1).

Periodic orbits have stable and unstable manifolds just as equilibrium points do; cf. Section 2.7 in Chapter 2. Let Γ be a (hyperbolic) periodic orbit and let N be a neighborhood of Γ. As in Section 2.7, the local stable and unstable manifolds of Γ are given by

$$S(\Gamma) = \{\mathbf{x} \in N \mid d(\phi_t(\mathbf{x}), \Gamma) \to 0 \text{ as } t \to \infty \text{ and } \phi_t(\mathbf{x}) \in N \text{ for } t \geq 0\}$$

and

$$U(\Gamma) = \{\mathbf{x} \in N \mid d(\phi_t(\mathbf{x}), \Gamma) \to 0 \text{ as } t \to -\infty \text{ and } \phi_t(\mathbf{x}) \in N \text{ for } t \leq 0\}.$$

The global stable and unstable manifolds of Γ are then defined by

$$W^s(\Gamma) = \bigcup_{t \leq 0} \phi_t(S(\Gamma))$$

and

$$W^u(\Gamma) = \bigcup_{t \geq 0} \phi_t(U(\Gamma)).$$

These manifolds are invariant under the flow ϕ_t of (1). The stability of periodic orbits as well as the existence and dimension of the manifolds $S(\Gamma)$ and $U(\Gamma)$ will be discussed more fully in the next two section.

Example 1. The system

$$\dot{x} = -y + x(1 - x^2 - y^2)$$
$$\dot{y} = x + y(1 - x^2 - y^2)$$
$$\dot{z} = z$$

has an isolated periodic orbit in the x, y plane given by $\mathbf{x}(t) = (\cos t, \sin t, 0)^T$. There is an equilibrium point at the origin. The z-axis, the cylinder $x^2 + y^2 = 1$ and the x, y plane are invariant manifolds of this system. The phase portrait for this system is shown in Figure 1 together with a cross-section in the x, z plane; the dashed curves in that cross-section are projections onto the x, z plane of orbits starting in the x, z plane. In this example, the stable manifold of Γ, $W^s(\Gamma)$, is the x, y plane excluding the origin and the unstable manifold of Γ, $W^u(\Gamma)$, is the unit cylinder. The unit cylinder is an attracting set for this system.

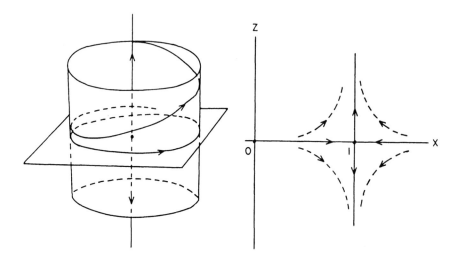

Figure 1. Some invariant manifolds for the system of Example 1 and a cross-section in the x, z plane.

A periodic orbit Γ of the type shown in Figure 1 where $W^s(\Gamma) \neq \Gamma$ and $W^u(\Gamma) \neq \Gamma$ is called a *periodic orbit of saddle type*. Any periodic orbit of saddle type is unstable.

We next consider periodic orbits of a planar system, (1) with $\mathbf{x} \in \mathbf{R}^2$.

Definition 2. A *limit cycle* Γ of a planar system is a cycle of (1) which is the α or ω-limit set of some trajectory of (1) other than Γ. If a cycle Γ is the ω-limit set of every trajectory in some neighborhood of Γ, then Γ is called an ω-*limit cycle* or *stable limit cycle*; if Γ is the α-limit set of every trajectory in some neighborhood of Γ, then Γ is called an α-*limit cycle* or an *unstable limit cycle*; and if Γ is the ω-limit set of one trajectory other than Γ and the α-limit set of another trajectory other than Γ, then Γ is called a *semi-stable limit cycle*.

Note that a stable limit cycle is actually an asymptotically stable cycle in the sense of Definition 1 and any stable limit cycle is an attractor. Example 1 in Section 3.2 exhibits a stable limit cycle and if we replace t by $-t$ in that example (thereby reversing the direction of the flow), we would obtain an unstable limit cycle. In Problem 2 below we see an example of a semi-stable limit cycle.

The following theorem is proved in [C/L], p. 396. In stating this theorem, we are using the Jordan curve theorem which states that any simple closed curve Γ separates the plane $\mathbf{R}^2$ into two disjoint open connected sets having

3.3. Periodic Orbits, Limit Cycles and Separatrix Cycles

Γ as their boundary. One of these sets, called the interior of Γ, is bounded and simply connected. The other, called the exterior of Γ, is unbounded and is not simply connected.

Theorem 1. *If one trajectory in the exterior of a limit cycle Γ of a planar C^1-system (1) has Γ as its ω-limit set, then every trajectory in some exterior neighborhood U of Γ has Γ as its ω-limit set. Moreover, any trajectory in U spirals around Γ as $t \to \infty$ in the sense that it intersects any straight line perpendicular to Γ an infinite number of times at $t = t_n$ where $t_n \to \infty$.*

The same sort of result holds for interior neighborhoods of Γ and also when Γ is the α-limit set of some trajectory.

The next example shows that limit cycles can accumulate at an equilibrium point and Problem 1 below shows that they can accumulate on a cycle; cf. Example 6 in Section 2.10 of Chapter 2.

Example 2. The system

$$\dot{x} = -y + x(x^2 + y^2) \sin \frac{1}{\sqrt{x^2 + y^2}}$$

$$\dot{y} = x + y(x^2 + y^2) \sin \frac{1}{\sqrt{x^2 + y^2}}$$

for $x^2 + y^2 \neq 0$ with $\dot{x} = \dot{y} = 0$ at $(0, 0)$ defines a C^1-system on $\mathbf{R}^2$ which can be written in polar coordinates as

$$\dot{r} = r^3 \sin \frac{1}{r}$$

$$\dot{\theta} = 1.$$

The origin is an equilibrium point and there are limit cycles Γ_n lying on the circles $r = 1/(n\pi)$. These limit cycles accumulate at the origin; i.e.,

$$\lim_{n \to \infty} d(\Gamma_n, 0) = 0.$$

Each of the limit cycles Γ_{2n} is stable while Γ_{2n+1} is unstable.

The next theorem, stated by Dulac [D] in 1923, shows that a planar analytic system, (1) with $\mathbf{f}(\mathbf{x})$ analytic in $E \subset \mathbf{R}^2$, cannot have an infinite number of limit cycles accumulating at a critical point as in the above example. Errors were recently found in Dulac's original proof; however, these errors were corrected in 1988 by a group of French mathematicians, Ecalle, Martinet, Moussu and Ramis, and independently by the Russian mathematician Y. Ilyashenko, who modified and extended Dulac's use of mixed Laurent-logarithmic type series. Cf. [8].

Theorem (Dulac). *In any bounded region of the plane, a planar analytic system* (1) *with* $\mathbf{f}(\mathbf{x})$ *analytic in* $\mathbf{R}^2$ *has at most a finite number of limit cycles. Any polynomial system has at most a finite number of limit cycles in* $\mathbf{R}^2$.

The next examples describe two important types of orbits that can occur in a dynamical system: homoclinic orbits and heteroclinic orbits. They also furnish examples of separatrix cycles and compound separatrix cycles for planar dynamical systems.

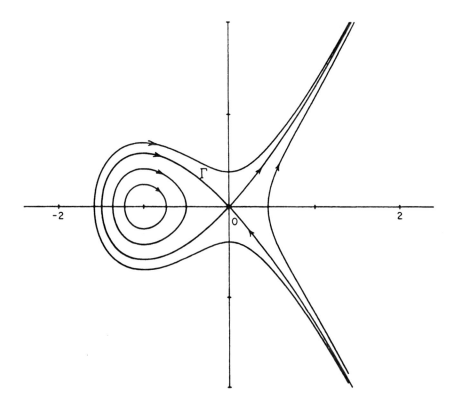

Figure 2. A homoclinic orbit Γ which defines a separatrix cycle.

Example 3. The Hamiltonian system
$$\dot{x} = y$$
$$\dot{y} = x + x^2$$
with Hamiltonian $H(x,y) = y^2/2 - x^2/2 - x^3/3$ has solution curves defined by
$$y^2 - x^2 - \frac{2}{3}x^3 = C.$$

3.3. Periodic Orbits, Limit Cycles and Separatrix Cycles 207

The phase portrait for this system is shown in Figure 2. The curve $y^2 = x^2 + 2x^3/3$, corresponding to $C = 0$, goes through the point $(-3/2, 0)$ and has the saddle at the origin (which also corresponds to $C = 0$) as its α and ω-limit sets. The solution curve $\Gamma \subset W^s(\mathbf{0}) \cap W^u(\mathbf{0})$ is called a *homoclinic orbit* and the flow on the simple closed curve determined by the union of this homoclinic orbit and the equilibrium point at the origin is called a *separatrix cycle*.

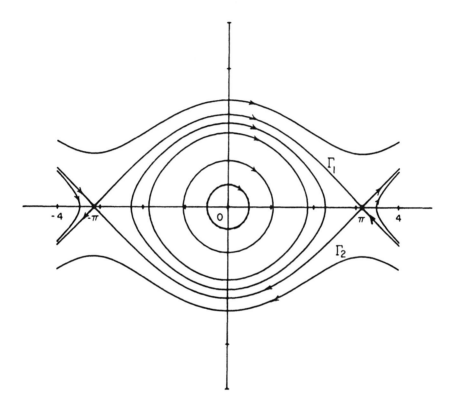

Figure 3. Heteroclinic orbits Γ_1 and Γ_2 defining a separatrix cycle.

Example 1 in Section 2.14 of Chapter 2, the undamped pendulum, furnishes an example of heteroclinic orbits; the trajectory Γ_1 in Figure 1 of that section, having the saddle $(-\pi, 0)$ as its α-limit set and the saddle at $(\pi, 0)$ as its ω-limit set is called a *heteroclinic orbit*. The trajectory Γ_2 in that figure is also a heteroclinic orbit; cf. Figure 3 above. The flow on the simple closed curve $S = \Gamma_1 \cup \Gamma_2 \cup \{(\pi, 0)\} \cup \{(-\pi, 0)\}$ defines a separatrix cycle.

A finite union of compatibly oriented separatrix cycles is called a *compound separatrix cycle* or *graphic*. An example of a compound separatrix cycle is given by the Hamiltonian system in Example 2 in Section 2.14 of Chapter 2.

Consider the following system.
$$\dot{x} = -2y$$
$$\dot{y} = 4x(x-1)(x-1/2).$$

The phase portrait for this system is given in Figure 2 in Section 2.14 of Chapter 2. The two trajectories Γ_1 and Γ_2 having the saddle at $(1/2, 0)$ as their α and ω-limit sets are homoclinic orbits and the flow on the (nonsimple) closed curve $S = \Gamma_1 \cup \Gamma_2 \cup \{(1/2, 0)\}$ is a compound separatrix cycle. S is the union of two positively oriented separatrix cycles. Other examples of compound separatrix cycles can be found in [A-II]. Several examples of compound separatrix cycles are shown in Figure 4 below.

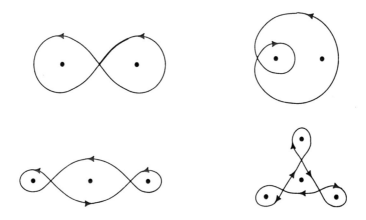

Figure 4. Examples of compound separatrix cycles.

PROBLEM SET 3

1. (a) Given
$$\dot{x} = -y + x(1 - x^2 - y^2)\sin|1 - x^2 - y^2|^{-1/2}$$
$$\dot{y} = x + y(1 - x^2 - y^2)\sin|1 - x^2 - y^2|^{-1/2}$$
for $x^2 + y^2 \neq 1$ with $\dot{x} = -y$ and $\dot{y} = x$ for $x^2 + y^2 = 1$. Show that there is a sequence of limit cycles
$$\Gamma_n^{\pm}: r = \sqrt{1 \pm \frac{1}{n^2\pi^2}}$$
for $n = 1, 2, 3, \ldots$ which accumulates on the cycle
$$\Gamma: \mathbf{x} = \gamma(t) = (\cos t, \sin t)^T.$$
What can you say about the stability of $\Gamma_n^{\pm}$?

3.3. Periodic Orbits, Limit Cycles and Separatrix Cycles

(b) Given

$$\dot{x} = -y + x(1 - x^2 - y^2)\sin(1 - x^2 - y^2)^{-1}$$
$$\dot{y} = x + y(1 - x^2 - y^2)\sin(1 - x^2 - y^2)^{-1}$$

for $x^2 + y^2 \neq 1$ with $\dot{x} = -y$ and $\dot{y} = x$ for $x^2 + y^2 = 1$. Show that there is a sequence of limit cycles

$$\Gamma_n: r = \sqrt{1 - \frac{1}{n\pi}}$$

for $n = \pm 1, \pm 2, \pm 3, \ldots$ which accumulates on the cycle Γ given in part (a). What can you say about the stability of Γ_n?

2. Show that the system

$$\dot{x} = -y + x(1 - x^2 - y^2)^2$$
$$\dot{y} = x + y(1 - x^2 - y^2)^2$$

has a semi-stable limit cycle Γ. Sketch the phase portrait for this system.

3. (a) Show that the linear system $\dot{x} = Ax$ with

$$A = \begin{bmatrix} 0 & -4 \\ 1 & 0 \end{bmatrix}$$

has a continuous band of cycles

$$\Gamma_\alpha: \gamma_\alpha(t) = (\alpha \cos 2t, \alpha/2 \sin 2t)^T$$

for $\alpha \in (0, \infty)$; cf. the example in Section 1.5 of Chapter 1. What is the period of each of these cycles?

(b) Show that the nonlinear system

$$\dot{x} = -y\sqrt{x^2 + y^2}$$
$$\dot{y} = x\sqrt{x^2 + y^2}$$

has a continuous band of cycles

$$\Gamma_\alpha: \gamma_\alpha(t) = (\alpha \cos \alpha t, \alpha \sin \alpha t)^T$$

for $\alpha \in (0, \infty)$. Note that the period $T_\alpha = 2\pi/\alpha$ of the cycle Γ_α is a continuous function of α for $\alpha \in (0, \infty)$ and that $T_\alpha \to 0$ as $\alpha \to 0$ while $T_\alpha \to \infty$ as $\alpha \to 0$.

4. Show that

$$\dot{x} = y$$
$$\dot{y} = x + x^2$$

is a Hamiltonian system with

$$H(x,y) = y^2/2 - x^2/2 - x^3/3$$

and that this system is symmetric with respect to the x-axis. Show that the origin is a saddle for this system and that $(-1, 0)$ is a center. Sketch the homoclinic orbit given by

$$y^2 = x^2 + \frac{2}{3}x^3.$$

Also, sketch all four trajectories given by this equation and sketch the phase portrait for this system.

5. Show that

$$\dot{x} = y + y(x^2 + y^2)$$
$$\dot{y} = x - x(x^2 + y^2)$$

is a Hamiltonian system with $4H(x,y) = (x^2 + y^2)^2 - 2(x^2 - y^2)$. Show that $\frac{dH}{dt} = 0$ along solution curves of this system and therefore that solution curves of this system are given by

$$(x^2 + y^2)^2 - 2(x^2 - y^2) = C.$$

Show that the origin is a saddle for this system and that $(\pm 1, 0)$ are centers for this system. (Note the symmetry with respect to the x-axis.) Sketch the two homoclinic orbits corresponding to $C = 0$ and sketch the phase portrait for this system noting the occurrence of a compound separatrix cycle for this system.

6. As in problem 5, sketch the compound separatrix cycle and the phase portrait for the Hamiltonian system

$$\dot{x} = y$$
$$\dot{y} = x - x^3$$

with $H(x,y) = y^2/2 - x^2/2 + x^4/4$.

7. (a) Write the system

$$\dot{x} = -y + x(1 - r^2)(4 - r^2)$$
$$\dot{y} = x + y(1 - r^2)(4 - r^2)$$

3.4. The Poincaré Map

with $r^2 = x^2+y^2$ in polar coordinates and show that this system has two limit cycles Γ_1 and Γ_2. Give solutions representing Γ_1 and Γ_2 and determine their stability. Sketch the phase portrait for this system and determine all limit sets of trajectories of this system.

(b) Consider the system

$$\dot{x} = -y + x(1 - x^2 - y^2 - z^2)(4 - x^2 - y^2 - z^2)$$
$$\dot{y} = x + y(1 - x^2 - y^2 - z^2)(4 - x^2 - y^2 - z^2)$$
$$\dot{z} = 0.$$

For $z = z_0$, a constant, write the resulting system in polar coordinates and deduce that this system has two invariant spheres. Sketch the phase portrait for this system and describe all limit sets of trajectories of this system. Does this system have an attracting set?

8. Consider the system

$$\dot{x} = -y + x(1 - x^2 - y^2)(4 - x^2 - y^2)$$
$$\dot{y} = x + y(1 - x^2 - y^2)(4 - x^2 - y^2)$$
$$\dot{z} = z.$$

Show that there are two periodic orbits Γ_1 and Γ_2 in the x, y plane represented by $\gamma_1(t) = (\cos t, \sin t)^T$ and $\gamma_2(t) = (2\cos t, 2\sin t)^T$, and determine their stability. Show that there are two invariant cylinders for this system given by $x^2+y^2 = 1$ and $x^2+y^2 = 4$; and describe the invariant manifolds $W^s(\Gamma_j)$ and $W^u(\Gamma_j)$ for $j = 1, 2$.

9. Reverse the direction of the flow in problem 8 (i.e., let $t \to -t$) and show that the resulting system has the origin and the periodic orbit Γ_2: $\mathbf{x} = (2\cos t, 2\sin t)^T$ as attractors.

3.4 The Poincaré Map

Probably the most basic tool for studying the stability and bifurcations of periodic orbits is the Poincaré map or first return map, defined by Henri Poincaré in 1881; cf. [P]. The idea of the Poincaré map is quite simple: If Γ is a periodic orbit of the system

$$\dot{\mathbf{x}} = \mathbf{f}(\mathbf{x}) \tag{1}$$

through the point $\mathbf{x}_0$ and Σ is a hyperplane perpendicular to Γ at $\mathbf{x}_0$, then for any point $\mathbf{x} \in \Sigma$ sufficiently near $\mathbf{x}_0$, the solution of (1) through $\mathbf{x}$ at

$t = 0$, $\phi_t(\mathbf{x})$, will cross Σ again at a point $\mathbf{P}(\mathbf{x})$ near $\mathbf{x}_0$; cf. Figure 1. The mapping $\mathbf{x} \to \mathbf{P}(\mathbf{x})$ is called the Poincaré map.

The Poincaré map can also be defined when Σ is a smooth surface, through a point $\mathbf{x}_0 \in \Gamma$, which is not tangent to Γ at $\mathbf{x}_0$. In this case, the surface Σ is said to intersect the curve Γ *transversally* at $\mathbf{x}_0$. The next theorem establishes the existence and continuity of the Poincaré map $\mathbf{P}(\mathbf{x})$ and of its first derivative $D\mathbf{P}(\mathbf{x})$.

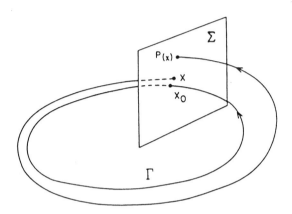

Figure 1. The Poincaré map.

Theorem 1. *Let E be an open subset of $\mathbf{R}^n$ and let $\mathbf{f} \in C^1(E)$. Suppose that $\phi_t(\mathbf{x}_0)$ is a periodic solution of (1) of period T and that the cycle*

$$\Gamma = \{\mathbf{x} \in \mathbf{R}^n \mid \mathbf{x} = \phi_t(\mathbf{x}_0), \quad 0 \le t \le T\}$$

is contained in E. Let Σ be the hyperplane orthogonal to Γ at $\mathbf{x}_0$; i.e., let

$$\Sigma = \{\mathbf{x} \in \mathbf{R}^n \mid (\mathbf{x} - \mathbf{x}_0) \cdot \mathbf{f}(\mathbf{x}_0) = 0\}.$$

Then there is a $\delta > 0$ and a unique function $\tau(\mathbf{x})$, defined and continuously differentiable for $\mathbf{x} \in N_\delta(\mathbf{x}_0)$, such that $\tau(\mathbf{x}_0) = T$ and

$$\phi_{\tau(\mathbf{x})}(\mathbf{x}) \in \Sigma$$

for all $\mathbf{x} \in N_\delta(\mathbf{x}_0)$.

Proof. The proof of this theorem is an immediate application of the implicit function theorem. For a given point $\mathbf{x}_0 \in \Gamma \subset E$, define the function

$$F(t, \mathbf{x}) = [\phi_t(\mathbf{x}) - \mathbf{x}_0] \cdot \mathbf{f}(\mathbf{x}_0).$$

It then follows from Theorem 1 in Section 2.5 of Chapter 2 that $F \in C^1(\mathbf{R} \times E)$ and it follows from the periodicity of $\phi_t(\mathbf{x}_0)$ that $F(T, \mathbf{x}_0) = 0$.

3.4. The Poincaré Map

Furthermore, since $\phi(t, \mathbf{x}_0) = \phi_t(\mathbf{x}_0)$ is a solution of (1) which satisfies $\phi(T, \mathbf{x}_0) = \mathbf{x}_0$, it follows that

$$\frac{\partial F(T, \mathbf{x}_0)}{\partial t} = \frac{\partial \phi(T, \mathbf{x}_0)}{\partial t} \cdot \mathbf{f}(\mathbf{x}_0)$$
$$= \mathbf{f}(\mathbf{x}_0) \cdot \mathbf{f}(\mathbf{x}_0) = |\mathbf{f}(\mathbf{x}_0)|^2 \neq 0$$

since $\mathbf{x}_0 \in \Gamma$ is not an equilibrium point of (1). Thus, it follows from the implicit function theorem, cf., e.g., Theorem 9.28 in [R], that there exists a $\delta > 0$ and a unique function $\tau(\mathbf{x})$ defined and continuously differentiable for all $\mathbf{x} \in N_\delta(\mathbf{x}_0)$ such that $\tau(\mathbf{x}_0) = T$ and such that

$$F(\tau(\mathbf{x}), \mathbf{x}) = 0$$

for all $\mathbf{x} \in N_\delta(\mathbf{x}_0)$. Thus, for all $\mathbf{x} \in N_\delta(\mathbf{x}_0)$, $[\phi(\tau(\mathbf{x}), \mathbf{x}) - \mathbf{x}_0] \cdot \mathbf{f}(\mathbf{x}_0) = 0$, i.e.,

$$\phi_{\tau(\mathbf{x})}(\mathbf{x}) \in \Sigma.$$

Definition 1. Let Γ, Σ, δ and $\tau(\mathbf{x})$ be defined as in Theorem 1. Then for $\mathbf{x} \in N_\delta(\mathbf{x}_0) \cap \Sigma$, the function

$$\mathbf{P}(\mathbf{x}) = \phi_{\tau(\mathbf{x})}(\mathbf{x})$$

is called *the Poincaré map* for Γ at $\mathbf{x}_0$.

Remark. It follows from Theorem 1 that $\mathbf{P} \in C^1(U)$ where $U = N_\delta(\mathbf{x}_0) \cap \Sigma$. And the same proof as in the proof of Theorem 1, using the implicit function theorem for analytic functions, implies that if $\mathbf{f}$ is analytic in E then $\mathbf{P}$ is analytic in U. Fixed points of the Poincaré map, i.e., points $\mathbf{x} \in \Sigma$ satisfying $\mathbf{P}(\mathbf{x}) = \mathbf{x}$, correspond to periodic orbits $\phi(\cdot, \mathbf{x})$ of (1). And there is no loss in generality in assuming that the origin has been translated to the point $\mathbf{x}_0 \in \Sigma$ in which case $\mathbf{x}_0 = \mathbf{0}$, $\Sigma \simeq \mathbf{R}^{n-1}$, $\mathbf{P}: \mathbf{R}^{n-1} \cap N_\delta(\mathbf{0}) \to \mathbf{R}^{n-1}$ and $D\mathbf{P}(\mathbf{0})$ is represented by an $(n-1) \times (n-1)$ matrix. By considering the system (1) with $t \to -t$, we can show that the Poincaré map $\mathbf{P}$ has a C^1 inverse, $\mathbf{P}^{-1}(\mathbf{x}) = \phi_{-\tau(\mathbf{x})}(\mathbf{x})$. Thus, $\mathbf{P}$ is a *diffeomorphism*; i.e., a smooth function with a smooth inverse.

Example 1. In example 1 of Section 3.2 it was shown that the system

$$\dot{x} = -y + x(1 - x^2 - y^2)$$
$$\dot{y} = x + y(1 - x^2 - y^2)$$

had a limit cycle Γ represented by $\boldsymbol{\gamma}(t) = (\cos t, \sin t)^T$. The Poincaré map for Γ can be found by solving this system written in polar coordinates

$$\dot{r} = r(1 - r^2)$$
$$\dot{\theta} = 1$$

with $r(0) = r_0$ and $\theta(0) = \theta_0$. The first equation can be solved either as a separable differential equation or as a Bernoulli equation. The solution is given by

$$r(t, r_0) = \left[1 + \left(\frac{1}{r_0^2} - 1\right) e^{-2t}\right]^{-1/2}$$

and

$$\theta(t, \theta_0) = t + \theta_0.$$

If Σ is the ray $\theta = \theta_0$ through the origin, then Σ is perpendicular to Γ and the trajectory through the point $(r_0, \theta_0) \in \Sigma \cap \Gamma$ at $t = 0$ intersects the ray $\theta = \theta_0$ again at $t = 2\pi$; cf. Figure 2. It follows that the Poincaré map is given by

$$P(r_0) = \left[1 + \left(\frac{1}{r_0^2} - 1\right) e^{-4\pi}\right]^{-1/2}.$$

Clearly $P(1) = 1$ corresponding to the cycle Γ and we see that

$$P'(r_0) = e^{-4\pi} r_0^{-3} \left[1 + \left(\frac{1}{r_0^2} - 1\right) e^{-4\pi}\right]^{-3/2}$$

and that $P'(1) = e^{-4\pi} < 1$.

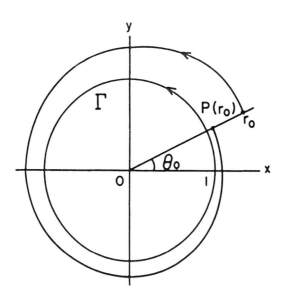

Figure 2. The Poincaré map for the system in Example 1.

We next cite some specific results for the Poincaré map of planar systems. We return to a more complete discussion of the Poincaré map of higher

3.4. The Poincaré Map

dimensional systems in the next section. For planar systems, if we translate the origin to the point $\mathbf{x}_0 \in \Gamma \cap \Sigma$, the normal line Σ will be a line through the origin; cf. Figure 3 below. The point $\mathbf{0} \in \Gamma \cap \Sigma$ divides the line Σ into two open segments Σ^+ and Σ^- where Σ^+ lies entirely in the exterior of Γ. Let s be the signed distance along Σ with $s > 0$ for points in Σ^+ and $s < 0$ for points in Σ^-.

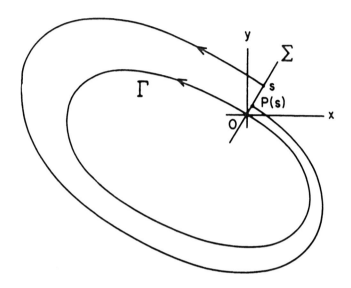

Figure 3. The straight line Σ normal to Γ at $\mathbf{0}$.

According to Theorem 1, the Poincaré map $P(s)$ is then defined for $|s| < \delta$ and we have $P(0) = 0$. In order to see how the stability of the cycle Γ is determined by $P'(0)$, let us introduce *the displacement function*

$$d(s) = P(s) - s.$$

Then $d(0) = 0$ and $d'(s) = P'(s) - 1$; and it follows from the mean value theorem that for $|s| < \delta$

$$d(s) = d'(\sigma)s$$

for some σ between 0 and s. Since $d'(s)$ is continuous, the sign of $d'(s)$ will be the same as the sign of $d'(0)$ for $|s|$ sufficiently small as long as $d'(0) \neq 0$. Thus, if $d'(0) < 0$ it follows that $d(s) < 0$ for $s > 0$ and that $d(s) > 0$ for $s < 0$; i.e., the cycle Γ is a stable limit cycle or an ω-limit cycle. Cf. Figure 3. Similarly, if $d'(0) > 0$ then Γ is an unstable limit cycle or an α-limit cycle. We have the corresponding results that if $P(0) = 0$ and $P'(0) < 1$, then Γ is a stable limit cycle and if $P(0) = 0$ and $P'(0) > 1$, then Γ is an unstable limit cycle. Thus, the stability of Γ is determined by the derivative of the Poincaré map.

In this regard, the following theorem which gives a formula for $P'(0)$ in terms of the right-hand side $\mathbf{f}(\mathbf{x})$ of (1), is very useful; cf. [A-II], p. 118.

Theorem 2. *Let E be an open subset of $\mathbf{R}^2$ and suppose that $\mathbf{f} \in C^1(E)$. Let $\gamma(t)$ be a periodic solution of (1) of period T. Then the derivative of the Poincaré map $P(s)$ along a straight line Σ normal to $\Gamma = \{\mathbf{x} \in \mathbf{R}^2 \mid \mathbf{x} = \gamma(t) - \gamma(0), 0 \leq t \leq T\}$ at $\mathbf{x} = 0$ is given by*

$$P'(0) = \exp \int_0^T \nabla \cdot \mathbf{f}(\gamma(t))\, dt.$$

Corollary. *Under the hypotheses of Theorem 2, the periodic solution $\gamma(t)$ is a stable limit cycle if*

$$\int_0^T \nabla \cdot \mathbf{f}(\gamma(t))\, dt < 0$$

and it is an unstable limit cycle if

$$\int_0^T \nabla \cdot \mathbf{f}(\gamma(t))\, dt > 0.$$

It may be a stable, unstable or semi-stable limit cycle or it may belong to a continuous band of cycles if this quantity is zero.

For Example 1 above, we have $\gamma(t) = (\cos t, \sin t)^T$, $\nabla \cdot \mathbf{f}(x,y) = 2 - 4x^2 - 4y^2$ and

$$\int_0^{2\pi} \nabla \cdot \mathbf{f}(\gamma(t))\, dt = \int_0^{2\pi} (2 - 4\cos^2 t - 4\sin^2 t)\, dt = -4\pi.$$

Thus, with $s = r - 1$, it follows from Theorem 2 that

$$P'(0) = e^{-4\pi}$$

which agrees with the result found in Example 1 by direct computation. Since $P'(0) < 1$, the cycle $\gamma(t)$ is a stable limit cycle in this example.

Definition 2. Let $P(s)$ be the Poincaré map for a cycle Γ of a planar analytic system (1) and let

$$d(s) = P(s) - s$$

be the displacement function. Then if

$$d(0) = d'(0) = \cdots = d^{(k-1)}(0) = 0 \quad \text{and} \quad d^{(k)}(0) \neq 0,$$

Γ is called a *multiple limit cycle* of multiplicity k. If $k = 1$ then Γ is called a *simple limit cycle*.

3.4. The Poincaré Map

We note that $\Gamma = \{\mathbf{x} \in \mathbf{R}^2 \mid \mathbf{x} = \gamma(t), 0 \leq t \leq T\}$ is a simple limit cycle of (1) iff

$$\int_0^T \nabla \cdot \mathbf{f}(\gamma(t))\, dt \neq 0.$$

It can be shown that if k is even then Γ is a semi-stable limit cycle and if k is odd then Γ is a stable limit cycle if $d^{(k)}(0) < 0$ and Γ is an unstable limit cycle if $d^{(k)}(0) > 0$. Furthermore, we shall see in Section 4.5 of Chapter 4 that if Γ is a multiple limit cycle of multiplicity k, then k limit cycles can be made to bifurcate from Γ under a suitable small perturbation of (1). Finally, it can be shown that in the analytic case, $d^{(k)}(0) = 0$ for $k = 0, 1, 2, \ldots$ iff Γ belongs to a continuous band of cycles.

Part of Dulac's Theorem for planar analytic systems, given in Section 3.3, follows immediately from the analyticity of the Poincaré map for analytic systems: Suppose that there are an infinite number of cycles Γ_n which accumulate on a cycle Γ. Then the Poincaré map $P(s)$ for Γ is an analytic function with an infinite number of zeros in a neighborhood of $s = 0$. It follows that $P(s) \equiv 0$ in a neighborhood of $s = 0$; i.e., the cycles Γ_n belong to a continuous band of cycles and are therefore not limit cycles. This result was established by Poincaré [P] in 1881:

Theorem (Poincaré). *A planar analytic system* (1) *cannot have an infinite number of limit cycles which accumulate on a cycle of* (1).

For planar analytic systems, it is convenient at this point to discuss the Poincaré map in the neighborhood of a focus and to define what we mean by a multiple focus. These results will prove useful in Chapter 4 where we discuss the bifurcation of limit cycles from a multiple focus.

Suppose that the planar analytic system (1) has a focus at the origin and that $\det D\mathbf{f}(0) \neq 0$. Then (1) is linearly equivalent to the system

$$\begin{aligned} \dot{x} &= ax - by + p(x,y) \\ \dot{y} &= bx + ay + q(x,y) \end{aligned} \quad (2)$$

with $b \neq 0$ where the power series expansions of p and q begin with second or higher degree terms. In polar coordinates, this system has the form

$$\begin{aligned} \dot{r} &= ar + 0(r^2) \\ \dot{\theta} &= b + 0(r). \end{aligned}$$

Let $r(t, r_0, \theta_0)$, $\theta(t, r_0, \theta_0)$ be the solution of this system satisfying $r(0, r_0, \theta_0) = r_0$ and $\theta(0, r_0, \theta_0) = \theta_0$. Then for $r_0 > 0$ sufficiently small and $b > 0$, $\theta(t, r_0, \theta_0)$ is a strictly increasing function of t. Let $t(\theta, r_0, \theta_0)$ be the inverse of this strictly increasing function and for a fixed θ_0, define the function

$$P(r_0) = r(t(\theta_0 + 2\pi, r_0, \theta_0), r_0, \theta_0).$$

Then for all sufficiently small $r_0 > 0$, $P(r_0)$ is an analytic function of r_0 which is called *the Poincaré map for the focus* at the origin of (2). Similarly, for $b < 0$, $\theta(t, r_0, \theta_0)$ is a strictly decreasing function of t and the formula

$$P(r_0) = r(t(\theta_0 - 2\pi, r_0, \theta_0), r_0, \theta_0)$$

is used to define the Poincaré map for the focus at the origin in this case. Cf. Figure 4.

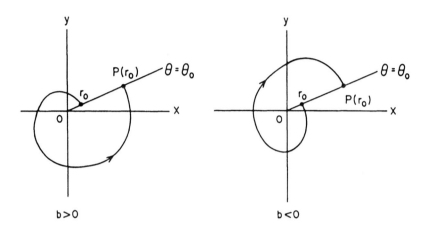

Figure 4. The Poincaré map for a focus at the origin.

The following theorem is proved in [A-II], p. 241.

Theorem 3. *Let $P(s)$ be the Poincaré map for a focus at the origin of the planar analytic system (2) with $b \neq 0$ and suppose that $P(s)$ is defined for $0 < s < \delta_0$. Then there is a $\delta > 0$ such that $P(s)$ can be extended to an analytic function defined for $|s| < \delta$. Furthermore, $P(0) = 0$, $P'(0) = \exp(2\pi a/|b|)$, and if $d(s) = P(s) - s$ then $d(s)d(-s) < 0$ for $0 < |s| < \delta$.*

The fact that $d(s)d(-s) < 0$ for $0 < |s| < \delta$ can be used to show that if

$$d(0) = d'(0) = \cdots = d^{(k-1)}(0) = 0 \quad \text{and} \quad d^{(k)}(0) \neq 0$$

then k is odd; i.e., $k = 2m + 1$. The integer $m = (k-1)/2$ is called the *multiplicity* of the focus. If $m = 0$ the focus is called a *simple focus* and it follows from the above theorem that the system (2), with $b \neq 0$, has a simple focus at the origin iff $a \neq 0$. The sign of $d'(0)$, i.e., the sign of a determines the stability of the origin in this case. If $a < 0$, the origin is a stable focus and if $a > 0$, the origin is an unstable focus. If $d'(0) = 0$, i.e., if $a = 0$, then (2) has a multiple focus or center at the origin. If $d'(0) = 0$ then the first nonzero derivative $\sigma \equiv d^{(k)}(0) \neq 0$ is called the *Liapunov*

3.4. The Poincaré Map

number for the focus. If $\sigma < 0$ then the focus is stable and if $\sigma > 0$ it is unstable. If $d'(0) = d''(0) = 0$ and $d'''(0) \neq 0$ then the Liapunov number for the focus at the origin of (2) is given by the formula

$$\sigma = d'''(0) = \frac{3\pi}{2b}\left\{[3(a_{30}+b_{03})+(a_{12}+b_{21})]\right.$$
$$\left. - \frac{1}{b}[2(a_{20}b_{20}-a_{02}b_{02}) - a_{11}(a_{02}+a_{20})+b_{11}(b_{02}+b_{20})]\right\} \quad (3)$$

where

$$p(x,y) = \sum_{i+j \geq 2} a_{ij} x^i y^j \quad \text{and} \quad q(x,y) = \sum_{i+j \geq 2} b_{ij} x^i y^j$$

in (2); cf. [A-II], p. 252.

This information will be useful in Section 4.4 of Chapter 4 where we shall see that m limit cycles can be made to bifurcate from a multiple focus of multiplicity m under a suitable small perturbation of the system (2).

Problem Set 4

1. Show that $\gamma(t) = (2\cos 2t, \sin 2t)^T$ is a periodic solution of the system

$$\dot{x} = -4y + x\left(1 - \frac{x^2}{4} - y^2\right)$$
$$\dot{y} = x + y\left(1 - \frac{x^2}{4} - y^2\right)$$

that lies on the ellipse $(x/2)^2 + y^2 = 1$; i.e., $\gamma(t)$ represents a cycle Γ of this system. Then use the corollary to Theorem 2 to show that Γ is a stable limit cycle.

2. Show that $\gamma(t) = (\cos t, \sin t, 0)^T$ represents a cycle Γ of the system

$$\dot{x} = -y + x(1 - x^2 - y^2)$$
$$\dot{y} = x + y(1 - x^2 - y^2)$$
$$\dot{z} = z.$$

Rewrite this system in cylindrical coordinates (r, θ, z); solve the resulting system and determine the flow $\phi_t(r_0, \theta_0, z_0)$; for a fixed $\theta_0 \in [0, 2\pi)$, let the plane

$$\Sigma = \{\mathbf{x} \in \mathbf{R}^3 \mid \theta = \theta_0, r > 0, z \in \mathbf{R}\}$$

and determine the Poincaré map $\mathbf{P}(r_0, z_0)$ where $\mathbf{P}\colon \Sigma \to \Sigma$. Compute $D\mathbf{P}(r_0, z_0)$ and show that $D\mathbf{P}(1, 0) = e^{2\pi B}$ where the eigenvalues of B are $\lambda_1 = -2$ and $\lambda_2 = 1$.

3. (a) Solve the linear system $\dot{\mathbf{x}} = A\mathbf{x}$ with

$$A = \begin{bmatrix} a & -b \\ b & a \end{bmatrix}$$

and show that at any point $(x_0, 0)$, on the x-axis, the Poincaré map for the focus at the origin is given by $P(x_0) = x_0 \exp(2\pi a/|b|)$. For $d(x) = P(x) - x$, compute $d'(0)$ and show that $d(-x) = -d(x)$.

(b) Write the system

$$\dot{x} = -y + x(1 - x^2 - y^2)$$
$$\dot{y} = x + y(1 - x^2 - y^2)$$

in polar coordinates and use the Poincaré map $P(r_0)$ determined in Example 1 for $r_0 > 0$ to find the function $P(s)$ of Theorem 3 which is analytic for $|s| < \delta$. (Why is $P(s)$ analytic? What is the domain of analyticity of $P(s)$?) Show that $d'(0) = e^{2\pi} - 1 > 0$ and hence that the origin is a simple focus which is unstable.

4. Show that the system

$$\dot{x} = -y + x(1 - x^2 - y^2)^2$$
$$\dot{y} = x + y(1 - x^2 - y^2)^2$$

has a limit cycle Γ represented by $\gamma(t) = (\cos t, \sin t)^T$. Use Theorem 2 to show that Γ is a multiple limit cycle. Since Γ is a semi-stable limit cycle, cf. Problem 2 in Section 3.3, we know that the multiplicity k of Γ is even. Can you show that $k = 2$?

5. Show that a quadratic system, (2) with $a = 0$, $b \neq 0$,

$$p(x,y) = \sum_{i+j=2} a_{ij} x^i y^j \quad \text{and} \quad q(x,y) = \sum_{i+j=2} b_{ij} x^i y^j,$$

either has a center or a focus of multiplicity $m \geq 2$ at the origin if $a_{20} + a_{02} = b_{20} + b_{02} = 0$.

3.5 The Stable Manifold Theorem for Periodic Orbits

In Section 3.4 we saw that the stability of a limit cycle Γ of a planar system is determined by the derivative of the Poincaré map, $P'(\mathbf{x}_0)$, at a point $\mathbf{x}_0 \in \Gamma$; in fact, if $P'(\mathbf{x}_0) < 1$ then the limit cycle Γ is (asymptotically) stable.

3.5. The Stable Manifold Theorem for Periodic Orbits

In this section we shall see that similar results, concerning the stability of periodic orbits, hold for higher dimensional systems

$$\dot{x} = f(x) \tag{1}$$

with $f \in C^1(E)$ where E is an open subset of $\mathbf{R}^n$. Assume that (1) has a periodic orbit of period T

$$\Gamma: x = \gamma(t), \quad 0 \leq t \leq T,$$

contained in E. In this case, according to the remark in Section 3.4, the derivative of the Poincaré map, $DP(x_0)$, at a point $x_0 \in \Gamma$ is an $(n-1) \times (n-1)$ matrix and we shall see that if $\|DP(x_0)\| < 1$ then the periodic orbit Γ is asymptotically stable.

We first of all show that the $(n-1) \times (n-1)$ matrix $DP(x_0)$ is determined by a fundamental matrix for the linearization of (1) about the periodic orbit Γ. *The linearization of (1) about Γ is defined as the nonautonomous linear system*

$$\dot{x} = A(t)x \tag{2}$$

with

$$A(t) = Df(\gamma(t)).$$

The $n \times n$ matrix $A(t)$ is a continuous, T-periodic function of t for all $t \in \mathbf{R}$. A *fundamental matrix for* (2) is a nonsingular $n \times n$ matrix $\Phi(t)$ which satisfies the matrix differential equation

$$\dot{\Phi} = A(t)\Phi$$

for all $t \in \mathbf{R}$. Cf. Problem 4 in Section 2.2 of Chapter 2. The columns of $\Phi(t)$ consist of n linearly independent solutions of (2) and the solution of (2) satisfying the initial condition $x(0) = x_0$ is given by

$$x(t) = \Phi(t)\Phi^{-1}(0)x_0.$$

Cf. [W], p. 77. For a periodic matrix $A(t)$, we have the following result known as *Floquet's Theorem* which is proved for example in [H] on pp. 60–61 or in [C/L] on p. 79.

Theorem 1. *If $A(t)$ is a continuous, T-periodic matrix, then for all $t \in \mathbf{R}$ any fundamental matrix solution for (2) can be written in the form*

$$\Phi(t) = Q(t)e^{Bt} \tag{3}$$

where $Q(t)$ is a nonsingular, differentiable, T-periodic matrix and B is a constant matrix. Furthermore, if $\Phi(0) = I$ then $Q(0) = I$.

Thus, at least in principle, a study of the nonautonomous linear system (2) can be reduced to a study of an autonomous linear system (with constant coefficients) studied in Chapter 1.

Corollary. *Under the hypotheses of Theorem 1, the nonautonomous linear system (2), under the linear change of coordinates*

$$\mathbf{y} = Q^{-1}(t)\mathbf{x},$$

reduces to the autonomous linear system

$$\dot{\mathbf{y}} = B\mathbf{y}. \tag{4}$$

Proof. According to Theorem 1, $Q(t) = \Phi(t)e^{-Bt}$. It follows that

$$\begin{aligned} Q'(t) &= \Phi'(t)e^{-Bt} - \Phi(t)e^{-Bt}B \\ &= A(t)\Phi(t)e^{-Bt} - \Phi(t)e^{-Bt}B \\ &= A(t)Q(t) = Q(t)B, \end{aligned}$$

since e^{-Bt} and B commute. And if

$$\mathbf{y}(t) = Q^{-1}(t)\mathbf{x}(t)$$

or equivalently, if

$$\mathbf{x}(t) = Q(t)\mathbf{y}(t),$$

then

$$\begin{aligned} \mathbf{x}'(t) &= Q'(t)\mathbf{y}(t) + Q(t)\mathbf{y}'(t) \\ &= A(t)Q(t)\mathbf{y}(t) - Q(t)B\mathbf{y}(t) + Q(t)\mathbf{y}'(t) \\ &= A(t)\mathbf{x}(t) + Q(t)[\mathbf{y}'(t) - B\mathbf{y}(t)]. \end{aligned}$$

Thus, since $Q(t)$ is nonsingular, $\mathbf{x}(t)$ is a solution of (2) if and only if $\mathbf{y}(t)$ is a solution of (4).

However, determining the matrix $Q(t)$ which reduces (2) to (4) or determining a fundamental matrix for (2) is in general a difficult problem which requires series methods and the theory of special functions.

As we shall see, if $\Phi(t)$ is a fundamental matrix for (2) which satisfies $\Phi(0) = I$, then $\|D\mathbf{P}(\mathbf{x}_0)\| = \|\Phi(T)\|$ for any point $\mathbf{x}_0 \in \Gamma$. It then follows from Theorem 1 that $\|D\mathbf{P}(\mathbf{x}_0)\| = \|e^{BT}\|$. The eigenvalues of e^{BT} are given by $e^{\lambda_j T}$ where λ_j, $j = 1, \ldots, n$, are the eigenvalues of the matrix B. The eigenvalues of B, λ_j, are called *characteristic exponents* of $\gamma(t)$ and the eigenvalues of e^{BT}, $e^{\lambda_j T}$, are called the *characteristic multipliers* of $\gamma(t)$. Even though the characteristic exponents, λ_j, are only determined modulo $2\pi i$, they suffice to uniquely determine the magnitudes of the characteristic multipliers $e^{\lambda_j T}$ which determine the stability of the periodic orbit Γ. This is made precise in what follows.

For $\mathbf{x} \in E$, let $\phi_t(\mathbf{x}) = \phi(t, \mathbf{x})$ be the flow of the system (1) and let $\gamma(t) = \phi_t(\mathbf{x}_0)$ be a periodic orbit of (1) through the point $\mathbf{x}_0 \in E$. Then since $\phi(t, \mathbf{x})$ satisfies the differential equation (1) for all $t \in \mathbf{R}$, the matrix

$$H(t, \mathbf{x}) = D\phi_t(\mathbf{x})$$

3.5. The Stable Manifold Theorem for Periodic Orbits

satisfies
$$\frac{\partial H(t, \mathbf{x})}{\partial t} = D\mathbf{f}(\phi_t(\mathbf{x}))H(t, \mathbf{x})$$
for all $t \in \mathbf{R}$. Thus, for $\mathbf{x} = \mathbf{x}_0$ and $t \in \mathbf{R}$, the function $\Phi(t) = H(t, \mathbf{x}_0)$ satisfies
$$\dot{\Phi} = D\mathbf{f}(\gamma(t))\Phi$$
and
$$\Phi(0) = I.$$
(Cf. the corollary in Section 2.3 of Chapter 2.) That is, $H(t, \mathbf{x}_0)$ is the fundamental matrix for (2) which satisfies $\Phi(0) = H(0, \mathbf{x}_0) = I$. Thus, by Theorem 1,
$$H(t, \mathbf{x}_0) = Q(t)e^{Bt}$$
where $Q(t)$ is T-periodic and satisfies $Q(0) = I$. It follows that
$$H(T, \mathbf{x}_0) = e^{BT}.$$

The next theorem shows that one of the characteristic exponents of $\gamma(t)$ is always zero, i.e., one of the characteristic multipliers is always 1, and if we suppose that $\lambda_n = 0$ and choose the basis for $\mathbf{R}^n$ so that the last column of e^{BT} is $(0, \ldots, 0, 1)^T$, then $D\mathbf{P}(\mathbf{x}_0)$ is the $(n-1) \times (n-1)$ matrix obtained by deleting the nth row and column in e^{BT}. As in the remark in Section 3.4, there is no loss in generality in assuming that the origin has been translated to the point $\mathbf{x}_0 \in \Gamma$.

Theorem 2. *Suppose that* $\mathbf{f} \in C^1(E)$ *where* E *is an open subset of* $\mathbf{R}^n$ *and that* $\gamma(t) = \phi_t(\mathbf{0})$ *is a periodic orbit of* (1) *contained in* E. *For* $\delta > 0$ *sufficiently small and* $\mathbf{x} \in \Sigma \cap N_\delta(\mathbf{0})$, *let* $\mathbf{P}(\mathbf{x})$ *be the Poincaré map for* $\gamma(t)$ *at* $\mathbf{0}$ *where* Σ *is the* $(n-1)$-*dimensional hyperplane orthogonal to* Γ *at* $\mathbf{0}$. *If* $\lambda_1, \ldots, \lambda_n$ *are the characteristic exponents of* $\gamma(t)$, *then one of them, say* λ_n, *is zero and the characteristic multipliers* $e^{\lambda_j T}$, $j = 1, \ldots, n-1$, *are the eigenvalues of* $D\mathbf{P}(\mathbf{0})$. *In fact, if the basis for* $\mathbf{R}^n$ *is chosen so that* $\mathbf{f}(\mathbf{0}) = (0, \ldots, 0, 1)^T$ *then the last column of* $H(T, \mathbf{0}) = D\phi_T(\mathbf{0})$ *is* $(0, \ldots, 0, 1)^T$ *and* $D\mathbf{P}(\mathbf{0})$ *is obtained by deleting the last row and column in* $H(T, \mathbf{0})$.

Proof. Since the periodic orbit $\gamma(t) = \phi_t(\mathbf{0})$ satisfies
$$\gamma'(t) = \mathbf{f}(\gamma(t)),$$
it follows that
$$\gamma''(t) = D\mathbf{f}(\gamma(t))\gamma'(t).$$
Thus, $\gamma'(t)$ is a solution of the nonautonomous linear system (2) and the initial condition $\gamma'(0) = \mathbf{f}(\mathbf{0})$. Then since the matrix
$$\Phi(t) = H(t, \mathbf{0}),$$

defined above, is the fundamental matrix for (2) satisfying $\Phi(0) = I$, it follows that
$$\gamma'(t) = \Phi(t)\mathbf{f}(0).$$
For $t = T$, $\gamma'(T) = \mathbf{f}(0)$ and we see that
$$H(T,0)\mathbf{f}(0) = \mathbf{f}(0);$$
i.e., 1 is an eigenvalue of $H(T,0)$ with eigenvector $\mathbf{f}(0)$. Since $H(T,0) = e^{BT}$, this implies that one of the eigenvalues of B is zero. Suppose that $\lambda_n = 0$ and that the basis for $\mathbf{R}^n$ has been chosen so that $\mathbf{f}(0) = (0,\ldots,0,1)^T$. Then $(0,\ldots,0,1)^T$ is an eigenvector of $H(T,0)$ corresponding to the eigenvalue $e^{\lambda_n T} = 1$. It follows that the last column of $H(T,0)$ is $(0,\ldots,0,1)^T$.

For $\delta > 0$ sufficiently small and $\mathbf{x} \in N_\delta(0)$, define the function
$$\mathbf{h}(\mathbf{x}) = \phi(\tau(\mathbf{x}),\mathbf{x})$$
where $\tau(\mathbf{x})$ is the function defined by Theorem 1 in Section 3.4. Then the Poincaré map $\mathbf{P}$ is the restriction of $\mathbf{h}$ to the subspace Σ. Since
$$D\mathbf{h}(\mathbf{x}) = D\phi(\tau(\mathbf{x}),\mathbf{x})$$
$$= \frac{\partial \phi}{\partial t}(\tau(\mathbf{x}),\mathbf{x})D\tau(\mathbf{x}) + D\phi_{\tau(\mathbf{x})}(\mathbf{x})$$
$$= \frac{\partial \phi}{\partial t}(\tau(\mathbf{x}),\mathbf{x})D\tau(\mathbf{x}) + H(\tau(\mathbf{x}),\mathbf{x}),$$
for $\mathbf{x} = 0$ we obtain
$$D\mathbf{h}(0) = \mathbf{f}(0)D\tau(0) + H(T,0)$$
$$= \begin{bmatrix} 0 & \cdots & 0 & 0 \\ & & & \vdots \\ 0 & \cdots & 0 & 0 \\ \frac{\partial \tau}{\partial x_1}(0) & \cdots & \frac{\partial \tau}{\partial x_n}(0) \end{bmatrix} + \begin{bmatrix} & & 0 \\ \tilde{H}(T,0) & & \vdots \\ & & 0 \\ \cdots & & 1 \end{bmatrix}.$$

And since with respect to the above basis $D\mathbf{P}(0)$ consists of the first $(n-1)$ rows and columns of $D\mathbf{h}(0)$, it follows that
$$D\mathbf{P}(0) = \tilde{H}(T,0)$$
where $\tilde{H}(T,0)$ consists of the first $(n-1)$ rows and columns of $H(T,0)$.

Remark. Even though it is generally very difficult to determine a fundamental matrix $\Phi(t)$ for (2), this theorem gives us a means of computing $D\mathbf{P}(\mathbf{x}_0)$ numerically; i.e., $D\mathbf{P}(\mathbf{x}_0)$ can be found by determining the effect

3.5. The Stable Manifold Theorem for Periodic Orbits

on the periodic orbit $\gamma(t) = \phi_t(\mathbf{x}_0)$ of small variations in the initial conditions $\mathbf{x}_0 \in \mathbf{R}^n$ defining the periodic orbit $\gamma(t)$. In fact, in a coordinate system with its origin at a point on the periodic orbit Γ and the x_n-axis tangent to Γ at that point and pointing in the same direction as the motion along Γ,

$$DP(0) = \left[\frac{\partial \phi_i}{\partial x_j}(T, 0)\right]$$

where $i, j = 1, \ldots, (n-1)$ and $\phi(t, \mathbf{x}) = \phi_t(\mathbf{x})$ is the flow defined by (1).

As was noted earlier, the stability of the periodic orbit $\gamma(t)$ is determined by the characteristic exponents $\lambda_1, \ldots, \lambda_{n-1}$ or by the characteristic multipliers $e^{\lambda_1 T}, \ldots, e^{\lambda_{n-1} T}$. This is made precise in the next theorem. The proof of this theorem is similar to the proof of the Stable Manifold Theorem in Section 2.7 in Chapter 2; cf. [H], p. 255.

Theorem (The Stable Manifold Theorem for Periodic Orbits). *Let $\mathbf{f} \in C^1(E)$ where E is an open subset of $\mathbf{R}^n$ containing a periodic orbit*

$$\Gamma \colon \mathbf{x} = \gamma(t)$$

of (1) *of period T. Let ϕ_t be the flow of* (1) *and $\gamma(t) = \phi_t(\mathbf{x}_0)$. If k of the characteristic exponents of $\gamma(t)$ have negative real part where $0 \leq k \leq n-1$ and $n-k-1$ of them have positive real part then there is a $\delta > 0$ such that the stable manifold of Γ,*

$$S(\Gamma) = \{\mathbf{x} \in N_\delta(\Gamma) \mid d(\phi_t(\mathbf{x}), \Gamma) \to 0 \text{ as } t \to \infty$$
$$\text{and } \phi_t(\mathbf{x}) \in N_\delta(\Gamma) \text{ for } t \geq 0\}$$

is a $(k+1)$-dimensional, differentiable manifold which is positively invariant under the flow ϕ_t and the unstable manifold of Γ,

$$U(\Gamma) = \{\mathbf{x} \in N_\delta(\Gamma) \mid d(\phi_t(\mathbf{x}), \Gamma) \to 0 \text{ as } t \to -\infty$$
$$\text{and } \phi_t(\mathbf{x}) \in N_\delta(\Gamma) \text{ for } t \leq 0\}$$

is an $(n-k)$-dimensional, differentiable manifold which is negatively invariant under the flow ϕ_t. Furthermore, the stable and unstable manifolds of Γ intersect transversally in Γ.

Remark 1. The local stable and unstable manifolds of Γ, $S(\Gamma)$ and $U(\Gamma)$, in the Stable Manifold Theorem can be used to define the global stable and unstable manifolds of Γ, $W^s(\Gamma)$ and $W^u(\Gamma)$, as in Section 3.3. It can be shown that $W^s(\Gamma)$ and $W^u(\Gamma)$ are unique, invariant, differentiable manifolds which have the same dimensions as $S(\Gamma)$ and $U(\Gamma)$ respectively. Also, if $\mathbf{f} \in C^r(E)$, then it can be shown that $W^s(\Gamma)$ and $W^u(\Gamma)$ are manifolds of class C^r.

Remark 2. Suppose that the origin has been translated to the point $\mathbf{x}_0$ so that $\gamma(t) = \phi_t(0)$. Let $\lambda_j = a_j + ib_j$ be the characteristic exponents of the periodic orbit $\gamma(t)$ and let $e^{\lambda_j T}$ be the characteristic multipliers of $\gamma(t)$, i.e., $e^{\lambda_j T}$ are the eigenvalues of the real $n \times n$ matrix

$$\Phi(T) = H(T, 0).$$

Furthermore, suppose that $\mathbf{u}_1 = \mathbf{f}(0)$ and that we have a basis of generalized eigenvectors of $\Phi(T)$ for $\mathbf{R}^n$ given by $\{\mathbf{u}_1, \ldots, \mathbf{u}_k, \mathbf{u}_{k+1}, \mathbf{v}_{k+1}, \ldots, \mathbf{u}_m, \mathbf{v}_m\}$ as in Section 1.9 of Chapter 1. We then define *the stable, unstable and center subspaces of the periodic orbit* Γ at the point $\mathbf{0} \in \Gamma$ as

$$E^s(\Gamma) = \text{Span}\{\mathbf{u}_j, \mathbf{v}_j \mid a_j < 0\}$$
$$E^c(\Gamma) = \text{Span}\{\mathbf{u}_j, \mathbf{v}_j \mid a_j = 0\}$$
$$E^u(\Gamma) = \text{Span}\{\mathbf{u}_j, \mathbf{v}_j \mid a_j > 0\}.$$

It can then be shown that the stable and unstable manifolds $W^s(\Gamma)$ and $W^u(\Gamma)$ of Γ are tangent to the stable and unstable subspaces $E^s(\Gamma)$ and $E^u(\Gamma)$ of Γ at the point $\mathbf{0} \in \Gamma$ respectively.

If $\text{Re}(\lambda_j) \neq 0$ for $j = 1, \ldots, n-1$, this theorem determines the behavior of trajectories near Γ. If $\text{Re}(\lambda_j) \neq 0$ for $j = 1, \ldots, n-1$ then Γ is called a *hyperbolic periodic orbit*. The periodic orbit in Example 1 of Section 3.3 is a hyperbolic periodic orbit with a two-dimensional stable manifold and two-dimensional unstable manifold. The characteristic exponents for that example are determined later in this section. If two or more of the characteristic exponents have zero real part then the periodic orbit Γ has a nontrivial center manifold, denoted by $W^c(\Gamma)$. In Example 2 in Section 3.2, the unit sphere S^2 is the center manifold for the periodic orbit $\gamma(t) = (\cos t, \sin t, 0)^T$.

The next theorem shows that not only do trajectories in the stable manifold $S(\Gamma)$ approach Γ as $t \to \infty$, but that the motion along trajectories in $S(\Gamma)$ is synchronized with the motion along Γ. We assume that the characteristic exponents have been ordered so that $\text{Re}(\lambda_j) < 0$ for $j = 1, \ldots, k$ and $\text{Re}(\lambda_j) > 0$ for $j = k+1, \ldots, n-1$. This theorem is proved for example, in Hartman [H], p. 254.

Theorem 3. *Under the hypotheses of the above theorem, there is an $\alpha > 0$ and a $K > 0$ such that $\text{Re}(\lambda_j) \leq -\alpha$ for $j = 1, \ldots, k$ and $\text{Re}(\lambda_j) \geq \alpha$ for $j = k+1, \ldots, n-1$ and for each $\mathbf{x} \in S(\Gamma)$ there exists an asymptotic phase t_0 such that for all $t \geq 0$*

$$|\phi_t(\mathbf{x}) - \gamma(t - t_0)| < K e^{-\alpha t/T};$$

and similarly for each $\mathbf{x} \in U(\Gamma)$ there exists an asymptotic phase t_0 such that for all $t \leq 0$

$$|\phi_t(\mathbf{x}) - \gamma(t - t_0)| < K e^{\alpha t/T};$$

Example 1. Consider the system

$$\dot{x} = x - y - x^3 - xy^2$$

3.5. The Stable Manifold Theorem for Periodic Orbits

$$\dot{y} = x + y - x^2y - y^3$$
$$\dot{z} = \lambda z.$$

Cf. Example 1 in Section 3.3. There is a periodic orbit

$$\Gamma: \gamma(t) = (\cos t, \sin t, 0)^T$$

of period $T = 2\pi$. We compute

$$D\mathbf{f}(\mathbf{x}) = \begin{bmatrix} 1 - 3x^2 - y^2 & -1 - 2xy & 0 \\ 1 - 2xy & 1 - x^2 - 3y^2 & 0 \\ 0 & 0 & \lambda \end{bmatrix}$$

where $\mathbf{x} = (x, y, z)^T$. The linearization of the above system about the periodic orbit $\gamma(t)$ is then given by

$$\dot{\mathbf{x}} = A(t)\mathbf{x}$$

where the periodic matrix

$$A(t) = \begin{bmatrix} -2\cos^2 t & -1 - \sin 2t & 0 \\ 1 - \sin 2t & -2\sin^2 t & 0 \\ 0 & 0 & \lambda \end{bmatrix}$$

This nonautonomous linear system has the fundamental matrix

$$\Phi(t) = \begin{bmatrix} e^{-2t}\cos t & -\sin t & 0 \\ e^{-2t}\sin t & \cos t & 0 \\ 0 & 0 & e^{\lambda t} \end{bmatrix}$$

which satisfies $\Phi(0) = I$. The serious student should verify that $\Phi(t)$ satisfies $\dot{\Phi} = A(t)\Phi$; cf. Problem 3 in Section 1.10 of Chapter 1. Once we have found the fundamental matrix for the linearized system, it is easy to see that

$$\Phi(t) = Q(t)e^{Bt}$$

where the periodic matrix

$$Q(t) = \begin{bmatrix} \cos t & -\sin t & 0 \\ \sin t & \cos t & 0 \\ 0 & 0 & 1 \end{bmatrix}$$

and

$$B = \begin{bmatrix} -2 & 0 & 0 \\ 0 & 0 & 0 \\ 0 & 0 & \lambda \end{bmatrix}.$$

As in Theorem 2, deleting the row and column containing the zero eigenvalue determines

$$D\mathbf{P}(\mathbf{x}_0) = \begin{bmatrix} e^{-2T} & 0 \\ 0 & e^{\lambda T} \end{bmatrix}$$

where $\mathbf{x}_0 = (1,0,0)^T$, $T = 2\pi$, and the characteristic exponents $\lambda_1 = -2$ and $\lambda_2 = \lambda$. For $\lambda > 0$, as in Example 1 in Section 3.3, there is a two-dimensional stable manifold $W^s(\Gamma)$ and a two-dimensional unstable manifold $W^u(\Gamma)$ which intersect orthogonally in γ; cf. Figure 1 in Section 3.3. For $\lambda < 0$, there is a three-dimensional stable manifold $\mathbf{R}^3 \sim \{0\}$, and a one-dimensional unstable manifold, Γ. If $\lambda = 0$, there is a two-dimensional stable manifold and a two-dimensional center manifold; the center manifold $W^c(\Gamma)$ is the unit cylinder.

This example is more easily treated by writing the above system in cylindrical coordinates as

$$\dot{r} = r(1-r^2)$$
$$\dot{\theta} = 1$$
$$\dot{z} = \lambda z.$$

The Poincaré map can then be computed directly by solving the above system as in Example 1 of Section 3.4:

$$\mathbf{P}(r,z) = \begin{pmatrix} \left[1 + \left(\frac{1}{r^2} - 1\right)e^{-4\pi}\right]^{-1/2} \\ ze^{2\pi\lambda} \end{pmatrix}.$$

It then follows that

$$D\mathbf{P}(r,z) = \begin{bmatrix} r^3 e^{-4\pi}\left[1 + \left(\frac{1}{r^2} - 1\right)e^{-4\pi}\right]^{-3/2} & 0 \\ 0 & e^{2\pi\lambda} \end{bmatrix}$$

and that

$$D\mathbf{P}(1,0) = \begin{bmatrix} e^{-4\pi} & 0 \\ 0 & e^{2\pi\lambda} \end{bmatrix}$$

as above. However, converting to polar coordinates does not always simplify the problem and allow us to compute the Poincaré map; cf. Problems 2 and 8 below. Just as there is a center manifold tangent to the center subspace E^c at a nonhyperbolic equilibrium point of (1), there is a center manifold tangent to the center subspace $E^c(\Gamma)$ of a nonhyperbolic periodic orbit Γ. We cite this result for future reference; cf., e.g., [Ru], p. 32.

Theorem 4 (The Center Manifold Theorem for Periodic Orbits). *Let $\mathbf{f} \in C^r(E)$ with $r \geq 1$ where E is an open subset of $\mathbf{R}^n$ containing a periodic orbit*

$$\Gamma: \mathbf{x} = \gamma(t)$$

of (1) of period T. Let ϕ_t be the flow of (1) and let $\gamma(t) = \phi_t(\mathbf{x}_0)$. If k of the characteristic exponents have negative real part, j have positive real part and $m = n-k-j$ have zero real part, then there is an m-dimensional center

3.5. The Stable Manifold Theorem for Periodic Orbits

manifold of Γ, $W^c(\Gamma)$, of class C^r which is invariant under the flow ϕ_t. Furthermore, $W^s(\Gamma), W^u(\Gamma)$ and $W^c(\Gamma)$ intersect transversally in Γ and if the origin has been translated to the point $\mathbf{x}_0$ so that $\gamma(t) = \phi_t(\mathbf{0})$, then $W^c(\Gamma)$ is tangent to the center subspace of Γ, $E^c(\Gamma)$, at the point $\mathbf{0} \in \Gamma$.

We next give some geometrical examples of stable manifolds, unstable manifolds and center manifolds of periodic orbits that can occur in $\mathbf{R}^3$ (Figures 1–4):

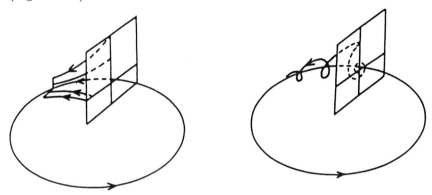

Figure 1. Periodic orbits with a three-dimensional stable manifold. These periodic orbits are asymptotically stable.

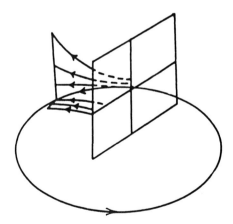

Figure 2. A periodic orbit with two-dimensional stable and unstable manifolds. This periodic orbit is unstable.

We conclude this section with one last result concerning the stability of periodic orbits for systems in $\mathbf{R}^n$ which is similar to the corollary in Section 3.4 for limit cycles of planar systems. This theorem is proved in [H] on p. 256.

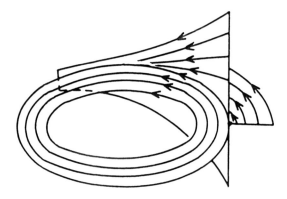

Figure 3. A periodic orbit with two-dimensional stable and center manifolds. This periodic orbit is stable, but not asymptotically stable.

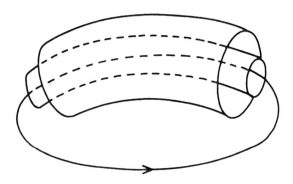

Figure 4. A periodic orbit with a three-dimensional center manifold. This periodic orbit is stable, but not asymptotically stable.

Theorem 5. Let $\mathbf{f} \in C^1(E)$ where E is an open subset of $\mathbf{R}^n$ containing a periodic orbit $\gamma(t)$ of (1) of period T. Then $\gamma(t)$ is not asymptotically stable unless

$$\int_0^T \nabla \cdot \mathbf{f}(\gamma(t))\, dt \leq 0.$$

Note that in Example 1, $\nabla \cdot \mathbf{f}(\mathbf{x}) = 2 - 4r^2 + \lambda$ and

$$\int_0^T \nabla \cdot \mathbf{f}(\gamma(t))\, dt = (\lambda - 2)2\pi > 0$$

if $\lambda > 2$. And $\lambda > 2$ certainly implies that the periodic orbit $\gamma(t)$ in that example is not asymptotically stable. In fact, we saw that $\lambda > 0$ implies

3.5. The Stable Manifold Theorem for Periodic Orbits

that $\gamma(t)$ is a periodic orbit of saddle type which is unstable. This example shows that for dimension $n \geq 3$, the condition

$$\int_0^T \nabla \cdot \mathbf{f}(\gamma(t))\, dt < 0$$

does not imply that $\gamma(t)$ is an asymptotically stable periodic orbit as it does for $n = 2$; cf. the corollary in Section 3.4.

Problem Set 5

1. Show that the nonlinear system

$$\dot{x} = -y + xz^2$$
$$\dot{y} = x + yz^2$$
$$\dot{z} = -z(x^2 + y^2)$$

has a periodic orbit $\gamma(t) = (\cos t, \sin t, 0)^T$. Find the linearization of this system about $\gamma(t)$, the fundamental matrix $\Phi(t)$ for this (autonomous) linear system which satisfies $\Phi(0) = I$, and the characteristic exponents and multipliers of $\gamma(t)$. What are the dimensions of the stable, unstable and center manifolds of $\gamma(t)$?

2. Consider the nonlinear system

$$\dot{x} = x - 4y - \frac{x^3}{4} - xy^2$$
$$\dot{y} = x + y - \frac{x^2 y}{4} - y^3$$
$$\dot{z} = z.$$

Show that $\gamma(t) = (2\cos 2t, \sin 2t, 0)^T$ is a periodic solution of this system of period $T = \pi$; cf. Problem 1 in Section 3.4. Determine the linearization of this system about the periodic orbit $\gamma(t)$,

$$\dot{\mathbf{x}} = A(t)\mathbf{x},$$

and show that

$$\Phi(t) = \begin{bmatrix} e^{-2t}\cos 2t & -2\sin 2t & 0 \\ \frac{1}{2}e^{-2t}\sin 2t & \cos 2t & 0 \\ 0 & 0 & e^t \end{bmatrix}$$

is the fundamental matrix for this nonautonomous linear system which satisfies $\Phi(0) = I$. Write $\Phi(t)$ in the form of equation (3), determine the characteristic exponents and the characteristic multipliers of $\gamma(t)$, and determine the dimensions of the stable and unstable manifolds of $\gamma(t)$. Sketch the periodic orbit $\gamma(t)$ and a few trajectories in the stable and unstable manifolds of $\gamma(t)$.

3. If $\Phi(t)$ is the fundamental matrix for (2) which satisfies $\Phi(0) = I$, show that for all $\mathbf{x}_0 \in \mathbf{R}^n$ and $t \in \mathbf{R}$, $\mathbf{x}(t) = \Phi(t)\mathbf{x}_0$ is the solution of (2) which satisfies the initial condition $\mathbf{x}(0) = \mathbf{x}_0$.

4. (a) Solve the linear system
$$\dot{x} = -y$$
$$\dot{y} = x$$
$$\dot{z} = y$$
with the initial condition $\mathbf{x}(0) = \mathbf{x}_0 = (x_0, y_0, z_0)^T$.

(b) Let $\mathbf{u}(t, \mathbf{x}_0) = \phi_t(\mathbf{x}_0)$ be the solution of this system and compute
$$\Phi(t) = D\phi_t(\mathbf{x}_0) = \frac{\partial \mathbf{u}}{\partial \mathbf{x}_0}(t, \mathbf{x}_0).$$

(c) Show that $\gamma(t) = (\cos t, \sin t, 1 - \cos t)^T$ is a periodic solution of this system and that $\Phi(t)$ is the fundamental matrix for (2) which satisfies $\Phi(0) = I$.

5. (a) Let $\Phi(t)$ be the fundamental matrix for (2) which satisfies $\Phi(0) = I$. Use Liouville's Theorem, (cf. [H], p. 46) which states that
$$\det \Phi(t) = \exp \int_0^t \mathrm{tr} A(s) ds,$$
to show that if $m_j = e^{\lambda_j T}$, $j = 1, \ldots, n$ are the characteristic multipliers of $\gamma(t)$ then
$$\sum_{j=1}^n m_j = \mathrm{tr}\Phi(T)$$
and
$$\prod_{j=1}^n m_j = \exp \int_0^T \mathrm{tr} A(t) dt.$$

(b) For a two-dimensional system, (2) with $\mathbf{x} \in \mathbf{R}^2$, use the above result and the fact that one of the multipliers is equal to one, say $m_2 = 1$, to show that the characteristic exponent
$$\lambda_1 = \frac{1}{T} \int_0^T \mathrm{tr} A(t) dt = \frac{1}{T} \int_0^T \nabla \cdot \mathbf{f}(\gamma(t)) dt$$
(cf. [A-II], p. 118) and that
$$\mathrm{tr}\Phi(T) = 1 + \exp \int_0^T \nabla \cdot \mathbf{f}(\gamma(t)) dt.$$

3.5. The Stable Manifold Theorem for Periodic Orbits

6. Use Liouville's Theorem (given in Problem 5) and the fact that $H(t, \mathbf{x}_0) = \Phi(t)$ is the fundamental matrix for (2) which satisfies $\Phi(0) = I$ to show that for all $t \in \mathbf{R}$

$$\det H(t, \mathbf{x}_0) = \exp \int_0^t \nabla \cdot \mathbf{f}(\gamma(s)) \, ds$$

where $\gamma(t) = \phi_t(\mathbf{x}_0)$ is a periodic solution of (1); cf. Problem 5 in Section 2.3 of Chapter 2.

7. (a) Suppose that the linearization of (1) about a periodic orbit Γ of (1) has a fundamental matrix solution given by

$$\Phi(t) = \begin{bmatrix} \cos t & -\sin t & 0 & 0 & 0 \\ \sin t & \cos t & 0 & 0 & 0 \\ 0 & 0 & \dfrac{4e^t - e^{-2t}}{3} & \dfrac{2(e^{-2t} - e^t)}{3} & 0 \\ 0 & 0 & \dfrac{2(e^t - e^{-2t})}{3} & \dfrac{4e^{-2t} - e^t}{3} & 0 \\ 0 & 0 & 0 & 0 & 1 \end{bmatrix}.$$

Find the characteristic exponents of the periodic orbit Γ and the dimensions of $W^s(\Gamma)$, $W^u(\Gamma)$ and $W^c(\Gamma)$.

(b) Same thing for

$$\Phi(t) = \begin{bmatrix} e^{-3t} \cos 2t & -e^{-3t} \sin 2t & 0 & 0 & 0 \\ e^{-3t} \sin 2t & e^{-3t} \cos 2t & 0 & 0 & 0 \\ 0 & 0 & e^{3t} & 0 & 0 \\ 0 & 0 & te^{3t} & e^{3t} & 0 \\ 0 & 0 & 0 & 0 & 1 \end{bmatrix}.$$

8. Consider the nonlinear system

$$\dot{x} = -2y + ax(4 - 4x^2 - y^2 + z)$$
$$\dot{y} = 8x + ay(4 - 4x^2 - y^2 + z)$$
$$\dot{z} = z(x - a^2)$$

where a is a parameter. Show that $\gamma(t) = (\cos 4t, 2 \sin 4t, 0)^T$ is a periodic solution of this system of period $T = \pi/2$. Determine the linearization of this system about the periodic orbit $\gamma(t)$,

$$\dot{\mathbf{x}} = A(t)\mathbf{x},$$

and show that

$$\Phi(t) = \begin{bmatrix} e^{-8at} \cos 4t & -\dfrac{1}{2} \sin 4t & \alpha(t) e^{-a^2 t + \frac{1}{4} \sin 4t} \\ 2e^{-8at} \sin 4t & \cos 4t & \beta(t) e^{-a^2 t + \frac{1}{4} \sin 4t} \\ 0 & 0 & e^{-a^2 t + \frac{1}{4} \sin 4t} \end{bmatrix}$$

is the fundamental matrix for this nonautonomous linear system which satisfies $\Phi(0) = I$, where $\alpha(t)$ and $\beta(t)$ are $\pi/2$-periodic functions which satisfy the nonhomogeneous linear system

$$\begin{pmatrix} \dot\alpha \\ \dot\beta \end{pmatrix} = \begin{bmatrix} a(a-4) - \cos 4t - 4a\cos 8t & -2 - 2a\sin 8t \\ 8 - 8a\sin 8t & a(a-4) - \cos 4t + 4a\cos 8t \end{bmatrix}$$

$$\cdot \begin{pmatrix} \alpha \\ \beta \end{pmatrix} + \begin{pmatrix} a\cos 4t \\ 2a\sin 4t \end{pmatrix}$$

and the initial conditions $\alpha(0) = \beta(0) = 0$. (This latter system can be solved using Theorem 1 and Remark 1 in Section 1.10 of Chapter 1 and the result of Problem 4 in Section 2.2 of Chapter 2; however, this is not necessary for our purposes.) Write $\Phi(t)$ in the form of equation (3), show that the characteristic exponents $\lambda_1 = -8a$ and $\lambda_2 = -a^2$, determine the characteristic multipliers of $\gamma(t)$, and determine the dimensions of the stable and unstable manifolds of $\gamma(t)$ for $a > 0$ and for $a < 0$.

3.6 Hamiltonian Systems with Two-Degrees of Freedom

Just as the Hamiltonian systems with one-degree of freedom offered some interesting examples which illustrated the general theory developed in Chapter 2, Hamiltonian systems with two-degrees of freedom give us some interesting examples which further illustrate the nature of the invariant manifolds $W^s(\Gamma)$, $W^u(\Gamma)$ and $W^c(\Gamma)$, of a periodic orbit Γ, discussed in this chapter. Before presenting these examples, we first show how projective geometry can be used to project the flow on an n-dimensional manifold in $\mathbf{R}^{n+1}$ onto a flow in $\mathbf{R}^n$. We begin with a simple example where we project the unit sphere $S^2 = \{\mathbf{x} \in \mathbf{R}^3 \mid |\mathbf{x}| = 1\}$ in $\mathbf{R}^3$ onto the (X, Y)-plane as indicated in Figure 1.

It follows from the similar triangles shown in Figure 2 that the equations defining (X, Y) in terms of (x, y, z) are given by

$$X = \frac{x}{1-z}, \qquad Y = \frac{y}{1-z}.$$

These equations set up a one-to-one correspondence between points (x, y, z) on the unit sphere S^2 with the north pole deleted and points $(X, Y) \in \mathbf{R}^2$. Points inside the circle $X^2 + Y^2 = 1$ correspond to points on the lower hemisphere, the origin in the (X, Y)-plane corresponds to the point $(0, 0, -1)$, and points outside the circle $X^2 + Y^2 = 1$ correspond to points on the upper hemisphere; the point $(0, 0, 1) \in S^2$ corresponds to the "point

3.6. Hamiltonian Systems with Two-Degrees of Freedom

at infinity." We can use this type of projective geometry to visualize flows on n-dimensional manifolds. The sphere in Figure 1 is referred to as the Bendixson sphere.

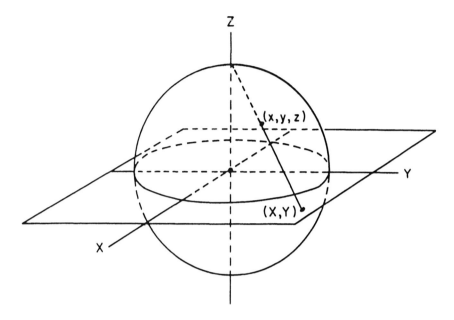

Figure 1. Stereographic projection of S^2 onto the (X,Y)-plane.

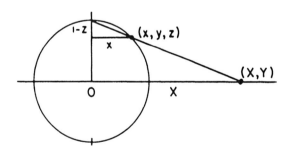

Figure 2. A cross-section of the sphere in Figure 1.

Example 1. Consider the system

$$\dot{x} = -y + xz^2$$
$$\dot{y} = x + yz^2$$
$$\dot{z} = -z(x^2 + y^2).$$

It is easy to see that for $V(x,y,z) = x^2 + y^2 + z^2$ we have $\dot{V} = 0$ along trajectories of this system. Thus, trajectories lie on the spheres $x^2 + y^2 +$

$z^2 = k^2$. The flow on any one of these spheres is topologically equivalent to the flow on the unit two-dimensional sphere S^2 shown in Figure 3. The

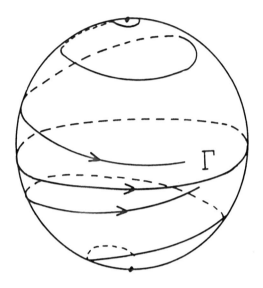

Figure 3. A flow on the unit sphere S^2.

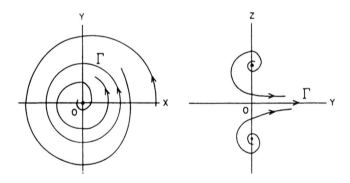

Figure 4. Various planar representations of the flow on S^2 shown in Figure 3.

flow on the equator of S^2 represents a periodic orbit
$$\Gamma: \gamma(t) = (\cos t, \sin t, 0)^T.$$
The periodic orbit Γ has a two dimensional stable manifold $W^s(\Gamma) = S^2 \sim \{(0, 0, \pm 1)\}$. If we project from the north pole of S^2 onto the (X, Y)-plane as in Figure 1, we obtain the flow in the (X, Y)-plane shown in Figure 4(a).

3.6. Hamiltonian Systems with Two-Degrees of Freedom

The periodic orbit Γ gets mapped onto the unit circle. We cannot picture the flow in a neighborhood of the north pole $(0,0,1)$ in the planar phase portrait in Figure 4(a); however, if we wish to picture the flow near $(0,0,1)$, we could project from the point $(1,0,0)$ onto the (Y, Z)-plane and obtain the planar phase portrait shown in Figure 4(b). Note that the periodic orbit Γ gets mapped onto the Y-axis which is "connected at the point at infinity" in Figure 4(b).

It can be shown that the periodic orbit Γ has characteristic exponents $\lambda_1 = \lambda_2 = 0$ and $\lambda_3 = -1$; cf. Problem 1 in Section 3.5. Thus, Γ has a two-dimensional stable manifold $W^s(\Gamma) = S^2 \sim \{(0,0,\mp 1)\}$ and a two-dimensional center manifold $W^c(\Gamma) = \{\mathbf{x} \in \mathbf{R}^3 \mid z = 0\}$. Of course, we do not see the center manifold $W^c(\Gamma)$ in either of the projections in Figure 4 since we are only projecting the sphere S^2 onto $\mathbf{R}^2$ in Figure 4.

Example 2. We now consider the Hamiltonian system with two-degrees of freedom

$$\dot{x} = y$$
$$\dot{y} = -x$$
$$\dot{z} = w$$
$$\dot{w} = -z$$

with Hamiltonian $H(x, y, z, w) = (x^2 + y^2 + z^2 + w^2)/2$. Trajectories of this system lie on the hyperspheres $x^2 + y^2 + z^2 + w^2 = k^2$ and the flow on each hypersphere is topologically equivalent to the flow on the unit three-sphere $S^3 = \{\mathbf{x} \in \mathbf{R}^4 \mid |\mathbf{x}| = 1\}$. On S^3, the total energy $H = 1/2$ is divided between the two harmonic oscillators $\ddot{x} + x = 0$ and $\ddot{z} + z = 0$; i.e., it follows from the above equations that

$$x^2 + y^2 = h^2 \quad \text{and} \quad z^2 + w^2 = 1 - h^2 \tag{1}$$

for some constant $h \in [0, 1]$. There are two periodic orbits on S^3,

$$\Gamma_1: \gamma_1(t) = (\cos t, -\sin t, 0, 0)^T$$

and

$$\Gamma_0: \gamma_0(t) = (0, 0, \cos t, -\sin t)^T,$$

corresponding to $h = 1$ and $h = 0$ respectively. If we project S^3 onto $\mathbf{R}^3$ from the point $(0, 0, 0, 1) \in S^3$, we obtain a one-to-one correspondence between points $(x, y, z, w) \in S^3 \sim \{(0,0,0,1)\}$ and points $(X, Y, Z) \in \mathbf{R}^3$ given by

$$X = \frac{x}{1-w}, \quad Y = \frac{y}{1-w}, \quad Z = \frac{z}{1-w}. \tag{2}$$

If $h = 1$, then from (1), $z = w = 0$ and $x^2 + y^2 = X^2 + Y^2 = 1$; i.e., the periodic orbit Γ_1 gets mapped onto the unit circle $X^2 + Y^2 = 1$ in the (X, Y)-plane in $\mathbf{R}^3$. And if $h = 0$, then from (1), $x = y = 0$ and

$$Z = \frac{z}{1 \mp \sqrt{1-z^2}};$$

i.e., the periodic orbit Γ_0 gets mapped onto the Z-axis "connected at infinity." These two periodic orbits in (X, Y, Z)-space are shown in Figure 5. The remaining trajectories of this system lie on two-dimensional tori, T_h^2, obtained as the cross product of two circles:

$$T_h^2 = \{\mathbf{x} \in \mathbf{R}^4 \mid x^2 + y^2 = h^2, \; z^2 + w^2 = 1 - h^2\}.$$

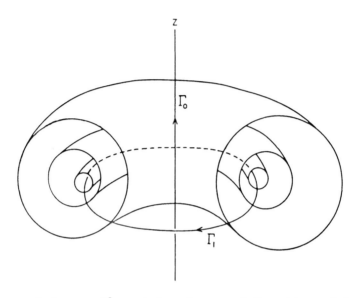

Figure 5. A flow on S^3 consisting of two periodic orbits and flows on invariant tori.

The equations for the projections of these two-dimensional tori onto $\mathbf{R}^3$ can be found by substituting (2) into (1). This yields

$$X^2 + Y^2 = \frac{h^2}{(1-w)^2} \quad \text{and} \quad Z^2(1-w)^2 + w^2 = 1 - h^2.$$

Solving the second equation for w and substituting into the first equation yields

$$\sqrt{X^2 + Y^2} = \frac{h(Z^2 + 1)}{1 \pm \sqrt{1 - h^2 - h^2 Z^2}}.$$

In cylindrical coordinates (R, θ, Z), this simplifies to the equations of the two-dimensional tori

$$T_h^2: \; Z^2 + \left(R - \frac{1}{h}\right)^2 = \frac{1 - h^2}{h^2}; \tag{3}$$

i.e., the tori T_h^2 are obtained by rotating the circles in the (X, Z)-plane, centered at $(1/h, 0, 0)$ with radius $\sqrt{1 - h^2}/h$, about the Z-axis. These invariant tori are shown in Figure 5. In this example, the periodic orbit Γ_1

3.6. Hamiltonian Systems with Two-Degrees of Freedom 239

has a four-dimensional center manifold

$$W^c(\Gamma_1) = \bigcup_{r>0} S_r^3$$

where on each three-sphere S_r^3 of radius r, Γ_1 has a three-dimensional center manifold as pictured in Figure 5. It can be shown that Γ_1 has four zero characteristic exponents; cf. Problem 1.

Example 3. Consider the Hamiltonian system

$$\dot{x} = y$$
$$\dot{y} = -x$$
$$\dot{z} = -z$$
$$\dot{w} = w$$

with Hamiltonian $H(x,y,z,w) = (x^2 + y^2)/2 - zw$. Trajectories of this system lie on the three-dimensional hypersurfaces

$$S: x^2 + y^2 - 2wz = k.$$

The flow on each of these hypersurfaces is topologically equivalent to the flow on the hypersurface $x^2 + y^2 - 2wz = 1$. On this hypersurface there is a periodic orbit

$$\Gamma: \gamma(t) = (\cos t, -\sin t, 0, 0)^T.$$

The linearization of this system about $\gamma(t)$ is given by

$$\dot{\mathbf{x}} = \begin{bmatrix} 0 & 1 & 0 & 0 \\ -1 & 0 & 0 & 0 \\ 0 & 0 & -1 & 0 \\ 0 & 0 & 0 & 1 \end{bmatrix} \mathbf{x}.$$

The fundamental matrix for this linear system satisfying $\Phi(0) = I$ is given by

$$\Phi(t) = \begin{bmatrix} R_t & 0 \\ 0 & I \end{bmatrix} e^{Bt}$$

where R_t is a rotation matrix and $B = \text{diag}[0, 0, -1, 1]$. The characteristic exponents $\lambda_1 = \lambda_2 = 0$, $\lambda_3 = -1$ and $\lambda_4 = 1$, and the periodic orbit Γ has a two-dimensional stable manifold $W^s(\Gamma)$ and a two-dimensional unstable manifold $W^u(\Gamma)$ on S, as well as a two-dimensional center manifold $W^c(\Gamma)$ (which does not lie on S). If we project the hypersurface S from the point $(1,0,0,0) \in S$ onto $\mathbf{R}^3$, the periodic orbit Γ gets mapped onto the Y-axis;

$W^s(\Gamma)$ is mapped onto the (Y, Z)-plane and $W^u(\Gamma)$ is mapped onto the (Y, W)-plane; cf. Figure 6.

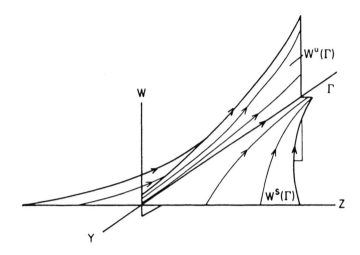

Figure 6. The two-dimensional stable and unstable manifolds of the periodic orbit Γ.

Example 4. Consider the pendulum oscillator

$$\dot{x} = y$$
$$\dot{y} = -x$$
$$\dot{z} = w$$
$$\dot{w} = -\sin z$$

with Hamiltonian $H(x, y, z, w) = (x^2 + y^2)/2 + (1 - \cos z) + w^2/2$. Trajectories of this system lie on the three-dimensional hypersurfaces

$$S: x^2 + y^2 + w^2 + 2(1 - \cos z) = k^2.$$

For $k > 2$ there are three periodic orbits on the hypersurface S:

$$\Gamma_0: \gamma_0(t) = (k \cos t, -k \sin t, 0, 0)$$
$$\Gamma_\pm: \gamma_\pm(t) = \left(\sqrt{k^2 - 4} \cos t, -\sqrt{k^2 - 4} \sin t, \pm \pi, 0 \right).$$

If we project from the point $(k, 0, 0, 0) \in S$ onto $\mathbf{R}^3$, the periodic orbit Γ_0 gets mapped onto the Y-axis and the periodic orbits $\Gamma_\pm$ get mapped onto the ellipses

$$\pi Y^2 + 4 \left(Z \pm \frac{k\pi}{4} \right)^2 = \left(\frac{\pi}{2} \right)^2;$$

3.6. Hamiltonian Systems with Two-Degrees of Freedom

cf. Problem 3. Since

$$Df(\mathbf{x}) = \begin{bmatrix} 0 & 1 & 0 & 0 \\ -1 & 0 & 0 & 0 \\ 0 & 0 & 0 & 1 \\ 0 & 0 & -\cos z & 0 \end{bmatrix},$$

we find that Γ_0 has all zero characteristic exponents. Thus Γ_0 has a three-dimensional center manifold on S. Similarly, $\Gamma_\pm$ have characteristic exponents $\lambda_1 = \lambda_2 = 0$, $\lambda_3 = 1$ and $\lambda_4 = -1$. Thus, $\Gamma_\pm$ have two-dimensional stable and unstable manifolds on S and two-dimensional center manifolds (which do not lie on S); cf. Figure 7.

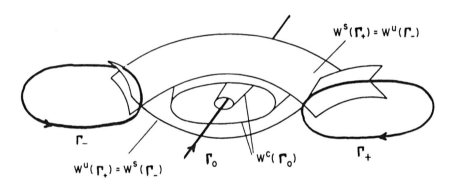

Figure 7. The flow of the pendulum-oscillator in projective space.

We see that for Hamiltonian systems with two-degrees of freedom, the trajectories of the system lie on three-dimensional hypersurfaces S given by $H(\mathbf{x}) = \text{constant}$. At any point $\mathbf{x}_0 \in S$ there is a two-dimensional hypersurface Σ normal to the flow. If $\mathbf{x}_0$ is a point on a periodic orbit then according to Theorem 1 in Section 3.4, there is an $\epsilon > 0$ and a Poincaré map

$$\mathbf{P}\colon N_\epsilon(\mathbf{x}_0) \cap \Sigma \to \Sigma.$$

Furthermore, we see that (i) a fixed point of $\mathbf{P}$ corresponds to a periodic orbit Γ of the system, (ii) if the iterates of $\mathbf{P}$ lie on a smooth curve, then this smooth curve is the cross-section of an invariant differentiable manifold of the system such as $W^s(\Gamma)$ or $W^u(\Gamma)$, and (iii) if the iterates of $\mathbf{P}$ lie on a closed curve, then this closed curve is the cross-section of an invariant torus of the system belonging to $W^c(\Gamma)$; cf. [G/H], pp. 212–216.

Problem Set 6

1. Show that the fundamental matrix for the linearization of the system in Example 2 about the periodic orbit Γ_1 which satisfies $\Phi(0) = I$ is equal to
$$\Phi(t) = \begin{bmatrix} R_t & 0 \\ 0 & R_t \end{bmatrix}$$
where R_t is the rotation matrix
$$R_t = \begin{bmatrix} \cos t & \sin t \\ -\sin t & \cos t \end{bmatrix}.$$
Thus $\Phi(t)$ can be written in the form of equation (3) in Section 3.5 with $B = 0$. What does this tell you about the dimension of the center manifold $W^c(\Gamma)$? Carry out the details in obtaining equation (3) for the invariant tori T_h^2.

2. Consider the Hamiltonian system with two-degrees of freedom
$$\dot{x} = \beta y$$
$$\dot{y} = -\beta x$$
$$\dot{z} = w$$
$$\dot{w} = -z$$
with $\beta > 0$.

(a) Show that for $H = 1/2$, the trajectories of this system lie on the three-dimensional ellipsoid
$$S:\ \beta(x^2 + y^2) + z^2 + w^2 = 1$$
and furthermore that for each $h \in (0,1)$ the trajectories lie on two-dimensional invariant tori
$$T_h^2 = \{\mathbf{x} \in \mathbf{R}^4 \mid x^2 + y^2 = h^2/\beta,\ z^2 + w^2 = 1 - h^2\} \subset S.$$

(b) Use the projection of S onto $\mathbf{R}^3$ given by equation (2) to show that these invariant tori are given by rotating the ellipsoids
$$Z^2 + \beta\left(X + \frac{1}{h\sqrt{\beta}}\right)^2 = \frac{1-h^2}{h^2}$$
about the Z-axis.

(c) Show that the flow is dense in each invariant tori, T_h^2, if β is irrational and that it consists of a one-parameter family of periodic orbits which lie on T_h^2 if β is rational; cf. Problem 2 in Section 3.2.

3.6. Hamiltonian Systems with Two-Degrees of Freedom

3. In Example 4, use the projective transformation

$$Y = \frac{y}{k-x}, \quad Z = \frac{z}{k-x}, \quad W = \frac{w}{k-x}$$

and show that the periodic orbit

$$\Gamma_+: \gamma(t) = (\cos t, -\sin t, \pi, 0)$$

gets mapped onto the ellipse

$$\pi Y^2 + 4\left(Z - \frac{k\pi}{4}\right)^2 = \left(\frac{\pi}{2}\right)^2.$$

4. Show that the Hamiltonian system

$$\dot{x} = y$$
$$\dot{y} = -x$$
$$\dot{z} = w$$
$$\dot{w} = -z + z^2$$

with Hamiltonian $H(x, y, z, w) = (x^2 + y^2 + z^2 + w^2)/2 - z^3/3$ has two periodic orbits

$$\Gamma_0: \gamma_0(t) = (k\cos t, -k\sin t, 0, 0)$$
$$\Gamma_1: \gamma_1(t) = \left(\sqrt{k^2 - 1/3}\cos t, -\sqrt{k^2 - 1/3}\sin t, 1, 0\right)$$

which lie on the surface $x^2 + y^2 + w^2 + z^2 - \frac{2}{3}z^3 = k^2$ for $k^2 > 1/3$. Show that under the projective transformation defined in Problem 3, Γ_0 gets mapped onto the Y-axis and Γ_1 gets mapped onto an ellipse. Show that Γ_0 has four zero characteristic exponents and that Γ_1 has characteristic exponents $\lambda_1 = \lambda_2 = 0$, $\lambda_3 = 1$ and $\lambda_4 = -1$. Sketch a local phrase portrait for this system in the projective space including Γ_0 and Γ_1 and parts of the invariant manifolds $W^c(\Gamma_0)$, $W^s(\Gamma_1)$ and $W^u(\Gamma_1)$.

5. Carry out the same sort of analysis as in Problem 4 for the Duffing-oscillator with Hamiltonian

$$H(x, y, z, w) = (x^2 + y^2 - z^2 - w^2)/2 + z^4/4,$$

and periodic orbits

$$\Gamma_0: \gamma_0(t) = (k\cos t, -k\sin t, 0, 0)$$
$$\Gamma_\pm: \gamma_\pm(t) = \left(\sqrt{k^2 - 1/4}\cos t, \sqrt{k^2 - 1/4}\sin t, \pm 1, 0\right).$$

6. Define an atlas for S^2 by using the stereographic projections onto $\mathbf{R}^2$ from the north pole $(0,0,1)$ of S^2, as in Figure 1, and from the south pole $(0,0,-1)$ of S^2; i.e., let $U_1 = \mathbf{R}^2$ and for $(x,y,z) \in S^2 \sim \{(0,0,1)\}$ define

$$\mathbf{h}_1(x,y,z) = \left(\frac{x}{1-z}, \frac{y}{1-z}\right).$$

Similarly let $U_2 = \mathbf{R}^2$ and for $(x,y,z) \in S^2 \sim \{(0,0,-1)\}$ define

$$\mathbf{h}_2(x,y,z) = \left(\frac{x}{1+z}, \frac{y}{1+z}\right).$$

Find $\mathbf{h}_2 \circ \mathbf{h}_1^{-1}$ (using Figures 1 and 2), find $D\mathbf{h}_2 \circ \mathbf{h}_1^{-1}(X,Y)$ and show that $\det D\mathbf{h}_2 \circ \mathbf{h}_1^{-1}(X,Y) \neq 0$ for all $(X,Y) \in \mathbf{h}_1(U_1 \cap U_2)$. (Note that at least two charts are needed in any atlas for S^2.)

3.7 The Poincaré–Bendixson Theory in R^2

In section 3.2, we defined the α and ω-limit sets of a trajectory Γ and saw that they were closed invariant sets of the system

$$\dot{\mathbf{x}} = \mathbf{f}(\mathbf{x}) \tag{1}$$

We also saw in the examples of Sections 3.2 and 3.3 that the α or ω-limit set of a trajectory could be a critical point, a limit cycle, a surface in $\mathbf{R}^3$ or a strange attractor consisting of an infinite number of interleaved branched surfaces in $\mathbf{R}^3$. For two-dimensional analytic systems, (1) with $\mathbf{x} \in \mathbf{R}^2$, the α and ω-limit sets of a trajectory are relatively simple objects: The α or ω-limit set of any trajectory of a two-dimensional, relatively-prime, analytic system is either a critical point, a cycle, or a compound separatrix cycle. A compound separatrix cycle or graphic of (1) is a finite union of compatibly oriented separatrix cycles of (1). Several examples of graphics were given in Section 3.3.

Let us first give a precise definition of separatrix cycles and graphics of (1) and then state and prove the main theorems in the Poincaré–Bendixson theory for planar dynamical systems. This theory originated with Henri Poincaré [P] and Ivar Bendixson [B] at the turn of the century. Recall that in Section 2.11 of Chapter 2 we defined a separatrix as a trajectory of (1) which lies on the boundary of a hyperbolic sector of (1). A more precise definition of a separatrix is given in Section 3.11.

Definition 1. *A separatrix cycle of* (1), S, *is a continuous image of a circle which consists of the union of a finite number of critical points and compatibly oriented separatrices of* (1), $\mathbf{p}_j, \Gamma_j, j = 1, \ldots, m$, *such that for* $j = 1, \ldots, m$, $\alpha(\Gamma_j) = \mathbf{p}_j$ *and* $\omega(\Gamma_j) = \mathbf{p}_{j+1}$ *where* $\mathbf{p}_{m+1} = \mathbf{p}_1$. *A*

3.7. The Poincaré–Bendixson Theory in R^2

compound separatrix cycle or graphic of (1), S, *is the union of a finite number of compatibly oriented separatrix cycles of* (1).

In the proof of the generalized Poincaré–Bendixson theorem for analytic systems given below, it is shown that if a graphic S is the limit set of a trajectory of (1) then the Poincaré map is defined on at least one side of S. The fact that the Poincaré map is defined on one side of a graphic S of (1) implies that for each separatrix $\Gamma_j \in S$, at least one of the sectors adjacent to Γ_j is a hyperbolic sector; it also implies that all of the separatrix cycles contained in S are compatibly oriented.

Theorem 1 (The Poincaré–Bendixson Theorem). *Suppose that* f $\in$ $C^1(E)$ *where E is an open subset of* $\mathbf{R}^2$ *and that* (1) *has a trajectory* Γ *with* Γ^+ *contained in a compact subset F of E. Then if $\omega(\Gamma)$ contains no critical point of* (1), $\omega(\Gamma)$ *is a periodic orbit of* (1).

Theorem 2 (The Generalized Poincaré–Bendixson Theorem). *Under the hypotheses of Theorem 1 and the assumption that* (1) *has only a finite number of critical points in F, it follows that $\omega(\Gamma)$ is either a critical point of* (1), *a periodic orbit of* (1), *or that $\omega(\Gamma)$ consists of a finite number of critical points,* $\mathbf{p}_1, \ldots, \mathbf{p}_m$, *of* (1) *and a countable number of limit orbits of* (1) *whose α and ω limit sets belong to* $\{\mathbf{p}_1, \ldots, \mathbf{p}_m\}$.

This theorem is proved on pp. 15–18 in [P/d] and, except for the finiteness of the number of limit orbits, it also follows as in the proof of the generalized Poincaré–Bendixson theorem for analytic systems given below. On p. 19 in [P/d] it is noted that the ω-limit set, $\omega(\Gamma)$, may consist of a "rose"; i.e., a single critical point $\mathbf{p}_1$, with a countable number of petals, consisting of elliptic sectors, whose boundaries are homoclinic loops at $\mathbf{p}_1$. In general, this theorem shows that $\omega(\Gamma)$ is either a single critical point of (1), a limit cycle of (1), or that $\omega(\Gamma)$ consists of a finite number of critical points $\mathbf{p}_1, \ldots, \mathbf{p}_m$, of (1) connected by a finite number of compatibly oriented limit orbits of (1) together with a finite number of "roses" at some of the critical points $\mathbf{p}_1, \ldots, \mathbf{p}_m$. Note that it follows from Lemma 1.7 in Chapter 1 of [P/d] that if $\mathbf{p}_1$ and $\mathbf{p}_2$ are distinct critical points of (1) which belong to the ω-limit set $\omega(\Gamma)$, then there exists at most one limit orbit $\Gamma_1 \subset \omega(\Gamma)$ such that $\alpha(\Gamma_1) = \mathbf{p}_1$ and $\omega(\Gamma_1) = \mathbf{p}_2$. For analytic or polynomial systems, $\omega(\Gamma)$ is somewhat simpler. In particular, it follows from Theorem VIII on p. 31 of [B] that any rose of an analytic system has only a finite number of petals.

Theorem 3 (The Generalized Poincaré–Bendixson Theorem for Analytic Systems). *Suppose that* (1) *is a relatively prime analytic system in an open set E of* $\mathbf{R}^2$ *and that* (1) *has a trajectory Γ with Γ^+ contained in a compact subset F of E. Then it follows that $\omega(\Gamma)$ is either a critical point of* (1), *a periodic orbit of* (1), *or a graphic of* (1).

Remark 1. It is well known that any relatively-prime analytic system (1) has at most a finite number of critical points in any bounded region of the plane; cf. [B], p. 30. The author has recently published a proof of this statement since it is difficult to find in the literature; cf. [24]. Also, it is important to note that under the hypotheses of the above theorems with Γ^- in place of Γ^+, the same conclusions hold for the α-limit set of Γ, $\alpha(\Gamma)$, as for the ω-limit set of Γ.

Furthermore, the above theorem, describing the α and ω-limit sets of trajectories of relatively-prime, analytic systems on compact subsets of $\mathbf{R}^2$, can be extended to all of $\mathbf{R}^2$ if we include graphics which contain the point at infinity on the Bendixson sphere (described in Figure 1 of Section 3.6); cf. Remark 3 and Theorem 4 below. In this regard, we note that any trajectory, $\mathbf{x}(t)$, of a planar analytic system either (i) is *bounded* (if $|\mathbf{x}(t)| \leq M$ for some constant M and for all $t \in \mathbf{R}$) or (ii) *escapes to infinity* (if $|\mathbf{x}(t)| \to \infty$ as $t \to \pm\infty$) or (iii) is *an unbounded oscillation* (if neither (i) nor (ii) hold). And it is exactly when $\mathbf{x}(t)$ is an unbounded oscillation that either the α or the ω-limit set of $\mathbf{x}(t)$ is a graphic containing the point at infinity on the Bendixson sphere. Cf. Problem 8.

The proofs of the above theorems follow from the lemmas established below. We first define what is meant by a transversal for (1).

Definition 2. A finite closed segment of a straight line, ℓ, contained in E, is called a *transversal for* (1) if there are no critical points of (1) on ℓ and if the vector field defined by (1) is not tangent to ℓ at any point of ℓ. A point $\mathbf{x}_0$ in E is a *regular point* of (1) if it is not a critical point of (1).

Lemma 1. *Every regular point $\mathbf{x}_0$ in E is an interior point of some transversal ℓ. Every trajectory which intersects a transversal ℓ at a point $\mathbf{x}_0$ must cross it. Let $\mathbf{x}_0$ be an interior point of a transversal ℓ; then for all $\epsilon > 0$ there is a $\delta > 0$ such that every trajectory passing through a point in $N_\delta(\mathbf{x}_0)$ at $t = 0$ crosses ℓ at some time t with $|t| < \epsilon$.*

Proof. The first statement follows from the definition of a regular point $\mathbf{x}_0$ by taking ℓ to be the straight line perpendicular to the vector defined by $\mathbf{f}(\mathbf{x}_0)$ at $\mathbf{x}_0$. The second statement follows from the fundamental existence-uniqueness theorem in Section 2.2 of Chapter 2 by taking $\mathbf{x}(0) = \mathbf{x}_0$; i.e., the solution $\mathbf{x}(t)$, with $\mathbf{x}(0) = \mathbf{x}_0$ defined for $-a < t < a$, defines a curve which crosses ℓ at $\mathbf{x}_0$. In order to establish the last statement, let $\mathbf{x} = (x,y)^T$, let $\mathbf{x}_0 = (x_0, y_0)^T$, and let ℓ be the straight line given by the equation $ax + by + c = 0$ with $ax_0 + by_0 + c = 0$. Then since $\mathbf{x}_0$ is a regular point of (1), there exists a neighborhood of $\mathbf{x}_0$, $N(\mathbf{x}_0)$, containing only regular points of (1). This follows from the continuity of $\mathbf{f}$. The solution $\mathbf{x}(t, \xi, \eta)$ passing through a point $(\xi, \eta) \in N(\mathbf{x}_0)$ at $t = 0$ is continuous in (t, ξ, η); cf. Section 2.3 in Chapter 2. Let

$$L(t, \xi, \eta) = ax(t, \xi, \eta) + by(t, \xi, \eta) + c.$$

3.7. The Poincaré–Bendixson Theory in R^2

Then $L(0, x_0, y_0) = 0$ and at any point (x_0, y_0) on ℓ

$$\frac{\partial L}{\partial t} = a\dot{x} + b\dot{y} \neq 0$$

since ℓ is a transversal. Thus it follows from the implicit function theorem that there is a continuous function $t(\xi, \eta)$ defined in some neighborhood of $\mathbf{x}_0$ such that $t(x_0, y_0) = 0$ and $L(t(\xi, \eta), \xi, \eta) = 0$ in that neighborhood. By continuity, for $\epsilon > 0$ there exists a $\delta > 0$ such that for all $(\xi, \eta) \in N_\delta(\mathbf{x}_0)$ we have $|t(\xi, \eta)| < \epsilon$. Thus the trajectory through any point $(\xi, \eta) \in N_\delta(\mathbf{x}_0)$ at $t = 0$ will cross the transversal ℓ at time $t = t(\xi, \eta)$ where $|t(\xi, \eta)| < \epsilon$.

Lemma 2. *If a finite closed arc of any trajectory Γ intersects a transversal ℓ, it does so in a finite number of points. If Γ is a periodic orbit, it intersects ℓ in only one point.*

Proof. Let the trajectory $\Gamma = \{\mathbf{x} \in E \mid \mathbf{x} = \mathbf{x}(t), t \in \mathbf{R}\}$ where $\mathbf{x}(t)$ is a solution of (1), and let A be the finite closed arc $A = \{\mathbf{x} \in E \mid \mathbf{x} = \mathbf{x}(t), a \leq t \leq b\}$. If A meets ℓ in infinitely many distinct points $\mathbf{x}_n = \mathbf{x}(t_n)$, then the sequence t_n will have a limit point $t^* \in [a, b]$. Thus, there is a subsequence, call it t_n, such that $t_n \to t^*$. Then $\mathbf{x}(t_n) \to \mathbf{y} = \mathbf{x}(t^*) \in \ell$ as $n \to \infty$. But

$$\frac{\mathbf{x}(t_n) - \mathbf{x}(t^*)}{t_n - t^*} \to \dot{\mathbf{x}}(t^*) = \mathbf{f}(\mathbf{x}(t^*))$$

as $n \to \infty$. And since $t_n, t^* \in [a.b]$ and $\mathbf{x}(t_n), \mathbf{x}(t^*) \in \ell$, it follows that

$$\frac{\mathbf{x}(t_n) - \mathbf{x}(t^*)}{t_n - t^*} \to \mathbf{v},$$

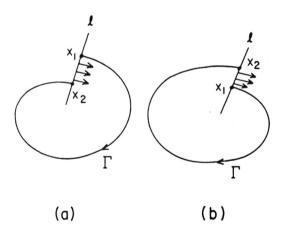

Figure 1. A Jordan curve defined by Γ and ℓ.

a vector tangent to ℓ at $\mathbf{y}$, as $n \to \infty$. This is a contradiction since ℓ is a transversal of (1). Thus, A meets ℓ in at most a finite number of points.

Now let $x_1 = x(t_1)$ and $x_2 = x(t_2)$ be two successive points of intersection of A with ℓ and assume that $t_1 < t_2$. Suppose that x_1 is distinct from x_2. Then the arc $A_{12} = \{x \in \mathbf{R}^2 \mid x = x(t), t_1 < t < t_2\}$ together with the closed segment $\overline{x_1 x_2}$ of ℓ comprises a Jordan curve J which separates the plane into two regions: cf. Figure 1. Then points $q = x(t)$ on Γ with $t < t_1$ (and near t_1) will be on the opposite side of J from points $p = x(t)$ with $t > t_2$ (and near t_2). Suppose that p is inside J as in Figure 1(a). Then to have Γ outside J for $t > t_2$, Γ must cross J. But Γ cannot cross A_{12} by the uniqueness theorem and Γ cannot cross $\overline{x_1 x_2} \subset \ell$ since the flow is inward on $\overline{x_1 x_2}$; otherwise, there would be a point on the segment $\overline{x_1 x_2}$ tangent to the vector field (1). Hence, Γ remains inside J for all $t > t_2$. Therefore, Γ cannot be periodic. A similar argument for p outside J as in Figure 1(b) also shows that Γ cannot be periodic. Thus, if Γ is a periodic orbit, it cannot meet ℓ in two or more points.

Remark 2. This same argument can be used to show that $\omega(\Gamma)$ intersects ℓ in only one point.

Lemma 3. *If Γ and $\omega(\Gamma)$ have a point in common, then Γ is either a critical point or a periodic orbit.*

Proof. Let $x_1 = x(t_1) \in \Gamma \cap \omega(\Gamma)$. If x_1 is a critical point of (1) then $x(t) = x_1$ for all $t \in \mathbf{R}$. If x_1 is a regular point of (1), then, by Lemma 1, it is an interior point of a transversal ℓ of (1). Since $x_1 \in \omega(\Gamma)$, it follows from the definition of the ω-limit set of Γ that any circle C with x_1 as center must contain in its interior a point $x = x(t^*)$ with $t^* > t_1 + 2$. If C is the circle with $\epsilon = 1$ in Lemma 1, then there is an $x_2 = x(t_2) \in \Gamma$ where $|t_2 - t^*| < 1$ and $x_2 \in \ell$. Assume that x_2 is distinct from x_1. Then the arc $\overline{x_1 x_2}$ of Γ intersects ℓ in a finite number of points by Lemma 2. Also, the successive intersections of Γ with ℓ form a monotone sequence which tends away from x_1. Hence, x_1 cannot be an ω-limit point of Γ, a contradiction. Thus, $x_1 = x_2$ and Γ is a periodic orbit of (1).

Lemma 4. *If $\omega(\Gamma)$ contains no critical points and $\omega(\Gamma)$ contains a periodic orbit Γ_0, then $\omega(\Gamma) = \Gamma_0$.*

Proof. Let $\Gamma_0 \subset \omega(\Gamma)$ be a periodic orbit with $\Gamma_0 \neq \omega(\Gamma)$. Then, by the connectedness of $\omega(\Gamma)$ in Theorem 1 of Section 3.2, Γ_0 contains a limit point y_0 of the set $\omega(\Gamma) \sim \Gamma_0$; otherwise, we could separate the sets Γ_0 and $\omega(\Gamma) \sim \Gamma_0$ by open sets and this would contradict the connectedness of $\omega(\Gamma)$. Let ℓ be a transversal through y_0. Then it follows from the fact that y_0 is a limit point of $\omega(\Gamma) \sim \Gamma_0$ that every circle with y_0 as center contains a point y of $\omega(\Gamma) \sim \Gamma_0$; and, by Lemma 1, for y sufficiently close to y_0, the trajectory Γ_y through the point y will cross ℓ at a point y_1. Since $y \in \omega(\Gamma) \sim \Gamma_0$ is a regular point of (1), the trajectory Γ is a limit orbit of (1) which is distinct from Γ_0 since $\Gamma \subset \omega(\Gamma) \sim \Gamma_0$. Hence, ℓ contains two distinct points $y_0 \in \Gamma_0 \subset \omega(\Gamma)$ and $y_1 \in \Gamma \subset \omega(\Gamma)$. But this contradicts Remark 2. Thus $\Gamma_0 = \omega(\Gamma)$.

3.7. The Poincaré–Bendixson Theory in R^2

Proof (of the Poincaré–Bendixson Theorem). If Γ is a periodic orbit, then $\Gamma \subset \omega(\Gamma)$ and by Lemma 4, $\Gamma = \omega(\Gamma)$. If Γ is not a periodic orbit, then since $\omega(\Gamma)$ is nonempty and consists of regular points only, there is a limit orbit Γ_0 of Γ such that $\Gamma_0 \subset \omega(\Gamma)$. Since Γ^+ is contained in a compact set $F \subset E$, the limit orbit $\Gamma_0 \subset F$. Thus Γ_0 has an ω-limit point $\mathbf{y}_0$ and $\mathbf{y}_0 \in \omega(\Gamma)$ since $\omega(\Gamma)$ is closed. If ℓ is a transversal through $\mathbf{y}_0$, then, since Γ_0 and $\mathbf{y}_0$ are both in $\omega(\Gamma)$, ℓ can intersect $\omega(\Gamma)$ only at $\mathbf{y}_0$ according to Remark 2. Since $\mathbf{y}_0$ is a limit point of Γ_0, it follows from Lemma 1 that ℓ must intersect Γ_0 in some point which, according to Lemma 2, must be $\mathbf{y}_0$. Hence Γ_0 and $\omega(\Gamma_0)$ have the point $\mathbf{y}_0$ in common. Thus, by Lemma 3, Γ_0 is a periodic orbit; and, by Lemma 4, $\Gamma_0 = \omega(\Gamma)$.

Proof (of the Generalized Poincaré–Bendixson Theorem for Analytic Systems). By hypothesis, $\omega(\Gamma)$ contains at most a finite number of critical points of (1) and they are isolated. (i) If $\omega(\Gamma)$ contains no regular points of (1) then $\omega(\Gamma) = \mathbf{x}_0$, a critical point of (1), since $\omega(\Gamma)$ is connected. (ii) If $\omega(\Gamma)$ consists entirely of regular points, then either Γ is a periodic orbit, in which case $\Gamma = \omega(\Gamma)$, or $\omega(\Gamma)$ is a periodic orbit by the Poincaré–Bendixson Theorem. (iii) If $\omega(\Gamma)$ consists of both regular points and a finite number of critical points, then $\omega(\Gamma)$ consists of limit orbits and critical points. Let $\Gamma_0 \subset \omega(\Gamma)$ be a limit orbit. Then, as in the proof of the Poincaré–Bendixson Theorem, Γ_0 cannot have a regular ω-limit point; if it did, we would have $\omega(\Gamma) = \Gamma_0$, a periodic orbit, and $\omega(\Gamma)$ would contain no critical points. Thus, each limit orbit in $\omega(\Gamma)$ has one of the critical points in $\omega(\Gamma)$ at its ω-limit set since $\omega(\Gamma)$ is connected. Similarly, each limit orbit in $\omega(\Gamma)$ has one of the critical points in $\omega(\Gamma)$ as its α-limit set. Thus, with an appropriate ordering of the critical points $\mathbf{p}_j$, $j = 1, \ldots, m$ (which may not be distinct) and the limit orbits $\Gamma_j \subset \omega(\Gamma)$, $j = 1, \ldots, m$, we have

$$\alpha(\Gamma_j) = \mathbf{p}_j \quad \text{and} \quad \omega(\Gamma_j) = \mathbf{p}_{j+1}$$

for $j = 1, \ldots, m$, where $\mathbf{p}_{m+1} = \mathbf{p}_1$. The finiteness of the number of limit orbits, Γ_j, follows from Theorems VIII and IX in [B] and Lemma 1.7 in [P/d]. And since $\omega(\Gamma)$ consists of limit orbits, Γ_j, and their α and ω-limit sets, $\mathbf{p}_j$, it follows that the trajectory Γ either spirals down to or out toward $\omega(\Gamma)$ as $t \to \infty$; cf. Theorem 3.2 on p. 396 in [C/L]. Therefore, as in Theorem 1 in Section 3.4, we can construct the Poincaré map at any point $\mathbf{p}$ sufficiently close to $\omega(\Gamma)$ which is either in the exterior of $\omega(\Gamma)$ or in the interior of one of the components of $\omega(\Gamma)$ respectively; i.e., $\omega(\Gamma)$ is a graphic of (1) and we say that the Poincaré map is defined on one side of $\omega(\Gamma)$. This completes the proof of the generalized Poincaré–Bendixson Theorem.

We next present a version of the generalized Poincaré–Bendixson theorem for flows on compact, two-dimensional manifolds (defined in Section 3.10); cf. Proposition 2.3 in Chapter 4 of [P/d]. In order to present this result, it is first necessary to define what we mean by a recurrent motion or recurrent trajectory.

Definition 3. Let Γ be a trajectory of (1). Then Γ is *recurrent* if $\Gamma \subset \alpha(\Gamma)$ or $\Gamma \subset \omega(\Gamma)$. A recurrent trajectory or orbit is called *trivial* if it is either a critical point or a periodic orbit of (1).

We note that critical points and periodic orbits of (1) are always trivial recurrent orbits of (1) and that for planar flows (or flows on S^2) these are the only recurrent orbits; however, flows on other two-dimensional surfaces can have more complicated recurrent motions. For example, every trajectory Γ of the irrational flow on the torus, T^2, described in Problem 2 of Section 3.2, is recurrent and nontrivial and its ω-limit limit set $\omega(\Gamma) = T^2$. The Cherry flow, described in Example 13 on p. 137 in [P/d], gives us an example of an analytic flow on T^2 which has one source and one saddle, the unstable separatrices of the saddle being nontrivial recurrent trajectories which intersect a transversal to the flow in a Cantor set. Also, we can construct vector fields with nontrivial recurrent motions on any two-dimensional, compact manifold except for the sphere, the projective plane and the Klein bottle. The fact that all recurrent motions are trivial on the sphere and on the projective plane follows from the Poincaré–Bendixson theorem and it was proved in 1969 by Markley [54] for the Klein bottle.

Theorem 4 (The Poincaré–Bendixson Theorem for Two-Dimensional Manifolds). *Let M be a compact two-dimensional manifold of class C^2 and let ϕ_t be the flow defined by a C^1 vector field on M which has only a finite number of critical points. If all recurrent orbits are trivial, then the ω-limit set of any trajectory,*

$$\gamma = \{\mathbf{x} \in M \mid \mathbf{x} = \phi_t(\mathbf{x}_0), \mathbf{x}_0 \in M, t \in \mathbf{R}\}$$

is either (i) *an equilibrium point of ϕ_t, i.e., a point $\mathbf{x}_0 \in M$ such that $\phi_t(\mathbf{x}_0) = \mathbf{x}_0$ for all $t \in \mathbf{R}$;* (ii) *a periodic orbit, i.e., a trajectory*

$$\Gamma_0 = \{\mathbf{x} \in M \mid \mathbf{x} = \phi_t(\mathbf{x}_0), \mathbf{x}_0 \in M, 0 \leq t \leq T, \phi_T(\mathbf{x}_0) = \mathbf{x}_0\};$$

or (iii) $\omega(\Gamma)$ *consists of a finite number of equilibrium points $\mathbf{p}_1, \ldots, \mathbf{p}_m$, of ϕ_t and a countable number of limit orbits whose α and ω limit sets belong to $\{\mathbf{p}_1, \ldots, \mathbf{p}_m\}$.*

Remark 3. Suppose that ϕ_t is the flow defined by a relatively-prime, analytic vector field on an analytic compact, two-dimensional manifold M; then if ϕ_t has only a finite number of equilibrium points on M and if all recurrent motions are trivial, it follows that $\omega(\Gamma)$ is either (i) an equilibrium point of ϕ_t, (ii) a periodic orbit, or (iii) a graphic on M.

We note that the ω-limit set $\omega(\Gamma)$ of a trajectory Γ of a flow on the sphere, the projective plane, or the Klein bottle is always one of the types listed in Theorem 4 (or in Remark 3 for analytic flows), but that $\omega(\Gamma)$ is generally more complicated for flows on other two-dimensional manifolds unless the recurrent motions are all trivial.

3.7. The Poincaré–Bendixson Theory in R^2

We cite one final theorem for periodic orbits of planar systems in this section. This theorem is proved for example on p. 252 in [H/S]. It can also be proved using index theory as is done in Section 3.12; cf. Corollary 2 in Section 3.12.

Theorem 5. *Suppose that* $\mathbf{f} \in C^1(E)$ *where* E *is an open subset of* $\mathbf{R}^2$ *which contains a periodic orbit* Γ *of* (1) *as well as its interior* U. *Then* U *contains at least one critical point of* (1).

Remark 4. For quadratic systems (1) where the components of $\mathbf{f}(\mathbf{x})$ consist of quadratic polynomials, it can be shown that U is a convex region which contains exactly one critical point of (1).

Problem Set 7

1. Consider the system
$$\dot{x} = -y + x(r^4 - 3r^2 + 1)$$
$$\dot{y} = x + y(r^4 - 3r^2 + 1)$$
where $r^2 = x^2 + y^2$.

   (a) Show that $\dot{r} < 0$ on the circle $r = 1$ and that $\dot{r} > 0$ on the circle $r = 2$. Use the Poincaré–Bendixson Theorem and the fact that the only critical point of this system is at the origin to show that there is a periodic orbit in the annular region $A_1 = \{\mathbf{x} \in \mathbf{R}^2 \mid 1 < |\mathbf{x}| < 2\}$.

   (b) Show that the origin is an unstable focus for this system and use the Poincaré–Bendixson Theorem to show that there is a periodic orbit in the annular region $A_2 = \{\mathbf{x} \in \mathbf{R}^2 \mid 0 < |\mathbf{x}| < 1\}$.

   (c) Find the unstable and stable limit cycles of this system.

2. (a) Use the Poincaré–Bendixson Theorem and the fact that the planar system
$$\dot{x} = x - y - x^3 \qquad \dot{y} = x + y - y^3$$
has only the one critical point at the origin to show that this system has a periodic orbit in the annular region $A = \{\mathbf{x} \in \mathbf{R}^2 \mid 1 < |\mathbf{x}| < \sqrt{2}\}$. **Hint:** Convert to polar coordinates and show that for all $\epsilon > 0$, $\dot{r} < 0$ on the circle $r = \sqrt{2} + \epsilon$ and $\dot{r} > 0$ on $r = 1 - \epsilon$; then use the Poincaré–Bendixson theorem to show that this implies that there is a limit cycle in
$$\bar{A} = \{\mathbf{x} \in \mathbf{R}^2 \mid 1 \leq |\mathbf{x}| \leq \sqrt{2}\};$$
and then show that no limit cycle can have a point in common with either one of the circles $r = 1$ or $r = \sqrt{2}$.

(b) Show that there is at least one stable limit cycle in A. (In fact, this system has exactly one limit cycle in A and it is stable. Cf. Problem 3 in Section 3.9.) This limit cycle and the annular region A are shown in Figure 2.

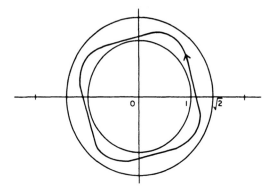

Figure 2. The limit cycle for Problem 2.

3. Let $\mathbf{f}$ be a C^1 vector field in an open set $E \subset \mathbf{R}^2$ containing an annular region A with a smooth boundary. Suppose that $\mathbf{f}$ has no zeros in $\bar{A}$, the closure of A, and that $\mathbf{f}$ is transverse to the boundary of A, pointing inward.

   (a) Prove that A contains a periodic orbit.

   (b) Prove that if A contains a finite number of cycles, then A contains at least one stable limit cycle of (1).

4. Let $\mathbf{f}$ be a C^1 vector field in an open set $E \subset \mathbf{R}^2$ containing the closure of the annular region $A = \{\mathbf{x} \in \mathbf{R}^2 \mid 1 < |\mathbf{x}| < 2\}$. Suppose that $\mathbf{f}$ has no zeros on the boundary of A and that at each boundary point $\mathbf{x} \in \dot{A}$, $\mathbf{f}(\mathbf{x})$ is tangent to the boundary of A.

   (a) Under the further assumption that A contains no critical points or periodic orbits of (1), sketch the possible phase portraits in A. (There are two topologically distinct phase portraits in A.)

   (b) Suppose that the boundary trajectories are oppositely oriented and that the flow defined by (1) preserves area. Show that A contains at least two critical points of the system (1). (This is reminiscent of Poincaré's Theorem for area preserving mappings of an annulus; cf. p. 220 in [G/H]. Recall that the flow defined by a Hamiltonian system with one-degree of freedom preserves area; cf. Problem 12 in Section 2.14 of Chapter 2.)

3.8. Lienard Systems

5. Show that
$$\dot{x} = y \qquad \dot{y} = -x + (1 - x^2 - y^2)y$$
has a unique stable limit cycle which is the ω-limit set of every trajectory except the critical point at the origin. **Hint:** Compute $\dot{r}$.

6. (a) Let Γ_0 be a periodic orbit of a C^1 dynamical system on an open set $E \subset \mathbf{R}^2$ with $\Gamma_0 \subset E$. Let T_0 be the period of Γ_0 and suppose that there is a sequence of periodic orbits $\Gamma_n \subset E$ of periods T_n containing points $\mathbf{x}_n$ which approach $\mathbf{x}_0 \in \Gamma_0$ as $n \to \infty$. Prove that $T_n \to T_0$. **Hint:** Use Lemma 2 to show that the function $\tau(\mathbf{x})$ of Theorem 1 in Section 3.4 satisfies $\tau(\mathbf{x}_n) = T_n$ for n sufficiently large.

    (b) This result does not hold for higher dimensional systems. It is true, however, that if $T_n \to T$, then T is a multiple of T_0. Sketch a periodic orbit Γ_0 in $\mathbf{R}^3$ and one neighboring orbit Γ_n of period T_n where for $\mathbf{x}_n \in \Gamma_n$ we have $\tau(\mathbf{x}_n) = T_n/2$.

7. Show that the C^1-system $\dot{x} = x - rx - ry + xy$, $\dot{y} = y - ry + rx - x^2$ can be written in polar coordinates as $\dot{r} = r(1-r)$, $\dot{\theta} = r(1 - \cos\theta)$. Show that it has an unstable node at the origin and a saddle node at $(1,0)$. Use this information and the Poincaré–Bendixson Theorem to sketch the phase portrait for this system and then deduce that for all $\mathbf{x} \neq 0$, $\phi_t(\mathbf{x}) \to (1,0)$ as $t \to \infty$, but that $(1,0)$ is not stable.

8. Show that the analytic system
$$\dot{x} = y \qquad \dot{y} = \left[\frac{y}{(1+y^2)} - x\right](1+y)$$
has an unbounded oscillation and that the ω-limit set of any trajectory starting on the positive y-axis is the invariant line $y = -1$. Sketch the phase portrait for this system on $\mathbf{R}^2$ and on the Bendixson sphere. **Hint:** Show that the line $y = -1$ is a trajectory of this system, that the only critical point is an unstable focus at the origin and that trajectories in the half plane $y < -1$ escape to infinity along parabolas $y = y_0 - x^2/2$ as $t \to \pm\infty$, i.e., show that $\frac{dy}{dx} = \frac{\dot{y}}{\dot{x}} \to -x$ as $y \to -\infty$.

3.8 Lienard Systems

In the previous section we saw that the Poincaré–Bendixson Theorem could be used to establish the existence of limit cycles for certain planar systems. It is a far more delicate question to determine the exact number of limit cycles of a certain system or class of systems depending on parameters. In this section we present a proof of a classical result on the uniqueness of the limit cycle for systems of the form

$$\dot{x} = y - F(x)$$
$$\dot{y} = -g(x) \qquad (1)$$

under certain conditions on the functions F and g. This result was first established by the French physicist A. Lienard in 1928 and the system (1) is referred to as a Lienard system. Lienard studied this system in the different, but equivalent form

$$\ddot{x} + f(x)\dot{x} + g(x) = 0$$

where $f(x) = F'(x)$ in a paper on sustained oscillations. This second-order differential equation includes the famous van der Pol equation

$$\ddot{x} + \mu(x^2 - 1)\dot{x} + x = 0 \qquad (2)$$

of vacuum-tube circuit theory as a special case.

We present several other interesting results on the number of limit cycles of Lienard systems and polynomial systems in this section which we conclude with a brief discussion of Hilbert's 16th Problem for planar polynomial systems. In the proof of Lienard's Theorem and in the statements of some of the other theorems in this section it will be useful to define the functions

$$F(x) = \int_0^x f(s)\,ds \quad \text{and} \quad G(x) = \int_0^x g(s)\,ds$$

and the energy function

$$u(x, y) = \frac{y^2}{2} + G(x).$$

Theorem 1 (Lienard's Theorem). *Under the assumptions that $F, g \in C^1(\mathbf{R})$, F and g are odd functions of x, $xg(x) > 0$ for $x \neq 0$, $F(0) = 0$, $F'(0) < 0$, F has single positive zero at $x = a$, and F increases monotonically to infinity for $x \geq a$ as $x \to \infty$, it follows that the Lienard system (1) has exactly one limit cycle and it is stable.*

The proof of this theorem makes use of the diagram below where the points P_j have coordinates (x_j, y_j) for $j = 0, 1, \ldots, 4$ and Γ is a trajectory of the Lienard system (1). The function $F(x)$ shown in Figure 1 is typical of functions which satisfy the hypotheses of Theorem 1. Before presenting the proof of this theorem, we first of all make some simple observations: Under the assumptions of the above theorem, the origin is the only critical point of (1); the flow on the positive y-axis is horizontal and to the right, and the flow on the negative y-axis is horizontal and to the left; the flow on the curve $y = F(x)$ is vertical, downward for $x > 0$ and upward for $x < 0$; the system (1) is invariant under $(x, y) \to (-x, -y)$ and therefore if $(x(t), y(t))$ describes a trajectory of (1) so does $(-x(t), -y(t))$; it follows

3.8. Lienard Systems

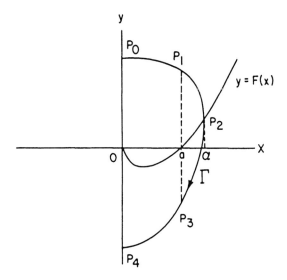

Figure 1. The function $F(x)$ and a trajectory Γ of Lienard's system.

that that if Γ is a closed trajectory of (1), i.e., a periodic orbit of (1), then Γ is symmetric with respect to the origin.

Proof. Due to the nature of the flow on the y-axis and on the curve $y = F(x)$, any trajectory Γ starting at a point P_0 on the positive y-axis crosses the curve $y = F(x)$ vertically at a point P_2 and then it crosses the negative y-axis horizontally at a point P_4; cf. Theorem 1.1, p. 202 in [H].

Due to the symmetry of the equation (1), it follows that Γ is a closed trajectory of (1) if and only if $y_4 = -y_0$; and for $u(x, y) = y^2/2 + G(x)$, this is equivalent to $u(0, y_4) = u(0, y_0)$. Now let A be the arc $\overline{P_0 P_4}$ of the trajectory Γ and consider the function $\phi(\alpha)$ defined by the line integral

$$\phi(\alpha) = \int_A du = u(0, y_4) - u(0, y_0)$$

where $\alpha = x_2$, the abscissa of the point P_2. It follows that Γ is a closed trajectory of (1) if and only if $\phi(\alpha) = 0$. We shall show that the function $\phi(\alpha)$ has exactly one zero $\alpha = \alpha_0$ and that $\alpha_0 > a$. First of all, note that along the trajectory Γ

$$du = g(x)\,dx + y\,dy = F(x)\,dy.$$

And if $\alpha \leq a$ then both $F(x) < 0$ and $dy = -g(x)\,dt < 0$. Therefore, $\phi(\alpha) > 0$; i.e., $u(0, y_4) > u(0, y_0)$. Hence, any trajectory Γ which crosses the curve $y = F(x)$ at a point P_2 with $0 < x_2 = \alpha \leq a$ is not closed.

Lemma. *For $\alpha \geq a$, $\phi(\alpha)$ is a monotone decreasing function which decreases from the positive value $\phi(a)$ to $-\infty$ as α increases in the interval $[a, \infty)$.*

For $\alpha > a$, as in Figure 1, we split the arc A into three parts $A_1 = \overline{P_0 P_1}$, $A_2 = \overline{P_1 P_3}$ and $A_3 = \overline{P_3 P_4}$ and define the functions

$$\phi_1(\alpha) = \int_{A_1} du, \quad \phi_2(\alpha) = \int_{A_2} du \quad \text{and} \quad \phi_3(\alpha) = \int_{A_3} du.$$

It follows that $\phi(\alpha) = \phi_1(\alpha) + \phi_2(\alpha) + \phi_3(\alpha)$. Along Γ we have

$$\begin{aligned} du &= \left[g(x) + y \frac{dy}{dx} \right] dx \\ &= \left[g(x) - \frac{y g(x)}{y - F(x)} \right] dx \\ &= \frac{-F(x) g(x)}{y - F(x)} dx. \end{aligned}$$

Along the arcs A_1 and A_3 we have $F(x) < 0$, $g(x) > 0$ and $dx/[y - F(x)] = dt > 0$. Therefore, $\phi_1(\alpha) > 0$ and $\phi_3(\alpha) > 0$. Similarly, along the arc A_2, we have $F(x) > 0$, $g(x) > 0$ and $dx/[y - F(x)] = dt > 0$ and therefore $\phi_2(\alpha) < 0$. Since trajectories of (1) do not cross, it follows that increasing α raises the arc A_1 and lowers the arc A_3. Along A_1, the x-limits of integration remain fixed at $x = x_0 = 0$, and $x = x_1 = a$; and for each fixed x in $[0, a]$, increasing α raises A_1 which increases y which in turn decreases the above integrand and therefore decreases $\phi_1(\alpha)$. Along A_3, the x-limits of integration remain fixed at $x_3 = a$ and $x_4 = 0$; and for each fixed $x \in [0, a]$, increasing α lowers A_3 which decreases y which in turn decreases the magnitude of the above integrand and therefore decreases $\phi_3(\alpha)$ since

$$\phi_3(\alpha) = \int_a^0 \frac{-F(x) g(x)}{y - F(x)} dx = \int_0^a \left| \frac{F(x) g(x)}{y - F(x)} \right| dx.$$

Along the arc A_2 of Γ we can write $du = F(x) dy$. And since trajectories of (1) do not cross, it follows that increasing α causes the arc A_2 to move to the right. Along A_2 the y-limits of integration remain fixed at $y = y_1$ and $y = y_3$; and for each fixed $y \in [y_3, y_1]$, increasing x increases $F(x)$ and since

$$\phi_2(x) = -\int_{y_3}^{y_1} F(x) \, dy,$$

this in turn decreases $\phi_2(\alpha)$. Hence for $\alpha \geq a$, ϕ is a monotone decreasing function of α. It remains to show that $\phi(\alpha) \to -\infty$ as $\alpha \to \infty$. It suffices to show that $\phi_2(\alpha) \to -\infty$ as $\alpha \to \infty$. But along A_2, $du = F(x) dy = -F(x) g(x) dt < 0$, and therefore for any sufficiently small $\epsilon > 0$

$$\begin{aligned} |\phi_2(\alpha)| &= -\int_{y_1}^{y_3} F(x) \, dy = \int_{y_3}^{y_1} F(x) \, dy > \int_{y_3+\epsilon}^{y_1-\epsilon} F(x) \, dy \\ &> F(\epsilon) \int_{y_3+\epsilon}^{y_1-\epsilon} dy \\ &= F(\epsilon)[y_1 - y_3 - 2\epsilon] \\ &> F(\epsilon)[y_1 - 2\epsilon]. \end{aligned}$$

3.8. Lienard Systems

But $y_1 > y_2$ and $y_2 \to \infty$ as $x_2 = \alpha \to \infty$. Therefore, $|\phi_2(\alpha)| \to \infty$ as $\alpha \to \infty$; i.e., $\phi_2(\alpha) \to -\infty$ as $\alpha \to \infty$.

Finally, since the continuous function $\phi(\alpha)$ decreases monotonically from the positive value $\phi(a)$ to $-\infty$ as α increases in $[a, \infty)$, it follows that $\phi(\alpha) = 0$ at exactly one value of α, say $\alpha = \alpha_0$, in (a, ∞). Thus, (1) has exactly one closed trajectory Γ_0 which goes through the point $(\alpha_0, F(\alpha_0))$. Furthermore, since $\phi(\alpha) < 0$ for $\alpha > \alpha_0$, it follows from the symmetry of the system (1) that for $\alpha \neq \alpha_0$, successive points of intersection of trajectory Γ through the point $(\alpha, F(\alpha))$ with the y-axis approach Γ_0; i.e., Γ_0 is a stable limit cycle of (1). This completes the proof of Lienard's Theorem.

Corollary. *For $\mu > 0$, van der Pol's equation (2) has a unique limit cycle and it is stable.*

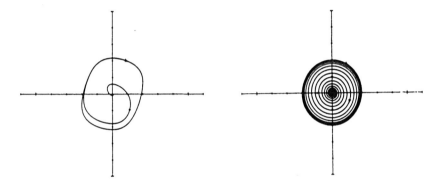

Figure 2. The limit cycle for the van der Pol equation for $\mu = 1$ and $\mu = .1$.

Figure 2 shows the limit cycle for the van der Pol equation (2) with $\mu = 1$ and $\mu = .1$. It can be shown that the limit cycle of (2) is asymptotic to the circle of radius 2 centered at the origin as $\mu \to 0$.

Example 1. It is not difficult to show that the functions $F(x) = (x^3 - x)/(x^2 + 1)$ and $g(x) = x$ satisfy the hypotheses of Lienard's Theorem; cf. Problem 1. It therefore follows that the system (1) with these functions has exactly one limit cycle which is stable. This limit cycle is shown in Figure 3.

In 1958 the Chinese mathematician Zhang Zhifen proved the following useful result which complements Lienard's Theorem. Cf. [35].

Theorem 2 (Zhang). *Under the assumptions that $a < 0 < b$, $F, g \in C^1(a, b)$, $xg(x) > 0$ for $x \neq 0$, $G(x) \to \infty$ as $x \to a$ if $a = -\infty$ and $G(x) \to \infty$ as $x \to b$ if $b = \infty$, $f(x)/g(x)$ is monotone increasing on $(a, 0) \cap (0, b)$ and is not constant in any neighborhood of $x = 0$, it follows that the system (1) has at most one limit cycle in the region $a < x < b$ and if it exists it is stable.*

258   3. Nonlinear Systems: Global Theory

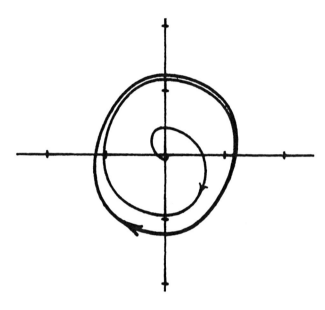

Figure 3. The limit cycle for the Lienard system in Example 1.

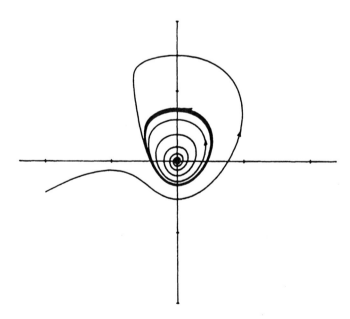

Figure 4. The limit cycle for the Lienard system in Example 2 with $\alpha = .02$.

3.8. Lienard Systems

Example 2. The author recently used this theorem to show that for $\alpha \in (0, 1)$, the quadratic system

$$\dot{x} = -y(1+x) + \alpha x + (\alpha+1)x^2$$
$$\dot{y} = x(1+x)$$

has exactly one limit cycle and it is stable. It is easy to see that the flow is horizontal and to the right on the line $x = -1$. Therefore, any closed trajectory lies in the region $x > -1$. If we define a new independent variable τ by $d\tau = -(1+x)dt$ along trajectories $\mathbf{x} = \mathbf{x}(t)$ of this system, it then takes the form of a Lienard system

$$\frac{dx}{d\tau} = y - \frac{\alpha x + (\alpha+1)x^2}{(1+x)}$$

$$\frac{dy}{d\tau} = -x.$$

Even though the hypotheses of Lienard's Theorem are not satisfied, it can be shown that the hypotheses of Zhang's theorem are satisfied. Therefore, this system has exactly one limit cycle and it is stable. The limit cycle for this system with $\alpha = .02$ is shown in Figure 4.

In 1981, Zhang proved another interesting theorem concerning the number of limit cycles of the Lienard system (1). Cf. [36]. Also, cf. Theorem 7.1 in [Y].

Theorem 3 (Zhang). *Under the assumptions that $g(x) = x$, $F \in C^1(\mathbf{R})$, $f(x)$ is an even function with exactly two positive zeros $a_1 < a_2$ with $F(a_1) > 0$ and $F(a_2) < 0$, and $f(x)$ is monotone increasing for $x > a_2$, it follows that the system (1) has at most two limit cycles.*

Example 3. Consider the Lienard system (1) with $g(x) = x$ and $f(x) = 1.6x^4 - 4x^2 + .8$. It is not difficult to show that the hypotheses of Theorem 3 are satisfied; cf. Problem 2. It therefore follows that the system (1) with $g(x) = x$ and $F(x) = .32x^5 - 4x^3/3 + .8x$ has at most two limit cycles. In fact, this system has exactly two limit cycles (cf. Theorem 6 below) and they are shown in Figure 5.

Of course, the more specific we are about the functions $F(x)$ and $g(x)$ in (1), the more specific we can be about the number of limit cycles that (1) has. For example, if $g(x) = x$ and $F(x)$ is a polynomial, we have the following results; cf. [18].

Theorem 4 (Lins, de Melo and Pugh). *The system (1) with $g(x) = x$, $F(x) = a_1 x + a_2 x^2 + a_3 x^3$, and $a_1 a_3 < 0$ has exactly one limit cycle. It is stable if $a_1 < 0$ and unstable if $a_1 > 0$.*

Remark. The Russian mathematician Rychkov showed that the system (1) with $g(x) = x$ and $F(x) = a_1 x + a_3 x^3 + a_3 x^5$ has at most two limit cycles.

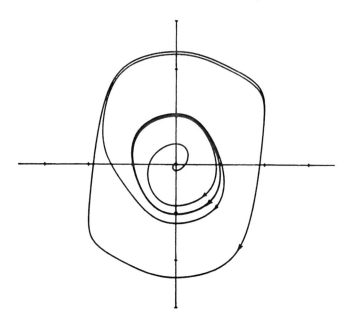

Figure 5. The two limit cycle of the Lienard system in Example 3.

In Section 3.4, we mentioned that m limit cycles can be made to bifurcate from a multiple focus of multiplicity m. This concept is discussed in some detail in Sections 4.4 and 4.5 of Chapter 4. Limit cycles which bifurcate from a multiple focus are called *local limit cycles*. The next theorem is proved in [3].

Theorem 5 (Blows and Lloyd). *The system* (1) *with* $g(x) = x$ *and* $F(x) = a_1 x + a_2 x^2 + \cdots + a_{2m+1} x^{2m+1}$ *has at most* m *local limit cycles and there are coefficients with* $a_1, a_3, a_5, \ldots, a_{2m+1}$ *alternating in sign such that* (1) *has* m *local limit cycles.*

Theorem 6 (Perko). *For* $\epsilon \neq 0$ *sufficiently small, the system* (1) *with* $g(x) = x$ *and* $F(x) = \epsilon[a_1 x + a_2 x^2 + \cdots + a_{2m+1} x^{2m+1}]$ *has at most* m *limit cycles; furthermore, for* $\epsilon \neq 0$ *sufficiently small, this system has exactly* m *limit cycles which are asymptotic to circles of radius* r_j, $j = 1, \ldots, m$, *centered at the origin as* $\epsilon \to 0$ *if and only if the* mth *degree equation*

$$\frac{a_1}{2} + \frac{3a_3}{8}\rho + \frac{5a_5}{16}\rho^2 + \frac{35a_7}{128}\rho^3 + \cdots + \binom{2m+2}{m+1}\frac{a_{2m+1}}{2^{2m+2}}\rho^m = 0 \quad (3)$$

has m *positive roots* $\rho = r_j^2$, $j = 1, \ldots, m$.

This last theorem is proved using Melnikov's Method in Section 4.10. It is similar to Theorem 76 on p. 414 in [A-II].

Example 4. Theorem 6 allows us to construct polynomial systems with as many limit cycles as we like. For example, suppose that we wish to find a

3.8. Lienard Systems

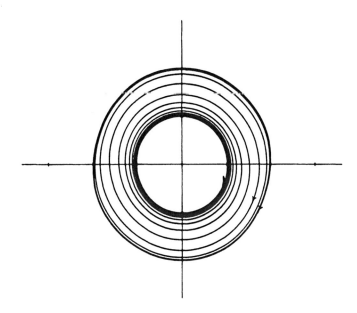

Figure 6. The limit cycles of the Lienard system in Example 4 with $\epsilon = .01$.

polynomial system of the form in Theorem 6 with exactly two limit cycles asymptotic to circles of radius $r = 1$ and $r = 2$. To do this, we simply set the polynomial $(\rho - 1)(\rho - 4)$ equal to the polynomial in equation (3) with $m = 2$ in order to determine the coefficients a_1, a_3 and a_5; i.e., we set

$$\rho^2 - 5\rho + 4 = \frac{5}{16} a_5 \rho^2 + \frac{3}{8} a_3 \rho + \frac{1}{2} a_1.$$

This implies that $a_5 = 16/5$, $a_3 = -40/3$ and $a_1 = 8$. For $\epsilon \neq 0$ sufficiently small, Theorem 6 then implies that the system

$$\dot{x} = y - \epsilon(8x - 40x^3/3 + 16x^5/5)$$
$$\dot{y} = -x$$

has exactly two limit cycles. For $\epsilon = .01$ these limit cycles are shown in Figure 6. They are very near the circles $r = 1$ and $r = 2$. For $\epsilon = .1$ these two limit cycles are shown in Figure 5. They are no longer near the two circles $r = 1$ and $r = 2$ for this larger value of ϵ. Arbitrary even-degree terms such as $a_2 x^2$ and $a_4 x^4$ may be added in the ϵ-term in this system without affecting the results concerning the number and geometry of the limit cycles of this example.

Example 5. As in Example 4, it can be shown that for $\epsilon \neq 0$ sufficiently small, the Lienard system

$$\dot{x} = y + \epsilon(72x - 392x^3/3 + 224x^5/5 - 128x^7/35)$$
$$\dot{y} = -x$$

has exactly three limit cycles which are asymptotic to the circles $r = 1$, $r = 2$ and $r = 3$ as $\epsilon \to 0$; cf. Problem 3. The limit cycles for this system with $\epsilon = .01$ and $\epsilon = .001$ are shown in Figure 7.

At the turn of the century, the world-famous mathematician David Hilbert presented a list of 23 outstanding mathematical problems to the Second International Congress of Mathematicians. The 16th Hilbert Problem asks for a determination of the maximum number of limit cycles, H_n, of an nth degree polynomial system

$$\dot{x} = \sum_{i+j=0}^{n} a_{ij} x^i y^j$$

$$\dot{y} = \sum_{i+j=0}^{n} b_{ij} x^i y^j. \qquad (4)$$

For given $(\mathbf{a}, \mathbf{b}) \in \mathbf{R}^{(n+1)(n+2)}$, let $H_n(\mathbf{a}, \mathbf{b})$ denote the number of limit cycles of the nth degree polynomial system (4) with coefficients $(\mathbf{a}, \mathbf{b})$. Note that Dulac's Theorem asserts that $H_n(\mathbf{a}, \mathbf{b}) < \infty$. The Hilbert number H_n is then equal to the sup $H_n(\mathbf{a}, \mathbf{b})$ over all $(\mathbf{a}, \mathbf{b}) \in \mathbf{R}^{(n+1)(n+2)}$.

Since linear systems in $\mathbf{R}^2$ do not have any limit cycles, cf. Section 1.5 in Chapter 1, it follows that $H_1 = 0$. However, even for the simplest class of nonlinear systems, (4) with $n = 2$, the Hilbert number H_2 has not been determined. In 1962, the Russian mathematician N. V. Bautin [2] proved that any quadratic system, (4) with $n = 2$, has at most three local limit cycles. And for some time it was believed that $H_2 = 3$. However, in 1979, the Chinese mathematicians S. L. Shi, L. S. Chen and M. S. Wang produced examples of quadratic systems with four limit cycles; cf. [29]. Hence $H_2 \geq 4$. Based on all of the current evidence it is believed that $H_2 = 4$ and in 1984, Y. X. Chin claimed to have proved this result; however, errors were pointed out in his work by Y. L. Cao.

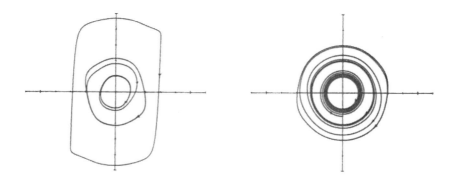

Figure 7. The limit cycles of the Lienard system in Example 5 with $\epsilon = .01$ and $\epsilon = .001$.

3.8. Lienard Systems

Regarding H_3, it is known that a cubic system can have at least eleven local limit cycles; cf. [42]. Also, in 1983, J. B. Li et al. produced an example of a cubic system with eleven limit cycles. Thus, all that can be said at this time is that $H_3 \geq 11$. Hilbert's 16th Problem for planar polynomial systems has generated much interesting mathematical research in recent years and will probably continue to do so for some time.

PROBLEM SET 8

1. Show that the functions $F(x) = (x^3 - x)/(x^2 + 1)$ and $g(x) = x$ satisfy the hypotheses of Lienard's Theorem.

2. Show that the functions $f(x) = 1.6x^4 - 4x^2 + .8$ and $F(x) = \int_0^x f(s)\,ds = .32x^5 - 4x^3/3 + .8x$ satisfy the hypotheses of Theorem 3.

3. Set the polynomial $(\rho - 1)(\rho - 4)(\rho - 9)$ equal to the polynomial in equation (3) with $m = 3$ and determine the coefficients a_1, a_3, a_5 and a_7 in the system of Example 5.

4. Construct a Lienard system with four limit cycles.

5. (a) Determine the phase portrait for the system
$$\dot{x} = y - x^2$$
$$\dot{y} = -x.$$

   (b) Determine the phase portrait for the Lienard system (1) with $F, g \in C^1(\mathbf{R})$, $F(x)$ an even function of x and g an odd function of x of the form $g(x) = x + 0(x^3)$. **Hint:** Cf. Theorem 6 in Section 2.10 of Chapter 2.

6. Consider the van der Pol system
$$\dot{x} = y + \mu(x - x^3/3)$$
$$\dot{y} = -x.$$

   (a) As $\mu \to 0^+$ show that the limit cycle L_μ of this system approaches the circle of radius two centered at the origin.

   (b) As $\mu \to \infty$ show that the limit cycle L_μ is asymptotic to the closed curve consisting of two horizontal line segments on $y = \pm 2\mu/3$ and two arcs of $y = \mu(x^3/3 - x)$. To do this, let $u = y/\mu$ and $\tau = t/\mu$ and show that as $\mu \to \infty$ the limit cycle of the resulting system approaches the closed curve consisting of the two horizontal line segments on $u = \pm 2/3$ and the two arcs of the cubic $u = x^3/3 - x$ shown in Figure 8.

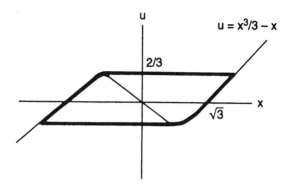

Figure 8. The limit of L_μ as $\mu \to \infty$.

7. Let F satisfy the hypotheses of Lienard's Theorem. Show that

$$\ddot{z} + F(\dot{z}) + z = 0$$

has a unique, asymptotically stable, periodic solution.

Hint: Let $x = \dot{z}$ and $y = -z$.

3.9 Bendixson's Criteria

Lienard's Theorem and the other theorems in the previous section establish the existence of exactly one or exactly m limit cycles for certain planar systems. Bendixson's Criteria and other theorems in this section establish conditions under which the planar system

$$\dot{\mathbf{x}} = \mathbf{f}(\mathbf{x}) \tag{1}$$

with $\mathbf{f} = (P, Q)^T$ and $\mathbf{x} = (x, y)^T \in \mathbf{R}^2$ has no limit cycles. In order to determine the global phase portrait of a planar dynamical system, it is necessary to determine the number of limit cycles around each critical point of the system. The theorems in this section and in the previous section make this possible for some planar systems. Unfortunately, it is generally not possible to determine the exact number of limit cycles of a planar system and this remains the single most difficult problem for planar systems.

Theorem 1 (Bendixson's Criteria). *Let $\mathbf{f} \in C^1(E)$ where E is a simply connected region in $\mathbf{R}^2$. If the divergence of the vector field $\mathbf{f}$, $\nabla \cdot \mathbf{f}$, is not identically zero and does not change sign in E, then (1) has no closed orbit lying entirely in E.*

3.9. Bendixson's Criteria

Proof. Suppose that Γ: $\mathbf{x} = \mathbf{x}(t)$, $0 \leq t \leq T$, is a closed orbit of (1) lying entirely in E. If S denotes the interior of Γ, it follows from Green's Theorem that

$$\int\int_S \nabla \cdot \mathbf{f}\, dx\, dy = \oint_\Gamma (P\, dy \quad Q\, dx)$$
$$= \int_0^T (P\dot{y} - Q\dot{x})\, dt$$
$$= \int_0^T (PQ - QP)\, dt = 0.$$

And if $\nabla \cdot \mathbf{f}$ is not identically zero and does not change sign in S, then it follows from the continuity of $\nabla \cdot \mathbf{f}$ in S that the above double integral is either positive or negative. In either case this leads to a contradiction. Therefore, there is no closed orbit of (1) lying entirely in E.

Remark 1. The same type of proof can be used to show that, under the hypotheses of Theorem 1, there is no separatrix cycle or graphic of (1) lying entirely in E; cf. Problem 1. And if $\nabla \cdot \mathbf{f} \equiv 0$ in E, it can be shown that, while there may be a center in E, i.e., a one-parameter family of cycles of (1), there is no limit cycle in E.

A more general result of this type, which is also proved using Green's Theorem, cf. Problem 2, is given by the following theorem:

Theorem 2 (Dulac's Criteria). *Let $\mathbf{f} \in C^1(E)$ where E is a simply connected region in $\mathbf{R}^2$. If there exists a function $B \in C^1(E)$ such that $\nabla \cdot (B\mathbf{f})$ is not identically zero and does not change sign in E, then (1) has no closed orbit lying entirely in E. If A is an annular region contained in E on which $\nabla \cdot (B\mathbf{f})$ does not change sign, then there is at most one limit cycle of (1) in A.*

As in the above remark, if $\nabla \cdot (B\mathbf{f})$ does not change sign in E, then it can be shown that there are no separatrix cycles or graphics of (1) in E and if $\nabla \cdot (B\mathbf{f}) \equiv 0$ in E, then (1) may have a center in E. Cf. [A–I], pp. 205–210. The next theorem, proved by the Russian mathematician L. Cherkas [5] in 1977, gives a set of conditions sufficient to guarantee that the Lienard system of Section 3.8 has no limit cycle.

Theorem 3 (Cherkas). *Assume that $a < 0 < b$, $F, g \in C^1(a, b)$, and $xg(x) > 0$ for $x \in (a, 0) \cup (0, b)$. Then if the equations*

$$F(u) = F(v)$$
$$G(u) = G(v)$$

have no solutions with $u \in (a, 0)$ and $v \in (0, b)$, the Lienard system (1) in Section 3.8 has no limit cycle in the region $a < x < b$.

Corollary 1. If $F, g \in C^1(\mathbf{R})$, g is an odd function of x and F is an odd function of x with its only zero at $x = 0$, then the Lienard system (1) in Section 3.8 has no limit cycles.

Finally, we cite one other result, due to Cherkas [4], which is useful in showing the nonexistence of limit cycles for certain quadratic systems.

Theorem 4. If $b < 0$, $ac \geq 0$ and $\alpha a \leq 0$, then the quadratic system

$$\dot{x} = \alpha x - y + ax^2 + bxy + cy^2$$
$$\dot{y} = x + x^2$$

has no limit cycle around the origin.

Example 1. Using this theorem, it is easy to show that for $\alpha \in [-1, 0]$, the quadratic system

$$\dot{x} = \alpha x - y + (\alpha + 1)x^2 - xy$$
$$\dot{y} = x + x^2$$

of Example 2 in Section 3.8 has no limit cycles. This follows since the origin is the only critical point of this system and therefore by Theorem 5 in Section 3.7 any limit cycle of this system must enclose the origin. But according to the above theorem with $b = -1 < 0$, $ac = 0$ and $\alpha a = \alpha(\alpha + 1) \leq 0$ for $\alpha \in [-1, 0]$, there is no limit cycle around the origin.

Problem Set 9

1. Show that, under the hypotheses of Theorem 1, there is no separatrix cycle S lying entirely in E. **Hint:** If such a separatrix cycle exists, then $S = \bigcup_{j=1}^m \Gamma_j$ where for $j = 1, \ldots, m$

$$\Gamma_j: \mathbf{x} = \gamma_j(t), \quad -\infty < t < \infty$$

   is a trajectory of (1). Apply Green's Theorem.

2. Use Green's Theorem to prove Theorem 2. **Hint:** In proving the second part of that theorem, assume that there are two limit cycles Γ_1 and Γ_2 in A, connect them with a smooth arc Γ_0 (traversed in both directions), and then apply Green's Theorem to the resulting simply connected region whose boundary is $\Gamma_1 + \Gamma_0 - \Gamma_2 - \Gamma_0$.

3. (a) Show that for the system

$$\dot{x} = x - y - x^3$$
$$\dot{y} = x + y - y^3$$

the divergence $\nabla \cdot \mathbf{f} < 0$ in the annular region $A = \{\mathbf{x} \in \mathbf{R} \mid 1 < |\mathbf{x}| < \sqrt{2}\}$ and yet there is a limit cycle in this region; cf. Problem 2 in Section 3.7. Why doesn't this contradict Bendixson's Theorem?

(b) Use the second part of Theorem 2 and the result of Problem 2 in Section 3.7 to show that there is exactly one limit cycle in A.

4. (a) Show that the limit cycle of the van der Pol equation

$$\dot{x} = y + x - x^3/3$$
$$\dot{y} = -x$$

must cross the vertical lines $x = \pm 1$; cf. Figure 2 in Section 3.8.

(b) Show that any limit cycle of the Lienard equation (1) in Section 3.8 with f and g odd C^1-functions must cross the vertical line $x = x_1$ where x_1 is the smallest zero of $f(x)$. If equation (1) in Section 3.8 has a limit cycle, use the Corollary to Theorem 3 to show that the function $f(x)$ has at least one positive zero.

5. (a) Use the Dulac function $B(x,y) = be^{-2\beta x}$ to show that the system

$$\dot{x} = y$$
$$\dot{y} = -ax - by + \alpha x^2 + \beta y^2$$

has no limit cycle in $\mathbf{R}^2$.

(b) Show that the system

$$\dot{x} = \frac{y}{1+x^2}$$
$$\dot{y} = \frac{-x + y(1 + x^2 + x^4)}{1+x^2}$$

has no limit cycle in $\mathbf{R}^2$.

3.10 The Poincaré Sphere and the Behavior at Infinity

In order to study the behavior of the trajectories of a planar system for large r, we could use the stereographic projection defined in Section 3.6; cf. Figure 1 in Section 3.6. In that case, the behavior of trajectories far from the origin could be studied by considering the behavior of trajectories near the "point at infinity," i.e., near the north pole of the unit sphere in Figure 1 of Section 3.6. However, if this type of projection is used, the point at infinity is typically a very complicated critical point of the flow induced on the sphere and it is often difficult to analyze the flow in a neighborhood of this critical point. The idea of analyzing the global behavior of a planar dynamical system by using a stereographic projection of the sphere onto the plane is due to Bendixson [B]. The sphere, including the critical point at infinity, is referred to as the Bendixson sphere.

A better approach to studying the behavior of trajectories "at infinity" is to use the so-called Poincaré sphere where we project from the center of the unit sphere $S^2 = \{(X, Y, Z) \in \mathbf{R}^3 \mid X^2 + Y^2 + Z^2 = 1\}$ onto the (x, y)-plane tangent to S^2 at either the north or south pole; cf. Figure 1. This type of central projection was introduced by Poincaré [P] and it has the advantage that the critical points at infinity are spread out along the equator of the Poincaré sphere and are therefore of a simpler nature than the critical point at infinity on the Bendixson sphere. However, some of the critical points at infinity on the Poincaré sphere may still be very complicated in nature.

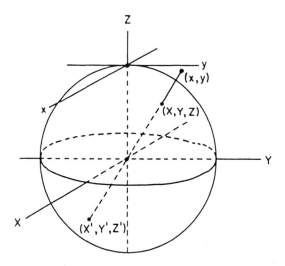

Figure 1. Central projection of the upper hemisphere of S^2 onto the (x, y)-plane.

Remark. The method of "blowing-up" a neighborhood of a complicated critical point uses a combination of central and stereographic projections onto the Poincaré and Bendixson spheres respectively. It allows one to reduce the study of a complicated critical point at the origin to the study of a finite number of hyperbolic critical points on the equator of the Poincaré sphere. Cf. Problem 13.

3.10. The Poincaré Sphere and the Behavior at Infinity

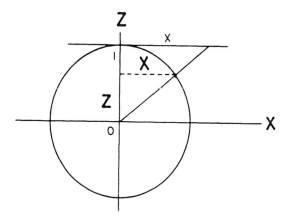

Figure 2. A cross-section of the central projection of the upper hemisphere.

If we project the upper hemisphere of S^2 onto the (x,y)-plane, then it follows from the similar triangles shown in Figure 2 that the equations defining (x,y) in terms of (X,Y,Z) are given by

$$x = \frac{X}{Z}, \quad y = \frac{Y}{Z}. \tag{1}$$

Similarly, it follows that the equations defining (X,Y,Z) in terms of (x,y) are given by

$$X = \frac{x}{\sqrt{1+x^2+y^2}}, \quad Y = \frac{y}{\sqrt{1+x^2+y^2}}, \quad X = \frac{1}{\sqrt{1+x^2+y^2}}.$$

These equations define a one-to-one correspondence between points (X,Y,Z) on the upper hemisphere of S^2 with $Z > 0$ and points (x,y) in the plane. The origin $0 \in \mathbf{R}^2$ corresponds to the north pole $(0,0,1) \in S^2$; points on the circle $x^2+y^2=1$ correspond to points on the circle $X^2+Y^2=1/2$, $Z=1/\sqrt{2}$ on S^2; and points on the equator of S^2 correspond to the "circle at infinity" or "points at infinity" of $\mathbf{R}^2$. Any two antipodal points (X,Y,Z) with (X',Y',Z') on S^2, but not on the equator of S^2, correspond to the same point $(x,y) \in \mathbf{R}^2$; cf. Figure 1. It is therefore only natural to regard any two antipodal points on the equator of S^2 as belonging to the same point at infinity. The hemisphere with the antipodal points on the equator identified is a model for the projective plane. However, rather than trying to visualize the flow on the projective plane induced by a dynamical system on $\mathbf{R}^2$, we shall visualize the flow on the Poincaré sphere induced by a dynamical system on $\mathbf{R}^2$ where the flow in neighborhoods of antipodal points is topologically equivalent, except that the direction of the flow may be reversed.

Consider a flow defined by a dynamical system on $\mathbf{R}^2$

$$\dot{x} = P(x,y)$$
$$\dot{y} = Q(x,y) \tag{2}$$

where P and Q are polynomial functions of x and y. Let m denote the maximum degree of the terms in P and Q. This system can be written in the form of a single differential equation

$$\frac{dy}{dx} = \frac{Q(x,y)}{P(x,y)}$$

or in differential form as

$$Q(x,y)\,dx - P(x,y)\,dy = 0. \tag{3}$$

Note that in either of these two latter forms we lose the direction of the flow along the solution curves of (2). It follows from (1) that

$$dx = \frac{Z\,dX - X\,dZ}{Z^2}, \quad dy = \frac{Z\,dY - Y\,dZ}{Z^2}. \tag{4}$$

Thus, the differential equation (3) can be written as

$$Q(Z\,dX - X\,dZ) - P(Z\,dY - Y\,dZ) = 0$$

where

$$P = P(x,y) = P(X/Z, Y/Z)$$

and

$$Q = Q(x,y) = Q(X/Z, Y/Z).$$

In order to eliminate Z in the denominators, multiply the above equation through by Z^m to obtain

$$ZQ^*\,dX - ZP^*\,dY + (YP^* - XQ^*)\,dZ = 0 \tag{5}$$

where

$$P^*(X,Y,Z) = Z^m P(X/Z, Y/Z)$$

and

$$Q^*(X,Y,Z) = Z^m Q(X/Z, Y/Z)$$

are polynomials in (X,Y,Z). This equation can be written in the form of the determinant equation

$$\begin{vmatrix} dX & dY & dZ \\ X & Y & Z \\ P^* & Q^* & 0 \end{vmatrix} = 0. \tag{5'}$$

3.10. The Poincaré Sphere and the Behavior at Infinity

Cf. [L], p. 202. The differential equation (5) then defines a family of solution curves or a flow on S^2. Each solution curve on the upper (or lower) hemisphere of S^2 defined by (5) corresponds to exactly one solution curve of the system (2) on $\mathbf{R}^2$. Furthermore, the flow on the Poincaré sphere S^2 defined by (5) allows us to study the behavior of the flow defined by (2) at infinity; i.e., we can study the flow defined by (5) in a neighborhood of the equator of S^2. The equator of S^2 consists of trajectories and critical points of (5). This follows since for $Z = 0$ in (5) we have $(YP^* - XQ^*)dZ = 0$. Thus, for $YP^* - XQ^* \neq 0$ we have $dZ = 0$; i.e., we have a trajectory through a regular point on the equator of S^2. And the critical points of (5) on the equator of S^2 where $Z = 0$ are given by the equation

$$YP^* - XQ^* = 0. \qquad (6)$$

If

$$P(x,y) = P_1(x,y) + \cdots + P_m(x,y)$$

and

$$Q(x,y) = Q_1(x,y) + \cdots + Q_m(x,y)$$

where P_j and Q_j are homogeneous jth degree polynomials in x and y, then

$$\begin{aligned}YP^* - XQ^* &= Z^m Y P_1(X/Z, Y/Z) + \cdots + Z^m Y P_m(X/Z, Y/Z) \\ &\quad - Z^m X Q_1(X/Z, Y/Z) - \cdots - Z^m X Q_m(X/Z, Y/Z) \\ &= Z^{m-1} Y P_1(X,Y) + \cdots + Y P_m(X,Y) \\ &\quad - Z^{m-1} X Q_1(X,Y) - \cdots - X Q_m(X,Y) \\ &= Y P_m(X,Y) - X Q_m(X,Y)\end{aligned}$$

for $Z = 0$. And for $Z = 0$, $X^2 + Y^2 = 1$. Thus, for $Z = 0$, (6) is equivalent to

$$\sin\theta P_m(\cos\theta, \sin\theta) - \cos\theta Q_m(\cos\theta, \sin\theta) = 0.$$

That is, the critical points at infinity are determined by setting the highest degree terms in $\dot\theta$, as determined by (2), with $r = 1$, equal to zero. We summarize these results in the following theorem.

Theorem 1. *The critical points at infinity for the mth degree polynomial system (2) occur at the points $(X, Y, 0)$ on the equator of the Poincaré sphere where $X^2 + Y^2 = 1$ and*

$$XQ_m(X,Y) - YP_m(X,Y) = 0 \qquad (6)$$

or equivalently at the polar angles θ_j and $\theta_j + \pi$ satisfying

$$G_{m+1}(\theta) \equiv \cos\theta Q_m(\cos\theta, \sin\theta) - \sin\theta P_m(\cos\theta, \sin\theta) = 0. \qquad (6')$$

This equation has at most $m+1$ pairs of roots θ_j and $\theta_j + \pi$ unless $G_{m+1}(\theta)$ is identically zero. If $G_{m+1}(\theta)$ is not identically zero, then the flow on the *equator of the Poincaré sphere is counter-clockwise at points corresponding to polar angles θ where $G_{m+1}(\theta) > 0$ and it is clockwise at points corresponding to polar angles θ where $G_{m+1}(\theta) < 0$.*

The behavior of the solution curves defined by (5) in a neighborhood of any critical point at infinity, i.e. any critical point of (5) on the equator of the Poincaré sphere S^2, can be determined by projecting that neighborhood onto a plane tangent to S^2 at that point; cf. [L], p. 205. Actually, it is only necessary to project the hemisphere with $X > 0$ onto the plane $X = 1$ and to project the hemisphere with $Y > 0$ onto the plane $Y = 1$ in order to determine the behavior of the flow in a neighborhood of any critical point on the equator of S^2. This follows because the flow on S^2 defined by (5) is topologically equivalent at antipodal points of S^2 if m is odd and it is topologically equivalent, with the direction of the flow reversed, if m is even; cf. Figure 1. We can project the flow on S^2 defined by (5) onto the plane $X = 1$ by setting $X = 1$ and $dX = 0$ in (5). Similarly we can project the flow defined by (5) onto the plane $Y = 1$ by setting $Y = 1$ and $dY = 0$ in (5). Cf. Figure 3. This leads to the results summarized in Theorem 2.

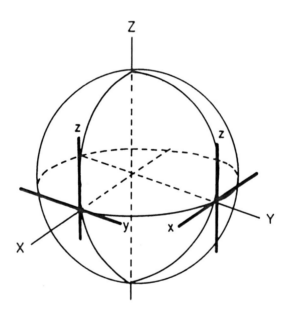

Figure 3. The projection of S^2 onto the planes $X = 1$ and $Y = 1$.

Theorem 2. *The flow defined by (5) in a neighborhood of any critical point of (5) on the equator of the Poincaré sphere S^2, except the points $(0, \pm 1, 0)$, is topologically equivalent to the flow defined by the system*

3.10. The Poincaré Sphere and the Behavior at Infinity

$$\pm \dot{y} = yz^m P\left(\frac{1}{z}, \frac{y}{z}\right) - z^m Q\left(\frac{1}{z}, \frac{y}{z}\right)$$
$$\pm \dot{z} = z^{m+1} P\left(\frac{1}{z}, \frac{y}{z}\right), \tag{7}$$

the signs being determined by the flow on the equator of S^2 as determined in Theorem 1. Similarly, the flow defined by (5) in a neighborhood of any critical point of (5) on the equator of S^2, except the points $(\pm 1, 0, 0)$, is topologically equivalent to the flow defined by the system

$$\pm \dot{x} = xz^m Q\left(\frac{x}{z}, \frac{1}{z}\right) - z^m P\left(\frac{x}{z}, \frac{1}{z}\right)$$
$$\pm \dot{z} = z^{m+1} Q\left(\frac{x}{z}, \frac{1}{z}\right), \tag{7'}$$

the signs being determined by the flow on the equator of S^2 as determined in Theorem 1.

Remark 2. A critical point of (7) at $(y_0, 0)$ corresponds to a critical point of (5) at the point

$$\left(\frac{1}{\sqrt{1+y_0^2}}, \frac{y_0}{\sqrt{1+y_0^2}}, 0\right)$$

on S^2; and a critical point of (7') at $(x_0, 0)$ corresponds to a critical point of (5) at the point

$$\left(\frac{x_0}{\sqrt{1+x_0^2}}, \frac{1}{\sqrt{1+x_0^2}}, 0\right)$$

on S^2.

Example 1. Let us determine the flow on the Poincaré sphere S^2 defined by the planar system

$$\dot{x} = x$$
$$\dot{y} = -y.$$

This system has a saddle at the origin and this is the only (finite) critical point of this system. The saddle at the origin of this system projects, under the central projection shown in Figure 1, onto saddles at the north and south poles of S^2 as shown in Figure 4. According to Theorem 1, the critical points at infinity for this system are determined by the solutions of

$$XQ_1(X,Y) - YP_1(X,Y) = -2XY = 0;$$

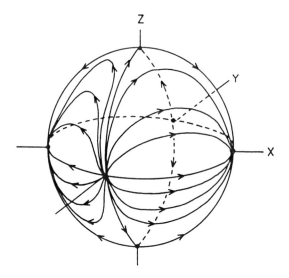

Figure 4. The flow on the Poincaré sphere S^2 defined by the system of Example 1.

i.e., there are critical points on the equator of S^2 at $\pm(1,0,0)$ and at $\pm(0,1,0)$. Also, we see from this expression that the flow on the equator of S^2 is clockwise for $XY > 0$ and counter-clockwise for $XY < 0$. According to Theorem 2, the behavior in a neighborhood of the critical point $(1,0,0)$ is determined by the behavior of the system

$$-\dot{y} = yz\left(\frac{1}{z}\right) + z\left(\frac{y}{z}\right)$$
$$-\dot{z} = z^2\left(\frac{1}{z}\right)$$

or equivalently

$$\dot{y} = -2y$$
$$\dot{z} = -z$$

near the origin. This system has a stable (improper) node at the origin of the type shown in Figure 2 in Section 1.5 of Chapter 1. The y-axis consists of trajectories of this system and all other trajectories come into the origin tangent to the z-axis. This completely determines the behavior at the critical point $(1,0,0)$; cf. Figure 4. The behavior at the antipodal point $(-1,0,0)$ is exactly the same as the behavior at $(1,0,0)$, i.e., there is also a stable (improper) node at $(-1,0,0)$, since $m = 1$ is odd in this example. Similarly, the behavior in a neighborhood of the critical point

3.10. The Poincaré Sphere and the Behavior at Infinity   275

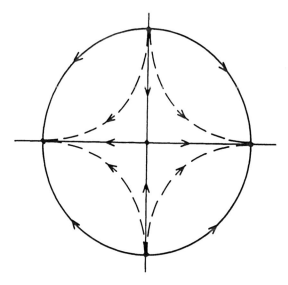

Figure 5. The global phase portrait for the system in Example 1.

$(0,1,0)$ is determined by the behavior of the system

$$\dot{x} = 2x$$
$$\dot{z} = z$$

near the origin. We see that there are unstable (improper) nodes at $(0,\pm 1,0)$ as shown in Figure 4. The fact that the x and y axes of the original system consist of trajectories implies that the great circles through the points $(\pm 1, 0, 0)$ and $(0, \pm 1, 0)$ consist of trajectories. Putting all of this information together and using the Poincaré–Bendixson Theorem yields the flow on S^2 shown in Figure 4.

If we project the upper hemisphere of the Poincaré sphere shown in Figure 4 orthogonally onto the unit disk in the (x, y)-plane, we capture all of the information about the behavior at infinity contained in Figure 4 in a planar figure that is much easier to draw; cf. Figure 5. The flow on the unit disk shown in Figure 5 is referred to as the global phase portrait for the system in Example 1 since Figure 5 describes the behavior of every trajectory of the system including the behavior of the trajectories at infinity. Note that the local behavior near the stable node at $(1, 0, 0)$ in Figure 4 is determined by the local behavior near the antipodal points $(\pm 1, 0)$ in Figure 5.

Notice that the separatrices partition the unit sphere in Figure 4 or the unit disk in Figure 5 into connected open sets, called components, and that the flow in each of these components is determined by the behavior of one typical trajectory (shown as a dashed curve in Figure 5). The separatrix configuration shown in Figure 5, and discussed more fully in the next section, completely determines the global phase portrait of the system of Example 1.

Example 2. Let us determine the global phase portrait of the quadratic system

$$\dot{x} = x^2 + y^2 - 1$$
$$\dot{y} = 5(xy - 1)$$

considered by Lefschetz [L] on p. 204 and originally considered by Poincaré [P] on p. 66. Since the circle $x^2 + y^2 = 1$ and the hyperbola $xy = 1$ do not intersect, there are no finite critical points of this system. The critical points at infinity are determined by

$$XQ_2(X,Y) - YP_2(X,Y) = Y(4X^2 - Y^2) = 0$$

together with $X^2 + Y^2 = 1$. That is, the critical points at infinity are at $\pm(1,0,0)$, $\pm(1,2,0)/\sqrt{5}$ and $\pm(1,-2,0)/\sqrt{5}$. Also, for $X = 0$ the above quantity is negative if $Y > 0$ and positive if $Y < 0$. (In fact, on the entire y-axis, $\dot{\theta} < 0$ if $y > 0$ and $\dot{\theta} > 0$ if $y < 0$.) This determines the direction of the flow on the equator of S^2. According to Theorem 2, the behavior near each of the critical points $(1,0,0)$, $(1,2,0)/\sqrt{5}$ and $(1,-2,0)/\sqrt{5}$ on S^2 is determined by the behavior of the system

$$\dot{y} = 4y - 5z^2 - y^3 + yz^2$$
$$\dot{z} = -z - zy^2 + z^3 \tag{8}$$

near the critical points $(0,0)$, $(2,0)$ and $(-2,0)$ respectively. Note that the critical points at infinity correspond to the critical points on the y-axis of (8) which are easily determined by setting $z = 0$ in (8). For the system (8) we have

$$D\mathbf{f}(0,0) = \begin{bmatrix} 4 & 0 \\ 0 & -1 \end{bmatrix} \quad \text{and} \quad D\mathbf{f}(\pm 2, 0) = \begin{bmatrix} -8 & 0 \\ 0 & -5 \end{bmatrix}.$$

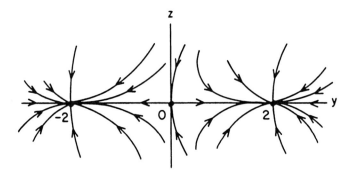

Figure 6. The behavior near the critical points of (8).

3.10. The Poincaré Sphere and the Behavior at Infinity   277

Thus, according to Theorems 3 and 4 in Section 2.10 of Chapter 2, $(0,0)$ is a saddle and $(\pm 2, 0)$ are stable (improper) nodes for the system (8); cf. Figure 6. Since $m = 2$ is even in this example, the behavior near the antipodal points $-(1,0,0)$, $-(1,2,0)/\sqrt{5}$ and $-(1,-2,0)/\sqrt{5}$ is topologically equivalent to the behavior near $(1,0,0)$, $(1,2,0)/\sqrt{5}$ and $(1,-2,0)/\sqrt{5}$ respectively with the directions of the flow reversed; i.e., $-(1,0,0)$ is a saddle and $-(1,\pm 2,0)/\sqrt{5}$ are unstable (improper) nodes. Finally, we note that along the entire x-axis $\dot{y} < 0$. Putting all of this information together and using the Poincaré–Bendixson Theorem leads to the global phase portrait shown in Figure 7; cf. Problem 1.

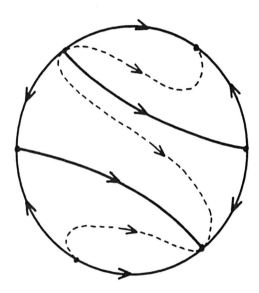

Figure 7. The global phase portrait for the system in Example 2.

The next example illustrates that the behavior at the critical points at infinity is typically more complicated than that encountered in the previous two examples where there were only hyperbolic critical points at infinity. In the next example, it is necessary to use the results in Section 2.11 of Chapter 2 for nonhyperbolic critical points in order to determine the behavior near the critical points at infinity. Note that if the right-hand sides of (7) or (7′) begin with quadratic or higher degree terms, then even the results in Section 2.11 of Chapter 2 will not suffice to determine the behavior at infinity and a more detailed analysis in the neighborhood of these degenerate critical points will be necessary; cf. [N/S]. The next theorem which follows from Theorem 1.1 and its corollaries on pp. 205 and 208 in [H] is often useful in determining the behavior of a planar system near a degenerate critical point.

Theorem 3. *Suppose that* $\mathbf{f} \in C^1(E)$ *where E is an open subset of $\mathbf{R}^2$ containing the origin and the origin is an isolated critical point of the system*

$$\dot{\mathbf{x}} = \mathbf{f}(\mathbf{x}). \tag{9}$$

For $\delta > 0$, let $\Omega = \{(r,\theta) \in \mathbf{R}^2 \mid 0 < r < \delta, \theta_1 < \theta < \theta_2\} \subset E$ be a sector at the origin containing no critical points of (9) and suppose that $\dot{\theta} > 0$ for $0 < r < \delta$ and $\theta = \theta_2$ and that $\dot{\theta} < 0$ for $0 < r < \delta$ and $\theta = \theta_1$. Then for sufficiently small $\delta > 0$

(i) *if $\dot{r} < 0$ for $r = \delta$ and $\theta_1 < \theta < \theta_2$, there exists at least one half-trajectory with its endpoint on $R = \{(r,\theta) \in \mathbf{R}^2 \mid r = \delta, \theta_1 < \theta < \theta_2\}$ which approaches the origin as $t \to \infty$; and*

(ii) *if $\dot{r} > 0$ for $r = \delta$ and $\theta_1 < \theta < \theta_2$, then every half-trajectory with its endpoint on R approaches the origin as $t \to -\infty$.*

Example 3. Let us determine the global phase portrait of the system

$$\dot{x} = -y(1+x) + \alpha x + (\alpha+1)x^2$$
$$\dot{y} = x(1+x)$$

depending on a parameter $\alpha \in (0,1)$. This system was considered in Example 2 in Section 3.8 where we showed that for $\alpha \in (0,1)$ there is exactly one stable limit cycle around the critical point at the origin. Also, the origin is the only (finite) critical point for this system. The critical points at infinity are determined by

$$XQ_2(X,Y) - YP_2(X,Y) = X[X^2 - (\alpha+1)XY + Y^2] = 0.$$

For $0 < \alpha < 1$, this equation has $X = 0$ as its only solution. Thus, the only critical points at infinity are at $\pm(0,1,0)$. And from (7'), the behavior at $(0,1,0)$ is determined by the behavior of the system

$$\begin{aligned} -\dot{x} &= x + z - \alpha xz - (\alpha+1)x^2 + x^2z + x^3 \\ -\dot{z} &= x^2z + xz^2 \end{aligned} \tag{10}$$

at the origin. In this case the matrix $A = D\mathbf{f}(0)$ for this system has one zero eigenvalue and Theorem 1 of Section 2.11 of Chapter 2 applies. In order to use that theorem, we must put the above equation in the form of equation (2) in Section 2.11 of Chapter 2. This can be done by letting $y = x + z$ and determining $\dot{y} = \dot{x} + \dot{z}$ from the above equations. We find that

$$\dot{y} = y + q_2(y,z) \equiv -y + z^2 - (\alpha+2)yz + (\alpha+1)y^2 + y^2z - y^3$$
$$\dot{z} = p_2(y,z) \equiv yz^2 - y^2z.$$

Solving $y + q_2(y,z) = 0$ in a neighborhood of the origin yields

$$y = \phi(z) = z^2[1 - (\alpha+2)z] + 0(z^4)$$

3.10. The Poincaré Sphere and the Behavior at Infinity

and then
$$p_2(z, \phi(z)) = z^4 - (\alpha + 3)z^5 + 0(z^6).$$
Thus, in Theorem 1 of Section 2.11 of Chapter 2 we have $m = 4$ (and $a_m = 1$). It follows that the origin of (10) is a saddle-node. Since $\dot{z} = 0$ on the x-axis (where $z = 0$) in equation (10), the x-axis is a trajectory. And for $z = 0$ we have $\dot{x} = x - (\alpha+1)x^2 + x^4$ in (10). Therefore, $\dot{x} > 0$ for $x > 0$ and $\dot{x} < 0$ for $x < 0$ on the x-axis. Next, on $z = -x$ we have $\dot{x} = x^2 + 0(x^3) > 0$ for small $x \neq 0$. And for sufficiently small $r > 0$, we have $\dot{r} > 0$ for $-z < x < 0$ in (10). On the z-axis we have $\dot{x} = -x < 0$ for $x > 0$. Thus, by Theorem 3, there is a trajectory of (10) in the sector $\pi/2 < \theta < 3\pi/4$ which approaches the origin as $t \to -\infty$; cf. Figure 8(a). It follows that the saddle-node at the origin of (10) has the local phase portrait shown in Figure 8(b). It then follows, using the Poincaré–Bendixson Theorem, that the global phase portrait for the system in this example is given in Figure 9; cf. Problem 2.

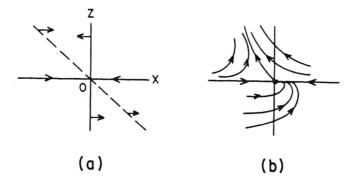

(a)   (b)

Figure 8. The saddle-node at the origin of (10).

The projective geometry and geometrical ideas presented in this section are by no means limited to flows in $\mathbf{R}^2$ which can be projected onto the upper hemisphere of S^2 by a central projection in order to study the behavior of the flow on the circle at infinity, i.e., on the equator of S^2. We illustrate how these ideas can be carried over to higher dimensions by presenting the theory and an example for flows in $\mathbf{R}^3$. The upper hemisphere of S^3 can be projected onto $\mathbf{R}^3$ using the transformation of coordinates given by

$$x = \frac{X}{W}, \quad y = \frac{Y}{W}, \quad z = \frac{Z}{W}$$

and

$$X = \frac{x}{\sqrt{1+|\mathbf{x}|^2}}, Y = \frac{y}{\sqrt{1+|\mathbf{x}|^2}}, Z = \frac{z}{\sqrt{1+|\mathbf{x}|^2}}, W = \frac{1}{\sqrt{1+|\mathbf{x}|^2}}$$

for $\mathbf{X} = (X, Y, Z, W) \in S^3$ with $|\mathbf{X}| = 1$ and for $\mathbf{x} = (x, y, z) \in \mathbf{R}^3$. If we consider a flow in $\mathbf{R}^3$ defined by

$$\dot{x} = P(x, y, z)$$
$$\dot{y} = Q(x, y, z) \quad (11)$$
$$\dot{z} = R(x, y, z)$$

where P, Q, and R are polynomial functions of x, y, z of maximum degree m, then this system can be equivalently written as

$$\frac{dy}{dx} = \frac{Q(x, y, z)}{P(x, y, z)} \quad \text{and} \quad \frac{dz}{dx} = \frac{R(x, y, z)}{P(x, y, z)}$$

or as

$$Q(x, y, z)dx - P(x, y, z)dy = 0 \quad \text{and} \quad R(x, y, z)dx - P(x, y, z)dz = 0$$

in which cases the direction of the flow along trajectories is no longer

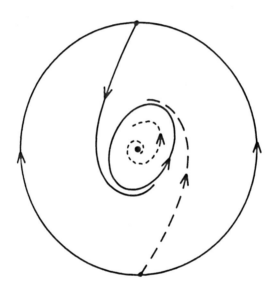

Figure 9. The global phase portrait for the system in Example 3.

determined. And then since

$$dx = \frac{WdX - XdW}{W^2}, \quad dy = \frac{WdY - YdW}{W^2}, \quad \text{and} \quad dz = \frac{WdZ - ZdW}{W^2},$$

we can write the system as

$$Q^*(WdX - WdW) - P^*(WdY - YdW) = 0$$

and

$$R^*(WdX - XdW) - P^*(WdZ - ZdW) = 0,$$

3.10. The Poincaré Sphere and the Behavior at Infinity

if we define the functions

$$P^*(\mathbf{X}) = Z^m P(\mathbf{X}/W)$$
$$Q^*(\mathbf{X}) = Z^m Q(\mathbf{X}/W)$$

and

$$R^*(\mathbf{X}) = Z^m R(\mathbf{X}/W).$$

Then corresponding to Theorems 1 and 2 above we have the following theorems.

Theorem 4. *The critical points at infinity for the mth degree polynomial system (11) occur at the points $(X, Y, Z, 0)$ on the equator of the Poincaré sphere S^3 where $X^2 + Y^2 + Z^2 = 1$ and*

$$XQ_m(X, Y, Z) - YP_m(X, Y, Z) = 0$$
$$XR_m(X, Y, Z) - ZP_m(X, Y, Z) = 0$$

and

$$YR_m(X, Y, Z) - ZQ_m(X, Y, Z) = 0$$

where P_m, Q_m and R_m denote the mth degree terms in P, Q, and R respectively.

Theorem 5. *The flow defined by the system (11) in a neighborhood*
(a) *of $(\pm 1, 0, 0, 0) \in S^3$ is topologically equivalent to the flow defined by the system*

$$\pm \dot{y} = yw^m P\left(\frac{1}{w}, \frac{y}{w}, \frac{z}{w}\right) - w^m Q\left(\frac{1}{w}, \frac{y}{w}, \frac{z}{w}\right)$$
$$\pm \dot{z} = zw^m P\left(\frac{1}{w}, \frac{y}{w}, \frac{z}{w}\right) - w^m R\left(\frac{1}{w}, \frac{y}{w}, \frac{z}{w}\right)$$
$$\pm \dot{w} = w^{m+1} P\left(\frac{1}{w}, \frac{y}{w}, \frac{z}{w}\right).$$

(b) *of $(0, \pm 1, 0, 0) \in S^3$ is topologically equivalent to the flow defined by the system*

$$\pm \dot{x} = xw^m Q\left(\frac{x}{w}, \frac{1}{w}, \frac{z}{w}\right) - w^m P\left(\frac{x}{w}, \frac{1}{w}, \frac{z}{w}\right)$$
$$\pm \dot{z} = zw^m Q\left(\frac{x}{w}, \frac{1}{w}, \frac{z}{w}\right) - w^m R\left(\frac{x}{w}, \frac{1}{w}, \frac{z}{w}\right)$$
$$\pm \dot{w} = w^{m+1} Q\left(\frac{x}{w}, \frac{1}{w}, \frac{z}{w}\right)$$

and

(c) of $(0, 0, \pm 1, 0) \in S^3$ is topologically equivalent to the flow defined by the system

$$\pm \dot{x} = xw^m R\left(\frac{x}{w}, \frac{y}{w}, \frac{1}{w}\right) - w^m P\left(\frac{x}{w}, \frac{y}{w}, \frac{1}{w}\right)$$

$$\pm \dot{y} = yw^m R\left(\frac{x}{w}, \frac{y}{w}, \frac{1}{w}\right) - w^m Q\left(\frac{x}{w}, \frac{y}{w}, \frac{1}{w}\right)$$

$$\pm \dot{w} = w^{m+1} R\left(\frac{x}{w}, \frac{y}{w}, \frac{1}{w}\right).$$

The direction of the flow, i.e., the arrows on the trajectories representing the flow, is not determined by Theorem 5. It follows from the original system (11). Let us apply this theory for flows in $\mathbf{R}^3$ to a specific example.

Example 4. Determine the flow on the Poincaré sphere S^3 defined by the system

$$\dot{x} = x$$
$$\dot{y} = y$$
$$\dot{z} = -z$$

in $\mathbf{R}^3$. From Theorem 4, we see that the critical points at infinity are determined by

$$XQ_1 - YP_1 = XY - YX = 0$$
$$XP_1 - XR_1 = ZX + XZ = 2XZ = 0$$

and

$$YR_1 - ZQ_1 = Y(-Z) - ZY = 2YZ = 0.$$

These equations are satisfied iff $X = Y = 0$ or $Z = 0$ and $X^2 + Y^2 + Z^2 = 1$; i.e., the critical points on S^3 are at $(0, 0, \pm 1, 0)$ and at points $(X, Y, 0, 0)$ with $X^2 + Y^2 = 1$. According to Theorem 5(c), the flow in a neighborhood of $(0, 0, \pm 1, 0)$ is determined by the system

$$\dot{x} = -xw\left(-\frac{1}{w}\right) + w\left(\frac{x}{w}\right) = 2x$$

$$\dot{y} = -yw\left(-\frac{1}{w}\right) + w\left(\frac{y}{w}\right) = 2y$$

$$\dot{w} = -w^2\left(-\frac{1}{w}\right) = w$$

where we have used the original system to select the minus sign in Theorem 5(c). We see that $(0, 0, \pm 1, 0)$ is an unstable node. The flow at any point

3.10. The Poincaré Sphere and the Behavior at Infinity

$(X, Y, 0, 0)$ with $X^2 + Y^2 = 1$ such as the point $(1, 0, 0, 0)$ is determined by the system in Theorem 5(a):

$$\dot{y} = -yw\left(\frac{1}{w}\right) + w\left(\frac{y}{w}\right) = 0$$

$$\dot{z} = -zw\left(\frac{1}{w}\right) - w\left(\frac{z}{w}\right) = -2z$$

$$\dot{w} = -w^2\left(\frac{1}{w}\right) = -w$$

where we have used the original system to select the minus sign in Theorem 5(a). We see that $(1, 0, 0, 0)$ is a nonisolated, stable critical point. We can summarize these results by projecting the upper hemisphere of the Poincaré sphere S^3 onto the unit ball in $\mathbf{R}^3$; this captures all of the information about the flow of the original system in $\mathbf{R}^3$ and it also includes the behavior on the sphere at infinity which is represented by the surface of the unit ball in $\mathbf{R}^3$. Cf. Figure 10.

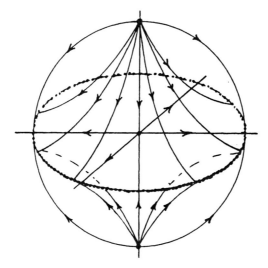

Figure 10. The "global phase portrait" for the system in Example 4.

Before ending this section, we shall define what we mean by a vector field on an n-dimensional manifold. In Example 1, we saw that a flow on $\mathbf{R}^2$ defined a flow on the Poincaré sphere S^2; cf. Figure 4. This flow on S^2 defines a corresponding vector field on S^2, namely, the set of all velocity vectors $\mathbf{v}$ tangent to the trajectories of the flow on S^2. Each of the velocity vectors $\mathbf{v}$ lies in the tangent plane $T_\mathbf{p} S^2$ to S^2 at a point $\mathbf{p} \in S^2$. We now give a more precise definition of a vector field on an n-dimensional differentiable manifold M. We shall use this definition for two-dimensional manifolds in Section 3.12 on index theory. Since any n-dimensional manifold

M can be embedded in $\mathbf{R}^N$ for some N with $n \leq N \leq 2n+1$ by Whitney's Theorem, we assume that $M \subset \mathbf{R}^N$ in the following definition.

Let M be an n-dimensional differentiable manifold with $M \subset \mathbf{R}^N$. A *smooth curve through a point* $\mathbf{p} \in M$ is a C^1-map $\gamma\colon (-a, a) \to M$ with $\gamma(0) = \mathbf{p}$. The velocity vector $\mathbf{v}$ tangent to γ at the point $\mathbf{p} = \gamma(0) \in M$,

$$\mathbf{v} = D\gamma(0).$$

The *tangent space to M at a point* $\mathbf{p} \in M$, $T_\mathbf{p} M$, is the set of all velocity vectors tangent to smooth curves passing through the point $\mathbf{p} \in M$. The tangent space $T_\mathbf{p} M$ is an n-dimensional linear space and we shall view it as a subspace of $\mathbf{R}^n$. A *vector field* $\mathbf{f}$ *on* M is a function $\mathbf{f}\colon M \to \mathbf{R}^N$ such that $\mathbf{f}(\mathbf{p}) \in T_\mathbf{p} M$ for all $\mathbf{p} \in M$. If we define the *tangent bundle of M*, TM, as the disjoint union of the tangent spaces $T_\mathbf{p} M$ to M for $\mathbf{p} \in M$, then a vector field on M, $\mathbf{f}\colon M \to TM$.

Let $\{(U_\alpha, \mathbf{h}_\alpha)\}_\alpha$ be an atlas for the n-dimensional manifold M and let $V_\alpha = \mathbf{h}_\alpha(U_\alpha)$. Recall that if the maps $\mathbf{h}_\alpha \circ \mathbf{h}_\beta^{-1}$ are all C^k-functions, then M is called a differentiable manifold of class C^k. For a vector field $\mathbf{f}$ on M there are functions $\mathbf{f}_\alpha\colon V_\alpha \to \mathbf{R}^n$ such that

$$\mathbf{f}(\mathbf{h}_\alpha^{-1}(\mathbf{x})) = D\mathbf{h}_\alpha^{-1}(\mathbf{x})\mathbf{f}_\alpha(\mathbf{x})$$

for all $\mathbf{x} \in V_\alpha$. Note that under the change of coordinates $\mathbf{x} \to \mathbf{h}_\alpha \circ \mathbf{h}_\beta^{-1}(\mathbf{x})$ we have

$$\mathbf{f}_\beta(\mathbf{h}_\beta \circ \mathbf{h}_\alpha^{-1}(\mathbf{x})) = D\mathbf{h}_\beta \circ \mathbf{h}_\alpha^{-1}(\mathbf{x})\mathbf{f}_\alpha(\mathbf{x})$$

for all $\mathbf{x}$ with $\mathbf{h}_\alpha^{-1}(\mathbf{x}) \in U_\alpha \cap U_\beta$. If the functions $\mathbf{f}_\alpha\colon V_\alpha \to \mathbf{R}^n$ are all in $C^1(V_\alpha)$ and the manifold M is at least of class C^2, then $\mathbf{f}$ is called a C^1-*vector field on* M.

A general theory, analogous to the local theory in Chapter 2, can be developed for the system

$$\dot{\mathbf{x}} = \mathbf{f}(\mathbf{x})$$

where $\mathbf{x} \in M$, an n-dimensional differentiable manifold of class C^2, and $\mathbf{f}$ is a C^1-vector field on M. In particular, we have local existence and uniqueness of solutions through any point $\mathbf{x}_0 \in M$. A solution or integral curve on M, $\phi_t(\mathbf{x}_0)$ is tangent to the vector field $\mathbf{f}$ at $\mathbf{x}_0$. If M is compact and $\mathbf{f}$ is a C^1-vector field on M then, by Chillingworth's Theorem, Theorem 4 in Section 3.1, $\phi(t, x_0) = \phi_t(x_0)$ is defined for all $t \in \mathbf{R}$ and $\mathbf{x}_0 \in M$ and it can be shown that $\phi \in C^1(\mathbf{R} \times M)$; $\phi_s \circ \phi_t = \phi_{s+t}$; and ϕ_t is called a *flow on the manifold* M. We have already seen several examples of flows on the two-dimensional manifolds S^2 and T^2 (and of course $\mathbf{R}^2$). In order to make these ideas more concrete, consider the following simple example of a vector field on the one-dimensional, compact manifold

$$C = S^1 = \{\mathbf{x} \in \mathbf{R}^2 \mid |\mathbf{x}| = 1\}.$$

3.10. The Poincaré Sphere and the Behavior at Infinity

Example 5. Let C be the unit circle in $\mathbf{R}^2$. Then C can be represented by points of the form $\mathbf{x}(\theta) = (\cos\theta, \sin\theta)^T$. C is a one-dimensional differentiable manifold of class C^∞. An atlas for C is given in Problem 7 in Section 2.7 of Chapter 2. For each $\theta \in [0, 2\pi)$, the tangent space to the manifold $M = C$ at the point $\mathbf{p}(\theta) = (\cos\theta, \sin\theta)^T \in M$

$$T_{\mathbf{p}(\theta)}M = \{\mathbf{x} \in \mathbf{R}^2 \mid x\cos\theta + y\sin\theta = 1\}$$

and the function

$$\mathbf{f}(\mathbf{p}(\theta)) = \begin{pmatrix} -\sin\theta \\ \cos\theta \end{pmatrix}$$

defines a C^1-vector field on M; cf. Figure 11. Note that we are viewing $T_\mathbf{p}M$ as a subspace of $\mathbf{R}^2$ and the vector $\mathbf{f}(\mathbf{p})$ as a free vector which can either be based at the origin $\mathbf{0} \in \mathbf{R}^2$ or at the point $\mathbf{p} \in M$.

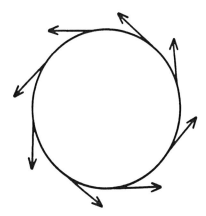

Figure 11. A vector field on the unit circle.

The solution to the system

$$\dot{\mathbf{x}} = \mathbf{f}(\mathbf{x})$$

through a point $\mathbf{p}(\theta) = (\cos\theta, \sin\theta)^T \in M$ is given by

$$\phi_t(\mathbf{p}(\theta)) = \begin{pmatrix} \cos(t+\theta) \\ \sin(t+\theta) \end{pmatrix}.$$

The flow ϕ_t represents the motion of a point moving counterclockwise around the unit circle with unit velocity $\mathbf{v} = \dot{\phi}_t(\mathbf{p})$ tangent to C at the point $\phi_t(\mathbf{p})$.

If the charts (U_1, h_1) and (U_2, h_2) are given by

$$U_1 = \left\{(x,y) \in \mathbf{R}^2 \mid y = \sqrt{1-x^2}, -1 < x < 1\right\}$$

$$h_1(x,y) = x, \quad h_1^{-1}(x) = \left(x, \sqrt{1-x^2}\right)^T$$

$$U_2 = \left\{(x,y) \in \mathbf{R}^2 \mid x = \sqrt{1-y^2}, -1 < y < 1\right\}$$

$$h_2(x,y) = y \quad \text{and} \quad h_2^{-1}(y) = \left(\sqrt{1-y^2}, y\right)^T,$$

then

$$Dh_1^{-1}(x) = \begin{pmatrix} 1 \\ \dfrac{-x}{\sqrt{1-x^2}} \end{pmatrix}, \quad Dh_2^{-1}(y) = \begin{pmatrix} \dfrac{-y}{\sqrt{1-y^2}} \\ 1 \end{pmatrix}$$

and for the functions

$$f_1(x) = -\sqrt{1-x^2} \quad \text{and} \quad f_2(y) = \sqrt{1-y^2}$$

where $-1 < x < 1$ and $-1 < y < 1$, we have

$$\mathbf{f}(x, \sqrt{1-x^2}) = \begin{pmatrix} -\sqrt{1-x^2} \\ x \end{pmatrix} = Dh_1^{-1}(x) f_1(x)$$

and

$$\mathbf{f}(\sqrt{1-y^2}, y) = \begin{pmatrix} -y \\ -\sqrt{1-y^2} \end{pmatrix} = Dh_2^{-1}(y) f_2(y).$$

It is also easily shown that

$$h_2 \circ h_1^{-1}(x) = \sqrt{1-x^2}, \quad Dh_2 \circ h_1^{-1}(x) = \dfrac{-x}{\sqrt{1-x^2}}$$

and that for $x > 0$

$$f_2(\sqrt{1-x^2}) = x = Dh_2 \circ h_1^{-1}(x) f_1(x).$$

Problem Set 10

1. Complete the details in obtaining the global phase portrait for Example 2 shown in Figure 7. In particular, show that

    (a) The separatrices approaching the saddle at the origin of system (8) do so in the first and fourth quadrants as shown in Figure 6.

    (b) In the global phase portrait shown in Figure 7, the separatrix having the critical point $(1,0,0)$ as its ω-limit set must have the unstable node at $(-1,2,0)/\sqrt{5}$ as its α-limit set. **Hint:** Use the Poincaré–Bendixson Theorem and the fact that $\dot{y} < 0$ on the x-axis for the system in Example 2.

3.10. The Poincaré Sphere and the Behavior at Infinity

2. Complete the details in obtaining the global phase portrait for Example 3 shown in Figure 9. In particular, show that the flow swirls counter-clockwise around the origin for the system in Example 3 and that the limit cycle of the system in Example 3 is the ω-limit set of the separatrix which has the saddle at $(0, 1, 0)$ as its α-limit set.

3. Draw the global phase portraits for the systems

   (a) $\begin{aligned} \dot{x} &= 2x \\ \dot{y} &= y \end{aligned}$

   (b) $\begin{aligned} \dot{x} &= x \\ \dot{y} &= y \end{aligned}$

   (c) $\begin{aligned} \dot{x} &= x - y \\ \dot{y} &= x + y \end{aligned}$

4. Draw the global phase portrait for the system

$$\dot{x} = -4y + 2xy - 8$$
$$\dot{y} = 4y^2 - x^2.$$

   Note that the nature of the finite critical points for this system was determined in Problem 4 in Section 2.10 of Chapter 2. Also, note that on the x-axis $\dot{y} < 0$ for $x \neq 0$.

5. Draw the global phase portrait for the system

$$\dot{x} = 2x - 2xy$$
$$\dot{y} = 2y - x^2 + y^2.$$

   Note that the nature of the finite critical points for this system was determined in Problem 4 in Section 2.10 of Chapter 2. Also, note the symmetry with respect to the y-axis for this system.

6. Draw the global phase portrait for the system

$$\dot{x} = -x^2 - y^2 + 1$$
$$\dot{y} = 2x.$$

   The same comments made in Problem 5 apply here.

7. Draw the global phase portrait for the system

$$\dot{x} = -x^2 - y^2 + 1$$
$$\dot{y} = 2xy.$$

   The same comments made in Problem 5 apply here and also, note that this system is symmetric with respect to both the x and y axes.

8. Draw the global phase portrait for this system
$$\dot{x} = x^2 - y^2 - 1$$
$$\dot{y} = 2y.$$

Note that the nature of the finite critical points for this system was determined in Problem 4 in Section 2.10 of Chapter 2. Also, note the symmetry with respect to the x-axis for this system. **Hint:** You will have to use Theorem 1 in Section 2.11 of Chapter 2 as in Example 3.

9. Let M be the two-dimensional manifold
$$S^2 = \{\mathbf{x} \in \mathbf{R}^3 \mid x^2 + y^2 + z^2 = 1\}.$$

(a) Show that any point $\mathbf{p}_0 = (x_0, y_0, z_0) \in S^2$, the tangent space to S^2 at the point $\mathbf{p}_0$
$$T_{\mathbf{p}_0} M = \{\mathbf{x} \in \mathbf{R}^3 \mid xx_0 + yy_0 + zz_0 = 1\}.$$

(b) What condition must the components of $\mathbf{f}$ satisfy in order that the function $\mathbf{f}$ defines a vector field on $M = S^2$?

(c) Show that the vector field on S^2 defined by the differentiable equation (5) is given by
$$\mathbf{f}(\mathbf{x}) = (xyQ^* - (y^2+z^2)P^*, xyP^* - (x^2+z^2)Q^*, xzP^* + yzQ^*)^T$$
where $P^*(x,y,z)$ and $Q^*(x,y,z)$ are defined as in equation (5); and show that this function satisfies the condition in part (b). **Hint:** Take the cross product of the vector of coefficients in equation (5) with the vector $(x,y,z)^T \in S^2$ to obtain $\mathbf{f}$.

10. Determine which of the global phase portraits shown in Figure 12 below correspond to the following quadratic systems. **Hint:** Two of the global phase portraits shown in Figure 12 correspond to the quadratic systems in Problems 2 and 3 in Section 3.11.

(a) $\dot{x} = -4y + 2xy - 8$
$\dot{y} = 4y^2 - x^2$

(b) $\dot{x} = 2x - 2xy$
$\dot{y} = 2y - x^2 + y^2$

(c) $\dot{x} = -x^2 - y^2 + 1$
$\dot{y} = 2x$

(d) $\dot{x} = -x^2 - y^2 + 1$
$\dot{y} = 2xy$

(e) $\dot{x} = x^2 - y^2 - 1$
$\dot{y} = 2y$

3.10. The Poincaré Sphere and the Behavior at Infinity   289

11. Set up a one-to-one correspondence between the global phase portraits (or separatrix configurations), shown in Figure 12 and the computer-drawn global phase portraits in Figure 13.

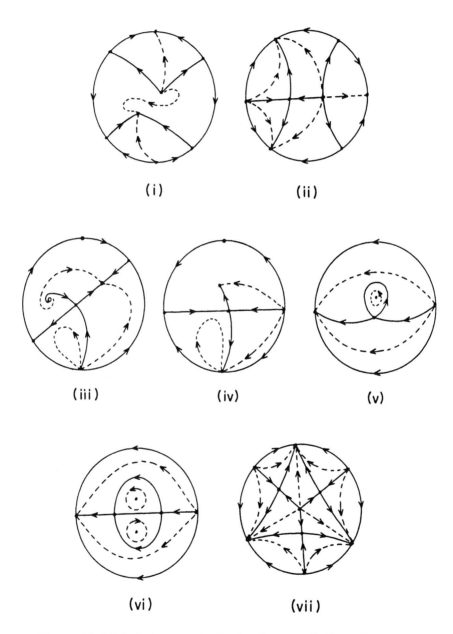

Figure 12. Global phase portrait of various quadratic systems.

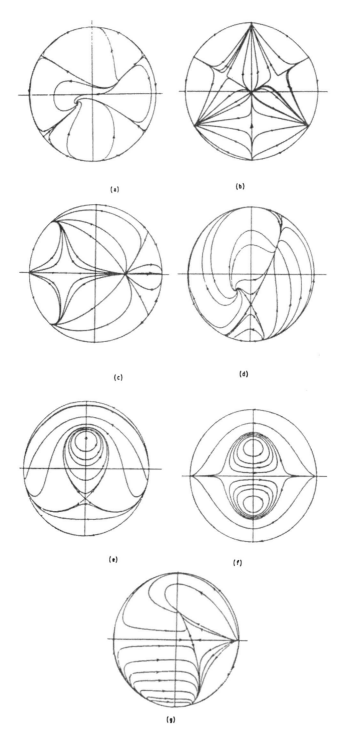

Figure 13. Computer-drawn global phase portraits.

3.10. The Poincaré Sphere and the Behavior at Infinity   291

12. The system of differential equations in Problem 10(c) above can be written as a first-order differential equation,

$$2xdx - (1 - x^2 - y^2)dy = 0.$$

Show that e^y is an integrating factor for this differential equation, i.e., show that when this differential equation is multiplied by e^y, it becomes an exact differential equation. Solve the resulting exact differential equation and use the solution to obtain a formula for the homoclinic loop at the saddle point $(0, -1)$ of the system 10(c).

13. Analyze the critical point at the origin of the system

$$\dot{x} = x^3 - 3xy^2$$
$$\dot{y} = 3x^2y - y^3$$

by the method of "blowing-up" outlined below:

(a) Write this system in polar coordinates to obtain

$$\frac{dr}{d\theta} = \frac{r(x^2 - y^2)}{2xy}.$$

(b) Project the x, y-plane onto the Bendixson sphere. This is most easily done by letting $\rho = 1/r$, $\xi = \rho \cos\theta$ and $\eta = \rho \sin\theta$. You should obtain

$$\frac{d\rho}{d\theta} = -\frac{1}{r^2}\frac{dr}{d\theta} = -\frac{\rho(\xi^2 - \eta^2)}{2\xi\eta}$$

or equivalently

$$\dot{\xi} = \xi$$
$$\dot{\eta} = -\eta.$$

(c) Now project the ξ, η-plane onto the Poincaré sphere as in Example 1. In this case, there are four hyperbolic critical points at infinity and they are all nodes. If we now project from the north pole of the Poincaré sphere (shown in Figure 4) onto the ξ, η-plane, we obtain the flow shown in Figure 14(a). This is called the "blow-up" of the degenerate critical point of the original system. By shrinking the circle in Figure 14(a) to a point, we obtain the flow shown in 14(b) which describes the flow of the original system in a neighborhood of the origin.

In general, by a finite number of "blow-ups," we can reduce the study of a complicated critical point at the origin to the study of a finite number of hyperbolic critical points on the equator of the Poincaré sphere. It is interesting to note that the flow on the Bendixson sphere for this problem is given in Figure 15.

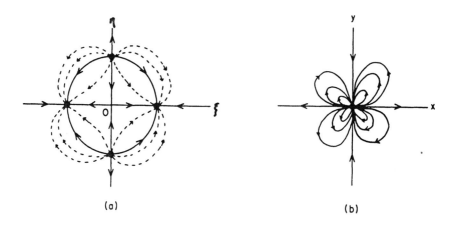

Figure 14. The blow-up of a degenerate critical point with four elliptic sectors.

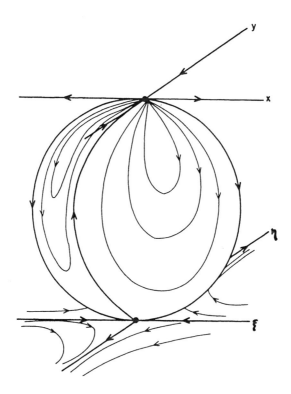

Figure 15. The flow on the Bendixson sphere defined by the system in Problem 13.

3.11 Global Phase Portraits and Separatrix Configurations

In this section, we present some ideas developed by Lawrence Markus [20] concerning the topological equivalence of two C^1-systems

$$\dot{\mathbf{x}} = \mathbf{f}(\mathbf{x}) \qquad (1)$$

and

$$\dot{\mathbf{x}} = \mathbf{g}(\mathbf{x}) \qquad (2)$$

on $\mathbf{R}^2$. Recall that (1) and (2) are said to be topologically equivalent on $\mathbf{R}^2$ if there exists a homeomorphism $H: \mathbf{R}^2 \to \mathbf{R}^2$ which maps trajectories of (1) onto trajectories of (2) and preserves their orientation by time; cf. Definition 2 in Section 3.1. The homeomorphism H need not preserve the parametrization by time.

Following Markus [20], we first of all extend our concept of what types of trajectories of (1) constitute a separatrix of (1) and then we give necessary and sufficient conditions for two relatively prime, planar, analytic systems to be topologically equivalent on $\mathbf{R}^2$. Markus established this result for C^1-systems having no "limit separatrices"; cf. Theorem 7.1 in [20]. His results apply directly to relatively prime, planar, analytic systems since they have no limit separatrices. In fact, the author [24] proved that if the components P and Q of $\mathbf{f} = (P, Q)^T$ are relatively prime, analytic functions, i.e., if (1) is a relatively prime, analytic system, then the critical points of (1) are isolated. And Bendixson [B] proved that at each isolated critical point $\mathbf{x}_0$ of a relatively prime, analytic system (1) there are at most a finite number of separatrices or trajectories of (1) which lie on the boundaries of hyperbolic sectors at $\mathbf{x}_0$; cf. Theorem IX, p. 32 in [B].

In order to make the concept of a separatrix as clear as possible, we first give a simplified definition of a separatrix for polynomial systems and then give Markus's more general definition for C^1-systems on $\mathbf{R}^2$. While the first definition below applies to any polynomial system, it does not in general apply to analytic systems on $\mathbf{R}^2$; cf. Problem 6.

Definition 1. A *separatrix* of a relatively prime, polynomial system (1) is a trajectory of (1) which is either

(i) a critical point of (1)

(ii) a limit cycle of (1)

(iii) a trajectory of (1) which lies on the boundary of a hyperbolic sector at a critical point of (1) on the Poincaré sphere.

In order to present Markus's definition of a separatrix of a C^1-system, it is first necessary to define what is meant by a parallel region. In [20] a

294   3. Nonlinear Systems: Global Theory

parallel region is defined as a collection of curves filling a plane region R which is topologically equivalent to either the plane filled with parallel lines, the punctured plane filled with concentric circles, or the punctured plane filled with rays from the deleted point; these three types of parallel regions are referred to as strip, annular, and spiral or radial regions respectively by Markus. These three basic types of parallel regions are shown in Figure 1 and several types of strip regions are shown in Figure 2.

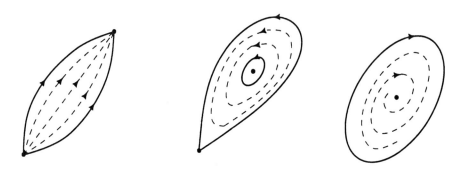

Figure 1. Various types of parallel or canonical regions of (1); strip, annular and spiral regions.

Figure 2. Various types of strip regions of (1); elliptic, parabolic and hyperbolic regions.

Definition 1'. A solution curve Γ of a C^1-system (1) with $\mathbf{f} \in C^1(\mathbf{R}^2)$ is a *separatrix* of (1) if Γ is not embedded in a parallel region N such that

(i) every solution of (1) in N has the same limit sets, $\alpha(\Gamma)$ and $\omega(\Gamma)$, as Γ and

(ii) N is bounded by $\alpha(\Gamma) \cup \omega(\Gamma)$ and exactly two solution curves Γ_1 and Γ_2 of (1) for which $\alpha(\Gamma_1) = \alpha(\Gamma_2) = \alpha(\Gamma)$ and $\omega(\Gamma_1) = \omega(\Gamma_2) = \omega(\Gamma)$.

3.11. Global Phase Portraits and Separatrix Configurations

Theorem 1. *If* (1) *is a* C^1-*system on* $\mathbf{R}^2$, *then the union of the set of separatrices of* (1) *is closed.*

This is Theorem 3.1 in [20]. Since the union of the set of separatrices of a relatively prime, analytic system (1), S, is closed, $\mathbf{R}^2 \sim S$ is open. The components of $\mathbf{R}^2 \sim S$, i.e., the open connected subsets of $\mathbf{R}^2 \sim S$, are called *the canonical regions of* (1). In Theorem 5.2 in [20], Markus shows that all of the canonical regions are parallel regions.

Definition 2. The *separatrix configuration* for a C^1-system (1) on $\mathbf{R}^2$ is the union S of the set of all separatrices of (1) together with one trajectory from each component of $\mathbf{R}^2 \sim S$. Two separatrix configurations S_1 and S_2 of a C^1-system (1) on $\mathbf{R}^2$ are said to be topologically equivalent if there is a homeomorphism $H \colon \mathbf{R}^2 \to \mathbf{R}^2$ which maps the trajectories in S_1 onto the trajectories in S_2 and preserves their orientation by time.

Theorem 2 (Markus). *Two relatively prime, planar, analytic systems are topologically equivalent on* $\mathbf{R}^2$ *if and only if their separatrix configurations are topologically equivalent.*

This theorem follows from Theorem 7.1 proved in [20] for C^1-systems having no limit separatrices. In regard to the global phase portraits of polynomial systems discussed in the previous section, this implies that on the Poincaré sphere S^2, it suffices to describe the set of separatrices S of (1), the flow on the equator E of S^2 and the behavior of one trajectory in each component of $S^2 \sim (S \cup E)$ in order to uniquely determine the global behavior of all trajectories of a polynomial system (1) for all time. However, in order for this statement to hold on the Poincaré sphere S^2, it is necessary that the critical points and separatrices of (1) do not accumulate at infinity, i.e., on the equator of S^2. This is certainly the case for relatively prime, polynomial systems as long as the equator of S^2 does not contain an infinite number of critical points as in Problem 3(b) in Section 3.10.

Several examples of separatrix configurations which determine the global behavior of all trajectories of certain polynomial systems on $\mathbf{R}^2$ for all time were given in the previous section; cf. Figures 5, 7, 9 and 12 in Section 3.10. We include one more example in this section in order to illustrate the idea of a Hopf bifurcation which we discuss in detail in Section 4.4.

Example 1. For $\alpha \in (-1, 1)$, the quadratic system

$$\dot{x} = \alpha x - y + (\alpha + 1)x^2 - xy$$
$$\dot{y} = x + x^2$$

was considered in Example 2 of Section 3.8, in Example 1 in Section 3.9, and in Example 3 of Section 3.10. The origin is the only finite critical point of this system; for $\alpha \in (-1, 1)$, there is a saddle-node at the critical point $(0, \pm 1, 0)$ at infinity as shown in Figure 9 of Section 3.10; for $\alpha \in (-1, 0]$ this system has no limit cycles; and for $\alpha \in (0, 1)$ there is a unique limit cycle around the origin. The global phase portrait for this system is determined by the separatrix configurations on (the upper hemisphere of) S^2 shown in Figure 3.

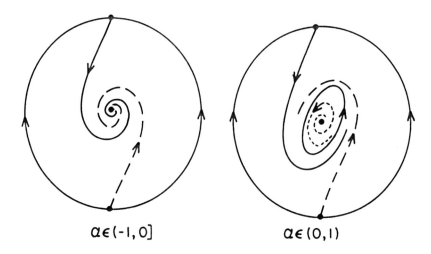

Figure 3. The separatrix configurations for the system of Example 1.

There is a significant difference between the two separatrix configurations shown in Figure 3. One of them contains a limit cycle while the other does not. Also, the stability of the focus at the origin is different. It will be shown in Chapter 4, using the theory of rotated vector fields, that a unique limit cycle is generated when the critical point at the origin changes its stability at that value of $\alpha = 0$. This is referred to as a Hopf bifurcation at the origin. The value $\alpha = 0$ at which the global phase portrait of this system changes its qualitative structure is called a bifurcation value. Various types of bifurcations are discussed in the next chapter. We note that for $\alpha \in (-1, 0) \cup (0, 1)$, the system of Example 1 is structurally stable on $\mathbf{R}^2$ (under strong C^1-perturbations, as defined in Section 4.1); however, it is not structurally stable on S^2 since it has a nonhyperbolic critical point at $(0, \pm 1, 0) \in S^2$. Cf. Section 4.1 where we discuss structural stability.

Problem Set 11

1. For the separatrix configurations shown in Figures 5, 7 and 9 in Section 3.10:

    (a) Determine the number of strip, annular and spiral canonical regions in each global phase portrait.

    (b) Classify each strip region as a hyperbolic, parabolic or elliptic region.

3.11. Global Phase Portraits and Separatrix Configurations

2. Determine the separatrix configuration on the Poincaré sphere for the quadratic system

$$\dot{x} = -4x - 2y + 4$$
$$\dot{y} = xy.$$

Hint: Show that the x-axis consists of separatrices. And use Theorems 1 and 2 in Section 2.11 of Chapter 2 to determine the nature of the critical points at infinity.

3. Determine the separatrix configuration on the Poincaré sphere for the quadratic system

$$\dot{x} = y - x^2 + 2$$
$$\dot{y} = 2x^2 - 2xy.$$

Hint: Use Theorem 2 in Section 2.11 of Chapter 2 in order to determine the nature of the critical points at infinity. Also, note that the straight line through the critical points $(2, 2)$ and $(0, -2)$ consists of separatrices.

4. Determine the separatrix configuration on the Poincaré sphere for the quadratic system

$$\dot{x} = \alpha x + x^2$$
$$\dot{y} = y.$$

Treat the cases $\alpha > 0$, $\alpha = 0$ and $\alpha < 0$ separately. Note that $\alpha = 0$ is a bifurcation value for this system.

5. Determine the separatrix configuration on the Poincaré sphere for the cubic system

$$\dot{x} = y$$
$$\dot{y} = -x^3 + 4xy$$

and show that the four trajectories determined by the invariant curves $y = x^2/(2\pm\sqrt{2})$, discussed in Problem 1 of Section 2.11, are separatrices according to Definition 1. **Hint:** Use equation $(7')$ in Theorem 2 of Section 3.10 to study the critical points at $(0, \pm 1, 0) \in S^2$. In particular, show that $z = x^2/(2 \pm \sqrt{2})$ are invariant curves of $(7')$ for this problem; that there are parabolic sectors between these two parabolas; that there is an elliptic sector above $z = x^2/(2 - \sqrt{2})$, a hyperbolic sector for $z < 0$ and two hyperbolic sectors between the x-axis and $z = x^2/(2+\sqrt{2})$. You should find the following separatrix configuration on S^2.

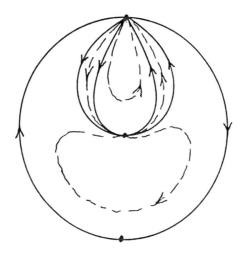

6. Determine the global phase portrait for Gauss' model of two competing species

$$\dot{x} = ax - bx^2 - rxy$$
$$\dot{y} = ay - by^2 - sxy$$

where $a > 0$, $s \geq r > b > 0$. Using the global phase portrait show that it is mathematically possible, but highly unlikely, for both species to survive, i.e., for $\lim_{t \to \infty} x(t)$ and $\lim_{t \to \infty} y(t)$ to both be non-zero.

7. (a) (Cf. Markus [20], p. 132.) Describe the flow on $\mathbf{R}^2$ determined by the analytic system

$$\dot{x} = \sin x$$
$$\dot{y} = \cos x$$

and show that, according to Definition 1', the separatrices of this system consist of those trajectories which lie on the straight lines $x = n\pi$, $n = 0, \pm 1, \pm 2, \ldots$. Note that we can map any strip $|x| \leq n\pi$ onto a portion of the Poincaré sphere using the central projection described in the previous section and that the separatrices of this system which lie in the strip $|x| \leq n\pi$ are exactly those trajectories which lie on boundaries of hyperbolic sectors at the point $(0, 1, 0)$ on this portion of the Poincaré sphere as in Definition 1.

(b) Same thing for the analytic system

$$\dot{x} = \sin^2 x$$
$$\dot{y} = \cos x.$$

3.12 Index Theory

As Markus points out, these two systems have the same separatrices, but they are not topologically equivalent. Of course, their separatrix configurations are not topologically equivalent.

3.12 Index Theory

In this section we define the index of a critical point of a C^1-vector field $\mathbf{f}$ on $\mathbf{R}^2$ or of a C^1-vector field $\mathbf{f}$ on a two-dimensional surface. By a two-dimensional surface, we shall mean a compact, two-dimensional differentiable manifold of class C^2. For a given vector field $\mathbf{f}$ on a two-dimensional surface S, if $\mathbf{f}$ has a finite number of critical points, $\mathbf{p}_1, \ldots, \mathbf{p}_m$, the index of the surface S relative to the vector field $\mathbf{f}$, $I_\mathbf{f}(S)$, is defined as the sum of the indices at each of the critical points, $\mathbf{p}_1, \ldots, \mathbf{p}_m$ in S. It is one of the most interesting facts of the index theory that the index of the surface S, $I_\mathbf{f}(S)$, is independent of the vector field $\mathbf{f}$ and, as we shall see, only depends on the topology of the surface S; in particular, $I_\mathbf{f}(S)$ is equal to the Euler–Poincaré characteristic of the surface S. This result is the famous Poincaré Index Theorem.

We begin this section with Poincaré's definition of the index of a Jordan curve C (i.e., a piecewise-smooth simple, closed curve C) relative to a C^1-vector field $\mathbf{f}$ on $\mathbf{R}^2$.

Definition 1. *The index $I_\mathbf{f}(C)$ of a Jordan curve C relative to a vector field $\mathbf{f} \in C^1(\mathbf{R}^2)$, where $\mathbf{f}$ has no critical point on C, is defined as the integer*

$$I_\mathbf{f}(C) = \frac{\Delta\Theta}{2\pi}$$

where $\Delta\Theta$ is the total change in the angle Θ that the vector $\mathbf{f} = (P, Q)^T$ makes with respect to the x-axis, i.e., $\Delta\Theta$ is the change in

$$\Theta(x, y) = \tan^{-1}\frac{Q(x, y)}{P(x, y)},$$

as the point (x, y) traverses C exactly once in the positive direction.

The index $I_\mathbf{f}(C)$ can be computer using the formula

$$I_\mathbf{f}(C) = \frac{1}{2\pi}\oint_C d\tan^{-1}\frac{dy}{dx} = \frac{1}{2\pi}\oint_C d\tan^{-1}\frac{Q(x,y)}{P(x,y)}$$
$$= \frac{1}{2\pi}\oint_C \frac{PdQ - QdP}{P^2 + Q^2}.$$

Cf. [A–I], p. 197.

Example 1. Let C be the circle of radius one, centered at the origin, and let us compute the index of C relative to the vector fields

$$\mathbf{f}(\mathbf{x}) = \begin{pmatrix} x \\ y \end{pmatrix}, \quad \mathbf{g}(\mathbf{x}) = \begin{pmatrix} -x \\ -y \end{pmatrix}, \quad \mathbf{h}(\mathbf{x}) = \begin{pmatrix} -y \\ x \end{pmatrix}, \quad \text{and} \quad \mathbf{k}(\mathbf{x}) = \begin{pmatrix} x \\ -y \end{pmatrix}.$$

These vector fields define flows having unstable and stable nodes, a center and a saddle at the origin respectively; cf. Section 1.5 in Chapter 1. The phase portraits for the flows generated by these four vector fields are shown in Figure 1. According to the above definition, we have

$$I_f(C) = 1, \quad I_g(C) = 1, \quad I_h(C) = 1, \quad \text{and} \quad I_k(C) = -1.$$

These indices can also be computed using the above formula; cf. Problem 1.

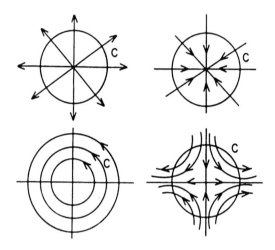

Figure 1. The flows defined by the vector fields in Example 1.

We now outline some of the basic ideas of index theory. We first need to prove a fundamental lemma.

Lemma 1. *If the Jordan curve C is decomposed into two Jordan curves, $C = C_1 + C_2$, as in Figure 2, then*

$$I_f(C) = I_f(C_1) + I_f(C_2)$$

with respect to any C^1-vector field $\mathbf{f} \in C^1(\mathbf{R}^2)$.

Proof. Let $\mathbf{p}_1$ and $\mathbf{p}_2$ be two distinct points on C which partition C into two arc A_1 and A_2 as shown in Figure 2. Let A_0 denote an arc in the interior of C from $\mathbf{p}_1$ to $\mathbf{p}_2$ and let $-A_0$ denote the arc from $\mathbf{p}_2$ to $\mathbf{p}_1$ traversed in the opposite direction. Let $C_1 = A_1 + A_0$ and let $C_2 = A_2 - A_0$.

It then follows that if $\Delta\Theta|_A$ denotes the change in the angle $\Theta(x, y)$ defined by the vector $\mathbf{f}(\mathbf{x})$ as the point $\mathbf{x} = (x, y)^T$ moves along the arc A in a well-defined direction, then $\Delta\Theta|_{-A} = -\Delta\Theta|_A$ and

$$I_f(C) = \frac{1}{2\pi}[\Delta\Theta|_{A_1} + \Delta\Theta|_{A_2}]$$

3.12. Index Theory

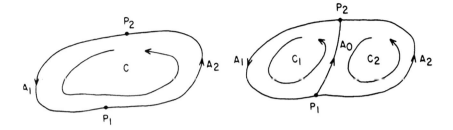

Figure 2. A decomposition of the Jordan curve C into $C_1 + C_2$.

$$= \frac{1}{2\pi}[\Delta\Theta|_{A_1} + \Delta\Theta|_{A_0} + \Delta\Theta|_{-A_0} + \Delta\Theta|_{A_2}]$$
$$= \frac{1}{2\pi}[\Delta\Theta|_{A_1+A_0} + \Delta\Theta|_{A_2-A_0}]$$
$$= \frac{1}{2\pi}[\Delta\Theta|_{C_1} + \Delta\Theta|_{C_2}]$$
$$= I_{\mathbf{f}}(C_1) + I_{\mathbf{f}}(C_2).$$

Theorem 1. *Suppose that* $\mathbf{f} \in C^1(E)$ *where E is an open subset of $\mathbf{R}^2$ which contains a Jordan curve C and that there are no critical points of $\mathbf{f}$ on C or in its interior. It then follows that* $I_{\mathbf{f}}(C) = 0$.

Proof. Since $\mathbf{f} = (P, Q)^T$ is continuous on E it is uniformly continuous on any compact subset of E. Thus, given $\epsilon = 1$, there is a $\delta > 0$ such that on any Jordan curve C_α which is contained inside a square of side δ in E, we have $0 \leq I_{\mathbf{f}}(C_\alpha) < \epsilon$; cf. Problem 2. Then, since $I_{\mathbf{f}}(C_\alpha)$ is a positive integer, it follows that $I_{\mathbf{f}}(C_\alpha) = 0$ for any Jordan curve C_α contained inside a square of side δ. We can cover the interior of C, Int C, as well as C with a square grid where the squares S_α in the grid have sides of length $\delta/2$. Choose $\delta > 0$ sufficiently small that any square S_α with $S_\alpha \cap \text{Int } C \neq \emptyset$ lies entirely in E and that $I_{\mathbf{f}}(C_\alpha) = 0$ where C_α is the boundary of $S_\alpha \cap \text{Int } C$. Since the closure of Int C is a compact set, a finite number of the squares S_α cover $\overline{\text{Int } C}$, say S_j, $j = 1, \ldots, N$. And by Lemma 1, we have

$$I_{\mathbf{f}}(C) = \sum_{j=1}^{N} I_{\mathbf{f}}(C_j) = 0.$$

Corollary 1. *Under the hypotheses of Theorem 1, if C_1 and C_2 are Jordan curves contained in E with $C_1 \subset \text{Int } C_2$, and if there are no critical points of $\mathbf{f}$ in* $\text{Int } C_2 \cap \text{Ext } C_1$, *then* $I_{\mathbf{f}}(C_1) = I_{\mathbf{f}}(C_2)$.

Definition 2. Let $\mathbf{f} \in C^1(E)$ where E is an open subset of $\mathbf{R}^2$ and let $\mathbf{x}_0 \in E$ be an isolated critical point of $\mathbf{f}$. Let C be a Jordan curve contained in E and containing $\mathbf{x}_0$ and no other critical point of $\mathbf{f}$ on its interior. Then the index of the critical point x_0 with respect to f

$$I_f(x_0) = I_f(C).$$

Theorem 2. *Suppose that* $f \in C^1(E)$ *where E is an open subset of* $\mathbf{R}^2$ *containing a Jordan curve C. Then if there are only a finite number of critical points, $x_1, \ldots, x_n$ of f in the interior of C, it follows that*

$$I_f(C) = \sum_{j=1}^{N} I_f(x_j).$$

This theorem is proved by enclosing each of the critical points x_j by a small circle C_j lying in the interior of C as in Figure 3. Let **a** and **b**

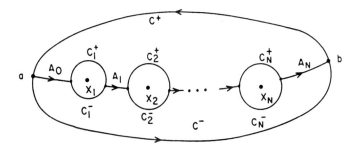

Figure 3. The decomposition of the Jordan curves $C, C_1, \ldots, C_N$.

be two distinct points on C. Since the interior of C, Int C, is arc-wise connected, we can construct arcs $\overline{ax_1}, \overline{x_1x_2}, \ldots, \overline{x_Nb}$ on the interior of C. Let $A_0, A_1, \ldots, A_N$ denote the part of these arcs on the exteriors of the circles $C_1, \ldots, C_N$ as in Figure 3. The curves $C, C_1, \ldots, C_N$ are then split into two parts, $C^+, C^-, C_1^+, C_1^-, \ldots, C_N^+, C_N^-$ by the arcs $A_0, \ldots, A_N$ as in Figure 3.

Define the Jordan curves $J_1 = C^+ + A_0 - C_1^+ - \cdots - C_N^+ + A_N$ and $J_2 = C^- - A_N - C_N^- - \cdots - C_1^- - A_0$. Then, by Theorem 1, $I_f(J_1) = I_f(J_2) = 0$. But

$$I_f(J_1) + I_f(J_2) = \frac{1}{2\pi}[\Delta\Theta|_C^+ + \Delta\Theta|_{A_0} - \Delta\Theta|_{C_1^+} - \cdots - \Delta\Theta|_{C_N^+} + \Delta\Theta|_{A_N}$$
$$+ \Delta\Theta|_{C^-} - \Delta\Theta|_{A_0} - \Delta\Theta|_{C_1^-} - \cdots - \Delta\Theta|_{C_N^-} - \Delta\Theta|_{A_N}]$$

$$= \frac{1}{2\pi}\left[\Delta\Theta|_{C^-} - \sum_{j=1}^{N}\Delta\Theta|_{C_j}\right].$$

Thus,

$$I_f(C) = \sum_{j=1}^{N} I_f(C_j) = \sum_{j=1}^{N} I_f(x_j).$$

3.12. Index Theory

Theorem 3. *Suppose that* $\mathbf{f} \in C^1(E)$ *where E is an open subset of $\mathbf{R}^2$ and that E contains a cycle Γ of the system*

$$\dot{\mathbf{x}} = \mathbf{f}(\mathbf{x}). \tag{1}$$

It follows that $I_\mathbf{f}(\Gamma) = 1$.

Proof. At any point $\mathbf{x} \in \Gamma$, define the unit vector $\mathbf{u}(\mathbf{x}) = \mathbf{f}(\mathbf{x})/|\mathbf{f}(\mathbf{x})|$. Then $I_\mathbf{u}(\Gamma) = I_\mathbf{f}(\Gamma)$ and we shall show that $I_\mathbf{u}(\Gamma) = 1$. We can rotate and translate the axes so that Γ is in the first quadrant and is tangent to the x-axis at some point $\mathbf{x}_0$ as in Figure 4. Let $\mathbf{x}(t) = (x(t), y(t))^T$ be the solution of (1) through the point $\mathbf{x}_0 = (x_0, y_0)^T$ at time $t = 0$. Then by normalizing the time as in Section 3.1, we may assume that the period of Γ is equal to 1; i.e.,

$$\Gamma = \{\mathbf{x} \in \mathbf{R}^2 \mid \mathbf{x} = \mathbf{x}(t), 0 \le t \le 1\}.$$

Now for points (s,t) in the triangular region

$$T = \{(s,t) \in \mathbf{R}^2 \mid 0 \le s \le t \le 1\},$$

we define the vector field $\mathbf{g}$ by

$$\mathbf{g}(s,s) = \mathbf{u}(\mathbf{x}(s)), \quad 0 \le s \le 1$$
$$\mathbf{g}(0,1) = -\mathbf{u}(\mathbf{x}_0)$$

and

$$\mathbf{g}(s,t) = \frac{\mathbf{x}(t) - \mathbf{x}(s)}{|\mathbf{x}(t) - \mathbf{x}(s)|}$$

for $0 \le s < t \le 1$ and $(s,t) \ne (0,1)$; cf. Figure 4. It then follows that $\mathbf{g}$ is continuous on T and that $\mathbf{g} \ne \mathbf{0}$ on T. Let $\tilde{\Theta}(s,t)$ be the angle that the vector $\mathbf{g}(s,t)$ makes with the x-axis. Then, assuming that the cycle Γ is positively oriented, $\tilde{\Theta}(0,0) = 0$ and since Γ is in the first quadrant, $\tilde{\Theta}(0,t) \in [0,\pi]$ for $0 \le t \le 1$, and therefore $\tilde{\Theta}(0,t)$ varies from 0 to π as t varies from 0 to 1. Similarly, it follows from the definition of $\mathbf{g}(s,t)$ that $\tilde{\Theta}(s,1)$ varies from π to 2π as s varies from 0 to 1. Let B denote the boundary of the region T. It then follows from Theorem 1 that $I_\mathbf{g}(B) = 0$. Thus the variation of $\tilde{\Theta}(s,s)$ as s varies from 0 to 1 to 1 is 2π. But this is exactly the variation of the angle that the vector $\mathbf{u}(\mathbf{x}(s))$ makes with the x-axis as s varies from 0 to 1. Thus $I_\mathbf{f}(\Gamma) = I_\mathbf{u}(\Gamma) = 1$. A similar argument yields the same result when Γ is negatively oriented. This completes the proof of Theorem 3.

Remark 1. For a separatrix cycle S of (1) such that the Poincaré map is defined on the interior or on the exterior of S, we can take a sequence of Jordan curves C_n approaching S and, using the fact that we only have hyperbolic sectors on either the interior or the exterior of S, we can show that $I_\mathbf{f}(C_n) = 1$. But for n sufficiently large, $I_\mathbf{f}(S) = I_\mathbf{f}(C_n)$. Thus, $I_\mathbf{f}(S) = 1$. It should be noted that if the Poincaré map is not defined on either side of S, then $I_\mathbf{f}(S)$ may not be equal to one.

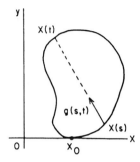 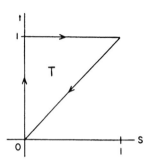

Figure 4. The vector field g on the triangular region T.

Corollary 2. *Under the hypotheses of Theorem 3, Γ contains at least one critical point of (1) on its interior. And, assuming that there are only a finite number of critical points of (1) on the interior of Γ, the sum of the indices at these critical points is equal to one.*

We next consider the relationship between the index of a critical point $\mathbf{x}_0$ of (1) with respect to the vector field $\mathbf{f}$ and with respect to its linearization $D\mathbf{f}(\mathbf{x}_0)\mathbf{x}$ at $\mathbf{x}_0$. The following lemma is useful in this regard. Its proof is left as an exercise; cf. Problem 3.

Lemma 2. *If $\mathbf{v}$ and $\mathbf{w}$ are two continuous vector fields defined on a Jordan curve C which never have opposite directions or are zero on C, then $I_\mathbf{v}(C) = I_\mathbf{w}(C)$.*

In proving the next theorem we write $\mathbf{x} = (x,y)^T$ and

$$\mathbf{f}(\mathbf{x}) = D\mathbf{f}(\mathbf{0})\mathbf{x} + \mathbf{g}(\mathbf{x}) = \begin{pmatrix} ax + by \\ cx + dy \end{pmatrix} + \begin{pmatrix} g_1(x,y) \\ g_2(x,y) \end{pmatrix}.$$

We say that $\mathbf{0}$ is a nondegenerate critical point of (1) if $\det D\mathbf{f}(\mathbf{0}) \neq 0$, i.e., if $ad - bc \neq 0$; and we say that $|\mathbf{g}(\mathbf{x})| = o(r)$ as $r \to 0$ if $|\mathbf{g}(\mathbf{x})|/r \to 0$ as $r \to 0$.

Theorem 5. *Suppose that $\mathbf{f} \in C^1(E)$ where E is an open subset of $\mathbf{R}^2$ containing the origin. If $\mathbf{0}$ is a nondegenerate critical point of (1) and $|\mathbf{g}(\mathbf{x})| = o(r)$ as $r \to 0$, then $I_\mathbf{f}(0) = I_\mathbf{v}(0)$ where $\mathbf{v}(\mathbf{x}) = D\mathbf{f}(\mathbf{0})\mathbf{x}$, the linearization of $\mathbf{f}$ at $\mathbf{0}$.*

This theorem is proved by showing that on a sufficiently small circle C centered at the origin the vector fields $\mathbf{v}$ and $\mathbf{f}$ are never in opposition and then applying Lemma 2. Suppose that $\mathbf{v}(\mathbf{x})$ and $\mathbf{f}(\mathbf{x})$ are in opposition at some point $\mathbf{x} \in C$. Then at the point $\mathbf{x} \in C$, $\mathbf{f} + s\mathbf{v} = 0$ for some $s > 0$. But $\mathbf{f} = \mathbf{v} + \mathbf{g}$ where $\mathbf{g} = \mathbf{f} - \mathbf{v}$ and therefore $(1+s)\mathbf{v} = -\mathbf{g}$ at the point

3.12. Index Theory

$\mathbf{x} \in C$; i.e., $(1+s)^2|\mathbf{v}|^2 = |\mathbf{g}^2|$ at the point $\mathbf{x} \in C$. Now

$$|\mathbf{v}^2| = r^2[(a\cos\theta + b\sin\theta)^2 + (c\cos\theta + d\sin\theta)^2].$$

And since $ad - bc \neq 0$, $\mathbf{v} = 0$ only at $\mathbf{x} = 0$. Thus, $\mathbf{v}$ is continuous and non-zero on the circle $r = 1$. Let

$$m = \min_{r=1} |\mathbf{v}|.$$

Then $m > 0$ and since $|\mathbf{v}|$ is homogeneous in r we have $|\mathbf{v}| \geq mr$ for $r \geq 0$. It then follows that at the point $\mathbf{x} \in C$

$$(1+s)^2 m^2 r^2 \leq |\mathbf{g}|^2.$$

But if the circle C is chosen sufficiently small, this leads to a contradiction since we then have

$$0 < m^2 \leq (1+s)^2 m^2 \leq |\mathbf{g}|^2/r^2$$

and $|\mathbf{g}|^2/r^2 \to 0$ as $r \to 0$. Thus, the vector fields $\mathbf{v}$ and $\mathbf{f}$ are never in opposition on a sufficiently small circle C centered at the origin. It then follows from Lemma 2 that $I_{\mathbf{f}}(0) = I_{\mathbf{v}}(0)$.

Since the index of a linear vector field is invariant under a nonsingular linear transformation, the following theorem is an immediate consequence of Theorem 5 and computation of the indices of generic linear vector fields such as those in Example 1; cf. [C/L], p. 401.

Theorem 6. *Under the hypotheses of Theorem 5, $I_{\mathbf{f}}(0)$ is -1 or $+1$ according to whether the origin is or is not a topological saddle for (1) or equivalently according to whether the origin is or is not a saddle for the linearization of (1) at the origin.*

According to Theorem 6, the index of any nondegenerate critical point of (1) is either ± 1. What can we say about the index of a critical point $\mathbf{x}_0$ of (1) when $\det D\mathbf{f}(\mathbf{x}_0) = 0$? The following theorem, due to Bendixson [B], answers this question for analytic systems. In Theorem 7, e denotes the number of elliptic sectors and h denotes the number of hyperbolic sectors of (1) at the origin. A proof of this theorem can be found on p. 511 in [A–I].

Theorem 7 (Bendixson). *Let the origin be an isolated critical point of the planar, analytic system (1). It follows that*

$$I_{\mathbf{f}}(0) = 1 + \left(\frac{e-h}{2}\right).$$

We see that for planar, analytic systems, this theorem implies the results of Theorem 6. It also implies that the number of elliptic sectors, e, and the number of hyperbolic sectors, h, have the same parity; i.e., $e = h \pmod 2$. Note that the index of a saddle-node is zero according to this theorem; cf. Figure 3 in Section 2.11 of Chapter 2.

We next outline the index theory for two-dimensional surfaces. By *a two-dimensional surface* S we mean a compact, two-dimensional, differentiable manifold of class C^2. First of all, let us define the index of S with respect to a vector field $\mathbf{f}$ on S; cf. Section 3.10. Suppose that the vector field $\mathbf{f}$ has only a finite number of critical points $\mathbf{p}_1, \ldots, \mathbf{p}_m$ on S. Then for $j = 1, \ldots, m$, each critical point $\mathbf{p}_j \in U_j$ for some chart $(U_j, \mathbf{h}_j)$. The corresponding function $\mathbf{f}_j \colon V_j \to \mathbf{R}^2$ then defines a C^1-vector field on $V_j = \mathbf{h}_j(U_j) \subset \mathbf{R}^2$; cf. Section 3.10. The point $\mathbf{q}_j = \mathbf{h}_j(\mathbf{p}_j)$ is then a critical point of the C^1-vector field $\mathbf{f}_j$ on the open set $V_j \subset \mathbf{R}^2$. *The index of the critical point* $\mathbf{p}_j \in S$ *with respect to the vector field* $\mathbf{f}$ *on* S, $I_\mathbf{f}(\mathbf{p}_j)$, is then defined to be equal to the index $I_{\mathbf{f}_j}(\mathbf{q}_j)$ of the critical point $\mathbf{q}_j \in V_j$ with respect to the vector field $\mathbf{f}_j$ on $V_j \subset \mathbf{R}^2$. *The index of the surface* S *with respect to the vector field* $\mathbf{f}$ *on* S is then defined to be equal to the sum of the indices at each of the critical points on S:

$$I_\mathbf{f}(S) = \sum_{j=1}^{m} I_\mathbf{f}(\mathbf{p}_j).$$

We next show that the index of a surface S with respect to a vector field $\mathbf{f}$ on the surface is actually independent of the vector field $\mathbf{f}$ and only depends on the topology of the surface. In fact, $I_\mathbf{f}(S) = \chi(S)$, the Euler–Poincaré characteristic of the surface. In order to define the Euler–Poincaré characteristic $\chi(S)$, we use the fact that any two-dimensional surface S can be decomposed into a finite number of curvilinear triangles. This is referred to as a triangulation of S. For a given triangulation of S, let v be the number of vertices, ℓ the number of edges, and T the number of triangles in the triangulation. *The Euler–Poincaré characteristic of* S is then defined to be the integer

$$\chi(S) = T + v - \ell.$$

It can be shown that $\chi(S)$ is a topological invariant which is independent of the triangulation and is related to the genus p of the surface S by the formula

$$\chi(S) = 2(1 - p). \tag{2}$$

For orientable surfaces, the genus p is equal to the number of "holes" in the surface. For example $p = 0$ for the two-dimensional sphere S^2, $p = 1$ for the two-dimensional torus T^2, and $p = 2$ for the two-dimensional anchor ring. Is is easy to see that the tetrahedron triangulation of the sphere (where we project the surface of a tetrahedron onto S^2) has $v = 4$ vertices, $\ell = 6$ edges, and $T = 4$ triangles. Thus, $\chi(S^2) = 2$. Any other triangulation of S^2 will yield the same result for the Euler–Poincaré characteristic of S^2. The torus T^2 can be triangulated to show that $\chi(T^2) = 0$; cf. Problem 5. Equation (2) for the characteristic also holds for non-orientable surfaces such as the projective plane P and the Klein bottle K. In this case,

3.12. Index Theory

the genus $p = q/2$ where the non-orientable surface S is topologically a two-dimensional sphere with q holes along whose boundaries the antipodal points are identified. Cf. the proof of Theorem 8 below. Thus, the characteristic of the projective plane $\chi(P) = 1$ and the characteristic of the Klein bottle $\chi(K) = 0$; cf. Problem 5.

Theorem 8 (The Poincaré Index Theorem). *The index $I_\mathbf{f}(S)$ of a two-dimensional surface S relative to any C^1-vector field $\mathbf{f}$ on S with at most a finite number of critical points is independent of the vector field $\mathbf{f}$ and is equal to the Euler–Poincaré characteristic of S; i.e.*

$$I_\mathbf{f}(S) = \chi(S).$$

We outline the proof of this interesting theorem which was first proved by Poincaré [P], p. 125, for orientable surfaces. The proof can be extended to non-orientable surfaces without difficulty. Following Gomory's proof in Lefschetz [L], p. 367, we first prove that for any C^1-vector field $\mathbf{f}$ on the sphere S^2 we have $I_\mathbf{f}(S^2) = 2$ and that for any vector field $\mathbf{f}$ on the projective plane P we have $I_\mathbf{f}(P) = 1$. We then use this information together with "surgery" on any other two-dimensional, topological surface S in order to complete the proof of the theorem.

Let $\mathbf{f}$ be any vector field on the two-dimensional sphere S^2 with a finite number of critical points. At any regular point on S^2 we can construct a sufficiently small circle C such that the flow is essentially parallel inside and on C. There will then be two points $\mathbf{p}$ and $\mathbf{q}$ on C where the vector field $\mathbf{f}$ is tangent to C. These two points divide C into two circular arcs C_1 and C_2; i.e., $C = C_1 + C_2$. The index of S^2 with respect to the vector field $\mathbf{f}$, $I_\mathbf{f}(S^2)$, is equal to the index of C in the negative direction with respect to the vector field $\mathbf{f}$; cf. Theorem 2. We now compute $I_\mathbf{f}(C)$ where C is traversed in the negative direction. Let γ be a smooth arc on S^2 joining $\mathbf{p}$ and $\mathbf{q}$ and containing no critical points of $\mathbf{f}$. We can open out the punctured sphere, S^2 with the interior of C removed, and lay it out flat to obtain a disk with the flow on the boundary shown in Figure 5. The details of opening out a neighborhood of the point $\mathbf{q}$ and viewing it from the exterior of the punctured sphere are shown in Figure 6. Note that the arc γ and the circle C divide the punctured sphere into two regions R_1 and R_2 and that on C and on γ near C, the flow is out of R_1 and into R_2. For points $\mathbf{p}'$ and $\mathbf{q}'$ on γ sufficiently near $\mathbf{p}$ and $\mathbf{q}$, the vector field will be essentially parallel to the vector field at $\mathbf{p}$ and $\mathbf{q}$ respectively. On the arc $\overline{\mathbf{p}'\mathbf{q}'}$ of γ define $\mathbf{v}$ to be the vector field perpendicular to the arc at each point of γ and pointing into R^2. Also on the circle C and on the small arcs $\overline{\mathbf{pp}'}$ and $\overline{\mathbf{qq}'}$ define the vector field $\mathbf{v} = \mathbf{f}/|\mathbf{f}|$. Then from Figure 5 we see that

$$I_\mathbf{v}(C_1 + \overline{\mathbf{pq}}) = 1 \quad \text{and} \quad I_\mathbf{v}(C_2 + \overline{\mathbf{qp}}) = 1.$$

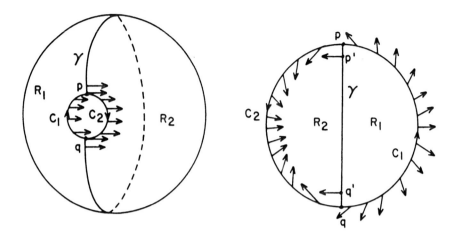

Figure 5. The flow in a neighborhood of the circle C on the sphere S^2.

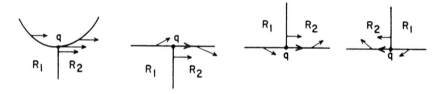

Figure 6. (a) A neighborhood of the point $q \in S$. (b) Stretching the neighborhood. (c) The neighborhood as viewed from the interior of S^2 after opening out the punctured sphere. (d) The neighborhood as viewed from the exterior of S^2 after opening out the punctured sphere.

But, for C traversed in the negative sense, we also have

$$I_f(S^2) = I_f(C) = I_f(C_1 + \overline{pq}) + I_f(C_2 + \overline{qp})$$

$$= \frac{1}{2\pi}[\Delta\Theta|_{C_1} + \Delta\Theta|_{\overline{pq}} + \Delta\Theta|_{C_2} + \Delta\Theta|_{\overline{qp}}]_f$$

$$= \frac{1}{2\pi}[\Delta\Theta|_{C_1}]_f + \frac{1}{2\pi}[\Delta\Theta|_{\overline{pq}}]_f + \frac{1}{2\pi}[\Delta\Theta|_{\overline{pq}}]_v$$

$$\quad + \frac{1}{2\pi}[\Delta\Theta|_{C_2}]_f + \frac{1}{2\pi}[\Delta\Theta|_{\overline{qp}}]_f - \frac{1}{2\pi}[\Delta\theta|_{\overline{pq}}]_v$$

$$= \frac{1}{2\pi}[\Delta\Theta|_{C_1} + \Delta\Theta|_{\overline{pq}} + \Delta\Theta|_{C_2} + \Delta\Theta|_{\overline{qp}}]_v$$

$$= I_v(C_1 + \overline{pq}) + I_v(C_2 + \overline{qp}) = 2$$

since $[\Delta\Theta|_{\overline{pq}}]_f + [\Delta\Theta|_{\overline{qp}}]_f = 0$ where $[\Delta\Theta|_A]_f$ denotes the change in the

3.12. Index Theory

angle Θ that the vector $\mathbf{f}$ makes with the x-axis as the arc A is traversed in the indicated direction. Thus, we have that $I_\mathbf{f}(S^2) = \chi(S^2) = 2$.

Next, let $\mathbf{f}$ be any vector field on the projective plane P with a finite number of critical points. We take the unit disk $\Omega = \{\mathbf{x} \in \mathbf{R}^2 \mid |\mathbf{x}| \leq 1\}$ with antipodal points identified as a model for P. Let Δ be the smooth surface which consists of two copies of Ω, say Ω_1 and Ω_2, joined at their boundaries, and let $\mathbf{g}$ be the vector field on Δ obtained by the continuous extension of the vector fields $\mathbf{f}$ on Ω_1 and Ω_2 across the boundaries of Ω_1 and Ω_2. The vector field $\mathbf{g}$ on Δ will then have twice as many critical points as the vector field $\mathbf{f}$ on Ω and diametrically opposite critical points on Δ will have the same index. Thus, since Δ is topologically a two-dimensional sphere, we have

$$I_\mathbf{f}(P) = \frac{1}{2} I_\mathbf{g}(\Delta) = \frac{1}{2} I_\mathbf{g}(S^2) = 1.$$

It follows that $I_\mathbf{f}(P) = \chi(P) = 1$.

Having shown that $I_\mathbf{f}(S^2) = 2$, we can determine the index of any orientable surface S, with respect to a vector field $\mathbf{f}$ on S, using surgery. Suppose that S is an orientable surface of genus p, i.e., topologically S is a donut with p holes or a sphere with p handles; cf. Figure 7. And suppose that $\mathbf{f}$ is a vector field on S with a finite number of critical points.

Figure 7. A surface of genus p.

At each hole in S we cut out a section S_j, $j = 1, \ldots, p$, containing no critical points of $\mathbf{f}$ and shrink the boundaries of each of the sections S_j to critical points P_j and P_j'. Also, shrink the boundary of the surface

$$\tilde{S} = S \sim \bigcup_{j=1}^{p} S_j$$

to critical points, Q_j and Q_j', $j = 1, \ldots, p$; cf. Figure 8. Let S' and S_j' denote the resulting surfaces. We then have $I_\mathbf{f}(Q_j) = I_\mathbf{f}(P_j)$ and $I_\mathbf{f}(Q_j') = I_\mathbf{f}(P_j')$ for $j = 1, \ldots, p$. Thus since the surfaces S_j', $j = 1, \ldots, p$, are topologically two-dimensional spheres, we have

$$I_\mathbf{f}(Q_j) + I_\mathbf{f}(Q_j') = I_\mathbf{f}(P_j) + I_\mathbf{f}(P_j') = 2$$

for $j = 1, \ldots, p$. But S' is also topologically a two-dimensional sphere and therefore

$$I_f(S) = I_f(S') - \sum_{j=1}^{p}[I_f(Q_j) + I_f(Q'_j)]$$
$$= 2 - 2p = \chi(S).$$

This completes the proof of Theorem 8 for orientable surfaces.

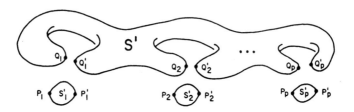

Figure 8. The surfaces $S', S'_1, \ldots, S'_p$.

Finally, suppose that S is a non-orientable surface of genus p and that $\mathbf{f}$ is a vector field with a finite number of critical points on S. Then topologically S is a two-dimensional sphere with $q = 2p$ holes along whose boundaries the antipodal points have been identified; cf. [L], p. 370. We cut out the q cross-caps which are topologically equivalent to the projective plane P; and then shrink the boundaries of the q resulting holes to q critical points $Q_j, j = 1, \ldots, q$ on S^2. But

$$I_f(Q_j) = I_f(P) = 1.$$

Thus,

$$I_f(S) = I_f(S^2) - \sum_{j=1}^{q} I_f(Q_j)$$
$$= 2 - q = 2 - 2p = \chi(S).$$

This completes the proof of the Poincaré Index Theorem.

The next corollary is an immediate consequence of the Poincaré Index Theorem and Theorem 6; cf. [P], p. 121.

Corollary 3. *Suppose that $\mathbf{f}$ is an analytic vector field on an analytic, two-dimensional surface S of genus p and that $\mathbf{f}$ has only hyperbolic critical points; i.e., isolated saddles, nodes and foci, on S. Then*

$$n + f - s = 2(1 - p)$$

where n, f and s are the number of nodes, foci and saddles on S respectively.

3.12. Index Theory

Example 2. In Example 1 in Section 3.10 we described the flow on the Poincaré sphere determined by the planar system

$$\dot{x} = x$$
$$\dot{y} = -y.$$

Cf. Figure 4 in Section 3.10. This simple planar system has a saddle at the origin. There are saddles at $\pm(0,0,1)$ and nodes at $\pm(1,0,0)$ and $\pm(0,1,0)$ on the Poincaré sphere. It follows from the above corollary that

$$\chi(S^2) = n + f - s = 4 - 2 = 2$$

as was to be expected. This same example also defines a vector field on the projective plane P, i.e., the vector field determined by the flow on P shown in Figure 5 in Section 3.10. We see that if the antipodal points of the disk shown in Figure 5 of Section 3.10 are identified, then we have a vector field on P with one saddle and two nodes and from the above corollary it follows that

$$\chi(P) = n + f - s = 2 - 1 = 1$$

as was to be expected.

Remark 2. As in the above example, the global phase portrait of any planar polynomial system of odd degree describes a flow or a vector field $\mathbf{f}$ on the projective plane. Therefore, the sum of the indices of all of the finite critical points and infinite critical points (where we have identified the antipodal points on the boundary of the unit disk) with respect to the vector field $\mathbf{f}$ must be equal to one.

Example 3. A model for the torus T^2 is the rectangle R with the opposite sides identified as in Figure 9. A flow on T^2 with two saddles and two nodes is defined by the flow on the rectangle R shown in Figure 9. The sum of the indices at the critical points of this flow is equal to zero, the Euler–Poincaré characteristic of T^2.

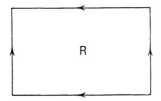
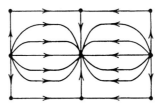

Figure 9. A flow on the torus T^2.

Problem Set 12

1. Use the index formula

$$I_\mathbf{f}(C) = \frac{1}{2\pi} \oint_C \frac{P dQ - Q dP}{P^2 + Q^2}$$

given at the beginning of this section to compute the indices $I_f(C)$, $I_g(C)$, $I_h(C)$ and $I_k(C)$ for the vector fields $\mathbf{f}, \mathbf{g}, \mathbf{h}$ and $\mathbf{k}$ of Example 1.

2. Let C_α be a Jordan curve as described in the proof of Theorem 1, and let $\mathbf{x}_0 = (x_0, y_0)^T$ be a point on C_α. Choose a coordinate system with the x-axis parallel to the vector $\mathbf{f}(\mathbf{x}_0)$ and in the direction of $\mathbf{f}(\mathbf{x}_0)$. Then for $\mathbf{f} = (P, Q)^T$ we have

$$Q(x_0, y_0) = 0 \quad \text{and} \quad P(x_0, y_0) = k_0$$

a positive constant. By the uniform continuity of P and Q in any compact subset of E, the $\delta > 0$ in the proof of Theorem 1 can be chosen sufficiently small that

$$\frac{-k_0}{2} < Q(x, y) < \frac{k_0}{2} \quad \text{and} \quad \frac{-k_0}{2} < P(x, y) - k_0 < \frac{k_0}{2}$$

for all points $(x, y) \in C_\alpha$. Prove that this implies that

$$\Delta\Theta/2\pi < 1/4$$

as the point (x, y) moves around C_α in the positive direction.

3. Prove Lemma 2. First of all, define the family of vector fields

$$\mathbf{v}_s = (1 - s)\mathbf{v} + s\mathbf{w}$$

and show that $\mathbf{v}_s \neq 0$ and that $\mathbf{v}_s$ is continuous for $0 \leq s \leq 1$. Then use the continuity of $I_{\mathbf{v}_s}(C)$ with respect to s and the fact that $I_{\mathbf{v}_s}(C)$ is an integer to show that

$$I_\mathbf{v}(C) = I_{\mathbf{v}_0}(C) = I_{\mathbf{v}_1}(C) = I_\mathbf{w}(C).$$

4. Use Bendixson's Index Theorem, i.e., Theorem 7, to determine the indices of the degenerate critical points shown in Figures 2–5 in Section 2.11 of Chapter 2.

5. (a) Use the triangulation shown on the top half of the torus T^2 in Figure 10 below, with an identical type of triangulation on the bottom half of T^2, in order to show that $\chi(T^2) = 0$. Describe a flow on T^2 with no critical points.

   (b) The Klein bottle K is a non-orientable surface of genus $p = q/2 = 1$. The usual model for the Klein bottle is the two-dimensional sphere S^2 with two holes along whose boundaries the antipodal points are identified; cf. Figure 11. If we make a cut C between the two holes and lay the surface out flat, we have a rectangle R with the opposite sides identified as shown in Figure 11. For the vector field $\mathbf{f}$ defined by the uniform parallel flow on R, shown in Figure 11, compute the index $I_f(K)$ and show that it is equal to $\chi(K)$.

3.12. Index Theory

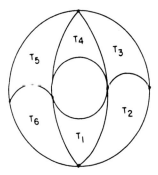

Figure 10. A triangulation of the torus T^2.

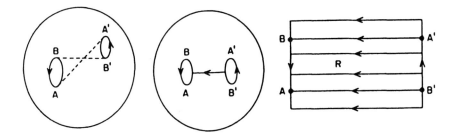

Figure 11. A flow on the Klein bottle K.

Remark 3. It follows from the index theory that the torus and the Klein bottle are the only two-dimensional surfaces on which there exist flows with no critical points. Furthermore, it can be shown that any flow on the Klein bottle with no critical points has a cycle.

6. Show that the sum of the indices at all of the finite and infinite critical points of the vector fields on the Poincaré sphere S^2 determined by the global phase portraits in Figures 4, 7, 9 and 12 in Section 3.10 are equal to two. **Hint:** Use Theorem 7 to determine the indices at the infinite critical points. Also, show that the sum of the indices at all of the critical points of the vector fields on the projective plane P determined by the global phase portraits in Figure 5 and those in Problem 3 in Section 3.10 are equal to one; cf. Remark 2.

7. Find the index of the degenerate critical points shown in Figure 12 below.

8. In sketching the following vector fields or flows, be sure that they are continuous; i.e., no opposing vectors can appear in any small neighborhood on the surface. Use Theorem 7 to compute the indices.

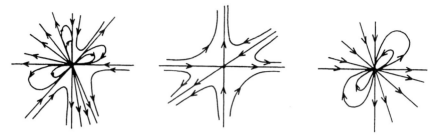

Figure 12. Degenerate critical points.

(a) Sketch a vector field or flow on the sphere S^2
  (i) with one critical point and compute the index at this critical point;
  (ii) with two critical points and compute the indices at these critical points;
  (iii) with three critical points and compute the indices at these critical points.

(b) Sketch a vector field or flow on the torus T^2
  (i) with no critical points;
  (ii) with at least one critical point and compute the index at each critical point.

(c) Sketch a vector field or flow on the anchor ring, i.e., an orientable surface of genus $p = 2$, and compute the sum of the indices at the critical points of this vector field.

(d) Sketch a vector field or flow on an orientable surface of genus $p = 3$ and compute the sum of the indices at the critical points of this vector field.

(e) Sketch a flow on the projective plane P
  (i) with one critical point
  (ii) with two critical points

(f) Sketch a flow on the Klein bottle K with one critical point.

9. Let $z = x + iy$, $\bar{z} = x - iy$ and show that the vector fields in the complex plane defined by

$$\dot{z} = z^k \quad \text{and} \quad \dot{z} = \bar{z}^k$$

have unique critical points at $z = 0$ with indices k and $-k$ respectively. **Hint:** Write $\dot{x} = \mathrm{Re}(z^k)$; $\dot{y} = \mathrm{Im}(z^k)$ and let $z = re^{i\theta}$. Sketch the phase portrait near the origin in those cases with indices ± 2 and ± 3.

4
Nonlinear Systems: Bifurcation Theory

In Chapters 2 and 3 we studied the local and global theory of nonlinear systems of differential equations

$$\dot{\mathbf{x}} = \mathbf{f}(\mathbf{x}) \tag{1}$$

with $\mathbf{f} \in C^1(E)$ where E is an open subset of $\mathbf{R}^n$. In this chapter we address the question of how the qualitative behavior of (1) changes as we change the function or vector field $\mathbf{f}$ in (1). If the qualitative behavior remains the same for all nearby vector fields, then the system (1) or the vector field $\mathbf{f}$ is said to be *structurally stable*. The idea of structural stability originated with Andronov and Pontryagin in 1937. Their work on planar systems culminated in Peixoto's Theorem which completely characterizes the structurally stable vector fields on a compact, two-dimensional manifold and establishes that they are generic. Unfortunately, no such complete results are available in higher dimensions ($n \geq 3$). If a vector field $\mathbf{f} \in C^1(E)$ is not structurally stable, it belongs to the bifurcation set in $C^1(E)$. The qualitative structure of the solution set or of the global phase portrait of (1) changes as the vector field $\mathbf{f}$ passes through a point in the bifurcation set. In this chapter, we study various types of bifurcations that occur in C^1-systems

$$\dot{\mathbf{x}} = \mathbf{f}(\mathbf{x}, \mu) \tag{2}$$

depending on a parameter $\mu \in \mathbf{R}$ (or on several parameters $\boldsymbol{\mu} \in \mathbf{R}^m$). In particular, we study bifurcations at nonhyperbolic equilibrium points and periodic orbits including bifurcations of periodic orbits from nonhyperbolic equilibrium points. These types of bifurcations are called local bifurcations since we focus on changes that take place near the equilibrium point or periodic orbit. We also consider global bifurcations in this chapter such as homoclinic loop bifurcations and bifurcations of limit cycles from a one-parameter family of periodic orbits such as those surrounding a center. This chapter is intended as an introduction to bifurcation theory and some of the simpler types of bifurcations that can occur in systems of the form (2). For the more general theory of bifurcations, the reader should consult Guckenheimer and Holmes [G/H], Wiggins [Wi], Chow and Hale [C/H], Golubitsky and Guillemin [G/G] and Ruelle [Ru].

Besides providing an introduction to the concept of structural stability and to bifurcation theory, this chapter also contains some interesting results on the global behavior of one-parameter families of periodic orbits defined by C^1 or analytic systems depending on a parameter. In particular, Duff's theory for limit cycles of planar families of rotated vector fields is presented in Section 4.6 and Wintner's Principle of Natural Termination for one-parameter families of periodic orbits of analytic systems in $\mathbf{R}^n$ is presented in Section 4.7. In Section 4.8, we see that for planar systems a limit cycle typically bifurcates from a homoclinic loop or separatrix cycle of (2) as the parameter μ is varied; however, in higher dimensions ($n \geq 3$), homoclinic loop bifurcations (or, more generally, bifurcations at a tangential homoclinic orbit) typically result in very wild or chaotic behavior. Establishing the existence of homoclinic tangencies or transverse homoclinic orbits and the resulting chaotic behavior is a very difficult task. However, Melnikov's Method which is presented in Sections 4.9–4.12 is one of the few analytic methods for studying homoclinic loop bifurcations and establishing the existence of transverse homoclinic orbits for perturbed dynamical systems. We conclude this chapter with a discussion of the various types of bifurcations that occur in planar systems and apply this theory in order to determine all of the possible bifurcations and the corresponding phase portraits that occur in the class of bounded quadratic systems in $\mathbf{R}^2$.

4.1 Structural Stability and Peixoto's Theorem

In this section, we present the concept of a structurally stable vector field or dynamical system and give necessary and sufficient conditions for a C^1-vector field $\mathbf{f}$ on a compact two-dimensional manifold to be structurally stable. The idea of structural stability originated with Andronov and Pontryagin in 1937; cf. [A–II], p. 56. We say that $\mathbf{f}$ is a structurally stable vector field if for any vector field $\mathbf{g}$ near $\mathbf{f}$, the vector fields $\mathbf{f}$ and $\mathbf{g}$ are topologically equivalent. Cf. Definition 2 in Section 3.1 of Chapter 3. The only concept that we need to make precise in this definition of structural stability is what it means for two C^1-vector fields $\mathbf{f}$ and $\mathbf{g}$ to be close.

If $\mathbf{f} \in C^1(E)$ where E is an open subset of $\mathbf{R}^n$, then the C^1-*norm* of $\mathbf{f}$

$$\|\mathbf{f}\|_1 = \sup_{\mathbf{x} \in E} |\mathbf{f}(\mathbf{x})| + \sup_{\mathbf{x} \in E} \|D\mathbf{f}(\mathbf{x})\| \tag{1}$$

where $|\cdot|$ denotes the Euclidean norm on $\mathbf{R}^n$ and $\|\cdot\|$ denotes the usual norm of the matrix $D\mathbf{f}(\mathbf{x})$ as defined in Section 1.3 of Chapter 1. The function $\|\cdot\|_1$ from $C^1(E)$ to $\mathbf{R}$ has all of the usual properties of a norm listed in Section 1.3 of Chapter 1. The set of functions in $C^1(E)$, bounded in the C^1-norm, is a Banach space, i.e., a complete normed linear space. We shall use the C^1-norm to measure the distance between any two functions

4.1. Structural Stability and Peixoto's Theorem

in $C^1(E)$. If K is a compact subset of E, then the C^1-*norm of* **f** *on* K is defined by

$$\|\mathbf{f}\|_1 = \max_{\mathbf{x} \in K} |\mathbf{f}(\mathbf{x})| + \max_{\mathbf{x} \in K} \|D\mathbf{f}(\mathbf{x})\| < \infty. \tag{1'}$$

Definition 1. Let E be an open subset of $\mathbf{R}^n$. A vector field $\mathbf{f} \in C^1(E)$ is said to be *structurally stable* if there is an $\varepsilon > 0$ such that for all $\mathbf{g} \in C^1(E)$ with

$$\|\mathbf{f} - \mathbf{g}\|_1 < \varepsilon$$

f and **g** are topologically equivalent on E; i.e., there is a homeomorphism $H: E \looparrowright E$ which maps trajectories of

$$\dot{\mathbf{x}} = \mathbf{f}(\mathbf{x}) \tag{2}$$

onto trajectories of

$$\dot{\mathbf{x}} = \mathbf{g}(\mathbf{x}) \tag{2'}$$

and preserves their orientation by time. In this case, we also say that *the dynamical system* (2) *is structurally stable.* If a vector field $\mathbf{f} \in C^1(E)$ is not structurally stable, then **f** is said to be *structurally unstable.* If K is a compact subset of E and $\mathbf{f} \in C^1(E)$, then if we use the C^1-norm (1') in Definition 1, we say that the vector field **f** is *structurally stable on* K.

Remark 1. If $E = \mathbf{R}^n$, then the ε-perturbations of **f** in the above definition, i.e., the functions $\mathbf{g} \in C^1(E)$ satisfying $\|\mathbf{f} - \mathbf{g}\|_1 < \varepsilon$, include the C^1, ε-perturbations of Guckenheimer and Holmes in Definition 1.7.1. on p. 38 in [G/H]. Also if K is a compact subset of E and if $\mathbf{g} \in C^1(K)$ satisfies

$$\max_{\mathbf{x} \in K} |\mathbf{f}(\mathbf{x}) - \mathbf{g}(\mathbf{x})| + \max_{\mathbf{x} \in K} \|D\mathbf{f}(\mathbf{x}) - D\mathbf{g}(\mathbf{x})\| < \varepsilon,$$

then there exists a compact subset $\tilde{K}$ of E containing K and a function $\tilde{\mathbf{g}} \in C^1(E)$ such that $\tilde{\mathbf{g}}(\mathbf{x}) = \mathbf{g}(\mathbf{x})$ for all $\mathbf{x} \in K$, $\tilde{\mathbf{g}}(\mathbf{x}) = \mathbf{f}(\mathbf{x})$ for all $\mathbf{x} \in E \sim \tilde{K}$ and $\|\mathbf{f} - \tilde{\mathbf{g}}\|_1 < \varepsilon$. Thus, in order to show that $\mathbf{f} \in C^1(\mathbf{R}^n)$ is not structurally stable on $\mathbf{R}^n$, it suffices to show that **f** is not structurally stable on some compact $K \subset \mathbf{R}^n$ with nonempty interior.

It was originally thought that structural stability was typical of any dynamical system modeling a physical problem. Consider, for example, a damped pendulum. If the mass, length or friction in the pendulum is changed by a sufficiently small amount, ε, the qualitative behavior of the solution will remain the same; i.e., the global phase portraits of the two systems (2) and (2') modeling the two pendula will be topologically equivalent. Thus, the dynamical system (2) modeling the physical system consisting of a damped pendulum is structurally stable. On the other hand, the dynamical system modeling an undamped pendulum in Example 1 in Section 2.14 of Chapter 2 is structurally unstable since the addition of any small amount of friction, i.e., damping, will change the undamped, periodic motion seen in Figure 1 of Section 2.14 in Chapter 2 to a damped motion; i.e., the centers in Figure 1 will become stable foci. Of course, a frictionless pendulum is not physically realizable. If we were to only consider physical problems which lead to systems of differential equations in $\mathbf{R}^2$, then we would not have to worry about arbitrarily small changes in the model leading to qualitatively different behavior of the system. However, there are higher dimensional systems (with $n \geq 3$) which are realistic models for certain physical problems (such as the three-body problem) and which are structurally unstable. Recently, many dynamical systems have been found which model physical problems and which have a strange attractor as part of their dynamics. These systems are not structurally stable and yet they are realistic models for certain physical systems; cf., e.g., [G/H], p. 259.

We next define what is meant by a structurally stable vector field on an n-dimensional compact manifold M: If $\mathbf{f}$ is a C^1-vector field on an n-dimensional, compact, differentiable manifold M, then for any (finite) atlas $\{U_j, \mathbf{h}_j\}_{j=1}^m$ for M we define the C^1-norm of $\mathbf{f}$ on M as

$$\|\mathbf{f}\|_1 = \max_j \|\mathbf{f}_j\|_1$$

where $\mathbf{f}_j\colon V_j \to \mathbf{R}^n$ and for $j = 1, \ldots, m$, $V_j = \mathbf{h}_j(U_j) \subset \mathbf{R}^n$ as in Section 3.10 of Chapter 3. Note that for different atlases for M we will get different norms on $C^1(M)$; however, all of these norms will be equivalent. Recall that two norms $\|\cdot\|_a$ and $\|\cdot\|_b$ on a linear space L are said to be equivalent if there are positive constants A and B such that

$$A\|\mathbf{x}\|_a \leq \|\mathbf{x}\|_b \leq B\|\mathbf{x}\|_a$$

for all $\mathbf{x} \in L$. Hence the resulting topologies on $C^1(M)$ will be equivalent. We say that two C^1-vector fields $\mathbf{f}, \mathbf{g} \in C^1(M)$ are *topologically equivalent on M* if there is a homeomorphism $H\colon M \to M$ which maps trajectories of (2) on M onto trajectories of (2') on M and preserves their orientation by time.

Definition 2. Let $\mathbf{f}$ be a C^1-vector field on a compact, n-dimensional, differentiable manifold M. Then $\mathbf{f} \in C^1(M)$ is *structurally stable on M* if there is an $\varepsilon > 0$ such that for all $\mathbf{g} \in C^1(M)$ with

$$\|\mathbf{f} - \mathbf{g}\|_1 < \varepsilon,$$

$\mathbf{g}$ is topologically equivalent to $\mathbf{f}$.

Remark 2. In 1962, Peixoto [22] gave a complete characterization of the structurally stable, C^1 vector fields on any compact, two-dimensional, differentiable manifold M (such as S^2) and he showed that they form a dense, open subset of $C^1(M)$; cf. Theorem 3 below. However, it was later shown

4.1. Structural Stability and Peixoto's Theorem

that on any open, two-dimensional, differentiable manifold E (such as $\mathbf{R}^2$), there is a subset of $C^1(E)$ which is open in the C^1-topology (defined by the C^1-norm) and which consists of structurally unstable vector fields. Nevertheless, in 1982, Kotus, Krych and Nitecki [16] showed how to control the behavior "at infinity" so as to guarantee the structural stability of a vector field on any two-dimensional, differentiable manifold under "strong C^1-perturbation" and they gave a complete characterization of the structurally stable vector fields on $\mathbf{R}^2$. Cf. Theorem 4 below.

Example 1. The vector field

$$\mathbf{f}(\mathbf{x}) = \begin{pmatrix} -y \\ x \end{pmatrix}$$

on $\mathbf{R}^2$ is not structurally stable on any compact set $K \subset \mathbf{R}^2$ containing the origin on its interior. To see this we let K be any compact subset of $\mathbf{R}^2$ which contains the origin on its interior and show that $\mathbf{f}$ is not structurally stable on K. Let $\|\cdot\|_1$ denote the C^1-norm on K and define the vector field

$$\mathbf{g}(\mathbf{x}) = \begin{pmatrix} -y + \mu x \\ x + \mu y \end{pmatrix}.$$

Then

$$\|\mathbf{f} - \mathbf{g}\|_1 = |\mu|(\max_{\mathbf{x} \in K} |\mathbf{x}| + 1)$$

and if d is the diameter of K, i.e. if

$$d = \max_{\mathbf{x},\mathbf{y} \in K} |\mathbf{x} - \mathbf{y}|,$$

it follows for all $\varepsilon > 0$ that if we choose $|\mu| = \varepsilon/(d+2)$ then $\|\mathbf{f}-\mathbf{g}\|_1 < \varepsilon$. The phase portraits for the system $\dot{\mathbf{x}} = \mathbf{g}(\mathbf{x})$ are shown in Figure 1. Clearly $\mathbf{f}$ is not topologically equivalent to $\mathbf{g}$; cf. Problem 1. Thus $\mathbf{f}$ is not structurally stable on $\mathbf{R}^2$. The number $\mu = 0$ is called a bifurcation value for the system $\dot{\mathbf{x}} = \mathbf{g}(\mathbf{x})$.

Example 2. The system

$$\dot{x} = -y + x(x^2 + y^2 - 1)^2$$
$$\dot{y} = x + y(x^2 + y^2 - 1)^2$$

is structurally unstable on any compact subset $K \subset \mathbf{R}^2$ which contains the unit disk on its interior. This can be seen by considering the system

$$\dot{x} = -y + x[(x^2 + y^2 - 1)^2 - \mu]$$
$$\dot{y} = x + y[(x^2 + y^2 - 1)^2 - \mu]$$

which is ε-close to the above system if $|\mu| = \varepsilon/(d+2)$ where d is the diameter of K. But writing this latter system in polar coordinates yields

$$\dot{r} = r[(r^2 - 1)^2 - \mu]$$
$$\dot{\theta} = 1.$$

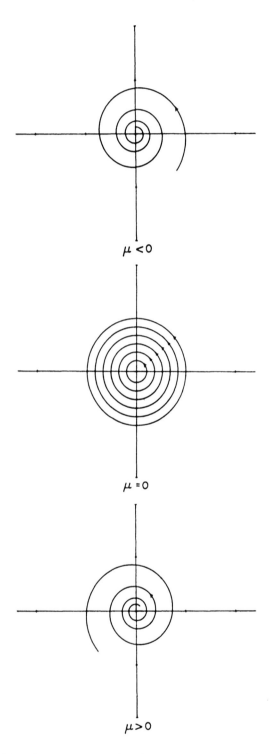

Figure 1. The phase portraits for the system $\dot{\mathbf{x}} = \mathbf{g}(\mathbf{x})$ in Example 1.

4.1. Structural Stability and Peixoto's Theorem

Hence, we have the phase portraits shown in Figure 2 below; and the above system with $\mu = 0$ is structurally unstable; cf. Problem 2. The number $\mu = 0$ is called a bifurcation value for the above system and for $\mu = 0$ this system has a limit cycle of multiplicity two represented by $\gamma(t) = (\cos t, \sin t)^T$.

Note that for $\mu = 0$, the origin is a nonhyperbolic critical point for the system in Example 1 and $\gamma(t)$ is a nonhyperbolic limit cycle of the system in Example 2. In general, dynamical systems with nonhyperbolic equilibrium points and/or nonhyperbolic periodic orbits are not structurally stable. This does not mean that dynamical systems with only hyperbolic equilibrium points and periodic orbits are structurally stable; cf., e.g., Theorem 3 below.

Before characterizing structurally stable planar systems, we cite some results on the persistance of hyperbolic equilibrium points and periodic orbits; cf., e.g., [H/S], pp. 305–312.

Theorem 1. *Let $\mathbf{f} \in C^1(E)$ where E is an open subset of $\mathbf{R}^n$ containing a hyperbolic critical point $\mathbf{x}_0$ of (2). Then for any $\varepsilon > 0$ there is a $\delta > 0$ such that for all $\mathbf{g} \in C^1(E)$ with*

$$\|\mathbf{f} - \mathbf{g}\|_1 < \delta$$

there exists a $\mathbf{y}_0 \in N_\varepsilon(\mathbf{x}_0)$ such that $\mathbf{y}_0$ is a hyperbolic critical point of (2'); furthermore, $D\mathbf{f}(\mathbf{x}_0)$ and $D\mathbf{g}(\mathbf{y}_0)$ have the same number of eigenvalues with negative (and positive) real parts.

Theorem 2. *Let $\mathbf{f} \in C^1(E)$ where E is an open subset of $\mathbf{R}^n$ containing a hyperbolic periodic orbit Γ of (2). Then for any $\varepsilon > 0$ there is a $\delta > 0$ such that for all $\mathbf{g} \in C^1(E)$ with*

$$\|\mathbf{f} - \mathbf{g}\|_1 < \delta$$

there exists a hyperbolic periodic orbit Γ' of (2') contained in an ε-neighborhood of Γ; furthermore, the stable manifolds $W^s(\Gamma)$ and $W^s(\Gamma')$, and the unstable manifolds $W^u(\Gamma)$ and $W^u(\Gamma')$, have the same dimensions.

One other important result for n-dimensional systems is that any linear system

$$\dot{\mathbf{x}} = A\mathbf{x}$$

where the matrix A has no eigenvalue with zero real part is structurally stable in $\mathbf{R}^n$. Besides nonhyperbolic equilibrium points and periodic orbits, there are two other types of behavior that can result in structurally unstable systems on two-dimensional manifolds. We illustrate these two types of behavior with some examples.

322　　　　　　　　　　　　　　　4. Nonlinear Systems: Bifurcation Theory

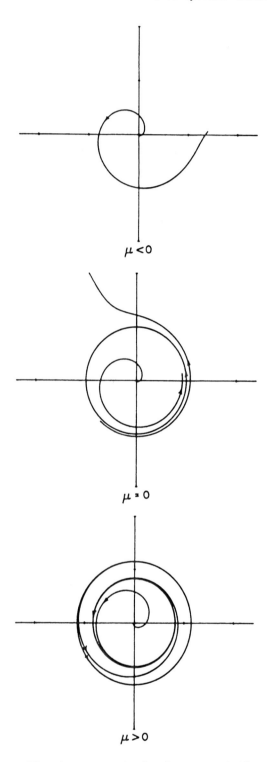

Figure 2. The phase portraits for the system in Example 2.

4.1. Structural Stability and Peixoto's Theorem

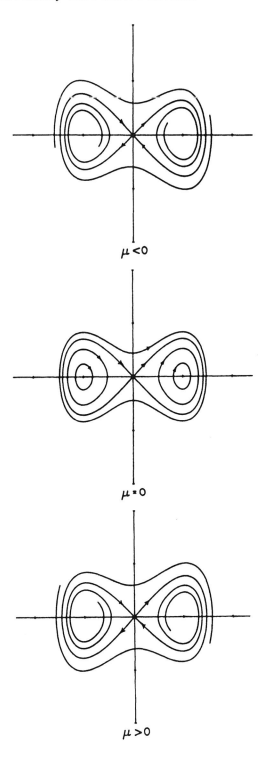

Figure 3. The phase portraits for the system in Example 3.

Example 3. Consider the system

$$\dot{x} = y$$
$$\dot{y} = \mu y + x - x^3.$$

For $\mu = 0$ this is a Hamiltonian system with Hamiltonian $H(x,y) = (y^2 - x^2)/2 + x^4/4$. The level curves for this function are shown in Figure 3. We see that for $\mu = 0$ there are two centers at $(\pm 1, 0)$ and two separatrix cycles enclosing these centers. For $\mu = 0$ this system is structurally unstable on any compact subset $K \subset \mathbf{R}^2$ containing the disk of radius 2 because the above system is ε-close to the system with $\mu = 0$ if $|\mu| = \varepsilon/(d+2)$ where d is the diameter of K. Also, the phase portraits for the above system are shown in Figure 3 and clearly the above system with $\mu \neq 0$ is not topologically equivalent to that system with $\mu = 0$; cf. Problem 3.

In Example 3, not only does the qualitative behavior near the nonhyperbolic critical points $(\pm 1, 0)$ change as μ varies through $\mu = 0$, but also there are no separatrix cycles for $\mu \neq 0$; i.e., separatrix cycles and more generally saddle-saddle connections do not persist under small perturbations of the vector field. Cf. Problem 4 for another example of a planar system with a saddle-saddle connection.

Definition 3. A point $\mathbf{x} \in E$ (or $\mathbf{x} \in M$) is a *nonwandering point of the flow ϕ_t* defined by (2) if for any neighborhood U of x and for any $T > 0$ there is a $t > T$ such that

$$\phi_t(U) \cap U \neq \emptyset.$$

The *nonwandering set Ω of the flow ϕ_t* is the set of all nonwandering points of ϕ_t in E (or in M). Any point $\mathbf{x} \in E \sim \Omega$ (or in $M \sim \Omega$) is called a *wandering point of ϕ_t*.

Equilibrium points and points on periodic orbits are examples of nonwandering points of a flow and for a relatively-prime, planar, analytic flow, the only nonwandering points are critical points, points on cycles and points on graphics that belong to the ω-limit set of a trajectory or the limit set of a sequence of periodic orbits of the flow (on $\mathbf{R}^2$ or on the Bendixson sphere; cf. Problem 8 in Section 3.7). This is not true in general as the next example shows; cf. Theorem 3 in Section 3.7 of Chapter 3.

Example 4. Let the unit square S with its opposite sides identified be a model for the torus and let (x, y) be coordinates on S which are identified (mod 1). Then the system

$$\dot{x} = \omega_1$$
$$\dot{y} = \omega_2$$

defines a flow on the torus; cf. Figure 4. The flow defined by this system is given by

$$\phi_t(x_0, y_0) = (\omega_1 t + x_0, \omega_2 t + y_0)^T.$$

4.1. Structural Stability and Peixoto's Theorem   325

If ω_1/ω_2 is irrational, then all points lie on orbits that never close, but densely cover S or the torus. If ω_1/ω_2 is rational then all points lie on periodic orbits. Cf. Problem 2 in Section 3.2 of Chapter 3. In either case all points of T^2 are nonwandering points, i.e., $\Omega = T^2$. And in either case, the system is structurally unstable since in either case, there is an arbitrarily small constant which, when added to ω_1, changes one case to the other.

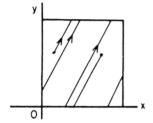

Figure 4. A flow on the unit square with its opposite sides identified and the corresponding flow on the torus.

We now state Peixoto's Theorem [22], proved in 1962, which completely characterizes the structurally stable C^1-vector fields on a compact, two-dimensional, differentiable manifold M.

Theorem 3 (Peixoto). *Let* **f** *be a C^1-vector field on a compact, two-dimensional, differentiable manifold M. Then* **f** *is structurally stable on M if and only if*

(i) *the number of critical points and cycles is finite and each is hyperbolic;*

(ii) *there are no trajectories connecting saddle points; and*

(iii) *the nonwandering set Ω consists of critical points and limit cycles only.*

Furthermore, if M is orientable, the set of structurally stable vector fields in $C^1(M)$ is an open, dense subset of $C^1(M)$.

If the set of all vector fields $\mathbf{f} \in C^r(M)$, with $r \geq 1$, having a certain property $\mathcal{P}$ contains an open, dense subset of $C^r(M)$, then the property $\mathcal{P}$ is called *generic*. Thus, according to Peixoto's Theorem, structural stability is a generic property of the C^1 vector fields on a compact, two-dimensional, differentiable manifold M. More generally, if V is a subset of $C^r(M)$ and the set of all vector fields $\mathbf{f} \in V$ having a certain property $\mathcal{P}$ contains an open, dense subset of V, then the property $\mathcal{P}$ is called *generic in V*.

If the phase space is planar, then by the Poincaré-Bendixson Theorem for analytic systems, the only possible limit sets are critical points, limit cycles and graphics and if there are no saddle-saddle connections, graphics are ruled out. The nonwandering set Ω will then consist of critical points and limit cycles only. Hence, if **f** is a vector field on the Poincaré sphere defined by a planar polynomial vector field as in Section 3.10 of Chapter 3, we have the following corollary of Peixoto's Theorem and Theorem 3 in Section 3.7 of Chapter 3.

Corollary 1. *Let* **f** *be a vector field on the Poincaré sphere defined by the differential equation*

$$\begin{vmatrix} dX & dY & dZ \\ X & Y & Z \\ P^* & Q^* & 0 \end{vmatrix} = 0$$

where

$$P^*(X,Y,Z) = Z^m P(X/Z, Y/Z),$$
$$Q^*(X,Y,Z) = Z^m Q(X/Z, Y/Z),$$

and P and Q are polynomials of degree m. Then **f** *is structurally stable on S^2 if and only if*

(i) *the number of critical points and cycles is finite and each is hyperbolic, and*

(ii) *there are no trajectories connecting saddle points on S^2.*

This corollary gives us an easy test for the structural stability of the global phase portrait of a planar polynomial system. In particular, the global phase portrait will be structurally unstable if there are nonhyperbolic critical points at infinity or if there is a trajectory connecting a saddle on the equator of the Poincaré sphere to another saddle on S^2. It can be shown that if the polynomial vector field **f** in Corollary 1 is structurally stable on S^2, then the corresponding polynomial vector field $(P,Q)^T$ is structurally stable on $\mathbf{R}^2$ under "strong C^1-perturbations". We say that a C^1-vector field **f** is structurally stable on $\mathbf{R}^2$ under strong C^1-perturbations (or that it is structurally stable with respect to the Whitney C^1-topology on $\mathbf{R}^2$) if it is topologically equivalent to all C^1-vector fields **g** satisfying

$$|\mathbf{f}(\mathbf{x}) - \mathbf{g}(\mathbf{x})| + \|D\mathbf{f}(\mathbf{x}) - D\mathbf{g}(\mathbf{x})\| < \varepsilon(\mathbf{x})$$

for some continuous, strictly positive function $\varepsilon(\mathbf{x})$ on $\mathbf{R}^2$. The fact that structural stability of the polynomial vector field **f** on S^2 in Corollary 1 implies that the corresponding polynomial vector field $(P,Q)^T$ is structurally

4.1. Structural Stability and Peixoto's Theorem

stable on $\mathbf{R}^2$ under strong C^1-perturbations follows from Corollary 1 and the next theorem proved in [16]; cf. Theorem 3.1 in [56]. Also, it follows from Theorem 23 in [A–II] and Theorem 4 below that structural stability of a polynomial vector field $\mathbf{f}$ on $\mathbf{R}^2$ under strong C^1-perturbations implies that $\mathbf{f}$ is structurally stable on any bounded region of $\mathbf{R}^2$; cf. Definition 10 in [A–II]. (The converses of the previous two statements are false as shown by the examples below.) In order to state the next theorem, we first define the concept of a saddle at infinity as defined in [56].

Definition 4. A **saddle at infinity** (SAI) of a vector field $\mathbf{f}$ defined on $\mathbf{R}^2$ is a pair $(\Gamma_\mathbf{p}^+, \Gamma_\mathbf{q}^-)$ of half-trajectories of $\mathbf{f}$, each escaping to infinity, such that there exist sequences $\mathbf{p}_n \to \mathbf{p}$ and $t_n \to \infty$ with $\phi(t_n, \mathbf{p}_n) \to \mathbf{q}$ in $\mathbf{R}^2$. $\Gamma_\mathbf{p}^+$ is called the stable separatrix of the SAI and $\Gamma_\mathbf{q}^-$ the unstable separatrix of the SAI. A **saddle connection** is a trajectory Γ of $\mathbf{f}$ with $\Gamma = \Gamma^+ \cup \Gamma^-$ where Γ^+ is a stable separatrix of a saddle or of a SAI while Γ^- is a separatrix of a saddle or of a SAI.

If we let $W^\pm(\mathbf{f})$ denote the union of all trajectories containing a stable or unstable separatrix of a saddle or SAI of $\mathbf{f}$, then there is a saddle connection only if $W^+(\mathbf{f}) \cap W^-(\mathbf{f}) \neq \emptyset$.

Theorem 4 (Kotus, Krych and Nitecki). *A polynomial vector field is structurally stable on $\mathbf{R}^2$ under strong C^1-perturbations iff*

(i) *all of its critical points and cycles are hyperbolic and*

(ii) *there are no saddle connections (where separatrices of saddles at infinity are taken into account).*

Furthermore, there is a dense open subset of the set of all mth-degree polynomials, every element of which is structurally stable on $\mathbf{R}^2$ under strong C^1-perturbations.

It is also shown in Proposition 2.3 in [56] that (i) and (ii) in Theorem 4 imply that the nonwandering set consists of equilibrium points and periodic orbits only. As in [56], we have stated Theorem 4 for polynomial vector fields on $\mathbf{R}^2$; however, it is important to note that Theorems A and B in [16] give necessary and sufficient conditions for any C^1 vector field on $\mathbf{R}^2$ to be structurally stable on $\mathbf{R}^2$ under strong C^1-perturbations and sufficient conditions for the structural stability of any C^1 vector field on a smooth two-dimensional open surface (i.e., a smooth two-dimensional differentiable manifold without boundary which is metrizable but not compact) under strong C^1-perturbations.

The following example has a saddle connection between two saddles at infinity on S^2 and also a saddle connection according to Definition 4. It is therefore not structurally stable on S^2, according to Corollary 1, or on $\mathbf{R}^2$ under strong C^1-perturbations, according to Theorem 4; however, it is structurally stable on any bounded region of $\mathbf{R}^2$, according to Theorem 23 in [A–II].

Example 5. The cubic system

$$\dot{x} = 1 - y^2$$
$$\dot{y} = xy + y^3$$

has the global phase portrait shown below:

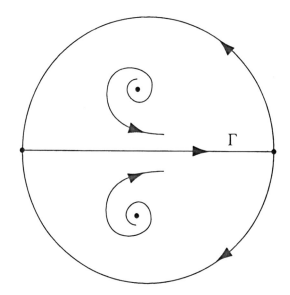

Clearly the trajectory Γ connects the two saddles at $(\pm 1, 0, 0)$ on the Poincaré sphere. And if we let p and q be any two points on Γ, then (Γ_p^+, Γ_q^-) is a saddle at infinity and (for p to the left of q on Γ) $\Gamma = \Gamma_p^+ \cup \Gamma_q^-$ is a saddle connection according to Definition 4. We also note that $W^+(\mathbf{f}) \cap W^-(\mathbf{f}) = \Gamma$ for the vector field given in this example.

The next example due to Chicone and Shafer, cf. (2.4) in [16], has a saddle connection according to Definition 4 and it is therefore not structurally stable on $\mathbf{R}^2$ under strong C^1-perturbations according to Theorem 4; although, it is structurally stable on any bounded region of $\mathbf{R}^2$ according to Theorem 23 in [A–II]. And since the corresponding vector field $\mathbf{f}$ in Corollary 1 has a nonhyperbolic critical point at $(0, \pm 1, 0)$ on S^2, $\mathbf{f}$ is not structurally stable on S^2.

4.1. Structural Stability and Peixoto's Theorem

Example 6. The quadratic system

$$\dot{x} = 2xy$$
$$\dot{y} = 2xy - x^2 + y^2 + 1$$

has the global phase portrait shown below:

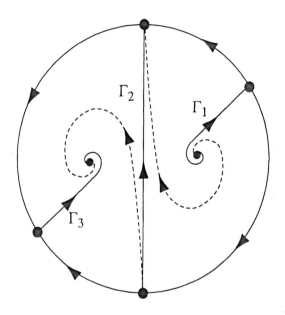

Let $p_i \in \Gamma_i$. Then $(\Gamma_{p_1}^+, \Gamma_{p_2}^-)$ and $(\Gamma_{p_2}^+, \Gamma_{p_3}^-)$ are two saddles at infinity and $\Gamma_2 = \Gamma_{p_2}^+ \cup \Gamma_{p_2}^-$ is a saddle connection according to Definition 4. We also note that $W^+(\mathbf{f}) \cap W^-(\mathbf{f}) = (\Gamma_1 \cup \Gamma_2) \cap (\Gamma_2 \cup \Gamma_3) = \Gamma_2$ for the vector field $\mathbf{f}$ given in this example.

The next example, which is (2.6) in [16], gives us a polynomial vector field which is structurally stable on $\mathbf{R}^2$ under strong C^1-perturbations according to Theorem 4, but whose projection onto S^2 is not structurally stable on S^2 according to Corollary 1 since there is a nonhyperbolic critical point at $(0, \pm 1, 0)$ on S^2.

Example 7. The cubic system

$$\dot{x} = x^3 - x$$
$$\dot{y} = 4x^2 - 1$$

has the global phase portrait shown below:

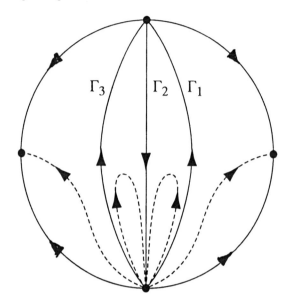

This system has no critical points or cycles in $\mathbf{R}^2$. Let $\mathbf{p}_i \in \Gamma_i$. Then $(\Gamma_{\mathbf{p}_1}^+, \Gamma_{\mathbf{p}_2}^-)$ and $(\Gamma_{\mathbf{p}_3}^+, \Gamma_{\mathbf{p}_2}^-)$ are saddles at infinity, but there is no saddle connection since $W^+(\mathbf{f}) \cap W^-(\mathbf{f}) = (\Gamma_1 \cup \Gamma_3) \cap \Gamma_2 = \emptyset$ for the vector field $\mathbf{f}$ in this example.

Shafer [56] has also given sufficient conditions for a polynomial vector field $\mathbf{f} \in \mathcal{P}_m$ to be structurally stable with respect to the coefficient topology on $\mathcal{P}_m$ (where $\mathcal{P}_m$ denotes the set of all polynomial vector fields of degree less than or equal to m on $\mathbf{R}^2$). It follows from Corollary 1 and the results in [56] that structural stability of the projection of the polynomial vector field on the Poincaré sphere implies structural stability of the polynomial vector field with respect to the coefficient topology on $\mathcal{P}_m$ which, in turn, implies structural stability of the polynomial vector field with respect to the Whitney C^1-topology.

In 1937 Andronov and Pontryagin showed that the conditions (i) and (ii) in Corollary 1 are necessary and sufficient for structural stability of a C^1-vector field on any **bounded** region of $\mathbf{R}^2$; cf. Definition 10 and Theorem 23 in [A–II]. And in higher dimensions, (i) together with the condition that the stable and unstable manifolds of any critical points and/or periodic orbits intersect transversally (cf. Definition 5 below) is necessary and sufficient for structural stability of a C^1-vector field on any **bounded** region of $\mathbf{R}^n$ whose boundary is transverse to the flow. However, there is no counterpart to Peixoto's Theorem for higher dimensional compact manifolds.

For a while, it was thought that conditions analogous to those in Peixoto's Theorem would completely characterize the structurally stable vector fields on a compact, n-dimensional, differentiable manifold; however, this proved

4.1. Structural Stability and Peixoto's Theorem

not to be the case. In order to formulate the analogous conditions for higher dimensional systems, we need to define what it means for two differentiable manifolds M and N to intersect transversally, i.e., nontangentially.

Definition 5. Let **p** be a point in $\mathbf{R}^n$. Then two differentiable manifolds M and N in $\mathbf{R}^n$ are said to *intersect transversally* at $\mathbf{p} \in M \cap N$ if $T_\mathbf{p}M \oplus T_\mathbf{p}N = \mathbf{R}^n$ where $T_\mathbf{p}M$ and $T_\mathbf{p}N$ denote the tangent spaces of M and N respectively at **p**. M and N are said to intersect transversally if they intersect transversally at every point $\mathbf{p} \in M \cap N$.

Definition 6. A *Morse–Smale system* is one for which

(i) the number of equilibrium points and periodic orbits is finite and each is hyperbolic;

(ii) all stable and unstable manifolds which intersect do so transversally; and

(iii) the nonwandering set consists of equilibrium points and periodic orbits only.

It is true that Morse–Smale systems on compact n-dimensional differentiable manifolds are structurally stable, but the converse is false in dimensions $n \geq 3$. As we shall see, there are structurally stable systems with strange attractors which are part of the nonwandering set. In dimensions $n \geq 3$, the structurally stable vector fields are not generic in $C^1(M)$. In fact, there are nonempty open subsets in $C^1(M)$ which consist of structurally unstable vector fields. Smale's work on differentiable dynamical systems and his construction of the horseshoe map were instrumental in proving that the structurally stable systems are not generic and that not all structurally stable systems are Morse–Smale; cf. [G/H], Chapter 5.

In the remainder of this chapter we consider the various types of bifurcations that can occur at nonhyperbolic equilibrium points and periodic orbits as well as the bifurcation of periodic orbits from equilibrium points and homoclinic loops. We also give a brief glimpse into what can happen at homoclinic loop bifurcations in higher dimensions ($n \geq 3$).

Problem Set 1

1. (a) In Example 1 show that $\|\mathbf{f} - \mathbf{g}\|_1 = |\mu|(\max_{\mathbf{x} \in K} |\mathbf{x}| + 1)$.

   (b) Show that for $\mu \neq 0$ the systems

   $$\begin{aligned} \dot{x} &= -y \\ \dot{y} &= x \end{aligned} \quad \text{and} \quad \begin{aligned} \dot{x} &= -y + \mu x \\ \dot{y} &= x + \mu y \end{aligned}$$

   are not topologically equivalent. **Hint:** Let ϕ_t and ψ_t be the flows defined by these two systems and assume that there is a homeomorphism $H: \mathbf{R}^2 \to \mathbf{R}^2$ and a strictly increasing, continuous function $t(\tau)$ mapping $\mathbf{R}$ onto $\mathbf{R}$ such that $\phi_{t(\tau)} = H^{-1} \circ \psi_\tau \circ H$. Use the fact that $\lim_{t \to \infty} \phi_t(1,0) \neq 0$ and that for $\mu < 0$, $\lim_{t \to \infty} \psi_t(\mathbf{x}) = \mathbf{0}$ for all $\mathbf{x} \in \mathbf{R}^2$ to arrive at a contradiction.

2. (a) In Example 2 show that $\|\mathbf{f} - \mathbf{g}\|_1 = |\mu|(\max_{\mathbf{x} \in K} |\mathbf{x}| + 1)$.

   (b) Show that the systems in Example 2 with $\mu = 0$ and $\mu \neq 0$ are not topologically equivalent. **Hint:** As in Problem 1, use the fact that for $|\mathbf{x}| < 1$
   $$\left| \lim_{t \to \infty} \phi_t(\mathbf{x}) \right| = 1 \text{ if } \mu = 0$$
   and
   $$\left| \lim_{t \to \infty} \psi_t(\mathbf{x}) \right| = \infty \text{ if } \mu < 0$$
   to arrive at a contradiction.

3. (a) In order to justify the phase portraits in Figure 3, show that the origin in Example 3 is a saddle and for $\mu < 0$ (or $\mu > 0$) the critical points $(\pm 1, 0)$ are stable (or unstable) foci; for $\mu \neq 0$, use Bendixson's Criteria to show that there are no cycles; and then use the Poincaré–Bendixson Theorem.

   (b) Show that the system in Example 3 with $\mu = 0$ is not structurally stable on the compact set $K = \{\mathbf{x} \in \mathbf{R}^2 \mid |\mathbf{x}| \leq 2\}$. **Hint:** As in Problems 1 and 2 use the fact that
   $$\lim_{t \to \infty} \phi_t(\sqrt{2}, 0) = (0, 0) \text{ for } \mu = 0$$
   and
   $$\lim_{t \to \infty} \psi_t(\sqrt{2}, 0) = (1, 0) \text{ for } \mu < 0$$
   to arrive at a contradiction.

4. (a) Draw the (local) phase portrait for the system
   $$\dot{x} = x(1-x)$$
   $$\dot{y} = -y(1-2x).$$

   (b) Show that this system (which has a saddle-saddle connection) is not structurally stable. **Hint:** For $\mu = \varepsilon/(d+2)$, show that the system
   $$\dot{x} = x(1-x)$$
   $$\dot{y} = -y(1-2x) + \mu x$$
   is ε-close to the system in part (a) on any compact set $K \subset \mathbf{R}^2$ of diameter d. Sketch the (local) phase portrait for this system with $\mu > 0$ and assuming that the systems in (a) and (b) are topologically equivalent for $\mu \neq 0$, arrive at a contradiction as in Problems 1–3.

4.1. Structural Stability and Peixoto's Theorem

5. Determine which of the global phase portraits in Figure 12 of Section 3.10 are structurally stable on S^2 and which are structurally stable on $\mathbf{R}^2$ under strong C^1-perturbations.

6. ([G/H], p. 42). Which of the following differential equations (considered as systems in $\mathbf{R}^2$) are structurally stable? Why or why not?

   (a) $\ddot{x} + 2\dot{x} + x = 0$
   (b) $\ddot{x} + \dot{x} + x^3 = 0$
   (c) $\ddot{x} + \sin x = 0$
   (d) $\ddot{x} + \dot{x}^2 + x = 0$.

7. Construct the (local) phase portrait for the system
$$\dot{x} = -y + xy$$
$$\dot{y} = x + (x^2 - y^2)/2$$
and show that it is structurally unstable.

8. Determine the nonwandering set Ω for the systems in Example 2 and Example 3 (for $\mu < 0$, $\mu = 0$, and $\mu > 0$).

9. Describe the nonwandering set on the Poincaré sphere for the global phase portraits in Problem 10 of Section 3.10 of Chapter 3.

10. Describe the nonwandering set for the phase portraits shown in Figure 5 below:

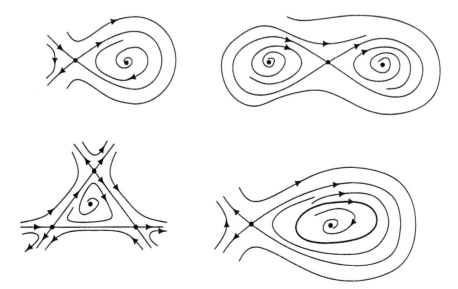

Figure 5. Some planar phase portraits with graphics.

4.2 Bifurcations at Nonhyperbolic Equilibrium Points

At the beginning of this chapter we mentioned that the qualitative behavior of the solution set of a system

$$\dot{\mathbf{x}} = \mathbf{f}(\mathbf{x}, \mu), \tag{1}$$

depending on a parameter $\mu \in \mathbf{R}$, changes as the vector field $\mathbf{f}$ passes through a point in the bifurcation set or as the parameter μ varies through a bifurcation value μ_0. A value μ_0 of the parameter μ in equation (1) for which the C^1-vector field $\mathbf{f}(\mathbf{x}, \mu_0)$ is not structurally stable is called a *bifurcation value*. We shall assume throughout this section that $\mathbf{f} \in C^1(E \times J)$ where E is an open set in $\mathbf{R}^n$ and $J \subset \mathbf{R}$ is an interval.

We begin our study of bifurcations of vector fields with the simplest kinds of bifurcations that occur in dynamical systems; namely, bifurcations at nonhyperbolic equilibrium points. In fact, we begin with a discussion of various types of critical points of one-dimensional systems

$$\dot{x} = f(x, \mu) \tag{1'}$$

with $x \in \mathbf{R}$ and $\mu \in \mathbf{R}$. The three simplest types of bifurcations that occur at a nonhyperbolic critical point of (1') are illustrated in the following examples.

Example 1. Consider the one-dimensional system

$$\dot{x} = \mu - x^2.$$

For $\mu > 0$ there are two critical points at $x = \pm\sqrt{\mu}$; $Df(x, \mu) = -2x$, $Df(\pm\sqrt{\mu}, \mu) = \mp 2\sqrt{\mu}$; and we see that the critical point at $x = \sqrt{\mu}$ is stable while the critical point at $x = -\sqrt{\mu}$ is unstable. (We continue to use D for the derivative with respect to x and the symbol $Df(x, \mu)$ will stand for the partial derivative of the function $f(x, \mu)$ with respect to x.) For $\mu = 0$, there is only one critical point at $x = 0$ and it is a nonhyperbolic critical point since $Df(0, 0) = 0$; the vector field $f(x) = -x^2$ is structurally unstable; and $\mu = 0$ is a bifurcation value. For $\mu < 0$ there are no critical points. The phase portraits for this differential equation are shown in Figure 1. For $\mu > 0$ the one-dimensional stable and unstable manifolds for the differential equation in Example 1 are given by $W^s(\sqrt{\mu}) = (-\sqrt{\mu}, \infty)$ and $W^u(-\sqrt{\mu}) = (-\infty, \sqrt{\mu})$. And for $\mu = 0$ the one-dimensional center manifold is given by $W^c(0) = (-\infty, \infty)$. All of the pertinent information concerning the bifurcation that takes place in this system at $\mu = 0$ is captured in the bifurcation diagram shown in Figure 2. The curve $\mu - x^2 = 0$ determines the position of the critical points of the system, a solid curve is used to indicate a family of stable critical points while a dashed curve is used to indicate a family of unstable critical points. This type of bifurcation is called a *saddle-node bifurcation*.

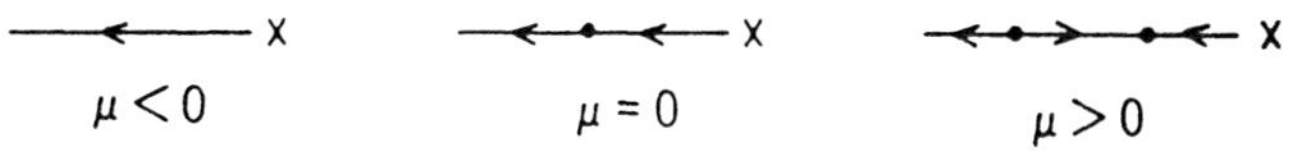

4.2. Bifurcations at Nonhyperbolic Equilibrium Points

Figure 1. The phase portraits for the differential equation in Example 1.

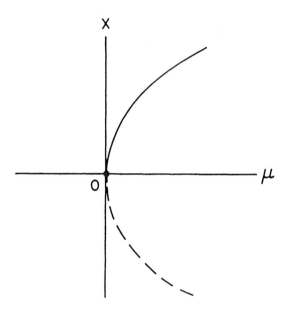

Figure 2. The bifurcation diagram for the saddle-node bifurcation in Example 1.

Example 2. Consider the one-dimensional system

$$\dot{x} = \mu x - x^2.$$

The critical points are at $x = 0$ and $x = \mu$. For $\mu = 0$ there is only one critical point at $x = 0$ and it is nonhyperbolic since $Df(0,0) = 0$; the vector field $f(x) = -x^2$ is structurally unstable; and $\mu = 0$ is a bifurcation value. The phase portraits for this differential equation are shown in Figure 3. For $\mu = 0$ we have $W^c(0) = (-\infty, \infty)$; the bifurcation diagram is shown in Figure 4. We see that there is an exchange of stability that takes place at the critical points of this system at the bifurcation value $\mu = 0$. This type of bifurcation is called a *transcritical bifurcation*.

336   4. Nonlinear Systems: Bifurcation Theory

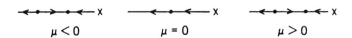

Figure 3. The phase portraits for the differential equation in Example 2.

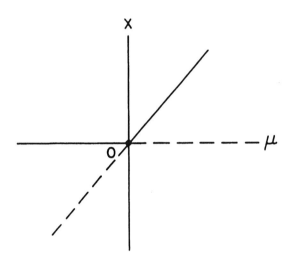

Figure 4. The bifurcation diagram for the transcritical bifurcation in Example 2.

Example 3. Consider the one-dimensional system

$$\dot{x} = \mu x - x^3.$$

For $\mu > 0$ there are critical points at $x = 0$ and at $x = \pm\sqrt{\mu}$. For $\mu \leq 0$, $x = 0$ is the only critical point. For $\mu = 0$ there is a nonhyperbolic critical point at $x = 0$ since $Df(0,0) = 0$; the vector field $f(x) = -x^3$ is structurally unstable; and $\mu = 0$ is a bifurcation value. The phase portraits are shown in Figure 5. For $\mu < 0$ we have $W^s(0) = (-\infty, \infty)$; however, for $\mu = 0$ we have $W^s(0) = \emptyset$ and $W^c(0) = (-\infty, \infty)$. The bifurcation diagram is shown in Figure 6 and this type of bifurcation is aptly called a *pitchfork bifurcation*.

Figure 5. The phase portraits for the differential equation in Example 3.

4.2. Bifurcations at Nonhyperbolic Equilibrium Points

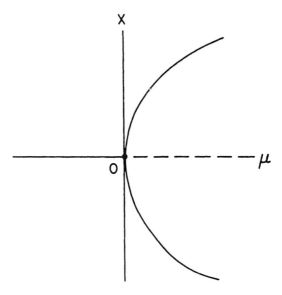

Figure 6. The bifurcation diagram for the pitchfork bifurcation in Example 3.

While the saddle-node, transcritical and pitchfork bifurcations in Examples 1–3 illustrate the most important types of bifurcations that occur in one-dimensional systems, there are certainly many other types of bifurcations that are possible in one-dimensional systems; cf., e.g., Problems 1 and 2. If $Df(0,0) = \cdots = D^{(m-1)}f(0,0) = 0$ and $D^m f(0,0) \neq 0$, then the one-dimensional system (1′) is said to have a critical point of multiplicity m at $x = 0$. In this case, at most m critical points can be made to bifurcate from the origin and there is a bifurcation which causes exactly m critical points to bifurcate from the origin. At the bifurcation value $\mu = 0$, the origin is a critical point of multiplicity two in Examples 1 and 2; it is a critical point of multiplicity three in Example 3; and it is a critical point of multiplicity four in Problem 1.

If $f(x_0, \mu_0) = Df(x_0, \mu_0) = 0$, then x_0 is a nonhyperbolic critical point of the system (1′) with $\mu = \mu_0$ and μ_0 is a bifurcation value of the system (1′). In this case, the type of bifurcation that occurs at the critical point $x = x_0$ at the bifurcation value $\mu = \mu_0$ in the one-dimensional system (1′) is determined by which of the higher order derivatives

$$\frac{\partial^m f(x_0, \mu_0)}{\partial x^j \partial \mu^k},$$

with $m \geq 2$, vanishes. This is also true in a sense for higher dimensional systems (1) and we have the following theorem proved by Sotomayor in 1976; cf. [G/H], p. 148. It is assumed that the function $\mathbf{f}(\mathbf{x}, \mu)$ is sufficiently differentiable so that all of the derivatives appearing in that theorem are continuous on $\mathbf{R}^n \times \mathbf{R}$. We use $D\mathbf{f}$ to denote the matrix of partial derivatives of the components of $\mathbf{f}$ with respect to the components of $\mathbf{x}$ and $\mathbf{f}_\mu$ to denote the vector of partial derivatives of the components of $\mathbf{f}$ with respect to the scalar μ.

Theorem 1 (Sotomayor). *Suppose that* $\mathbf{f}(\mathbf{x}_0, \mu_0) = \mathbf{0}$ *and that the* $n \times n$ *matrix* $A \equiv D\mathbf{f}(\mathbf{x}_0, \mu_0)$ *has a simple eigenvalue* $\lambda = 0$ *with eigenvector* $\mathbf{v}$ *and that* A^T *has an eigenvector* $\mathbf{w}$ *corresponding to the eigenvalue* $\lambda = 0$. *Furthermore, suppose that* A *has* k *eigenvalues with negative real part and* $(n - k - 1)$ *eigenvalues with positive real part and that the following conditions are satisfied*

$$\mathbf{w}^T \mathbf{f}_\mu(\mathbf{x}_0, \mu_0) \neq 0, \quad \mathbf{w}^T[D^2\mathbf{f}(\mathbf{x}_0, \mu_0)(\mathbf{v}, \mathbf{v})] \neq 0. \tag{2}$$

Then there is a smooth curve of equilibrium points of (1) *in* $\mathbf{R}^n \times \mathbf{R}$ *passing through* $(\mathbf{x}_0, \mu_0)$ *and tangent to the hyperplane* $\mathbf{R}^n \times \{\mu_0\}$. *Depending on the signs of the expressions in* (2), *there are no equilibrium points of* (1) *near* $\mathbf{x}_0$ *when* $\mu < \mu_0$ (*or when* $\mu > \mu_0$) *and there are two equilibrium points of* (1) *near* $\mathbf{x}_0$ *when* $\mu > \mu_0$ (*or when* $\mu < \mu_0$). *The two equilibrium points of* (1) *near* $\mathbf{x}_0$ *are hyperbolic and have stable manifolds of dimensions* k *and* $k + 1$ *respectively; i.e., the system* (1) *experiences a saddle-node bifurcation at the equilibrium point* $\mathbf{x}_0$ *as the parameter* μ *passes through the bifurcation value* $\mu = \mu_0$. *The set of* C^∞-*vector fields satisfying the above condition is an open, dense subset in the Banach space of all* C^∞, *one-parameter, vector fields with an equilibrium point at* $\mathbf{x}_0$ *having a simple zero eigenvalue.*

The bifurcation diagram for the saddle-node bifurcation in Theorem 1 is given by the one shown in Figure 2 with the x-axis in the direction of the eigenvector $\mathbf{v}$. (Actually, the diagram in Figure 2 might have to be rotated about the x or μ axes or both in order to obtain the correct bifurcation diagram for Theorem 1.)

If the conditions (2) are changed to

$$\begin{aligned} \mathbf{w}^T \mathbf{f}_\mu(\mathbf{x}_0, \mu_0) &= 0, \\ \mathbf{w}^T[D\mathbf{f}_\mu(\mathbf{x}_0, \mu_0)\mathbf{v}] &\neq 0 \quad \text{and} \\ \mathbf{w}^T[D^2\mathbf{f}(\mathbf{x}_0, \mu_0)(\mathbf{v}, \mathbf{v})] &\neq 0, \end{aligned} \tag{3}$$

then the system (1) experiences a transcritical bifurcation at the equilibrium point $\mathbf{x}_0$ as the parameter μ varies through the bifurcation value

4.2. Bifurcations at Nonhyperbolic Equilibrium Points

$\mu = \mu_0$ and the bifurcation diagram is given by Figure 4 with the x-axis in the direction of the eigenvector $\mathbf{v}$. And if the conditions (2) are changed to

$$\mathbf{w}^T \mathbf{f}_\mu(\mathbf{x}_0, \mu_0) = 0, \quad \mathbf{w}^T[D\mathbf{f}_\mu(\mathbf{x}_0, \mu_0)\mathbf{v}] \neq 0,$$
$$\mathbf{w}^T[D^2\mathbf{f}(\mathbf{x}_0, \mu_0)(\mathbf{v}, \mathbf{v})] = 0 \text{ and } \mathbf{w}^T[D^3\mathbf{f}(\mathbf{x}_0, \mu_0)(\mathbf{v}, \mathbf{v}, \mathbf{v})] \neq 0, \tag{1}$$

then the system (1) experiences a pitchfork bifurcation at the equilibrium point $\mathbf{x}_0$ as the parameter μ varies through the bifurcation value $\mu = \mu_0$ and the bifurcation diagram is given by Figure 6 with the x-axis in the direction of the eigenvector $\mathbf{v}$.

Sotomayor's theorem also establishes that in the class of C^∞, one-parameter, vector fields with an equilibrium point having one zero eigenvalue, the saddle-node bifurcations are generic in the sense that any such vector field can be perturbed to a saddle-node bifurcation. Transcritical and pitchfork bifurcations are not generic in this sense and further conditions on the one-parameter family of vector fields $\mathbf{f}(\mathbf{x}, \mu)$ are required before (1) can experience these types of bifurcations.

We next present some examples of saddle-node, transcritical and pitchfork bifurcations at nonhyperbolic critical points of planar systems which, once again, illustrate that the qualitative behavior near a nonhyperbolic critical point is determined by the behavior of the system on the center manifold; cf. Examples 4–6 and Examples 1–3 above.

Example 4. Consider the planar system

$$\dot{x} = \mu - x^2$$
$$\dot{y} = -y.$$

In the notation of Theorem 1, we have

$$A = D\mathbf{f}(\mathbf{0}, 0) = \begin{bmatrix} 0 & 0 \\ 0 & -1 \end{bmatrix}$$

$$\mathbf{f}_\mu(\mathbf{0}, 0) = \begin{pmatrix} 1 \\ 0 \end{pmatrix}$$

$\mathbf{v} = \mathbf{w} = (1, 0)^T$, $\mathbf{w}^T\mathbf{f}_\mu(\mathbf{0}, 0) = 1$ and $\mathbf{w}^T[D^2\mathbf{f}(\mathbf{0}, 0)(\mathbf{v}, \mathbf{v})] = -2$. There is a saddle-node bifurcation at the nonhyperbolic critical point $(0, 0)$ at the bifurcation value $\mu = 0$. For $\mu < 0$ there are no critical points. For $\mu = 0$ there is a critical point at the origin and, according to Theorem 1 in Section 2.11 of Chapter 2, it is a saddle-node. For $\mu > 0$ there are two critical points at $(\pm\sqrt{\mu}, 0)$; $(\sqrt{\mu}, 0)$ is a stable node and $(-\sqrt{\mu}, 0)$ is a saddle. The phase portraits for this system are shown in Figure 7 and the bifurcation diagram is the same as the one in Figure 2. Note that the

340   4. Nonlinear Systems: Bifurcation Theory

x-axis is in the $\mathbf{v}$ direction and that it is an analytic center manifold of the nonhyperbolic critical point $\mathbf{0}$ for $\mu = 0$.

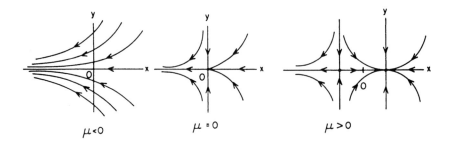

Figure 7. The phase portraits for the system in Example 4.

Remark. It follows from Lemma 2 in Section 3.12 that the index of a closed curve C relative to a vector field $\mathbf{f}$ (where $\mathbf{f}$ has no critical points on C) is preserved under small perturbations of the vector field. For example, we see that for sufficiently small μ, the index of a closed curve containing the origin on its interior is zero for any of the vector fields shown in Figure 7.

Example 5. Consider the planar system

$$\dot{x} = \mu x - x^2$$
$$\dot{y} = -y$$

which satisfies the conditions (3). There is a transcritical bifurcation at the origin at the bifurcation value $\mu = 0$. There are critical points at the origin and at $(\mu, 0)$. The phase portraits for this system are shown in Figure 8. The bifurcation diagram for this example is the same as the one in Figure 4.

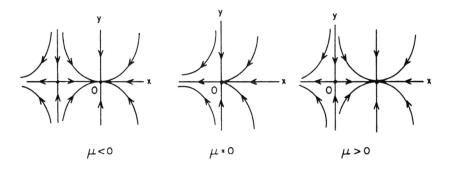

Figure 8. The phase portraits for the system in Example 5.

4.2. Bifurcations at Nonhyperbolic Equilibrium Points

Example 6. The system

$$\dot{x} = \mu x - x^3$$
$$\dot{y} = -y$$

satisfies the conditions (4) and there is a pitchfork bifurcation at $x_0 = 0$ and $\mu_0 = 0$. For $\mu \leq 0$ the only critical point is at the origin and for $\mu > 0$ there are critical points at the origin and at $(\pm\sqrt{\mu}, 0)$. For $\mu = 0$, the nonhyperbolic critical point at the origin is a node according to Theorem 1 in Section 2.11 of Chapter 2. The phase portraits for this system are shown in Figure 9 and the bifurcation diagram for this example is the same as the one shown in Figure 6.

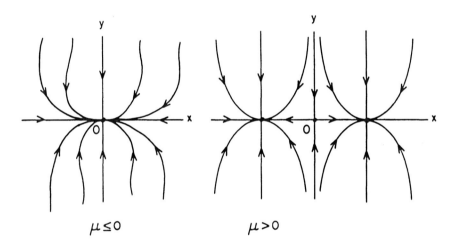

Figure 9. The phase portraits for the systems in Example 6.

Just as in the case of one-dimensional systems, we can have equilibrium points of multiplicity m for higher dimensional systems. An equilibrium point is of multiplicity m if any perturbation produces at most m nearby equilibrium points and if there is a perturbation which produces exactly m nearby equilibrium points. This is discussed in detail for planar systems in Chapter VIII of [A–II]. The origin in Examples 4 and 5 is a critical point of multiplicity two; it is of multiplicity three in Example 6; and it is of multiplicity four in Problem 7.

Problem Set 2

1. Consider the one-dimensional system

$$\dot{x} = -x^4 + 5\mu x^2 - 4\mu^2.$$

Determine the critical points and the bifurcation value for this differential equation. Draw the phase portraits for various values of the parameter μ and draw the bifurcation diagram.

2. Carry out the same analysis as in Problem 1 for the one-dimensional system
$$\dot{x} = x^2 - x\mu^2.$$

3. Define the function
$$f(x) = \begin{cases} x^3 \sin \dfrac{1}{x} & \text{for } x \neq 0 \\ 0 & \text{for } x = 0 \end{cases}$$

Show that $f \in C^1(\mathbf{R})$. Consider the one-dimensional system
$$\dot{x} = f(x) - \mu$$
with f defined above.

(a) Show that for $\mu = 0$ there are an infinite number of critical points in any neighborhood of the origin, that the nonzero critical points are hyperbolic and alternate in stability, and that the origin is a nonhyperbolic critical point.

(b) Show that $\mu = 0$ is a bifurcation value.

(c) Draw a bifurcation diagram and show that there are an infinite number of bifurcation values which accumulate at $\mu = 0$. What type of bifurcations occur at the nonzero bifurcation values?

4. Verify that the conditions (3) are satisfied by the system in Example 5. What are the dimensions of the various stable, unstable and center manifolds that occur in this system?

5. Verify that the conditions (4) are satisfied by the system in Example 6. What are the dimensions of the various stable, unstable and center manifolds that occur in this system?

6. If f satisfies the conditions of Theorem 1, what are the dimensions of the stable and unstable manifolds at the two hyperbolic critical points that occur near x_0 for $\mu > \mu_0$ (or $\mu < \mu_0$)? What are the dimensions if the conditions (2) are changed to (3) or (4)? Cf. Problems 4 and 5.

7. Consider the two-dimensional system
$$\dot{x} = -x^4 + 5\mu x^2 - 4\mu^2$$
$$\dot{y} = -y.$$

Determine the critical points and the bifurcation diagram for this system. Draw the phase portraits for various values of μ and draw the bifurcation diagram. Cf. Problem 1.

4.3 Higher Codimension Bifurcations at Nonhyperbolic Equilibrium Points

Let us continue our discussion of bifurcations at nonhyperbolic critical points and consider systems

$$\dot{\mathbf{x}} = \mathbf{f}(\mathbf{x}, \boldsymbol{\mu}), \qquad (1)$$

which depend on one or more parameters $\boldsymbol{\mu} \in \mathbf{R}^m$. The system (1) has a nonhyperbolic critical point $\mathbf{x}_0 \in \mathbf{R}^n$ for $\boldsymbol{\mu} = \boldsymbol{\mu}_0 \in \mathbf{R}^m$ iff $\mathbf{f}(\mathbf{x}_0, \boldsymbol{\mu}_0) = 0$ and the $n \times n$ matrix $A \equiv D\mathbf{f}(\mathbf{x}_0, \boldsymbol{\mu}_0)$ has at least one eigenvalue with zero real part. We continue our discussion of the simplest case when the matrix A has exactly one zero eigenvalue and relegate a discussion of the cases when A has a pair of pure imaginary eigenvalues or a pair of zero eigenvalues to the next section and to Section 4.13, respectively. The case when A has exactly one zero eigenvalue (and no other eigenvalues with zero real parts) is the simplest case to study since the behavior of the system (1) for $\boldsymbol{\mu}$ near the bifurcation value $\boldsymbol{\mu}_0$ is completely determined by the behavior of the associated one-dimensional system

$$\dot{x} = F(x, \mu) \qquad (2)$$

on the center manifold for μ near μ_0. Cf. Examples 1–6 in the previous section. On the other hand, a more in-depth study of this case will allow us to illustrate some ideas and terminology that are basic to an understanding of bifurcation theory. In particular, we shall use examples of single zero eigenvalue bifurcations to gain an understanding of the concepts of the codimension and the universal unfolding of a bifurcation.

If a structurally unstable vector field $\mathbf{f}_0(\mathbf{x})$ is embedded in an m-parameter family of vector fields (1) with $\mathbf{f}(\mathbf{x}, \boldsymbol{\mu}_0) = \mathbf{f}_0(\mathbf{x})$, then the m-parameter family of vector fields is called an *unfolding* of the vector field $\mathbf{f}_0(\mathbf{x})$ and (1) is called a *universal unfolding* of $\mathbf{f}_0(\mathbf{x})$ at a nonhyperbolic critical point $\mathbf{x}_0$ if it is an unfolding of $\mathbf{f}_0(\mathbf{x})$ and if every other unfolding of $\mathbf{f}_0(\mathbf{x})$ is topologically equivalent to a family of vector fields induced from (1), in a neighborhood of $\mathbf{x}_0$. The minimum number of parameters necessary for (1) to be a universal unfolding of the vector field $\mathbf{f}_0(\mathbf{x})$ at a nonhyperbolic critical point $\mathbf{x}_0$ is called the *codimension* of the bifurcation at $\mathbf{x}_0$. Cf. p. 123 in [G/H] and pp. 284–286 in [Wi-II], where we see that if M is a manifold in some infinite-dimensional vector space or Banach space B, then the codimension of M is the smallest dimension of a submanifold $N \subset B$ that intersects M transversally. Thus, if S is the set of all structurally stable vector fields in $B \equiv C^1(E)$ and $\mathbf{f}_0 \in S^c$ (the complement of S), then $\mathbf{f}_0$ belongs to the bifurcation set in $C^1(E)$ that is locally isomorphic to a manifold M in B and the codimension of the bifurcation that occurs at $\mathbf{f}_0$ is equal to the codimension of the manifold M. We illustrate these ideas by returning to the saddle-node and pitch-fork bifurcations studied in the previous section.

There is no loss of generality in assuming that $\mathbf{x}_0 = 0$ and that $\mu_0 = 0$, and this assumption will be made throughout the remainder of this section.

For the saddle-node bifurcation, the one-dimensional system (2) has the normal form $F(x, 0) = F_0(x) = ax^2$, and the constant a can be made equal to -1 by rescaling the time; i.e., we shall consider unfoldings of the normal form
$$\dot{x} = -x^2. \tag{3}$$
First of all, note that adding higher degree x-terms to (3) does not affect the behavior of the critical point at the origin; e.g., the system $\dot{x} = -x^2 + \mu_3 x^3$ has critical points at $x = 0$ and at $x = 1/\mu_3$; $x = 0$ is a nonhyperbolic critical point, and the hyperbolic critical point $x = 1/\mu_3 \to \infty$ as $\mu_3 \to 0$. Thus it suffices to consider unfoldings of (3) of the form
$$\dot{x} = \mu_1 + \mu_2 x - x^2.$$
Cf. [Wi-II], pp. 263 and 280. Furthermore, by translating the origin of this system to the point $x = \mu_2/2$, we obtain the system
$$\dot{x} = \mu - x^2 \tag{4}$$
with $\mu = \mu_1 + \mu_2^2/4$. Thus, all possible types of qualitative dynamical behavior that can occur in an unfolding of (3) are captured in (4), and the one-parameter family of vector fields (4) is a universal unfolding of the vector field (3) at the nonhyperbolic critical point at the origin. Cf. [Wi-II], pp. 280 and 300. Thus, all possible types of dynamical behavior for systems near (3) are exhibited in Figure 1 of the previous section, and the saddle-node bifurcation described in Figure 1 of Section 4.1 is a codimension-one bifurcation.

For the pitch-fork bifurcation, the one-dimensional system (2) has the normal form (after rescaling time) $F(x, 0) = F_0(x) = -x^3$, and we consider unfoldings of
$$\dot{x} = -x^3. \tag{5}$$
As we shall see, the one-parameter family of vector fields considered in Example 3 of the previous section is not a universal unfolding of the vector field (5) at the nonhyperbolic critical point at the origin. As in the case of the saddle-node bifurcation, we need not consider higher degree terms (of degree greater than three) in (5), and, by translating the origin, we can eliminate any second-degree terms. Therefore, a likely candidate for a universal unfolding of the vector field (5) at the nonhyperbolic critical point at the origin is the two-parameter family of vector fields
$$\dot{x} = \mu_1 + \mu_2 x - x^3. \tag{6}$$
And this is indeed the case; cf. [Wi-II], p. 301 or [G/H], p. 356; i.e., the pitch-fork bifurcation is a codimension-two bifurcation. In order to investigate the various types of dynamical behavior that occur in the system (6), we note that for $\mu_2 > 0$ the cubic equation
$$x^3 - \mu_2 x - \mu_1 = 0$$

4.3. Higher Codimension Bifurcations

has three roots iff $\mu_1^2 < 4\mu_2^3/27$, two roots (at $x = \pm\sqrt{\mu_2/3}$) iff $\mu_1^2 = 4\mu_2^3/27$ and one root if $\mu_1^2 > 4\mu_2^3/27$; it also has one root for all $\mu_2 \leq 0$ and $\mu_1 \in \mathbf{R}$. It then is easy to deduce the various types of dynamical behavior of (6) shown in Figure 1 below for $\mu_2 > 0$.

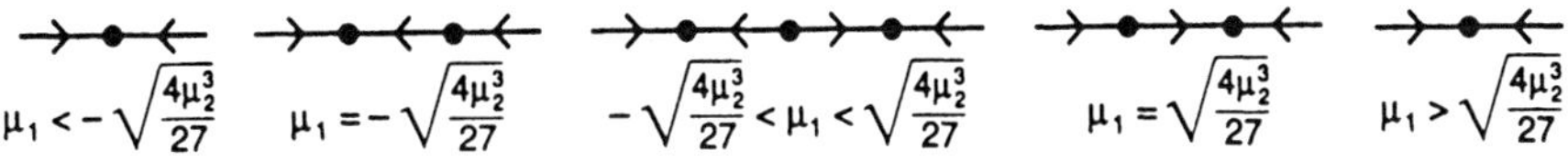

$$\mu_1 < -\sqrt{\frac{4\mu_2^3}{27}} \qquad \mu_1 = -\sqrt{\frac{4\mu_2^3}{27}} \qquad -\sqrt{\frac{4\mu_2^3}{27}} < \mu_1 < \sqrt{\frac{4\mu_2^3}{27}} \qquad \mu_1 = \sqrt{\frac{4\mu_2^3}{27}} \qquad \mu_1 > \sqrt{\frac{4\mu_2^3}{27}}$$

Figure 1. The phase portraits for the differential equation (6) with $\mu_2 > 0$.

The first three phase portraits in Figure 1 include all of the possible types of qualitative behavior for the differential equation (6) as well as the qualitative behavior for $\mu_2 \leq 0$ and $\mu_1 \in \mathbf{R}$, which is described by the first phase portrait in Figure 1. Notice that the second phase portrait in Figure 1 does not appear in the list of phase portraits in Figure 5 of the previous section for the unfolding of the normal form (5) given by the one-parameter family of differential equations in Example 3 of the previous section. Clearly, that unfolding of (5) is not a universal unfolding; i.e., the pitch-fork bifurcation is a codimension-two and not a codimension-one bifurcation. The bifurcation diagram for the vector field (6), i.e., the locus of points satisfying $\mu_1 + \mu_2 x - x^3 = 0$, which determines the location of the critical points of (6) for various values of the parameter $\mu \in \mathbf{R}^2$, is shown in Figure 2; cf. Figure 12.1, p. 169 in [G/S]. The bifurcation set or the set of bifurcation points in the μ-plane, i.e., the set of μ-points where (6) is structurally unstable, is also shown in Figure 2; it is the projection of the bifurcation points in the bifurcation diagram onto the μ-plane. The bifurcation set (in the μ-plane) consists of the two curves,

$$\mu_1 = \pm\sqrt{\frac{4\mu_2^3}{27}}$$

for $\mu_2 > 0$, at which points (6) undergoes a saddle-node bifurcation, and the origin. The two curves of saddle-node bifurcation points intersect in a cusp at the origin in the μ-plane, and the differential equation (6) is said to have a *cusp bifurcation* at $\mu = 0$. Notice that for parameter values μ in the shaded region in Figure 2, the system (6) has three hyperbolic critical points, and for point μ outside the closure of this region the system (6) has one hyperbolic critical point.

We conclude this section by describing the bifurcation set for universal unfolding of the normal form

$$\dot{x} = -x^4, \qquad (7)$$

which has a multiplicity-four critical point at the origin. As in the previous two examples, it seems that a likely candidate for the universal unfolding of

346								4. Nonlinear Systems: Bifurcation Theory

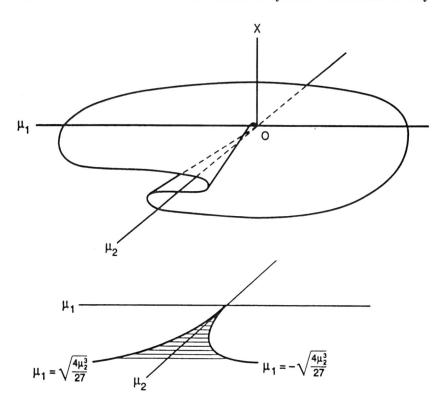

Figure 2. The bifurcation diagram and bifurcation set (in the μ-plane) for the differential equation (6).

(7) at the nonhyperbolic critical point at the origin is the three-parameter family of vector fields

$$\dot{x} = \mu_1 + \mu_2 x + \mu_3 x^2 - x^4. \tag{8}$$

And this is indeed the case; cf. [G/S], pp. 206–208. The bifurcation diagram for (8) is difficult to draw as it lies in $\mathbf{R}^4$; however, the bifurcation set for (8) in the three-dimensional parameter space is shown in Figure 3; cf. Figure 4.3, p. 208 in [G/S]. The shape of the bifurcation surface shown in Figure 3 gives this codimension-three bifurcation its name, the *swallow-tail bifurcation*.

The bifurcation set in $\mathbf{R}^3$ consists of two saddle-node bifurcation surfaces, SN_1 and SN_2, which intersect in two cusp bifurcation curves, C_1 and C_2, which intersect in a cusp at the origin. [The intersection of the surface SN_1 with itself describes the locus of μ-points where (8) has two distinct nonhyperbolic critical points at which saddle-node bifurcations occur.] For parameter values in the shaded region in Figure 3 between the saddle-node bifurcation surfaces, the differential equation (6) has four hyperbolic critical

4.3. Higher Codimension Bifurcations

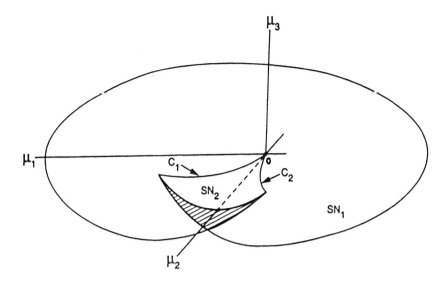

Figure 3. The bifurcation set (in three-dimensional parameter space) for the differential equation (8).

points: On SN_2 and on the part of SN_1 adjacent to the shaded region in Figure 3, it has two hyperbolic critical points and one nonhyperbolic critical point; on the remaining part of SN_1 and at the origin, it has one nonhyperbolic critical point; on the C_1 and C_2 curves, it has one hyperbolic and one nonhyperbolic critical point; for points above the surface shown in Figure 3, (8) has two hyperbolic critical points, and for points below this surface (8) has no critical points. The various phase portraits for the differential equation (8) are determined in Problem 1.

The examples discussed in this section were meant to illustrate the concepts of the codimension and universal unfolding of a structurally unstable vector field at a nonhyperbolic critical point. They also serve to illustrate the fact that for a single zero eigenvalue, a multiplicity m critical point results in a codimension-$(m-1)$ bifurcation. We close this section with an example that illustrates once again the power of the center manifold theory, not only in determining the qualitative behavior of a system at a nonhyperbolic critical point, but also in determining the possible types of qualitative dynamical behavior for nearby systems.

Example 1 (The Cusp Bifurcation for Planar Vector Fields). In light of the comments made earlier in this section, it is not surprising that the universal unfolding of the normal form

$$\dot{x} = -x^3$$
$$\dot{y} = -y$$

348   4. Nonlinear Systems: Bifurcation Theory

is given by the two-parameter family of vector fields

$$\dot{x} = \mu_1 + \mu_2 x - x^3$$
$$\dot{y} = -y. \tag{9}$$

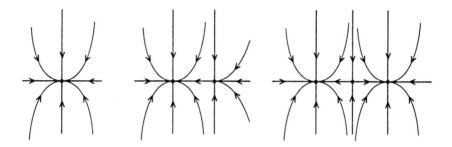

Figure 4. The phase portraits for the system (9) with $\mu_2 > 0$.

This system has a codimension-two, cusp bifurcation at $\mu = 0 \in \mathbf{R}^2$. The bifurcation diagram and the bifurcation set (in the μ-plane) are shown in Figure 2. The possible types of phase portraits for this system are shown in Figure 4, where we see that there is a saddle-node at the points $\mathbf{x} = (\pm\sqrt{\mu_2/3}, 0)$ for points on the saddle-node bifurcation curves $\mu_1 = \mp\sqrt{4\mu_2^3/27}$ for $\mu_2 > 0$; there is a saddle at the origin for μ in the shaded region in Figure 2, and the system (9) has exactly one critical point, a stable node, at any μ-point outside the closure of that region and also at $\mu = 0$ (in which case the origin of this system is a nonhyperbolic critical point). Notice that the middle phase portrait in Figure 4 does not appear in Figure 9 of the previous section.

PROBLEM SET 3

1. (a) Draw the phase portraits for the differential equation (8) with the parameter $\mu = (\mu_1, \mu_2, \mu_3)$ in the different regions of parameter space described in Figure 3.

   (b) Draw the phase portraits for the system

   $$\dot{x} = \mu_1 + \mu_2 x + \mu_3 x^2 - x^4$$
   $$\dot{y} = -y$$

   with the parameter $\mu = (\mu_1, \mu_2, \mu_3)$ in the different regions of parameter space described in Figure 3.

2. Determine a universal unfolding for the following system, and draw

the various types of phase portraits possible for systems near this system:

$$\dot{x} = xy$$
$$\dot{y} = y - x^2$$

Hint: Determine the flow on the center manifold as in Problem 3 in Section 2.12.

3. Same thing as in Problem 2 for the system

$$\dot{x} = x^2 - xy$$
$$\dot{y} = -y + x^2.$$

Hint: See Problem 4 in Section 2.12.

4. Same thing as in Problem 2 for the system

$$\dot{x} = x^2 y$$
$$\dot{y} = -y - x^2.$$

5. Same thing as in Problem 2 for the system

$$\dot{x} = -x^4$$
$$\dot{y} = -y.$$

6. Show that the universal unfolding of

$$\dot{x} = ax^2 + bxy + cy^2$$
$$\dot{y} = -y + dx^2 + exy + fy^2$$

(a) has a codimension-one saddle-node bifurcation at $\mu = 0$ if $a \neq 0$,

(b) has a codimension-two cusp bifurcation at $\mu = 0$ if $a = 0$ and $bd \neq 0$,

(c) has a codimension-three swallow-tail bifurcation at $\mu = 0$ if $a = b = 0$ and $cd \neq 0$.

Hint: See Problem 6 in Section 2.12.

4.4 Hopf Bifurcations and Bifurcations of Limit Cycles from a Multiple Focus

In the previous sections, we considered various types of bifurcations that can occur at a nonhyperbolic equilibrium point x_0 of a system

$$\dot{\mathbf{x}} = \mathbf{f}(\mathbf{x}, \mu) \tag{1}$$

depending on a parameter $\mu \in \mathbf{R}$ when the matrix $D\mathbf{f}(\mathbf{x}_0, \mu_0)$ had a simple zero eigenvalue. In particular, we saw that the saddle-node bifurcation was generic. In this section we consider various types of bifurcations that can occur when the matrix $D\mathbf{f}(\mathbf{x}_0, \mu_0)$ has a simple pair of pure imaginary eigenvalues and no other eigenvalues with zero real part. In this case, the implicit function theorem guarantees that for each μ near μ_0 there will be a unique equilibrium point $\mathbf{x}_\mu$ near $\mathbf{x}_0$; however, if the eigenvalues of $D\mathbf{f}(\mathbf{x}_\mu, \mu)$ cross the imaginary axis at $\mu = \mu_0$, then the dimensions of the stable and unstable manifolds of $\mathbf{x}_\mu$ will change and the local phase portrait of (1) will change as μ passes through the bifurcation value μ_0. In the generic case, a Hopf bifurcation occurs where a periodic orbit is created as the stability of the equilibrium point $\mathbf{x}_\mu$ changes. We illustrate this idea with a simple example and then present a general theory for planar systems. The reader should refer to [G/H], p. 150 or [Ru], p. 82 for the more general theory of Hopf bifurcations in higher dimensional systems which is summarized in Theorem 2 below. We also discuss other types of bifurcations for planar systems in this section where several limit cycles bifurcate from a critical point $\mathbf{x}_0$ when $D\mathbf{f}(\mathbf{x}_0, \mu_0)$ has a pair of pure imaginary eigenvalues.

Example 1 (A Hopf Bifurcation). Consider the planar system

$$\dot{x} = -y + x(\mu - x^2 - y^2)$$
$$\dot{y} = x + y(\mu - x^2 - y^2).$$

The only critical point is at the origin and

$$D\mathbf{f}(\mathbf{0}, \mu) = \begin{bmatrix} \mu & -1 \\ 1 & \mu \end{bmatrix}.$$

By Theorem 4 in Section 2.10 of Chapter 2, the origin is a stable or an unstable focus of this nonlinear system if $\mu < 0$ or if $\mu > 0$ respectively. For $\mu = 0$, $D\mathbf{f}(\mathbf{0}, 0)$ has a pair of pure imaginary eigenvalues and by Theorem 5 in Section 2.10 of Chapter 2 and Dulac's Theorem, the origin is either a center or a focus for this nonlinear system with $\mu = 0$. Actually, the structure of the phase portrait becomes apparent if we write this system in polar coordinates; cf. Example 2 in Section 2.2 of Chapter 2:

$$\dot{r} = r(\mu - r^2)$$
$$\dot{\theta} = 1.$$

We see that at $\mu = 0$ the origin is a stable focus and for $\mu > 0$ there is a stable limit cycle

$$\Gamma_\mu: \gamma_\mu(t) = \sqrt{\mu}(\cos t, \sin t)^T.$$

The curves Γ_μ represent a one-parameter family of limit cycles of this system. The phase portraits for this system are shown in Figure 1 and the bifurcation diagram is shown in Figure 2. The upper curve in the bifurcation

4.4. Hopf Bifurcations and Bifurcations of Limit Cycles

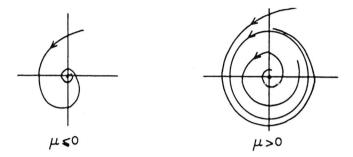

Figure 1. The phase portraits for the system in Example 1.

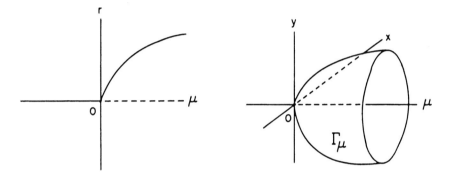

Figure 2. The bifurcation diagram and the one-parameter family of limit cycles Γ_μ resulting from the Hopf bifurcation in Example 1.

diagram shown in Figure 2 represents the one-parameter family of limit cycles Γ_μ which defines a surface in $\mathbf{R}^2 \times \mathbf{R}$; cf. Figure 2. The bifurcation of the limit cycle from the origin that occurs at the bifurcation value $\mu = 0$ as the origin changes its stability is referred to as a Hopf bifurcation.

Next, consider the planar analytic system

$$\begin{aligned} \dot{x} &= \mu x - y + p(x, y) \\ \dot{y} &= x + \mu y + q(x, y) \end{aligned} \tag{2}$$

where the analytic functions

$$p(x, y) = \sum_{i+j \geq 2} a_{ij} x^i y^j = (a_{20} x^2 + a_{11} xy + a_{02} y^2)$$
$$+ (a_{30} x^3 + a_{21} x^2 y + a_{12} xy^2 + a_{03} y^3) + \cdots$$

and

$$q(x,y) = \sum_{i+j\geq 2} b_{ij}x^iy^j = (b_{20}x^2 + b_{11}xy + b_{02}y^2)$$
$$+ (b_{30}x^3 + b_{21}x^2y + b_{12}xy^2 + b_{03}y^3) + \cdots.$$

In this case, for $\mu = 0$, $D\mathbf{f}(0,0)$ has a pair of pure imaginary eigenvalues and the origin is called a weak focus or a multiple focus. The multiplicity m of a multiple focus was defined in Section 3.4 of Chapter 3 in terms of the Poincaré map $P(s)$ for the focus. In particular, by Theorem 3 in Section 3.4 of Chapter 3, we have

$$P'(0) = e^{2\pi\mu}$$

for the system (2) and for $\mu = 0$ we have $P'(0) = 1$ or equivalently $d'(0) = 0$ where $d(s) = P(s) - s$ is the displacement function. For $\mu = 0$ in (2), the Liapunov number σ is given by equation (3) in Section 3.4 of Chapter 3 as

$$\sigma = \frac{3\pi}{2}[3(a_{30} + b_{03}) + (a_{12} + b_{21}) - 2(a_{20}b_{20} - a_{02}b_{02})$$
$$+ a_{11}(a_{02} + a_{20}) - b_{11}(b_{02} + b_{20})]. \tag{3}$$

In particular, if $\sigma \neq 0$ then the origin is a weak focus of multiplicity one, it is stable if $\sigma < 0$ and unstable if $\sigma > 0$, and a Hopf bifurcation occurs at the origin at the bifurcation value $\mu = 0$. The following theorem is proved in [A–II] on pp. 261–264.

Theorem 1 (The Hopf Bifurcation). *If $\sigma \neq 0$, then a Hopf bifurcation occurs at the origin of the planar analytic system (2) at the bifurcation value $\mu = 0$; in particular, if $\sigma < 0$, then a unique stable limit cycle bifurcates from the origin of (2) as μ increases from zero and if $\sigma > 0$, then a unique unstable limit cycle bifurcates from the origin of (2) as μ decreases from zero. If $\sigma < 0$, the local phase portraits for (2) are topologically equivalent to those shown in Figure 1 and there is a surface of periodic orbits which has a quadratic tangency with the (x,y)-plane at the origin in $\mathbf{R}^2 \times \mathbf{R}$; cf. Figure 2.*

In the first case ($\sigma < 0$) in Theorem 1 where the critical point generates a *stable* limit cycle as μ passes through the bifurcation value $\mu = 0$, we have what is called a *supercritical Hopf bifurcation* and in the second case ($\sigma > 0$) in Theorem 1 where the critical point generates an *unstable* limit cycle as μ passes through the bifurcation value $\mu = 0$, we have what is called a *subcritical Hopf bifurcation*.

Remark 1. For a general planar analytic system

$$\dot{x} = ax + by + p(x,y)$$
$$\dot{y} = cx + dy + q(x,y) \tag{2'}$$

with $\Delta = ad - bc > 0$, $a + d = 0$ and $p(x,y), q(x,y)$ given by the above series, the matrix

$$D\mathbf{f}(0) = \begin{bmatrix} a & b \\ c & d \end{bmatrix}$$

4.4. Hopf Bifurcations and Bifurcations of Limit Cycles

will have a pair of imaginary eigenvalues and the origin will be a weak focus; the Liapunov number σ is then given by the formula

$$\sigma = \frac{-3\pi}{2b\Delta^{3/2}}\{[ac(a_{11}^2 + a_{11}b_{02} + a_{02}b_{11}) + ab(b_{11}^2 + a_{20}b_{11} + a_{11}b_{02})$$
$$+ c^2(a_{11}a_{02} + 2a_{02}b_{02}) - 2ac(b_{02}^2 - a_{20}a_{02}) - 2ab(a_{20}^2 - b_{20}b_{02})$$
$$- b^2(2a_{20}b_{20} + b_{11}b_{20}) + (bc - 2a^2)(b_{11}b_{02} - a_{11}a_{20})]$$
$$- (a^2 + bc)[3(cb_{03} - ba_{30}) + 2a(a_{21} + b_{12}) + (ca_{12} - bb_{21})]\}. \quad (3')$$

Cf. [A–II], p. 253. For $\sigma \neq 0$ in equation $(3')$, Theorem 1 with $\mu = a + d$ also holds for the system $(2')$.

The addition of any higher degree terms to the linear system

$$\dot{x} = \mu x - y$$
$$\dot{y} = x + \mu y$$

will result in a Hopf bifurcation at the origin at the bifurcation value $\mu = 0$ provided the Liapunov number $\sigma \neq 0$. The hypothesis that **f** is analytic in Theorem 1 can be weakened to $\mathbf{f} \in C^3(E \times J)$ where E is an open subset of $\mathbf{R}^2$ containing the origin and $J \subset \mathbf{R}$ is an interval. For $\mathbf{f} \in C^1(E \times J)$, a one-parameter family of limit cycles is still generated at the origin at the bifurcation value $\mu = 0$, but the surface of periodic orbits will not necessarily be tangent to the (x,y)-plane at the origin; cf. Problem 2.

The following theorem, proved by E. Hopf in 1942, establishes the existence of the Hopf bifurcation for higher dimensional systems when $D\mathbf{f}(\mathbf{x}_0, \mu_0)$ has a pair of pure imaginary eigenvalues, λ_0 and $\bar{\lambda}_0$, and no other eigenvalues with zero real part; cf. [G/H], p. 151.

Theorem 2 (Hopf). *Suppose that the C^4-system (1) with $\mathbf{x} \in \mathbf{R}^n$ and $\mu \in \mathbf{R}$ has a critical point $\mathbf{x}_0$ for $\mu = \mu_0$ and that $D\mathbf{f}(\mathbf{x}_0, \mu_0)$ has a simple pair of pure imaginary eigenvalues and no other eigenvalues with zero real part. Then there is a smooth curve of equilibrium points $\mathbf{x}(\mu)$ with $\mathbf{x}(\mu_0) = \mathbf{x}_0$ and the eigenvalues, $\lambda(\mu)$ and $\bar{\lambda}(\mu)$ of $D\mathbf{f}(\mathbf{x}(\mu), \mu)$, which are pure imaginary at $\mu = \mu_0$, vary smoothly with μ. Furthermore, if*

$$\frac{d}{d\mu}[\mathrm{Re}\lambda(\mu)]_{\mu=\mu_0} \neq 0,$$

then there is a unique two-dimensional center manifold passing through the point $(\mathbf{x}_0, \mu_0)$ and a smooth transformation of coordinates such that the system (1) on the center manifold is transformed into the normal form

$$\dot{x} = -y + ax(x^2 + y^2) - by(x^2 + y^2) + O(|\mathbf{x}|^4)$$
$$\dot{y} = x + bx(x^2 + y^2) + ay(x^2 + y^2) + O(|\mathbf{x}|^4)$$

in a neighborhood of the origin which, for $a \neq 0$, has a weak focus of *multiplicity one at the origin and*

$$\dot{x} = \mu x - y + ax(x^2 + y^2) - by(x^2 + y^2)$$
$$\dot{y} = x + \mu y + bx(x^2 + y^2) + ay(x^2 + y^2)$$

is a universal unfolding of this normal form in a neighborhood of the origin on the center manifold.

In the higher dimensional case, if $a \neq 0$ there is a two-dimensional surface S of stable periodic orbits for $a < 0$ [or unstable periodic orbits for $a > 0$; cf. Problem 1(b)] which has a quadratic tangency with the eigenspace of λ_0 and $\bar{\lambda}_0$ at the point $(\mathbf{x}_0, \mu_0) \in \mathbf{R}^n \times \mathbf{R}$; i.e., the surface S is tangent to the center manifold $W^c(\mathbf{x}_0)$ of (1) at the nonhyperbolic equilibrium point $\mathbf{x}_0$ for $\mu = \mu_0$. Cf. Figure 3.

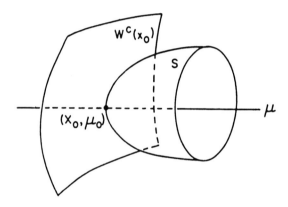

Figure 3. A one-parameter family of periodic orbits S resulting from a Hopf bifurcation at a nonhyperbolic equilibrium point $\mathbf{x}_0$ and a bifurcation value μ_0.

We next illustrate the use of formula (3) for the Liapunov number σ in determining whether a Hopf bifurcation is supercritical or subcritical.

Example 2. The quadratic system

$$\dot{x} = \mu x - y + x^2$$
$$\dot{y} = x + \mu y + x^2$$

has a weak focus of multiplicity one at the origin for $\mu = 0$ since by equation (3) the Liapunov number $\sigma = -3\pi \neq 0$. Furthermore, since $\sigma < 0$, it follows from Theorem 1 that a unique stable limit cycle bifurcates from the origin as the parameter μ increases through the bifurcation value $\mu = 0$; i.e. the Hopf bifurcation is supercritical. The limit cycle for this system at the parameter value $\mu = .1$ is shown in Figure 4. The bifurcation diagram is the same as the one shown in Figure 2.

4.4. Hopf Bifurcations and Bifurcations of Limit Cycles

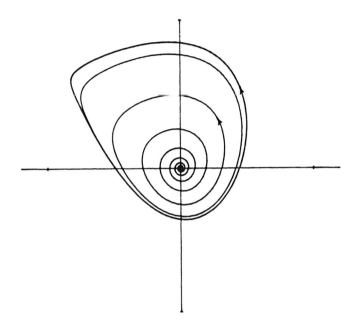

Figure 4. The limit cycle for the system in Example 2 with $\mu = .1$.

If $\mu = \sigma = 0$ in equation (2) where σ is given by equation (3), then the origin will be a weak focus of multiplicity $m > 1$ of the planar analytic system (2). The next theorem, proved in Chapter IX of [A–II] shows that at most m limit cycles can bifurcate from the origin as μ varies through the bifurcation value $\mu = 0$ and that there is an analytic perturbation of the vector field in

$$\dot{x} = -y + p(x, y)$$
$$\dot{y} = x + q(x, y)$$
(4)

which causes exactly m limit cycles to bifurcate from the origin at $\mu = 0$. In order to state this theorem, we need to extend the notion of the C^1-norm defined on the class of functions $C^1(E)$ to the C^k-norm defined on the class $C^k(E)$; i.e., for $\mathbf{f} \in C^k(E)$ where E is an open subset of $\mathbf{R}^n$, we define

$$\|\mathbf{f}\|_k = \sup_E |\mathbf{f}(\mathbf{x})| + \sup_E \|D\mathbf{f}(\mathbf{x})\| + \cdots + \sup_E \|D^k \mathbf{f}(\mathbf{x})\|$$

where for the norms $\|\cdot\|$ on the right-hand side of this equation we use

$$\|D^k \mathbf{f}(\mathbf{x})\| = \max \left| \frac{\partial^k \mathbf{f}(\mathbf{x})}{\partial x_{j_1} \cdots \partial x_{j_k}} \right|,$$

the maximum being taken over $j_1, \ldots, j_k = 1, \ldots, n$. Each of the spaces of functions in $C^k(E)$, bounded in the C^k-norm, is then a Banach space and $C^{k+1}(E) \subset C^k(E)$ for $k = 0, 1, 2, \ldots$.

Theorem 3 (The Bifurcation of Limit Cycles from a Multiple Focus). *If the origin is a multiple focus of multiplicity m of the analytic system (4) then for $k \geq 2m + 1$*

(i) *there is a $\delta > 0$ and an $\varepsilon > 0$ such that any system ε-close to (4) in the C^k-norm has at most m limit cycles in $N_\delta(0)$ and*

(ii) *for any $\delta > 0$ and $\varepsilon > 0$ there is an analytic system which is ε-close to (4) in the C^k-norm and has exactly m simple limit cycles in $N_\delta(0)$.*

In Theorem 5 in Section 3.8 of Chapter 3, we saw that the Lienard system

$$\dot{x} = y - [a_1 x + a_2 x^2 + \cdots + a_{2m+1} x^{2m+1}]$$
$$\dot{y} = -x$$

has at most m local limit cycles and that there are coefficients with $a_1, a_3, \ldots, a_{2m+1}$ alternating in sign such that this system has m local limit cycles. This type of system with $a_1 = \cdots = a_{2m} = 0$ and $a_{2m+1} \neq 0$ has a weak focus of multiplicity m at the origin. We consider one such system with $m = 2$ in the next example.

Example 3. Consider the system

$$\dot{x} = y - \varepsilon[\mu x + a_3 x^3 + a_5 x^5]$$
$$\dot{y} = -x$$

with $a_3 \leq 0, a_5 > 0$ and small $\varepsilon \neq 0$. By (3'), if $\mu = 0$ and $a_3 = 0$ then $\sigma = 0$. Therefore, the origin is a weak focus of multiplicity $m \geq 2$. And by Theorem 5 in Section 3.8 of Chapter 3, it is a weak focus of multiplicity $m \leq 2$. Thus, the origin is a weak focus of multiplicity $m = 2$. By Theorem 2, at most two limit cycles bifurcate from the origin as μ varies through the bifurcation value $\mu = 0$. To find coefficients a_3 and a_5 for which exactly two limit cycles bifurcate from the origin at $\mu = 0$, we use Theorem 6 in Section 3.8 of Chapter 3; cf. Example 4 in that section. We find that for $\mu > 0$ and for sufficiently small $\varepsilon \neq 0$ the system

$$\dot{x} = y - \varepsilon \left[\mu x - 4\mu^{1/2} x^3 + \frac{16}{5} x^5 \right]$$
$$\dot{y} = -x$$

has exactly two limit cycles around the origin that are asymptotic to circles of radius $r = \sqrt[4]{\mu}$ and $r = \sqrt[4]{\mu/4}$ as $\varepsilon \to 0$. For $\varepsilon = .01$, the limit cycles of this system are shown in Figure 5 for $\mu = .5$ and $\mu = 1$. The bifurcation diagram for this last system with small $\varepsilon \neq 0$ is shown in Figure 6.

A rich source of examples for planar systems having limit cycles are systems of the form

$$\dot{x} = -y + x\psi(r, \mu)$$
$$\dot{y} = x + y\psi(r, \mu)$$
(5)

4.4. Hopf Bifurcations and Bifurcations of Limit Cycles   357

where $r = \sqrt{x^2 + y^2}$. Many of Poincaré's examples in [P] are of this form. The bifurcation diagram, of curves representing one-parameter families of limit cycles, is given by the graph of the relation $\psi(r, \mu) = 0$ in the upper half of the (μ, r)-plane where $r > 0$. The system in Example 1 is of this form and we consider one other example of this type having a weak focus of multiplicity two at the origin.

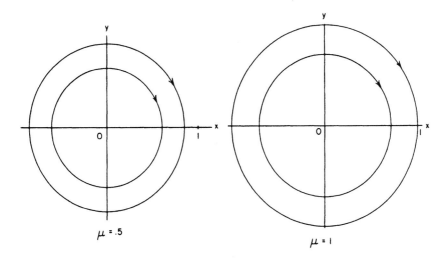

Figure 5. The limit cycles for the system in Example 3 with $\mu = .5$ and $\mu = 1$.

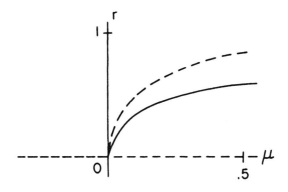

Figure 6. The bifurcation diagram for the limit cycles which bifurcate from the weak focus of multiplicity two of the system in Example 3 (with $\varepsilon < 0$, the stabilities being reversed for $\varepsilon > 0$).

Example 4. Consider the planar analytic system

$$\dot{x} = -y + x(\mu - r^2)(\mu - 2r^2)$$
$$\dot{y} = x + y(\mu - r^2)(\mu - 2r^2)$$

where $r^2 = x^2 + y^2$. According to the above comment, the bifurcation diagram is given by the curves $r = \sqrt{\mu}$ and $r = \sqrt{\mu/2}$ in the upper half of the (μ, r)-plane; cf. Figure 7. In fact, writing this system in polar coordinates shows explicitly that for $\mu > 0$ there are two limit cycles represented by $\gamma_1(t) = \sqrt{\mu}(\cos t, \sin t)^T$ and $\gamma_2(t) = \sqrt{\mu/2}(\cos t, \sin t)^T$. The inner limit cycle $\gamma_2(t)$ is stable and the outer limit cycle $\gamma_1(t)$ is unstable. The multiplicity of the weak focus at the origin of this system is two.

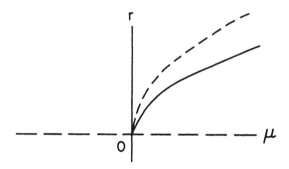

Figure 7. The bifurcation diagram for the system in Example 4.

As in the last two examples (and as in Theorem 5 in Section 3.8), we can obtain polynomial systems with a weak focus of arbitrarily large multiplicity; however, it is a much more delicate question to determine the maximum number of local limit cycles that are possible for a polynomial system of fixed degree. The answer to this question is currently known only for quadratic systems. Bautin [2] showed that a quadratic system can have at most three local limit cycles and that there exists a quadratic system with a weak focus of multiplicity three. The next theorem, established in [62], gives us a complete set of results for determining the Liapunov number, $W_k = d^{(2k+1)}(0)$, or the multiplicity of the weak focus at the origin for the quadratic system (6) below. We note that it follows from equation (3) that $\sigma = 3\pi W_1/2$ for the quadratic system (6); cf. Problem 7.

Theorem 4 (Li). *Consider the quadratic system*

$$\dot{x} = -y + ax^2 + bxy$$
$$\dot{y} = x + \ell x^2 + mxy + ny^2 \tag{6}$$

4.4. Hopf Bifurcations and Bifurcations of Limit Cycles

and define

$$W_1 = a(b - 2\ell) - m(\ell + n)$$
$$W_2 = a(2a + m)(3a - m)[a^2(b - 2\ell - n) + (\ell + n)^2(n - b)]$$
$$W_3 = a^7\ell(2a + m)(2\ell + n)[a^2(b - 2\ell - n) + (\ell + n)^2(n - b)].$$

Then the origin is

(i) a weak focus of multiplicity 1 iff $W_1 \neq 0$,

(ii) a weak focus of multiplicity 2 iff $W_1 = 0$ and $W_2 \neq 0$,

(iii) a weak focus of multiplicity 3 iff $W_1 = W_2 = 0$ and $W_3 \neq 0$, and

(iv) a center iff $W_1 = W_2 = W_3 = 0$.

Furthermore, if for $k = 1, 2$ or 3 the origin is a weak focus of multiplicity k and the Liapunov number $W_k < 0$ (or $W_k > 0$), then the origin of (6) is stable (or unstable).

Remark 2. It follows from Theorem 2 that the one-parameter family of vector fields in Example 1 is a universal unfolding of the normal form

$$\dot{x} = -y - x(x^2 + y^2)$$
$$\dot{y} = x - y(x^2 + y^2)$$

(cf. the normal form in Problem 1(b) with $a = -1$ and $b = 0$) whose linear part has a pair of pure imaginary eigenvalues $\pm i$ and which has $\sigma = -9\pi$ according to (3). Thus, the Hopf bifurcation described in Example 1 or in Theorem 2 is a codimension-one bifurcation. More generally, a bifurcation at a weak focus of multiplicity m is a codimension-m bifurcation. See Remark 4 at the end of Section 4.15.

Remark 3. The system in Example 1 defines a one-parameter family of (negatively) rotated vector fields with parameter $\mu \in \mathbf{R}$ (according to Definition 1 in Section 4.6). And according to Theorem 5 in Section 4.6, any one-parameter family of rotated vector fields can be used to obtain a universal unfolding of the normal form for a C^1 or polynomial system with a weak focus of multiplicity one.

PROBLEM SET 4

1. (a) Show that for $a + b \neq 0$ the system

$$\dot{x} = \mu x - y + a(x^2 + y^2)x - b(x^2 + y^2)y + 0(|\mathbf{x}|^4)$$
$$\dot{y} = x + \mu y + a(x^2 + y^2)x + b(x^2 + y^2)y + 0(|\mathbf{x}|^4)$$

has a Hopf bifurcation at the origin at the bifurcation value $\mu = 0$. Determine whether it is supercritical or subcritical.

(b) Show that for $a \neq 0$ the system in the last paragraph in Section 2.13 (or in Theorem 2 above),

$$\dot{x} = \mu x - y + a(x^2 + y^2)x - b(x^2 + y^2)y + 0(|\mathbf{x}|^4)$$
$$\dot{y} = x + \mu y + b(x^2 + y^2)x + a(x^2 + y^2)y + 0(|\mathbf{x}|^4),$$

has a Hopf bifurcation at the origin at the bifurcation value $\mu = 0$. Determine whether it is supercritical or subcritical. Note that for $0 < r \ll 1$ either one of the above systems defines a one-parameter family of (negatively) rotated vector fields with parameter μ (according to Definition 1 in Section 4.6). Cf. Problem 2(b) in Section 4.6.

2. Consider the C^1-system

$$\dot{x} = \mu x - y - x\sqrt{x^2 + y^2}$$
$$\dot{y} = x + \mu y - y\sqrt{x^2 + y^2}.$$

(a) Show that the vector field $\mathbf{f}$ defined by this system belongs to $C^1(\mathbf{R}^2 \times \mathbf{R})$; i.e., show that all of the first partial derivatives with respect to x, y and μ are continuous for all x, y and μ.

(b) Write this system in polar coordinates and show that for $\mu > 0$ there is a unique stable limit cycle around the origin and that for $\mu \leq 0$ there is no limit cycle around the origin. Sketch the phase portraits for these two cases.

(c) Draw the bifurcation diagram and sketch the conical surface generated by the one-parameter family of limit cycles of this system.

3. Write the differential equation

$$\ddot{x} + \mu\dot{x} + (x - x^3) = 0$$

as a planar system and use equation (3') to show that at the bifurcation value $\mu = 0$ the quantity $\sigma = 0$. Draw the phase portrait for the Hamiltonian system obtained by setting $\mu = 0$.

4. Write the system

$$\dot{x} = -y + x(\mu - r^2)(\mu - 2r^2)$$
$$\dot{y} = x + y(\mu - r^2)(\mu - 2r^2)$$

in polar coordinates and show that for $\mu > 0$ there are two limit cycles represented by $\gamma_1(t) = \sqrt{\mu}(\cos t, \sin t)^T$ and $\gamma_2(t) = \sqrt{\mu/2}(\cos t, \sin t)^T$. Draw the phase portraits for this system for $\mu \leq 0$ and $\mu > 0$.

4.4. Hopf Bifurcations and Bifurcations of Limit Cycles

5. Draw the bifurcation diagram in the (μ, r)-plane for the system
$$\dot{x} = -y + x(r^2 - \mu)(2r^2 - \mu)(3r^2 - \mu)$$
$$\dot{y} = x + y(r^2 - \mu)(2r^2 - \mu)(3r^2 - \mu).$$
What can you say about the multiplicity m of the weak focus at the origin of this system?

6. Use Theorem 6 in Section 3.8 of Chapter 3 to find coefficients $a_3(\mu)$, $a_5(\mu)$ and a_7 for which three limit cycles, asymptotic to circles of radii $r_1 = \mu^{1/6}/2$, $r_2 = \mu^{1/6}$ and $r_3 = \sqrt{2}\mu^{1/6}$ as $\varepsilon \to 0$, bifurcate from the origin of the Lienard system
$$\dot{x} = y - \varepsilon[\mu x + a_3 x^3 + a_5 x^5 + a_7 x^7]$$
$$\dot{y} = -x$$
at the bifurcation value $\mu = 0$.

7. Use equation (3) to show that $\sigma = 3\pi W_1/2$ for the system (6) in Theorem 4 with W_1 given by the formula in that theorem.

8. Consider the quadratic system
$$\dot{x} = \mu x - y + x^2 + xy$$
$$\dot{y} = x + \mu y + x^2 + mxy + ny^2.$$

   (a) For $\mu = 0$, derive a set of necessary and sufficient conditions for this system to have a weak focus of multiplicity one at the origin. If $m = 0$ or if $n = -1$, is the Hopf bifurcation at the origin supercritical or subcritical?

   (b) For $\mu = 0$, derive a set of necessary and sufficient conditions for this system to have a weak focus of multiplicity two at the origin. In this case, what happens as μ varies through the bifurcation value $\mu_0 = 0$? **Hint:** See Theorem 5 in Section 4.6. Also, see Problem 4(a) in Section 4.15.

   (c) For $\mu = 0$, show that there is exactly one point in the (m, n) plane for which this system has a weak focus of multiplicity three at the origin. In this case, what happens as μ varies through the bifurcation value $\mu_0 = 0$? (See the hint in part b.)

   (d) For $\mu = 0$, show that there are exactly three points in the (m, n) plane for which this system has a center. At which one of these points do we have a Hamiltonian system?

9. Use equation (3') to show that $\sigma = kF[cF^2 + (cF + 1)(F - E + 2c)]$, with the positive constant $k = 3\pi/\left[2|EF|\sqrt{|1 + EF|^3}\right]$, for the quadratic system
$$\dot{x} = -x + Ey + y^2$$
$$\dot{y} = Fx + y - xy + cy^2$$
with $1 + EF < 0$. (This result will be useful in doing some of the problems in Section 4.14).

4.5 Bifurcations at Nonhyperbolic Periodic Orbits

Several interesting types of bifurcations can take place at a nonhyperbolic periodic orbit; i.e., at a periodic orbit having two or more characteristic exponents with zero real part. As in Theorem 2 in Section 3.5 of Chapter 3, one of the characteristic exponents is always zero. In the simplest case of a nonhyperbolic periodic orbit when there is one other zero characteristic exponent, the periodic orbit Γ has a two-dimensional center manifold $W^c(\Gamma)$ and the simplest types of bifurcations that occur on this manifold are the saddle-node, transcritical and pitchfork bifurcations, the saddle-node bifurcation being generic. This is the case when the derivative of the Poincaré map, $D\mathbf{P}(\mathbf{x}_0)$, at a point $\mathbf{x}_0 \in \Gamma$, has one eigenvalue equal to one. If $D\mathbf{P}(\mathbf{x}_0)$ has one eigenvalue equal to -1, then generically a period-doubling bifurcation occurs which corresponds to a flip bifurcation for the Poincaré map. And if $D\mathbf{P}(\mathbf{x}_0)$ has a pair of complex conjugate eigenvalues on the unit circle, then generically Γ bifurcates into an invariant two-dimensional torus; this corresponds to a Hopf bifurcation for the Poincaré map. Cf. Chapter 2 in [Ru].

We shall consider C^1-systems

$$\dot{\mathbf{x}} = \mathbf{f}(\mathbf{x}, \mu) \tag{1}$$

depending on a parameter $\mu \in \mathbf{R}$ where $\mathbf{f} \in C^1(E \times J)$, E is an open subset in $\mathbf{R}^n$ and $J \subset \mathbf{R}$ is an interval. Let $\phi_t(\mathbf{x}, \mu)$ be the flow of the system (1) and assume that for $\mu = \mu_0$, the system (1) has a periodic orbit $\Gamma_0 \subset E$ given by $\mathbf{x} = \phi_t(\mathbf{x}_0, \mu_0)$. Let Σ be the hyperplane perpendicular to Γ_0 at point $\mathbf{x}_0 \in \Gamma_0$. Then, using the implicit function theorem as in Theorem 1 in Section 3.4 of Chapter 3, it can be shown that there is a C^1 function $\tau(\mathbf{x}, \mu)$ defined in a neighborhood $N_\delta(\mathbf{x}_0, \mu_0)$ of the point $(\mathbf{x}_0, \mu_0) \in E \times J$ such that

$$\phi_{\tau(\mathbf{x},\mu)}(\mathbf{x}, \mu) \in \Sigma$$

for all $(\mathbf{x}, \mu) \in N_\delta(\mathbf{x}_0, \mu_0)$. As in Section 3.4 of Chapter 3, it can be shown that for each $\mu \in N_\delta(\mu_0)$, $\mathbf{P}(\mathbf{x}, \mu)$ is a C^1-diffeomorphism on $N_\delta(\mathbf{x}_0)$. Also, if we assume that (1) has a one-parameter family of periodic orbits Γ_μ, i.e., if $\mathbf{P}(\mathbf{x}, \mu)$ has a one-parameter family of fixed points $\mathbf{x}_\mu$, then as in Theorem 2 in Section 3.4 of Chapter 3, we have the following convenient formula for computing the derivative of the Poincaré map of a planar C^1-system (1):

$$DP(\mathbf{x}_\mu, \mu) = \exp \int_0^{T_\mu} \nabla \cdot \mathbf{f}(\gamma_\mu(t), \mu) dt \tag{2}$$

at a point $\mathbf{x}_\mu \in \Gamma_\mu$ where the one-parameter family of periodic orbits

$$\Gamma_\mu: \mathbf{x} = \gamma_\mu(t), \quad 0 \le t \le T_\mu$$

and T_μ is the period of $\gamma_\mu(t)$. Before beginning our study of bifurcations at nonhyperbolic periodic orbits, we illustrate the dependence of the Poincaré map $P(\mathbf{x}, \mu)$ on the parameter μ with an example.

4.5. Bifurcations at Nonhyperbolic Periodic Orbits

Example 1. Consider the planar system

$$\dot{x} = -y + x(\mu - r^2)$$
$$\dot{y} = x + y(\mu - r^2)$$

of Example 1 in Section 4.4. As we saw in the previous section, a one-parameter family of limit cycles

$$\Gamma_\mu: \gamma_\mu(t) = \sqrt{\mu}(\cos t, \sin t)^T,$$

with $\mu > 0$, is generated in a supercritical Hopf bifurcation at the origin at the bifurcation value $\mu = 0$. The bifurcation diagram is shown in Figure 2 in Section 4.4. In polar coordinates, we have

$$\dot{r} = r(\mu - r^2)$$
$$\dot{\theta} = 1.$$

The first equation can be solved as a Bernoulli equation and for $r(0) = r_0$ we obtain the solution

$$r(t, r_0, \mu) = \left[\frac{1}{\mu} + \left(\frac{1}{r_0^2} - \frac{1}{\mu}\right)e^{-2\mu t}\right]^{-1/2}$$

for $\mu > 0$. On any ray from the origin, the Poincaré map $P(r_0, \mu) = r(2\pi, r_0, \mu)$; i.e.,

$$P(r_0, \mu) = \left[\frac{1}{\mu} + \left(\frac{1}{r_0^2} - \frac{1}{\mu}\right)e^{-4\mu\pi}\right]^{-1/2}.$$

It is not difficult to compute the derivative of this function with respect to r_0 and obtain

$$DP(r_0, \mu) = e^{-4\mu\pi} r_0^{-3} \left[\frac{1}{\mu} + \left(\frac{1}{r_0^2} - \frac{1}{\mu}\right)e^{-4\mu\pi}\right]^{-3/2}.$$

Solving $P(r, \mu) = r$, we obtain a one-parameter family of fixed points of $P(r, \mu), r_\mu = \sqrt{\mu}$ for $\mu > 0$ and this leads to

$$DP(r_\mu, \mu) = e^{-4\mu\pi}.$$

This formula can also be obtained using equation (2); cf. Problem 1. Since for $\mu > 0, DP(r_\mu, \mu) < 1$, the periodic orbits Γ_μ are all stable and hyperbolic.

The system (1) is said to have a nonhyperbolic periodic orbit Γ_0 through the point $\mathbf{x}_0$ at a bifurcation value μ_0 if $D\mathbf{P}(\mathbf{x}_0, \mu_0)$ has an eigenvalue of unit modulus. We begin our study of bifurcations at nonhyperbolic periodic orbits with some simple examples of the saddle-node, transcritical and pitchfork bifurcations that occur when $D\mathbf{P}(\mathbf{x}_0, \mu_0)$ has an eigenvalue equal to one.

Example 2 (A Saddle-Node Bifurcation at a Nonhyperbolic Periodic Orbit). Consider the planar system

$$\dot{x} = -y - x[\mu - (r^2 - 1)^2]$$
$$\dot{y} = x - y[\mu - (r^2 - 1)^2]$$

which is of the form of equation (5) in Section 4.4. Writing this system in polar coordinates yields

$$\dot{r} = -r[\mu - (r^2 - 1)^2]$$
$$\dot{\theta} = 1.$$

For $\mu > 0$ there are two one-parameter families of periodic orbits

$$\Gamma_\mu^\pm: \gamma_\mu^\pm(t) = \sqrt{1 \pm \mu^{1/2}}(\cos t, \sin t)^T$$

with parameter μ. Since the origin is unstable for $0 < \mu < 1$, the smaller limit cycle Γ_μ^- is stable and the larger limit cycle Γ_μ^+ is unstable. For $\mu = 0$ there is a semistable limit cycle Γ_0 represented by $\gamma_0(t) = (\cos t, \sin t)^T$. The phase portraits for this system are shown in Figure 1 and the bifurcation diagram is shown in Figure 2. Note that there is a supercritical Hopf bifurcation at the origin at the bifurcation value $\mu = 1$.

In Example 2 the points $r_\mu^\pm = \sqrt{1 \pm \mu^{1/2}}$ are fixed points of the Poincaré map $P(r, \mu)$ of the periodic orbit $\gamma_0(t)$ along any ray from the origin

$$\Sigma = \{\mathbf{x} \in \mathbf{R}^2 \mid r > 0, \theta = \theta_0\};$$

i.e., we have $d(\sqrt{1 \pm \mu^{1/2}}, \mu) = 0$ where $d(r, \mu) = P(r, \mu) - r$ is the displacement function. The bifurcation diagram is given by the graph of the relation $d(r, \mu) = 0$ in the (μ, r)-plane. Using equation (2), we can compute the derivative of the Poincaré map at $r_\mu^\pm = \sqrt{1 \pm \mu^{1/2}}$:

$$DP(\sqrt{1 \pm \mu^{1/2}}, \mu) = e^{\pm 8\mu^{1/2}(1 \pm \mu^{1/2})\pi},$$

cf. Problem 2. We see that for $0 < \mu < 1, DP(\sqrt{1 - \mu^{1/2}}, \mu) < 1$ and $DP(\sqrt{1 + \mu^{1/2}}, \mu) > 1$; the smaller limit cycle is stable and the larger limit cycle is unstable as illustrated in Figures 1 and 2. Furthermore, for $\mu = 0$ we have $DP(1, 0) = 1$, i.e., $\gamma_0(t)$ is a nonhyperbolic periodic orbit with both of its characteristic exponents equal to zero.

Remark 1. In general, for planar systems, the bifurcation diagram is given by the graph of the relation $d(s, \mu) = 0$ in the (μ, s)-plane where

$$d(s, \mu) = P(s, \mu) - s$$

is the displacement function along a straight line Σ normal to the nonhyper-

4.5. Bifurcations at Nonhyperbolic Periodic Orbits

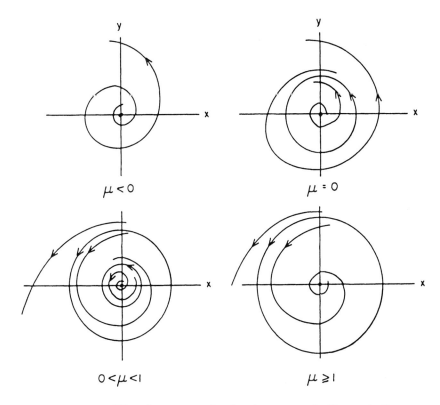

Figure 1. The phase portraits for the system in Example 2.

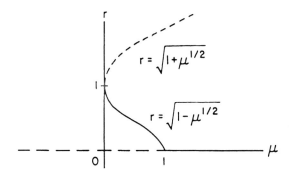

Figure 2. The bifurcation diagram for the saddle-node bifurcation at the nonhyperbolic periodic orbit $\gamma_0(t)$ of the system in Example 2.

bolic periodic orbit Γ_0 at $\mathbf{x}_0$. We take s to be the signed distance along the straight line Σ, with s positive at points on the exterior of Γ_0 and negative at points on the interior of Γ_0, as in Figure 3 in Section 3.4 of Chapter 3.

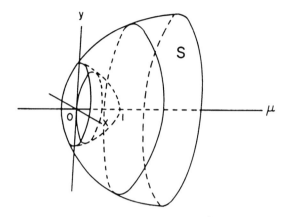

Figure 3. The one-parameter family of periodic orbits S of the system in Example 2.

In this context, it follows that for each fixed value of μ, the values of s for which $d(s, \mu) = 0$ define points $\mathbf{x}_j$ on Σ near the point $\mathbf{x}_0 \in \Gamma_0 \cap \Sigma$ through which the system (1) has periodic orbits $\gamma_j(t) = \phi_t(\mathbf{x}_j)$. For example, in Figure 2, each vertical line $\mu = $ constant, with $0 < \mu < 1$, intersects the curve $d(r, \mu) = 0$ in two points $(\mu, \sqrt{1 \pm \mu^{1/2}})$; and the system in Example 2 has periodic orbits $\gamma_\mu^\pm(t)$ through the points $(\sqrt{1 \pm \mu^{1/2}}, 0)$ on the x-axis in the phase plane. As in Figure 2 in Section 4.4, each one-parameter family of periodic orbits generates a surface S in $\mathbf{R}^2 \times \mathbf{R}$. For example, the periodic orbits of the system in Example 2 generate the surface S shown in Figure 3. Since in general there is only one surface generated at a saddle-node bifurcation at a nonhyperbolic periodic orbit, we regard the two one-parameter families of periodic orbits (with parameter μ) as belonging to one and the same family of periodic orbits. In this case, we can always define a new parameter β (such as the arc length along a path on the surface S) so that $\Gamma_{\mu(\beta)}$ defines a one-parameter family of periodic orbits with parameter β.

Example 3 (A Transcritical Bifurcation at a Nonhyperbolic Periodic Orbit). Consider the planar system

$$\dot{x} = -y - x(1 - r^2)(1 + \mu - r^2)$$
$$\dot{y} = x - y(1 - r^2)(1 + \mu - r^2).$$

In polar coordinates we have

$$\dot{r} = -r(1 - r^2)(1 + \mu - r^2)$$
$$\dot{\theta} = 1.$$

For all $\mu \in \mathbf{R}$, this system has a one-parameter family of periodic orbits

4.5. Bifurcations at Nonhyperbolic Periodic Orbits

represented by
$$\gamma_0(t) = (\cos t, \sin t)^T$$
and for $\mu > -1$, there is another one-parameter family of periodic orbits represented by
$$\gamma_\mu(t) = \sqrt{1+\mu}(\cos t, \sin t)^T.$$
The bifurcation diagram, showing the transcritical bifurcation that occurs at the nonhyperbolic periodic orbit $\gamma_0(t)$ at the bifurcation value $\mu = 0$, is shown in Figure 4. Note that a subcritical Hopf bifurcation occurs at the nonhyperbolic critical point at the origin at the bifurcation value $\mu = -1$.

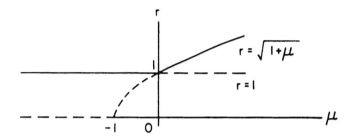

Figure 4. The bifurcation diagram for the transcritical bifurcation at the nonhyperbolic periodic orbit $\gamma_0(t)$ of the system in Example 3.

In Example 3 the points $r_\mu = \sqrt{1+\mu}$ and $r_\mu = 1$ are fixed points of the Poincaré map $P(r, \mu)$ of the nonhyperbolic periodic orbit Γ_0; i.e., we have
$$d(\sqrt{1+\mu}, \mu) = 0$$
for all $\mu > -1$ and
$$d(1, \mu) = 0$$
for all $\mu \in \mathbf{R}$ where $d(r, \mu) = P(r, \mu) - r$. Furthermore, using equation (2), we can compute
$$DP(\sqrt{1+\mu}, \mu) = e^{-4\mu(1+\mu)\pi}$$
and
$$DP(1, \mu) = e^{4\mu\pi},$$
cf. Problem 2. This determines the stability of the two families of periodic orbits as indicated in Figure 4. We see that $DP(1, 0) = 1$; i.e., there is a nonhyperbolic periodic orbit Γ_0 with both of its characteristic exponents equal to zero at the bifurcation value $\mu = 0$. In this example, there are two distinct surfaces of periodic orbits, a cylindrical surface and a parabolic surface, which intersect in the nonhyperbolic periodic orbit Γ_0; cf. Problem 3.

Example 4 (A Pitchfork Bifurcation at a Nonhyperbolic Periodic Orbit). Consider the planar system

$$\dot{x} = -y + x(1 - r^2)[\mu - (r^2 - 1)^2]$$
$$\dot{y} = x + y(1 - r^2)[\mu - (r^2 - 1)^2].$$

In polar coordinates we have

$$\dot{r} = r(1 - r^2)[\mu - (r^2 - 1)^2]$$
$$\dot{\theta} = 1.$$

For all $\mu \in \mathbf{R}$, this system has a one-parameter family of periodic orbits represented by

$$\gamma_0(t) = (\cos t, \sin t)^T$$

and for $\mu > 0$ there is another family (with two branches as in Remark 1) represented by

$$\gamma_\mu^\pm(t) = \sqrt{1 \pm \mu^{1/2}}(\cos t, \sin t)^T.$$

Using equation (2) we can compute

$$DP(\sqrt{1 \pm \mu^{1/2}}, \mu) = e^{4\mu(1 \pm \mu^{1/2})\pi}$$

and

$$DP(1, \mu) = e^{-4\pi\mu}.$$

This determines the stability of the two families of periodic orbits as indicated in Figure 5(a). We also see that $DP(1,0) = 1$; i.e., there is a

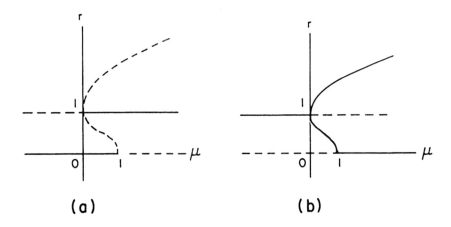

Figure 5. The bifurcation diagram for the pitchfork bifurcation at the nonhyperbolic periodic orbit $\gamma_0(t)$ of the system (a) in Example 4 and (b) in Example 4 with $t \to -t$.

4.5. Bifurcations at Nonhyperbolic Periodic Orbits

nonhyperbolic periodic orbit Γ_0 with both of its characteristic exponents equal to zero at the bifurcation value $\mu = 0$. Note that a Hopf bifurcation occurs at the nonhyperbolic critical point at the origin at the bifurcation value $\mu = 1$. Also, note that if we reverse the sign of t in this example, i.e., let $t \to -t$, then we reverse the stability of the periodic orbits and we would have the bifurcation diagram with a pitchfork bifurcation shown in Figure 5(b).

Using the implicit function theorem, conditions can be given on the derivatives of the Poincaré map which imply the existence of a saddle-node, transcritical or pitchfork bifurcation at a nonhyperbolic periodic orbit of (1). We shall only give these conditions for planar systems. In the next theorem $P(s, \mu)$ denotes the Poincaré map along a normal line Σ to a nonhyperbolic periodic orbit Γ_0 at a bifurcation value $\mu = \mu_0$ in (1). As in Section 4.2, D denotes the partial derivative of $P(s, \mu)$ with respect to the spatial variable s.

Theorem 1. *Suppose that* $\mathbf{f} \in C^2(E \times J)$ *where E is an open subset of* $\mathbf{R}^2$ *and $J \subset \mathbf{R}$ is an interval. Assume that for $\mu = \mu_0$ the system (1) has a periodic orbit $\Gamma_0 \subset E$ and that $P(s, \mu)$ is the Poincaré map for Γ_0 defined in a neighborhood $N_\delta(0, \mu_0)$. Then if $P(0, \mu_0) = 0$, $DP(0, \mu_0) = 1$,*

$$D^2 P(0, \mu_0) \neq 0 \quad \text{and} \quad P_\mu(0, \mu_0) \neq 0, \tag{3}$$

it follows that a saddle-node bifurcation occurs at the nonhyperbolic periodic orbit Γ_0 at the bifurcation value $\mu = \mu_0$; i.e., depending on the signs of the expressions in (3), there are no periodic orbits of (1) near Γ_0 when $\mu < \mu_0$ (or when $\mu > \mu_0$) and there are two periodic orbits of (1) near Γ_0 when $\mu > \mu_0$ (or when $\mu < \mu_0$). The two periodic orbits of (1) near Γ_0 are hyperbolic and of the opposite stability.

If the conditions (3) are changed to

$$\begin{aligned} P_\mu(0, \mu_0) = 0 \quad DP_\mu(0, \mu_0) \neq 0 \quad \text{and} \\ D^2 P(0, \mu_0) \neq 0, \end{aligned} \tag{4}$$

then a transcritical bifurcation occurs at the nonhyperbolic periodic orbit Γ_0 at the bifurcation value $\mu = \mu_0$. And if the conditions (3) are changed to

$$\begin{aligned} P_\mu(0, \mu_0) = 0, \quad DP_\mu(0, \mu_0) \neq 0 \\ D^2 P(0, \mu_0) = 0 \quad \text{and} \quad D^3 P(0, \mu_0) \neq 0, \end{aligned} \tag{5}$$

then a pitchfork bifurcation occurs at the nonhyperbolic periodic orbit Γ_0 at the bifurcation value $\mu = \mu_0$.

Remark 2. Under the conditions (3) in Theorem 1, the periodic orbit Γ_0 is a multiple limit cycle of multiplicity $m = 2$ and exactly two limit cycles bifurcate from the semi-stable limit cycle Γ_0 as μ varies from μ_0 in one sense or the other. In particular, if $D^2 P(0, \mu_0)$ and $P_\mu(0, \mu_0)$ have opposite signs, then there are two limit cycles near Γ_0 for all sufficiently small $\mu - \mu_0 > 0$ and if $D^2 P(0, \mu_0)$ and $P_\mu(0, \mu_0)$ have the same sign, then there are two limit cycles near Γ_0 for all sufficiently small $\mu_0 - \mu > 0$.

Remark 3. It follows from equation (2) that

$$DP(0, \mu_0) = e^{\int_0^{T_0} \nabla \cdot \mathbf{f}(\gamma_0(\tau), \mu_0) d\tau}$$

where T_0 is the period of the nonhyperbolic periodic orbit

$$\Gamma_0: \mathbf{x} = \gamma_0(t).$$

Furthermore, in this case we also have a formula for the partial derivative of the Poincaré map P with respect to the parameter μ in terms of the vector field $\mathbf{f}$ along the periodic orbit Γ_0:

$$P_\mu(0, \mu_0) = \frac{-\omega_0}{|\mathbf{f}(\gamma_0(0))|} \int_0^{T_0} e^{\int_t^{T_0} \nabla \cdot \mathbf{f}(\gamma_0(\tau), \mu_0) d\tau} \mathbf{f} \wedge \mathbf{f}_\mu(\gamma_0(t), \mu_0) dt \qquad (6)$$

where $\omega_0 = \pm 1$ according to whether Γ_0 is positively or negatively oriented, and the wedge product of two vectors $\mathbf{u} = (u_1, u_2)^T$ and $\mathbf{v} = (v_1, v_2)^T$ is given by the determinant

$$\mathbf{u} \wedge \mathbf{v} = \begin{vmatrix} u_1 & u_2 \\ v_1 & v_2 \end{vmatrix} = u_1 v_2 - v_1 u_2.$$

This formula was apparently first derived by Andronov et al.; cf. equation (36) on p. 384 in [A–II]. It is closely related to the Melnikov function defined in Section 4.9.

These same types of bifurcations also occur in higher dimensional systems when the derivative of the Poincaré map, $DP(\mathbf{x}_0, \mu_0)$, for the periodic orbit Γ_0 has a single eigenvalue equal to one and no other eigenvalues of unit modulus; cf., e.g. Theorem II.2, p. 65 in [Ru]. Furthermore, in this case the saddle-node bifurcation is generic; cf. pp. 58 and 64 in [Ru].

Before discussing some of the other types of bifurcations that can occur at nonhyperbolic periodic orbits of higher dimensional systems (1) with $n \geq 3$, we first discuss some of the other types of bifurcations that can occur at multiple limit cycles of planar systems. Recall that a limit cycle $\gamma_0(t)$ is a multiple limit cycle of (1) if and only if

$$\int_0^{T_0} \nabla \cdot \mathbf{f}(\gamma_0(t)) dt = 0.$$

Cf. Definition 2 in Section 3.4 of Chapter 3. Analogous to Theorem 2 in Section 4.4 for the bifurcation of m limit cycles from a weak focus of multiplicity m, we have the following theorem, proved on pp. 278–282 in [A–II], for the bifurcation of m limit cycles from a multiple limit cycle of multiplicity m of a planar analytic system

$$\begin{aligned} \dot{x} &= P(x, y) \\ \dot{y} &= Q(x, y). \end{aligned} \qquad (7)$$

4.5. Bifurcations at Nonhyperbolic Periodic Orbits

Theorem 2. *If Γ_0 is a multiple limit cycle of multiplicity m of the planar analytic system (7), then*

(i) *there is a $\delta > 0$ and an $\varepsilon > 0$ such that any system ε-close to (7) in the C^m-norm has at most m limit cycles in a δ-neighborhood, $N_\delta(\Gamma_0)$, of Γ_0 and*

(ii) *for any $\delta > 0$ and $\varepsilon > 0$ there is an analytic system which is ε-close to (7) in the C^m-norm and has exactly m simple limit cycles in $N_\delta(\Gamma_0)$.*

It can be shown that the nonhyperbolic limit cycles in Examples 2 and 3 are of multiplicity $m = 2$ and that the nonhyperbolic limit cycle in Example 4 is of multiplicity $m = 3$. It can also be shown that the system (5) in Section 4.4 with

$$\psi(r,\mu) = [\mu - (r^2 - 1)^2][\mu - 2(r^2 - 1)^2]$$

has a multiple limit cycle $\gamma_0(t) = (\cos t, \sin t)^T$ of multiplicity $m = 4$ at the bifurcation value $\mu = 0$ and that exactly four hyperbolic limit cycles bifurcate from $\gamma_0(t)$ as μ increases from zero; cf. Problem 4. Codimension-$(m-1)$ bifurcations occur at multiplicity-m limit cycles of planar systems. These bifurcations were studied by the author in [39].

We next consider some examples of period-doubling bifurcations which occur when $D\mathbf{P}(\mathbf{x}_0, \mu_0)$ has a simple eigenvalue equal to -1 and no other eigenvalues of unit modulus. Figure 6 shows what occurs geometrically at a period-doubling bifurcation at a nonhyperbolic periodic orbit Γ_0 (shown as a dashed curve in Figure 6). Since trajectories do not cross, it is geometrically impossible to have a period-doubling bifurcation for a planar system.

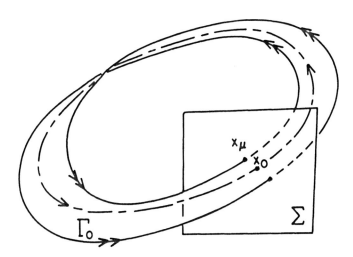

Figure 6. A period-doubling bifurcation at a nonhyperbolic periodic orbit Γ_0.

This also follows from the fact that, by equation (2), $DP(\mathbf{x}_0, \mu_0) = 1$ for any nonhyperbolic limit cycle Γ_0.

Suppose that $\mathbf{P}(\mathbf{x}, \mu)$ is the Poincaré map defined in a neighborhood of the point $(\mathbf{x}_0, \mu_0) \in E \times J$ where $\mathbf{x}_0$ is a point on the periodic orbit Γ_0 of (1) with $\mu = \mu_0$, and suppose that $D\mathbf{P}(\mathbf{x}_0, \mu_0)$ has an eigenvalue -1 and no other eigenvalues of unit modulus. Then generically a period-doubling bifurcation occurs at the nonhyperbolic periodic orbit Γ_0 and it is characterized by the fact that for all μ near μ_0, and on one side or the other of μ_0, there is a point $\mathbf{x}_\mu \in \Sigma$, the hyperplane normal to Γ_0 at $\mathbf{x}_0$, such that $\mathbf{x}_\mu$ is a fixed point of $\mathbf{P}^2 = \mathbf{P} \circ \mathbf{P}$; i.e.,

$$\mathbf{P}^2(\mathbf{x}_\mu, \mu) = \mathbf{x}_\mu.$$

Since the periodic orbits with periods approximately equal to $2T_0$ correspond to fixed points of the second iterate $\mathbf{P}^2$ of the Poincaré map, the bifurcation diagram, which can be obtained by using a center manifold reduction as on pp. 157–159 in [G/H], has the form shown in Figure 7. The curves in Figure 7 represent the locus of fixed points of $\mathbf{P}^2$ in the (μ, x) plane where x is the distance along the curve where the center manifold $W^c(\Gamma_0)$ intersects the hyperplane Σ. In Figure 7(a) the solid curve for $\mu > 0$ corresponds to a single periodic orbit whose period is approximately equal to $2T_0$ for small $\mu > 0$. Since there is only one periodic orbit whose period is approximately equal to $2T_0$ for small $\mu > 0$, the bifurcation diagram for a period-doubling bifurcation is often shown as in Figure 7(b).

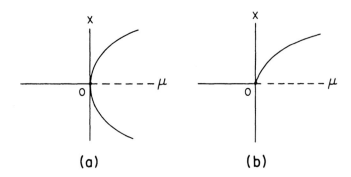

Figure 7. The bifurcation diagram for a period-doubling bifurcation.

We next look at some period-doubling bifurcations that occur in the Lorenz system introduced in Example 4 in Section 3.2 of Chapter 3. The Lorenz system was first studied by the meteorologist-mathematician E. N. Lorenz in 1963. Lorenz derived a relatively simple system of three nonlinear differential equations which captures many of the salient features of convective fluid motion. The Lorenz system offers a rich source of examples

4.5. Bifurcations at Nonhyperbolic Periodic Orbits

of various types of bifurcations that occur in dynamical systems. We discuss some of these bifurcations in the next example where we rely heavily on Sparrow's excellent numerical study of the Lorenz system [S]. Besides period-doubling bifurcations, the Lorenz system also exhibits a homoclinic loop bifurcation and the attendant chaotic motion in which the numerically computed solutions oscillate in a pseudo-random way, apparently forever; cf. [S] and Section 4.8.

Example 5 (The Lorenz System). Consider the Lorenz system

$$\dot{x} = 10(y - x)$$
$$\dot{y} = \mu x - y - xz \qquad (8)$$
$$\dot{z} = xy - 8z/3$$

depending on the parameter μ with $\mu > 0$. These equations are symmetric under the transformation $(x, y, z) \to (-x, -y, z)$. Thus, if (8) has a periodic orbit Γ (such as the ones shown in Figures 11 and 12 below), it will also have a corresponding periodic orbit Γ' which is the image of Γ under this transformation. Lorenz showed that there is an ellipsoid $E^2 \subset \mathbf{R}^3$ which all trajectories eventually enter and never leave and that there is a bounded attracting set of zero volume within E^2 toward which all trajectories tend. For $0 < \mu \leq 1$, this set is simply the origin; i.e., for $0 < \mu \leq 1$ there is only the one critical point at the origin and it is globally stable; cf. Problem 6 in Section 3.2 of Chapter 3. For $\mu = 1$, the origin is a nonhyperbolic critical point of (8) and there is a pitchfork bifurcation at the origin which occurs as μ increases through the bifurcation value $\mu = 1$. The two critical points which bifurcate from the origin are located at the points

$$C_{1,2} = (\pm 2\sqrt{2(\mu-1)/3}, \pm 2\sqrt{2(\mu-1)/3}, \mu - 1);$$

cf. Problem 6 in Section 3.2 of Chapter 3. For $\mu > 1$ the critical point at the origin has a one-dimensional unstable manifold $W^u(0)$ and a two-dimensional stable manifold $W^s(0)$. The eigenvalues $\lambda_{1,2}$ at the critical points $C_{1,2}$ satisfy a cubic equation and they all have negative real part for $1 < \mu < \mu_H$ where $\mu_H = 470/19 \simeq 24.74$; cf. [S], p. 10. Parenthetically, we remark that $\lambda_{1,2}$ are both real for $1 < \mu \leq 1.34$ and there are complex pairs of eigenvalues at $C_{1,2}$ for $\mu > 1.34$. The behavior near the critical points of (8) for relatively small μ (say $0 < \mu < 10$) is summarized in Figure 8. Note that the z-axis is invariant under the flow for all values of μ.
As μ increases, the trajectories in the unstable manifold at the origin $W^u(0)$ leave the origin on increasingly larger loops before they spiral down to the critical points $C_{1,2}$; cf. Figure 9. As μ continues to increase, Sparrow's numerical work shows that a very interesting phenomenon occurs in the Lorenz system at a parameter value $\mu' \simeq 13.926$: The trajectories in the unstable manifold $W^u(0)$ intersect the stable manifold $W^s(0)$ and form two homoclinic loops (which are symmetric images of each other); cf. Figure 10.

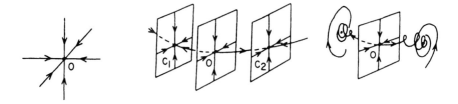

Figure 8. A pitchfork bifurcation occurs at the origin of the Lorenz system at the bifurcation value $\mu = 1$.

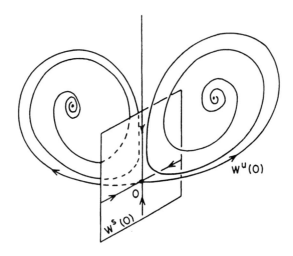

Figure 9. The stable and unstable manifolds at the origin for $10 \leq \mu < \mu' \simeq 13.926$.

Homoclinic loop bifurcations are discussed in Section 4.8, although most of the theoretical results in that section are for two-dimensional systems where generically a single periodic orbit bifurcates from a homoclinic loop. In this case, a pair of unstable periodic orbits $\Gamma_{1,2}$ also bifurcates from the two homoclinic loops as μ increases beyond μ' as shown in Figure 11; however, something much more interesting occurs for the three-dimensional Lorenz system (8) which has no counterpart for two-dimensional systems. An infinite number of periodic orbits of arbitrarily long period bifurcate from the homoclinic loops as μ increases beyond μ' and there is a bounded invariant set which contains all of these periodic orbits as well as an infinite number of nonperiodic motions; cf. Figure 6 in Section 3.2 and [S], p. 21. Sparrow refers to this type of homoclinic loop bifurcation as a "homoclinic explosion" and it is part of what makes the Lorenz system so interesting.

4.5. Bifurcations at Nonhyperbolic Periodic Orbits

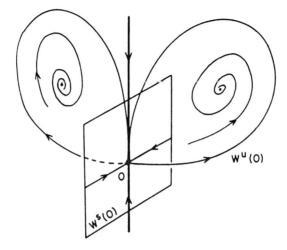

Figure 10. The homoclinic loop which occurs in the Lorenz system at the bifurcation value $\mu' \simeq 13.926$.

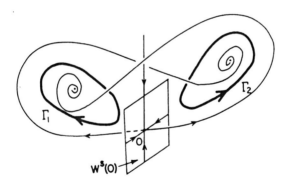

Figure 11. A symmetric pair of unstable periodic orbits which result from the homoclinic loop bifurcation at $\mu = \mu'$.

The critical points $C_{1,2}$ each have one negative and two pure imaginary eigenvalues at the parameter value $\mu = \mu_H \simeq 24.74$. A subcritical Hopf bifurcation occurs at the nonhyperbolic critical points $C_{1,2}$ at the bifurcation value $\mu = \mu_H$; cf. Section 4.4. Sparrow has computed the unstable periodic orbits $\Gamma_{1,2}$ for several parameter values in the range $13.926 \simeq \mu' < \mu < \mu_H \simeq 24.74$. The projection of Γ_2 on the (x, z)-plane is shown in Figure 12; cf. [S], p. 27.

We see that the periodic orbit Γ_2 approaches the critical point at the origin and forms a homoclinic loop as μ decreases to μ': For $\mu > \mu_H$, all

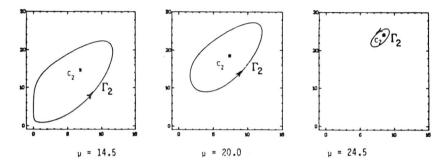

Figure 12. Some periodic orbits in the one-parameter family of periodic orbits generated by the subcritical Hopf bifurcation at the critical point C_2 at the bifurcation value $\mu = \mu_H \simeq 24.74$. Reprinted with permission from Sparrow (Ref. [S]).

three critical points are unstable and, at least for $\mu > \mu_H$ and μ near μ_H, there must be a strange invariant set within E^2 toward which all trajectories tend. In fact, from the numerical results in [S], it seems quite certain that, at least for μ near μ_H, the Lorenz system (8) has a strange attractor as described in Example 4 in Section 3.2 of Chapter 3. This strange attractor actually appears at a value $\mu = \mu_A \cong 24.06 < \mu_H$ at which value there is a heteroclinic connection of $W^u(0)$ and $W^s(\Gamma_{1,2})$; cf. [S], p. 32.

In order to describe some of the period-doubling bifurcations that occur in the Lorenz system and to see what happens to all of the periodic orbits born in the homoclinic explosion that occurs at $\mu = \mu' \simeq 13.926$, we next look at the behavior of (8) for large μ (namely for $\mu > \mu_\infty \simeq 313$); cf. [S], Chapter 7. For $\mu > 313$, Sparrow's work [S] indicates that there is only one periodic orbit Γ_∞ and it is stable and symmetric under the transformation $(x, y, z) \to (-x, -y, z)$. For $\mu > 313$, the stable periodic orbit Γ_∞ and the critical points $0, C_1$ and C_2 make up the nonwandering set Ω. The projection of the stable, symmetric periodic orbit Γ_∞ on the (x, z)-plane is shown in Figure 13 for $\mu = 350$. At a bifurcation value $\mu \simeq 313$, the nonhyperbolic periodic orbit undergoes a pitchfork bifurcation as described earlier in this section; cf. the bifurcation diagram in Figure 5(b). As μ decreases below $\mu_\infty \simeq 313$, the periodic orbit Γ_∞ becomes unstable and two stable (nonsymmetric) periodic orbits are born; cf. the bifurcation diagram in Figure 15 below. The projection of one of these stable, nonsymmetric, periodic orbits on the (x, z)-plane is shown in Figure 13 for $\mu = 260$; cf. [S], p. 67. Recall that for each nonsymmetric periodic orbit Γ in the Lorenz system, there exists a corresponding nonsymmetric periodic orbit Γ' obtained from Γ under the transformation $(x, y, z) \to (-x, -y, z)$.

As μ continues to decrease below $\mu = \mu_\infty$ a period-doubling bifurcation occurs in the Lorenz system at the bifurcation value $\mu = \mu_2 \simeq 224$. The

4.5. Bifurcations at Nonhyperbolic Periodic Orbits

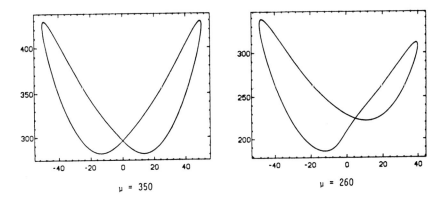

Figure 13. The symmetric and nonsymmetric periodic orbits of the Lorenz system which occur for large μ. Reprinted with permission from Sparrow (Ref. [S]).

projection of one of the resulting periodic orbits Γ_2 whose period is approximately equal to twice the period of the periodic orbit Γ_1 shown in Figure 13 is shown in Figure 14(a) for $\mu = 222$. There is another one Γ_2' which is the symmetric image of Γ_2. This is not the end of the period-doubling story in the Lorenz system! Another period-doubling bifurcation occurs at a nonhyperbolic periodic orbit of the family Γ_2 at the bifurcation value $\mu = \mu_4 \simeq 218$. The projection of one of the resulting periodic orbits Γ_4 whose period is approximately equal to twice the period of the periodic orbit Γ_2 (or four times the period of Γ_1) is shown in Figure 14(b) for $\mu = 216.2$. There is another one Γ_4' which is the symmetric image of Γ_4; cf. [S], p. 68.

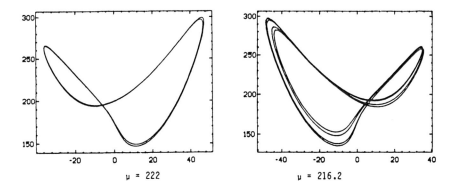

Figure 14. The nonsymmetric periodic orbits which result from period-doubling bifurcations in the Lorenz system. Reprinted with permission from Sparrow (Ref. [S]).

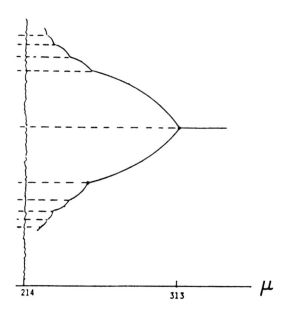

Figure 15. The bifurcation diagram showing the pitchfork bifurcation and period-doubling cascade that occurs in the Lorenz system. Reprinted with permission from Sparrow (Ref. [S]).

And neither is this the end of the story, for as μ continues to decrease below μ_4, more and more period-doubling bifurcations occur. In fact, there is an infinite sequence of period-doubling bifurcations which accumulate at a bifurcation value $\mu = \mu^* \simeq 214$. This is indicated in the bifurcation diagram shown in Figure 15; cf. [S], p. 69. This type of accumulation of period-doubling bifurcations is referred to as a period-doubling cascade. Interestingly enough, there are some universal properties common to all period-doubling cascades. For example the limit of the ratio

$$\frac{\mu_{n-1} - \mu_n}{\mu_n - \mu_{n+1}},$$

where μ_n is the bifurcation value at which the nth period-doubling bifurcation occurs, is equal to some universal constant $\delta = 4.6992\ldots$; cf., e.g., [S], p. 58.

Before leaving this interesting example, we mention one other period-doubling cascade that occurs in the Lorenz system. Sparrow's calculations [S] show that at $\mu \simeq 166$ a saddle-node bifurcation occurs at a nonhyperbolic periodic orbit. For $\mu < 166$, this results in two symmetric periodic orbits, one stable and one unstable. The stable, symmetric, periodic orbit is shown in Figure 16 for $\mu = 160$. As μ continues to decrease, first a pitchfork bifurcation and then a period-doubling cascade occurs, similar to

4.5. Bifurcations at Nonhyperbolic Periodic Orbits

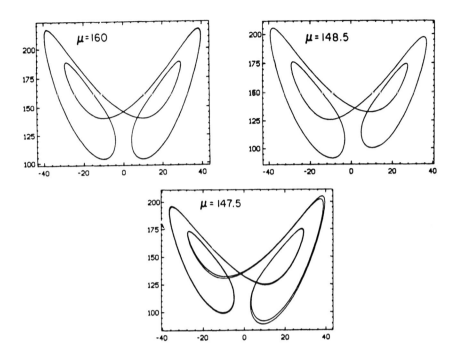

Figure 16. A stable, symmetric periodic orbit born in a saddle-node bifurcation at $\mu \simeq 166$; a stable, nonsymmetric, periodic orbit born in a pitchfork bifurcation at $\mu \simeq 154$; and a stable, nonsymmetric, periodic orbit born in a period-doubling bifurcation at $\mu \simeq 148$. Reprinted with permission from Sparrow (Ref. [S]).

that discussed above. One of the stable, nonsymmetric, periodic orbits resulting from the pitchfork bifurcation which occurs at $\mu \simeq 154.5$ is shown in Figure 16 for $\mu = 148.5$. One of the double-period, periodic orbits is also shown in Figure 16 for $\mu = 147.5$. The bifurcation diagram for the parameter range $145 < \mu < 166$ is shown in Figure 17; cf. [S], p. 62.

There are many other period-doubling cascades and homoclinic explosions that occur in the Lorenz system for the parameter range $25 < \mu < 145$ which we will not discuss here. Sparrow has studied several of these bifurcations in [S]; cf. his summary on p. 99 of [S]. Many of the periodic orbits born in the saddle-node and pitchfork bifurcations and in the period-doubling cascades in the Lorenz system (8) persist as μ decreases to the value $\mu = \mu' \simeq 13.926$ at which the first homoclinic explosion occurs and which is where they terminate as μ decreases. Others terminate in other homoclinic explosions as μ decreases (or as μ increases).

To really begin to understand the complicated dynamics that occur in higher dimensional systems (with $n \geq 3$) such as the Lorenz system, it is necessary to study dynamical systems defined by maps or diffeomorphisms such as the Poincaré map. While the numerical studies of Lorenz, Sparrow

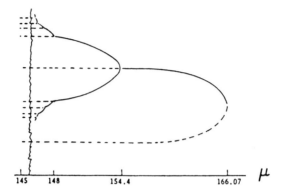

Figure 17. The bifurcation diagram showing the saddle-node and pitchfork bifurcations and the period-doubling cascade that occur in the Lorenz system for $145 < \mu < 167$. Reprinted with permission from Sparrow (Ref. [S]).

and others have made it clear that the Lorenz system has some complicated dynamics which include the appearance of a strange attractor, the study of dynamical systems defined by maps rather than flows has made it possible to mathematically establish the existence of strange attractors for maps which have transverse homoclinic orbits; cf. [G/H] and [Wi]. Much of the success of this approach is due to the program of study of differentiable dynamical systems begun by Stephen Smale in the sixties. In particular, the Smale Horseshoe map, which occurs whenever there is a transverse homoclinic orbit, motivated much of the development of the modern theory of dynamical systems; cf., e.g. [Ru]. We shall discuss some of these ideas more thoroughly in Section 4.8; however, this book focuses on dynamical systems defined by flows rather than maps.

Before ending this section on bifurcations at nonhyperbolic periodic orbits, we briefly mention one last type of generic bifurcation that occurs at a periodic orbit when $D\mathbf{P}(\mathbf{x}_0, \mu_0)$ has a pair of complex conjugate eigenvalues on the unit circle. In this case, an invariant, two-dimensional torus results from the bifurcation and this corresponds to a Hopf bifurcation at the fixed point $\mathbf{x}_0$ of the Poincaré map $\mathbf{P}(\mathbf{x}, \mu)$; cf. [G/H], pp. 160–165 or [Ru], pp. 63 and 82. The idea of what occurs in this type of bifurcation is illustrated in Figure 18; however, an analysis of this type of bifurcation is beyond the scope of this book.

Problem Set 5

1. Using equation (2), compute the derivative of the Poincaré map $DP(r_\mu, \mu)$ for the one-parameter family of periodic orbits

$$\gamma_\mu(t) = \sqrt{\mu}(\cos t, \sin t)^T$$

of the system in Example 1.

4.5. Bifurcations at Nonhyperbolic Periodic Orbits

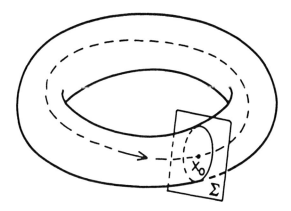

Figure 18. A bifurcation at a nonhyperbolic periodic orbit which results in the creation of an invariant torus and which corresponds to a Hopf bifurcation of its Poincaré map.

2. Verify the computations of the derivatives of the Poincaré maps obtained in Examples 2, 3 and 4 by using equation (2).

3. Sketch the surfaces of periodic orbits in Examples 3 and 4.

4. Write the planar system
$$\dot{x} = -y + x[\mu - (r^2 - 1)^2][\mu - 4(r^2 - 1)^2]$$
$$\dot{y} = x + y[\mu - (r^2 - 1)^2][\mu - 4(r^2 - 1)^2]$$
in polar coordinates and show that for all $\mu > 0$ there are one-parameter families of periodic orbits given by
$$\gamma_1^\pm(t) = \sqrt{1 \pm \mu^{1/2}}(\cos t, \sin t)^T$$
and
$$\gamma_2^\pm(t) = \sqrt{1 \pm \mu^{1/2}/2}(\cos t, \sin t)^T.$$
Using equation (2), compute the derivative of the Poincaré maps, $DP(\sqrt{1 \pm \mu^{1/2}}, \mu)$ and $DP(\sqrt{1 \pm \mu^{1/2}/2}, \mu)$, for these families. And draw the bifurcation diagram.

5. Show that the vector field **f** defined by the right-hand side of
$$\dot{x} = -y + x[\mu - (r - 1)^2]$$
$$\dot{y} = x + y[\mu - (r - 1)^2]$$
is in $C^1(\mathbf{R}^2)$ but that $\mathbf{f} \notin C^2(\mathbf{R}^2)$. Write this system in polar coordinates, determine the one-parameter families of periodic orbits, and draw the bifurcation diagram. How do the bifurcations for this system compare with those in Example 2?

6. For the functions $\psi(r,\mu)$ given below, write the system (5) in Section 4.4 in polar coordinates in order to determine the one-parameter families of periodic orbits of the system. Draw the bifurcation diagram in each case and determine the various types of bifurcations.

   (a) $\psi(r,\mu) = (r-1)(r-\mu-1)$
   (b) $\psi(r,\mu) = (r-1)(r-\mu-1)(r+\mu)$
   (c) $\psi(r,\mu) = (r-1)(r-\mu-1)(r+\mu+1)$
   (d) $\psi(r,\mu) = (\mu-1)(r^2-1)[\mu-1-(r^2-1)^2]$

7. (One-dimensional maps) A map $P\colon \mathbf{R} \to \mathbf{R}$ is said to have a nonhyperbolic fixed point at $x = x_0$ if $P(x_0) = x_0$ and $|DP(x_0)| = 1$.

   (a) Show that the map $P(x,\mu) = \mu - x^2$ has a nonhyperbolic fixed point $x = -1/2$ at the bifurcation value $\mu = -1/4$. Sketch the bifurcation diagram, i.e. sketch the locus of fixed points of P, where $P(x,\mu) = x$, in the (μ, x) plane and show that the map $P(x,\mu)$ has a saddle-node bifurcation at the point $(\mu, x) = (-1/4, -1/2)$.

   (b) Show that the map $P(x,\mu) = \mu x(1-x)$ has a nonhyperbolic fixed point at $x = 0$ at the bifurcation value $\mu = 1$. Sketch the bifurcation diagram in the (μ, x) plane and show that $P(x,\mu)$ has a transcritical bifurcation at the point $(\mu, x) = (1, 0)$.

8. (Two-dimensional maps) A map $\mathbf{P}\colon \mathbf{R}^2 \to \mathbf{R}^2$ is said to have a nonhyperbolic fixed point at $\mathbf{x} = \mathbf{x_0}$ if $\mathbf{P}(\mathbf{x_0}) = \mathbf{x_0}$ and $D\mathbf{P}(\mathbf{x_0})$ has an eigenvalue of unit modulus.

   (a) Show that the map $\mathbf{P}(\mathbf{x},\mu) = (\mu - x^2, 2y)^T$ has a nonhyperbolic fixed point at $\mathbf{x} = (-1/2, 0)^T$ at the bifurcation value $\mu = -1/4$. Sketch the bifurcation diagram in the (μ, x) plane and show that $\mathbf{P}(\mathbf{x},\mu)$ has a saddle-node bifurcation at the point $(\mu, x, y) = (-1/4, -1/2, 0)$.

   (b) Show that the map $\mathbf{P}(\mathbf{x},\mu) = (y, -x/2 + \mu y - y^3)^T$ has a nonhyperbolic fixed point at $\mathbf{x} = \mathbf{0}$ at the bifurcation value $\mu = 3/2$. Sketch the bifurcation diagram in the (μ, x) plane and show that $\mathbf{P}(\mathbf{x},\mu)$ has a pitchfork bifurcation at the point $(\mu, x, y) = (3/2, 0, 0)$.

   Hint: Follow the procedure outlined in Problem 7(a).

9. Consider the one-dimensional map $P(x,\mu) = \mu - x^2$ in Problem 7(a). Compute $DP(x,\mu)$ (where D stands for the partial derivative with respect to x) and show that along the upper branch of fixed points given by $x = (-1 + \sqrt{1+4\mu})/2$ we have
$$DP\left(\frac{-1+\sqrt{1+4\mu}}{2}, \mu\right) = -1$$

4.6. One-Parameter Families of Rotated Vector Fields

at the bifurcation value $\mu = 3/4$. Thus, for $\mu = 3/4$, the map $P(x, \mu)$ has a nonhyperbolic fixed point at $x = 1/2$ and $DP(1/2, 3/4) = -1$. We therefore expect a period-doubling bifurcation or a so-called flip bifurcation for the map $P(x, \mu)$ to occur at the point $(\mu_0, x_0) = (3/4, 1/2)$ in the (μ, x) plane. Show that this is indeed the case by showing that the iterated map

$$P^2(x, \mu) = \mu - (\mu - x^2)^2$$

has a pitchfork bifurcation at the point $(\mu_0, x_0) = (3/4, 1/2)$; i.e., show that the conditions (4) in Section 4.2 are satisfied for the map $F = P^2$. (These conditions reduce to $F(x_0, \mu_0) = x_0, DF(x_0, \mu_0) = 1, D^2F(x_0, \mu_0) = 0, D^3F(x_0, \mu_0) \neq 0, F_\mu(x_0, \mu_0) = 0$, and $DF_\mu(x_0, \mu_0) \neq 0$.) Show that the pitchfork bifurcation for $F = P^2$ is supercritical by showing that for $\mu = 1$, the equation $P^2(x, \mu) = x$ has four solutions. Sketch the bifurcation diagram for P^2, i.e. the locus of points where $P^2(x, \mu) = x$ and show that this implies that the map $P(x, \mu)$ has a period-doubling or flip bifurcation at the point $(\mu_0, \mathbf{x}_0) = (3/4, 1/2)$.

10. Show that the one-dimensional map $P(x, \mu) = \mu x(1 - x)$ of Problem 7(b) has a nonhyperbolic fixed point at $x = 2/3$ at the bifurcation value $\mu = 3$ and that $DP(2/3, 3) = -1$. Show that the map $P(x, \mu)$ has a flip bifurcation at the point $(\mu_0, x_0) = (3, 2/3)$ in the (μ, x) plane. Sketch the bifurcation diagram in the (μ, x) plane.

11. Why can't the one-dimensional maps in Problems 9 and 10 be the Poincaré maps of any two-dimensional system of differential equations? Note that they could be the Poincaré maps of a higher dimensional system (with $n \geq 3$) where x is the distance along the one-dimensional manifold $W^c(\Gamma) \subset \Sigma$ where the center manifold of a periodic orbit Γ of the system intersects a hyperplane Σ normal to Γ.

4.6 One-Parameter Families of Rotated Vector Fields

We next study planar analytic systems

$$\dot{\mathbf{x}} = \mathbf{f}(\mathbf{x}, \mu) \tag{1}$$

which depend on the parameter $\mu \in \mathbf{R}$ in a very specific way. We assume that as the parameter μ increases, the field vectors $\mathbf{f}(\mathbf{x}, \mu)$ or equivalently $(P(x, y, \mu), Q(x, y, \mu))^T$ all rotate in the same sense. If this is the case, then the system (1) is said to define a one-parameter family of rotated vector fields and we can prove some very specific results concerning the bifurcations and global behavior of the limit cycles and separatrix cycles of such a system. These results were established by G. D. F. Duff [7] in 1953 and were later extended by the author [23], [63].

Definition 1. The system (1) with $\mathbf{f} \in C^1(\mathbf{R}^2 \times \mathbf{R})$ is said to define a *one-parameter family of rotated vector fields* if the critical points of (1) are isolated and at all ordinary points of (1) we have

$$\begin{vmatrix} P & Q \\ P_\mu & Q_\mu \end{vmatrix} > 0. \tag{2}$$

If the sense of the above inequality is reversed, then the system (1) is said to define a *one-parameter family of negatively rotated vector fields*.

Note that the condition (2) is equivalent to $\mathbf{f} \wedge \mathbf{f}_\mu > 0$. Since the angle that the field vector $\mathbf{f} = (P, Q)^T$ makes with the x-axis

$$\Theta = \tan^{-1} \frac{Q}{P},$$

it follows that

$$\frac{\partial \Theta}{\partial \mu} = \frac{PQ_\mu - QP_\mu}{P^2 + Q^2}.$$

Hence, condition (2) implies that at each ordinary point $\mathbf{x}$ of (1), where $P^2 + Q^2 \neq 0$, the field vector $\mathbf{f}(\mathbf{x}, \mu)$ at $\mathbf{x}$ rotates in the positive sense as μ increases. If in addition to the condition (2) at each ordinary point of (1) we have $\tan \Theta(x, y, \mu) \to \pm\infty$ as $\mu \to \pm\infty$, or if $\Theta(x, y, \mu)$ varies through π radians as μ varies in $\mathbf{R}$, then (1) is said to define a *semicomplete family of rotated vector fields* or simply a semicomplete family. Any vector field

$$\mathbf{F}(\mathbf{x}) = \begin{pmatrix} X(x, y) \\ Y(x, y) \end{pmatrix}$$

can be embedded in a semicomplete family of rotated vector fields

$$\begin{aligned} \dot{x} &= X(x, y) - \mu Y(x, y) \\ \dot{y} &= \mu X(x, y) + Y(x, y) \end{aligned} \tag{3}$$

with parameter $\mu \in \mathbf{R}$. And as Duff pointed out, any vector field $\mathbf{F} = (X, Y)^T$ can also be embedded in a "complete" family of rotated vector fields

$$\begin{aligned} \dot{x} &= X(x, y) \cos\mu - Y(x, y) \sin\mu \\ \dot{y} &= X(x, y) \sin\mu + Y(x, y) \cos\mu \end{aligned} \tag{4}$$

with parameter $\mu \in (-\pi, \pi]$. The family of rotated vector fields (4) is called *complete* since each vector in the vector field defined by (4) rotates through 2π radians as the parameter μ varies in $(-\pi, \pi]$. Duff also showed that any nonsingular transformation of coordinates with a positive Jacobian determinant takes a semicomplete (or complete) family of rotated vector fields into a semicomplete (or complete) family of rotated vector fields; cf. Problem 1. We first establish the result that limit cycles of any one-parameter family of rotated vector fields expand or contract monotonically as the parameter μ increases. In order to establish this result, we first prove two lemmas which are of some interest in themselves.

4.6. One-Parameter Families of Rotated Vector Fields

Lemma 1. *Cycles of distinct fields of a semicomplete family of rotated vector fields do not intersect.*

Proof. Suppose that Γ_{μ_1} and Γ_{μ_2} are two cycles of the distinct vector fields $\mathbf{F}(\mu_1)$ and $\mathbf{F}(\mu_2)$ defined by (1). Suppose for definiteness that $\mu_1 < \mu_2$ and that the cycles Γ_{μ_1} and Γ_{μ_2} have a point in common. Then at that point

$$\operatorname{Arg} \mathbf{F}(\mu_1) < \operatorname{Arg} \mathbf{F}(\mu_2);$$

i.e., $\mathbf{F}(\mu_2)$ points into the interior of Γ_{μ_1}. But according to Definition 1, this is true at every point on Γ_{μ_1}. Thus, once the trajectory Γ_{μ_2} enters the interior of the Jordan curve Γ_{μ_1}, it can never leave it. This contradicts the fact that the cycle Γ_{μ_2} is a closed curve. Thus, Γ_{μ_1} and Γ_{μ_2} have no point in common.

Lemma 2. *Suppose that the system (1) defines a one-parameter family of rotated vector fields. Then there exists an outer neighborhood U of any externally stable cycle Γ_{μ_0} of (1) such that through every point of U there passes a cycle Γ_μ of (1) where $\mu < \mu_0$ if Γ_{μ_0} is positively oriented and $\mu > \mu_0$ if Γ_{μ_0} is negatively oriented. Corresponding statements hold regarding unstable cycles and inner neighborhoods.*

Proof. Let N_{ε_0} denote the outer, ε_0-neighborhood of Γ_{μ_0} and let $\mathbf{x}$ be any point of N_ε with $0 < \varepsilon \leq \varepsilon_0$. Let $\Gamma(\mathbf{x}, \mu)$ denote the trajectory of (1) passing through the point $\mathbf{x}$ at time $t = 0$. Let ℓ be a line segment normal to Γ_{μ_0} which passes through $\mathbf{x}$. If ε_0 and $|\mu - \mu_0|$ are sufficiently small, ℓ is a transversal to the vector field $\mathbf{F}(\mu)$ in N_{ε_0}. Let $P(\mathbf{x}, \mu)$ be the Poincaré map for the cycle Γ_{μ_0} with $\mathbf{x} \in \ell$. Then since Γ_{μ_0} is externally stable, ℓ meets $\mathbf{x}$, $P(\mathbf{x}, \mu_0)$ and Γ_{μ_0} in that order. Furthermore, by continuity, $P(\mathbf{x}, \mu)$ moves continuously along ℓ as μ varies in a small neighborhood of μ_0, and for $|\mu - \mu_0|$ sufficiently small, the arc of $\Gamma(\mathbf{x}, \mu)$ from $\mathbf{x}$ to $P(\mathbf{x}, \mu)$ is contained in N_{ε_0}. We shall show that for $\varepsilon > 0$ sufficiently small, there is a $\mu < \mu_0$ if Γ_{μ_0} is positively oriented and a $\mu > \mu_0$ if Γ_{μ_0} is negatively oriented such that $P(\mathbf{x}, \mu) = \mathbf{x}$. The result stated will then hold for the neighborhood $U = N_\varepsilon$. The trajectories are differentiable, rectifiable curves. Let s denote the arc length along $\Gamma(\mathbf{x}, \mu_0)$ measured in the direction of increasing t and let n denote the distance taken along the outer normals to this curve; cf. [A–II], p. 110. In N_ε, the angle function $\Theta(\mathbf{x}, \mu)$ satisfies a local Lipschitz condition

$$|\Theta(s, n_1, \mu) - \Theta(s, n_2, \mu)| < M|n_1 - n_2|$$

for some constant M independent of s, n, μ and $\varepsilon < \varepsilon_0$. Also the continuous positive function $\partial \Theta / \partial \mu$ has in N_ε a positive lower bound m independent of ε. Let $n = h(s, \mu)$ be the equation of $\Gamma(\mathbf{x}, \mu)$ so that $h(s, \mu_0) = 0$. Suppose for definiteness that Γ_{μ_0} is positively oriented and let μ decrease from μ_0. It then follows that for $\mu_0 - \mu$ sufficiently small

$$\frac{dh}{ds} > \frac{1}{2}[m(\mu_0 - \mu) - Mh].$$

Integrating this differential inequality (using Gronwall's Lemma) from 0 to L, the length of the arc $\Gamma(\mathbf{x}, \mu)$ from $\mathbf{x}$ to $P(\mathbf{x}, \mu)$, we get

$$h(L, \mu) \geq \frac{m}{M}(\mu_0 - \mu)[1 - e^{-ML/2}] \equiv K(\mu_0 - \mu).$$

Thus, we see that $h(L, \mu) > \varepsilon$ for $\mu_0 - \mu > \varepsilon/K$. But this implies that for such a value of μ the point $P(\mathbf{x}, \mu)$ has moved outward from Γ_{μ_0} on ℓ past $\mathbf{x}$. Since the motion of $P(\mathbf{x}, \mu)$ along ℓ is continuous, there exists an intermediate value μ_1 of μ such that $P(\mathbf{x}, \mu_1) = \mathbf{x}$. It follows that $|\mu_1 - \mu_0| \leq \varepsilon/K$. If $\varepsilon > 0$ is sufficiently small, $\Gamma(\mathbf{x}, \mu)$ remains in N_{ε_0} for $\mu_1 \leq \mu \leq \mu_0$. This proves the result. If Γ_{μ_0} is negatively oriented we need only write $\mu - \mu_0$ in place of $\mu_0 - \mu$ in the above proof.

The next theorem follows from Lemmas 1 and 2. Cf. [7].

Theorem 1. *Stable and unstable limit cycles of a one-parameter family of rotated vector fields* (1) *expand or contract monotonically as the parameter μ varies in a fixed sense and the motion covers an annular neighborhood of the initial position.*

The following table determines the variation of the parameter μ which causes the limit cycle Γ_μ to expand. The opposite variation of μ will cause Γ_μ to contract. $\Delta\mu > 0$ indicates increasing μ. The orientation of Γ_μ is denoted by ω and the stability of Γ_μ by σ where $(+)$ denotes an unstable limit cycle and $(-)$ denotes a stable limit cycle.

ω	+	+	−	−
σ	+	−	+	−
$\Delta\mu$	+	−	−	+

Figure 1. The variation of the parameter μ which causes the limit cycle Γ_μ to expand.

The next theorem, describing a saddle-node bifurcation at a semistable limit cycle of (1), can also be proved using Lemmas 1 and 2 as in [7]. And since $\mathbf{f} \wedge \mathbf{f}_\mu = PQ_\mu - QP_\mu > 0$, it follows from equation (6) in Section 4.5 that the partial derivative of the Poincaré map with respect to the parameter μ is not zero; i.e.,

$$\frac{\partial d(n, \mu)}{\partial \mu} \neq 0$$

where $d(n, \mu)$ is the displacement function along ℓ and n is the distance along the transversal ℓ. Thus, by the implicit function theorem, the relation

4.6. One-Parameter Families of Rotated Vector Fields

$d(n, \mu) = 0$ can be solved for μ as a function of n. It follows that *the only bifurcations that can occur at a multiple limit cycle of a one-parameter family of rotated vector fields are saddle-node bifurcations.*

Theorem 2. *A semistable limit cycle Γ_μ of a one-parameter family of rotated vector fields* (1) *splits into two simple limit cycles, one stable and one unstable, as the parameter μ is varied in one sense and it disappears as μ is varied in the opposite sense.*

The variation of μ which causes the bifurcation of Γ_μ into two hyperbolic limit cycles is determined by the table in Figure 1 where in this case σ denotes the external stability of the semistable limit cycle Γ_μ.

A lemma similar to Lemma 2 can also be proved for separatrix cycles and graphics of (1) and this leads to the following result. Cf. [7].

Theorem 3. *A separatrix cycle or graphic Γ_0 of a one-parameter family of rotated vector fields* (1) *which is isolated from other cycles of* (1) *generates a unique limit cycle on its interior or exterior (which is of the same stability as Γ_0 on its interior or exterior respectively) as the parameter μ is varied in a suitable sense as determined by the table in Figure 1.*

Recall that in the proof of Theorem 3 in Section 3.7 of Chapter 3, it was shown that the Poincaré map is defined on the interior or exterior of any separatrix cycle or graphic of (1) and this is the side on which a limit cycle is generated as μ varies in an appropriate sense. The variation of μ which causes a limit cycle to bifurcate from the separatrix cycle or graphic is determined by the table in Figure 1. In this case, σ denotes the external stability when the Poincaré map is defined on the exterior of the separatrix cycle or graphic and it denotes the negative of the internal stability, i.e., $(-)$ for an interiorly unstable separatrix cycle or graphic and $(+)$ for an interiorly stable separatrix cycle or graphic, when the Poincaré map is defined on the interior of the separatrix cycle or graphic.

For example, if for $\mu = \mu_0$ we have a simple, positively oriented ($\omega = +1$), separatrix loop at a saddle which is stable on its interior (so that $\sigma = +1$), a unique stable limit cycle is generated on its interior as μ increases from μ_0 (by Table 1); cf. Figure 2.

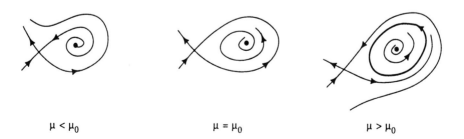

$\mu < \mu_0$ $\qquad\qquad\qquad$ $\mu = \mu_0$ $\qquad\qquad\qquad$ $\mu > \mu_0$

Figure 2. The generation of a limit cycle at a separatrix cycle of (1).

388 4. Nonlinear Systems: Bifurcation Theory

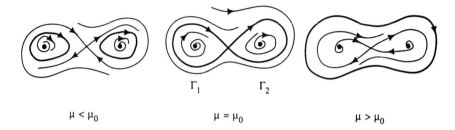

Figure 3. The generation of a limit cycle at a graphic of (1).

As another example, consider the case where for $\mu = \mu_0$ we have a graphic Γ_0 composed of two separatrix cycles Γ_1 and Γ_2 at a saddle point as shown in Figure 3. The orientation is negative so $\omega = -1$. The graphic is externally stable so $\sigma_0 = -1$. Thus, by Table 1 and Theorem 3, a unique, stable limit cycle is generated by the graphic Γ_0 on its exterior as μ increases from μ_0. Similarly, the separatrix cycles Γ_1 and Γ_2 are internally stable so that σ_1 and $\sigma_2 = +1$. Thus, by Table 1 and Theorem 3, a unique, stable limit cycle is generated by each of the separatrix cycles Γ_1 and Γ_2 on their interiors as μ decreases from μ_0. Cf. Figure 3.

Consider one last example of a graphic Γ_0 which is composed of two separatrix cycles Γ_1 and Γ_2 with Γ_1 on the interior of Γ_2 as shown in Figure 4 with $\mu = \mu_0$. In this case, Γ_1 is positively oriented (so that $\omega_1 = +1$) and is internally stable (so that $\sigma_1 = +1$). Thus, Γ_1 generates a unique, stable limit cycle on its interior as μ increases from μ_0. Similarly, Γ_2 is negatively oriented (so that $\omega_2 = -1$) and it is externally stable (so that $\sigma_2 = -1$). Thus, Γ_2 generates a unique, stable limit cycle on its exterior as μ increases from μ_0. And the graphic Γ_0 is negatively oriented (so that $\omega_0 = -1$) and it is internally stable (so that $\sigma_0 = +1$). Thus Γ_0 generates a unique, stable limit cycle on its interior as μ decreases from μ_0. Cf. Figure 4.

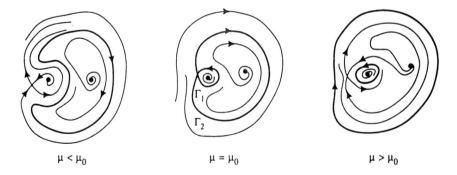

Figure 4. The generation of a limit cycle at a graphic of (1).

4.6. One-Parameter Families of Rotated Vector Fields

Specific examples of these types of homoclinic loop bifurcations are given in Section 4.8 where homoclinic loop bifurcations are discussed for more general vector fields. Of course, we do not have the specific results for more general vector fields that we do for the one-parameter families of rotated vector fields being discussed in this section.

Theorems 1 and 2 above describe the local behavior of any one-parameter family of limit cycles generated by a one-parameter family of rotated vector fields (1). The next theorem, proved by Duff [7] in 1953 and extended by the author in [24] and [63], establishes a result which describes the global behavior of any one-parameter family of limit cycles generated by a semicomplete analytic family of rotated vector fields. Note that in the next theorem, as in Section 4.5, the two limit cycles generated at a saddle-node bifurcation at a semistable limit cycle are considered as belonging to the same one-parameter family of limit cycles since they are both defined by the same branch of the relation $d(n, \mu) = 0$ where $d(n, \mu)$ is the displacement function.

Theorem 4. *Let Γ_μ be a one-parameter family of limit cycles of a semicomplete analytic family of rotated vector fields with parameter μ and let G be the annular region covered by Γ_μ as μ varies in $\mathbf{R}$. Then the inner and outer boundaries of G consist of either a single critical point or a graphic of (1) on the Bendixson sphere.*

We next cite a result describing a Hopf bifurcation at a weak focus of a one-parameter family of rotated vector fields (1). We assume that the origin of the system (1) has been translated to the weak focus, i.e., that the system (1) has been written in the form

$$\dot{\mathbf{x}} = A(\mu)\mathbf{x} + \mathbf{F}(\mathbf{x}, \mu) \tag{5}$$

as in Section 2.7 of Chapter 2. Duff [7] showed that if (5) defines a semicomplete family of rotated vector fields, then if $\det A(\mu_0) \neq 0$ at some $\mu_0 \in \mathbf{R}$, it follows that $\det A(\mu) \neq 0$ for all $\mu \in \mathbf{R}$.

Theorem 5. *Assume that $\mathbf{F} \in C^2(\mathbf{R}^2 \times \mathbf{R})$, that the system (5) defines a one-parameter family of rotated vector fields with parameter $\mu \in \mathbf{R}$, that $\det A(\mu_0) > 0$, trace $A(\mu_0) = 0$, trace $A(\mu) \not\equiv 0$, and that the origin is not a center of (5) for $\mu = \mu_0$. It then follows that the origin of (5) absorbs or generates exactly one limit cycle at the bifurcation value $\mu = \mu_0$.*

The variation of μ from $\mu_0, \Delta\mu$, which causes the bifurcation of a limit cycle, is determined by the table in Figure 1 where σ denotes the stability of the origin of (5) at $\mu = \mu_0$ (which is the same as the stability of the bifurcating limit cycle) and ω denotes the orientation of the bifurcating limit cycle which is the same as the θ-direction in which the flow swirls around the weak focus at the origin of (5) for $\mu = \mu_0$. Note that if the origin of (5) is a weak focus of multiplicity one, then the stability of the origin is determined by the sign of the Liapunov number σ which is given by equation (3) or (3') in Section 4.4.

We cite one final result, established recently by the author [25], which describes how a one-parameter family of limit cycles, Γ_μ, generated by a semicomplete analytic family of rotated vector fields with parameter $\mu \in \mathbf{R}$, terminates. The termination of one-parameter families of periodic orbits of more general vector fields is considered in the next section. In the statement of the next theorem, we use the quantity

$$\rho_\mu = \max_{\mathbf{x} \in \Gamma_\mu} |\mathbf{x}|$$

to measure the maximum distance of the limit cycle Γ_μ from the origin and we say that the orbits in the family become unbounded as $\mu \to \mu_0$ if $\rho_\mu \to \infty$ as $\mu \to \mu_0$.

Theorem 6. *Any one-parameter family of limit cycles Γ_μ generated by a semi-complete family of rotated vector fields (1) either terminates as the parameter μ or the orbits in the family become unbounded or the family terminates at a critical point or on a graphic of (1).*

Some typical bifurcation diagrams of global one-parameter families of limit cycles generated by a semicomplete family of rotated vector fields are shown in Figure 5 where we plot the quantity ρ_μ versus the parameter μ.

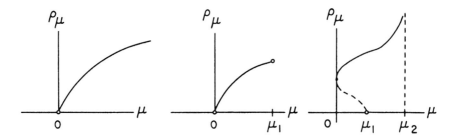

Figure 5. Some typical bifurcation diagrams of global one-parameter families of limit cycles of a one-parameter family of rotated vector fields.

In the first case in Figure 5, a one-parameter family of limit cycles is born in a Hopf bifurcation at a critical point of (1) at $\mu = 0$ and it expands monotonically as $\mu \to \infty$. In the second case, the family is born at a Hopf bifurcation at $\mu = 0$, it expands monotonically with increasing μ, and it terminates on a graphic of (1) at $\mu = \mu_1$. In the third case, there is a Hopf bifurcation at $\mu = \mu_1$, a saddle-node bifurcation at $\mu = 0$, and the family expands monotonically to infinity, i.e., $\rho_\mu \to \infty$, as $\mu \to \mu_2$.

We end this section with some examples that display various types of limit cycle behavior that occur in families of rotated vector fields. Note

4.6. One-Parameter Families of Rotated Vector Fields

that many of our examples are of the form of the system

$$\dot{x} = -y + x[\psi(r) - \mu]$$
$$\dot{y} = x + y[\psi(r) - \mu] \qquad (6)$$

which forms a one-parameter family of rotated vector fields since

$$\begin{vmatrix} P & Q \\ P_\mu & Q_\mu \end{vmatrix} = r^2 > 0$$

at all ordinary points of (6). The bifurcation diagram for (6) is given by the graph of the relation $r[\psi(r) - \mu] = 0$ in the region $r \geq 0$.

Example 1. The system

$$\dot{x} = -\mu x - y + xr^2$$
$$\dot{y} = x - \mu y + yr^2$$

has the form of the system (6) and therefore defines a one-parameter family of rotated vector fields. The only critical point is at the origin, $\det A(\mu) = 1 + \mu^2 > 0$, and trace $A(\mu) = -2\mu$. According to Theorem 5, there is a Hopf bifurcation at the origin at $\mu = 0$; and since for $\mu = 0$ the origin is unstable (since $\dot{r} = r^3$ for $\mu = 0$) and positively oriented, it follows from the table in Figure 1 that a limit cycle is generated as μ increases from zero. The bifurcation diagram in the (μ, r)-plane is given by the graph of $r[r^2 - \mu] = 0$; cf. Figure 6.

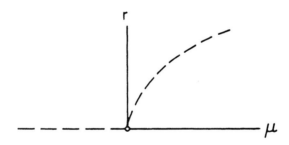

Figure 6. A Hopf bifurcation at the origin.

Example 2. The system

$$\dot{x} = -y + x[-\mu + (r^2 - 1)^2]$$
$$\dot{y} = x + y[-\mu + (r^2 - 1)^2]$$

forms a one-parameter family of rotated vector fields with parameter μ. The origin is the only critical point of this system, $\det A(\mu) = 1 + (1 - \mu)^2 > 0$, and trace $A(\mu) = 2(1 - \mu)$. According to Theorem 5, there is

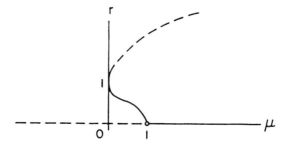

Figure 7. The bifurcation diagram for the system in Example 2.

a Hopf bifurcation at the origin at $\mu = 1$, and since the origin is stable (cf. equation (3) in Section 4.4) and positively oriented, according to the table in Figure 1, a one-parameter family of limit cycles is generated as μ decreases from one. The bifurcation diagram is given by the graph of the relation $r[-\mu + (r^2 - 1)^2] = 0$; cf. Figure 7. We see that there is a saddle-node bifurcation at the semistable limit cycle $\gamma_0(t) = (\cos t, \sin t)^T$ at the bifurcation value $\mu = 0$. The one-parameter family of limit cycles born at the Hopf bifurcation at $\mu = 1$ terminates as the parameter and the orbits in the family increase without bound.

The system considered in the next example satisfies the condition (2) in Definition 1 except on a curve $G(x, y) = 0$ which is not a trajectory of (1). The author has recently established that all of the results in this section hold for such systems which are referred to as *one-parameter families of rotated vector fields* (mod $G = 0$); cf. [23] and [63].

Example 3. Consider the system

$$\dot{x} = -x + y^2$$
$$\dot{y} = -\mu x + y + \mu y^2 - xy.$$

We have

$$\begin{vmatrix} P & Q \\ P_\mu & Q_\mu \end{vmatrix} = (-x + y^2)^2 > 0$$

except on the parabola $x = y^2$ which is not a trajectory of this system. The critical points are at the origin, which is a saddle, and at $(1, \pm 1)$. Since

$$D\mathbf{f}(1, \pm 1) = \begin{bmatrix} -1 & \pm 2 \\ -\mu \mp 1 & \pm 2\mu \end{bmatrix},$$

we have $\det A(\mu) = 2 > 0$ and $\operatorname{tr} A(\mu) = -1 \pm 2\mu$ at the critical points $(1, \pm 1)$. Since this system is invariant under the transformation $(x, y, \mu) \to (x, -y, -\mu)$, we need only consider $\mu \geq 0$. According to Theorem 5, there is a Hopf bifurcation at the critical point $(1, 1)$ at $\mu = 1/2$. By equation (3')

4.6. One-Parameter Families of Rotated Vector Fields

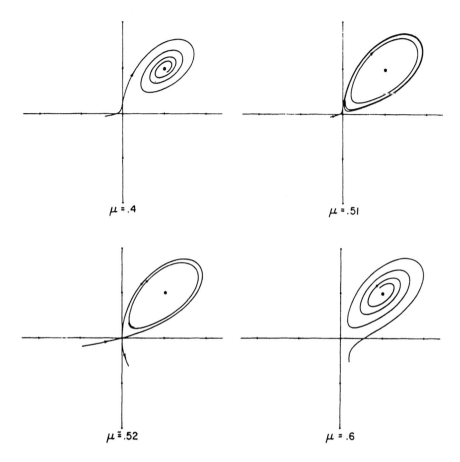

Figure 8. Phase portraits for the system in Example 3.

in Section 4.4, $\sigma < 0$ for $\mu = 1/2$ and therefore since $\omega = -1$, it follows from the table in Figure 1 that a unique limit cycle is generated at the critical point $(1, 1)$ as μ increases from $\mu = 1/2$. This stable, negatively oriented limit cycle expands monotonically with increasing μ until it intersects the saddle at the origin and forms a separatrix cycle at a bifurcation value $\mu = \mu_1$ which has been numerically determined to be approximately .52. The bifurcation diagram is the same as the one shown in the second diagram in Figure 5. Numerically drawn phase portraits for various values of $\mu \in [.4, .6]$ are shown in Figure 8.

Remark. It was stated by Duff in [7] and proved by the author in [63] that a rotation of the vector field causes the separatrices at a hyperbolic saddle to precess in such a way that, after a (positive) rotation of the field vectors through π radians, a stable separatrix has turned (in the positive sense about the saddle point) into the position of one of the unstable separatrices of the initial field. In particular, a rotation of a vector field with a saddle connection will cause the saddle connection to break.

PROBLEM SET 6

1. Show that any nonsingular transformation with a positive Jacobian determinant takes a one-parameter family of rotated vector fields into a one-parameter family of rotated vector fields.

2. (a) Show that the system
$$\dot{x} = \mu x + y - xr^2$$
$$\dot{y} = -x + \mu y - yr^2$$
defines a one-parameter family of rotated vector fields. Use Theorem 5 and the table in Figure 1 to determine whether the Hopf bifurcation at the origin of this system is subcritical or supercritical. Draw the bifurcation diagram.

   (b) Show that the system
$$\dot{x} = \mu x - y + (ax - by)(x^2 + y^2) + O(|\mathbf{x}|^4)$$
$$\dot{y} = x + \mu y + (ay + bx)(x^2 + y^2) + O(|\mathbf{x}|^4)$$
of Problem 1(b) in Section 4.4 defines a one-parameter family of negatively rotated vector fields with parameter $\mu \in \mathbf{R}$ in some neighborhood of the origin. Determine whether the Hopf bifurcation at the origin is subcritical or supercritical. **Hint:** From equation (3) in Section 4.4, it follows that $\sigma = 9\pi a$.

3. Show that the system
$$\dot{x} = y(1+x) + \mu x + (\mu - 1)x^2$$
$$\dot{y} = -x(1+x)$$
satisfies the condition (2) except on the vertical lines $x = 0$ and $x = -1$. Use the results of this section to show that there is a subcritical Hopf bifurcation at the origin at the bifurcation value $\mu = 0$. How must the one-parameter family of limit cycles generated at the Hopf bifurcation at $\mu = 0$ terminate according to Theorem 6? Draw the bifurcation diagram.

4. Draw the global phase portraits for the system in Problem 3 for $-3 < \mu < 1$. **Hint:** There is a saddle-node at the point $(0, \pm 1, 0)$ at infinity.

5. Draw the global phase portraits for the system in Example 3 for the parameter range $-2 < \mu < 2$. **Hint:** There is a saddle-node at the point $(\pm 1, 0, 0)$ at infinity.

6. Show that the system

$$\dot{x} = y + y^2$$
$$\dot{y} = -2x + \mu y - xy + (\mu + 1)y^2$$

satisfies condition (2) except on the horizontal lines $y = 0$ and $y = -1$. Use the results of this section to show that there is a supercritical Hopf bifurcation at the origin at the value $\mu = 0$. Discuss the global behavior of this limit cycle. Draw the global phase portraits for the parameter range $-1 < \mu < 1$.

7. Show that the system

$$\dot{x} = -x + y + y^2$$
$$\dot{y} = -\mu x + (\mu + 4)y + (\mu - 2)y^2 - xy$$

satisfies the condition (2) except on the curve $x = y + y^2$ which is not a trajectory of this system. Use the results of this section to show that there is a subcritical Hopf bifurcation at the upper critical point and a supercritical Hopf bifurcation at the lower critical point of this system at the bifurcation value $\mu = 1$. Discuss the global behavior of these limit cycles. Draw the global phase portraits for the parameter range $0 < \mu < 4$.

4.7 The Global Behavior of One-Parameter Families of Periodic Orbits

In this section we discuss the global behavior of one-parameter families of periodic orbits of a system of differential equations

$$\dot{\mathbf{x}} = \mathbf{f}(\mathbf{x}, \mu) \qquad (1)$$

depending on a parameter $\mu \in \mathbf{R}$. We assume that $\mathbf{f}$ is a real, analytic function of $\mathbf{x}$ and μ and that the components of $\mathbf{f}$ are relatively prime. The global behavior of families of periodic orbits has been a topic of recent research interest. We cite some of the recent results on this topic which generalize the corresponding results in Section 4.6 and, in particular, we cite a classical result established in 1931 by A. Wintner referred to as Wintner's Principle of Natural Termination.

In Section 4.6 we saw that the only kind of bifurcation that occurs at a nonhyperbolic limit cycle of a family of rotated vector fields is a saddle-node bifurcation. We regard the two limit cycles generated at a saddle-node bifurcation as belonging to the same one-parameter family of limit cycles; e.g., we can use the distance along a normal arc to the family as the parameter. For families of rotated vector fields, we have the result in Theorem 6 of Section 4.6 that any one-parameter family of limit cycles expands or contracts monotonically with the parameter until the parameter or the size of the orbits in the family becomes unbounded or until the family terminates at a critical point or on a graphic of (1).

The next examples show that we cannot expect this simple type of behavior in general. The first example shows that, in general, limit cycles do not expand or contract monotonically with the parameter. This permits "cyclic families" to occur as illustrated in the second example. And the third example shows that several one-parameter families of limit cycles can bifurcate from a nonhyperbolic limit cycle.

Example 1. Consider the system

$$\dot{x} = -y + x[(x-\mu)^2 + y^2 - 1]$$
$$\dot{y} = x + y[(x-\mu)^2 + y^2 - 1].$$

According to Theorem 1 and equation (3) in Section 4.4, there is a one-parameter family of limit cycles generated in a Hopf bifurcation at the origin as μ increases from the bifurcation value $\mu = -1$. This family terminates in a Hopf bifurcation at the origin as the parameter μ approaches the bifurcation value $\mu = 1$ from the left. Some of the limit cycles in this one-parameter family of limit cycles are shown in Figure 1 for the parameter range $0 \leq \mu < 1$. Since this system is invariant under the transformation $(x, y, \mu) \to (-x, -y, -\mu)$, the limit cycles for $\mu \in (-1, 0]$ are obtained by reflecting those in Figure 1 about the origin. We see that the orbits do not expand or contract monotonically with the parameter. The bifurcation diagram for this system is shown in Figure 2.

Remark 1. Even though the growth of the limit cycles in a one-parameter family of limit cycles is, in general, not monotone, we can still determine whether a hyperbolic limit cycle Γ_0 expands or contracts at a point $\mathbf{x}_0 \in \Gamma_0$ by computing the rate of change of the distance s along a normal ℓ to the limit cycle Γ_0: $\mathbf{x} = \gamma_0(t)$ at the point $\mathbf{x}_0 = \gamma_0(t_0)$; i.e.

$$\frac{ds}{d\mu} = -\frac{d_\mu(0, \mu_0)}{d_s(0, \mu_0)}$$

where $d(s, \mu) = P(s, \mu) - s$ is the displacement function along ℓ,

$$d_s(0, \mu_0) = e^{\int_0^{T_0} \nabla \cdot \mathbf{f}(\gamma_0(t), \mu_0) dt} - 1$$

by Theorem 2 in Section 3.4 of Chapter 3, and

$$d_\mu(0, \mu_0) = \frac{-\omega_0}{|\mathbf{f}(\mathbf{x}_0, \mu_0)|} \int_{t_0}^{T_0+t_0} e^{\int_{t_0}^{T_0+t_0} \nabla \cdot \mathbf{f}(\gamma_0(\tau), \mu_0) d\tau} \mathbf{f} \wedge \mathbf{f}_\mu(\gamma_0(t), \mu_0) dt$$

as in equation (6) in Section 4.5. All of the quantities which determine the sign of $ds/d\mu$, i.e., which determine whether Γ_0 expands or contracts at the point $\mathbf{x}_0 = \gamma(t_0) \in \Gamma_0$, have the same sign along Γ_0 except the integral

$$I_0 \equiv \int_{t_0}^{T_0+t_0} e^{\int_t^{T_0+t_0} \nabla \cdot \mathbf{f}(\gamma_0(\tau), \mu_0) d\tau} \mathbf{f} \wedge \mathbf{f}_\mu(\gamma_0(t), \mu_0) dt.$$

4.7. The Global Behavior of One-Parameter Families

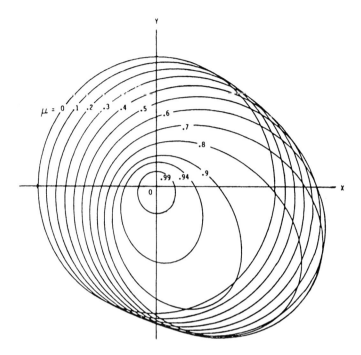

Figure 1. Part of the one-parameter family of limit cycles of the system in Example 1.

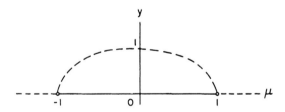

Figure 2. The bifurcation diagram for the system in Example 1.

And the sign of this integral determines whether the limit cycle Γ_0 in the one-parameter family of limit cycles is expanding or contracting with the parameter μ according to the following table which is similar to the table in Figure 1 of Section 4.6. The following table determines the change in $\Delta\mu = \mu - \mu_0$ which causes an expansion of the limit cycle at the point $x_0 = \gamma(t_0) \in \Gamma_0$. As in Section 4.6, ω denotes the orientation of Γ_0 and σ its stability.

ωI_0	+	+	−	−
σ	+	−	+	−
$\Delta \mu$	+	−	−	+

For a one-parameter family of rotated vector fields, $\mathbf{f} \wedge \mathbf{f}_\mu > 0$ and therefore the integral I_0 is positive at all points on Γ_0. The above formula for $ds/d\mu$ is then equivalent to a formula given by Duff [7] in 1953.

Example 2. Consider the system

$$\dot{x} = -y + x[(r^2 - 2)^2 + \mu^2 - 1]$$
$$\dot{y} = x + y[(r^2 - 2)^2 + \mu^2 - 1].$$

The bifurcation diagram is given by the graph of the relation $(r^2 - 2)^2 + \mu^2 - 1 = 0$ in the (μ, r)-plane. It is shown in Figure 3. We see that there are saddle-node bifurcations at the nonhyperbolic periodic orbits represented by $\gamma(t) = \sqrt{2}(\cos t, \sin t)^T$ at the bifurcation values $\mu = \pm 1$. This type of one-parameter family of periodic orbits is called a cyclic family. Loosely speaking a *cyclic family* is one that has a closed-loop bifurcation diagram.

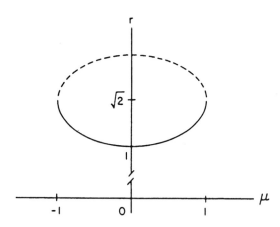

Figure 3. The bifurcation diagram for the system in Example 2.

Example 3. Consider the analytic system

$$\dot{x} = -y + x[\mu - (r^2 - 1)^2][\mu - (r^2 - 1)][\mu + (r^2 - 1)]$$
$$\dot{y} = x + y[\mu - (r^2 - 1)^2][\mu - (r^2 - 1)][\mu + (r^2 - 1)].$$

4.7. The Global Behavior of One-Parameter Families

The bifurcation diagram is given by the graph of the relation

$$[\mu - (r^2 - 1)^2][\mu - (r^2 - 1)][\mu + (r^2 - 1)] = 0$$

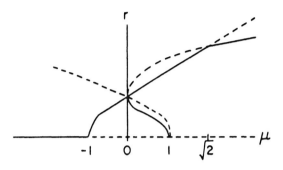

Figure 4. The bifurcation diagram for the system in Example 3.

in the (μ, r)-plane with $r \geq 0$. It is shown in Figure 4. The main point of this rather complicated example is to show that we can have several one-parameter families of limit cycles passing through a nonhyperbolic periodic orbit. In this case there are three one-parameter families passing through the point $(0, 1)$ in the bifurcation diagram. There is also a Hopf bifurcation at the origin at the bifurcation value $\mu = -1$.

As in the last example, the bifurcation diagram can be quite complicated; however, in 1931 A. Wintner [34] showed that, in the case of analytic systems (1), there are at most a finite number of one-parameter families of periodic orbits that bifurcate from a nonhyperbolic periodic orbit and any one-parameter family of periodic orbits can be continued through a bifurcation in a unique way. Winter used Puiseux series in order to establish this result. Any one-parameter family of periodic orbits generates a two-dimensional surface in $\mathbf{R}^n \times \mathbf{R}$ where each cross-section of the surface obtained by holding μ constant, $\mu = \mu_0$, is a periodic orbit Γ_0 of (1); cf. Figure 3 in Section 4.5. In this context, a cyclic family is simply a two-dimensional torus in $\mathbf{R}^n \times \mathbf{R}$. If a one-parameter family of periodic orbits of an analytic system (1) cannot be analytically extended to a larger one-parameter family of periodic orbits, it is called a *maximal family*. The question of how a maximal one-parameter family of periodic orbits terminates is answered by the following theorem proved by A. Wintner in 1931. Cf. [34].

Theorem 1 (Wintner's Principle of Natural Termination). *Any maximal, one-parameter family of periodic orbits of an analytic system* (1) *is either cyclic or it terminates as either the parameter μ, the periods T_μ, or the periodic orbits Γ_μ in the family become unbounded; or the family* terminates at an equilibrium point or at a period-doubling bifurcation orbit of (1).

As was pointed out in Section 4.5, period-doubling bifurcations do not occur in planar systems and it can be shown that the only way that the periods T_μ of the periodic orbits Γ_μ in a one-parameter family of periodic orbits of a planar system can become unbounded is when the orbits Γ_μ approach a graphic or degenerate critical point of the system. Hence, for planar analytic systems, we have the following more specific result, recently established by the author [25, 39], concerning the termination of a maximal one-parameter family of limit cycles.

Theorem 2 (Perko's Planar Termination Principle). *Any maximal, one-parameter family of limit cycles of a planar, relatively prime, analytic system (1) is either cyclic or it terminates as either the parameter μ or the limit cycles in the family become unbounded; or the family terminates at a critical point or on a graphic of (1).*

We see that, except for the possible occurrence of cyclic families, this planar termination principle for general analytic systems is exactly the same as the termination principle for analytic families of rotated vector fields given by Theorem 6 in Section 4.6. Some results on the termination of the various family branches of periodic orbits of nonanalytic systems (1) have recently been given by Alligood, Mallet-Paret and Yorke [1]. Their results extend Wintner's Principle of Natural Termination to C^1-systems provided that we include the possibility of a family terminating as the "virtual periods" become unbounded.

PROBLEM SET 7

1. Show that the system in Example 1 experiences a subcritical Hopf bifurcation at the origin at the bifurcation values $\mu = \pm 1$.

2. Show that the system in Example 3 experiences a Hopf bifurcation at the origin at the bifurcation value $\mu = -1$.

3. Draw the bifurcation diagram and describe the various bifurcations that take place in the system

$$\dot{x} = -y + x[(r^2 - 2)^2 + \mu^2 - 1][r^2 + 2\mu^2 - 2]$$
$$\dot{y} = x + y[(r^2 - 2)^2 + \mu^2 - 1][r^2 + 2\mu^2 - 2].$$

Describe the termination of all of the noncyclic, maximal, one-parameter families of limit cycles of this system.

4. Same as Problem 3 for the system

$$\dot{x} = -y + x[(r^2 - 2)^2 + \mu^2 - 1][r^2 + \mu^2 - 3]$$
$$\dot{y} = x + y[(r^2 - 2)^2 + \mu^2 - 1][r^2 + \mu^2 - 3].$$

4.8 Homoclinic Bifurcations

In Section 4.6, we saw that a separatrix cycle or homoclinic loop of a planar family of rotated vector fields

$$\dot{\mathbf{x}} = \mathbf{f}(\mathbf{x}, \mu) \tag{1}$$

generates a limit cycle as the parameter μ is varied in a certain sense, i.e., as the vector field $\mathbf{f}$ is rotated in a suitable sense, described by the table in Figure 1 in Section 4.6. In this section, we first of all consider homoclinic loop bifurcations for general planar vector fields

$$\dot{\mathbf{x}} = \mathbf{f}(\mathbf{x}) \tag{2}$$

and then we look at some of the very interesting phenomena that result in higher dimensional systems when the Poincaré map has a transverse homoclinic orbit. Transverse homoclinic orbits for the higher dimensional system (2) with $n \geq 3$ typically result from a tangential homoclinic bifurcation. These concepts are defined later in this section.

Even when the planar system (2) does not define a family of rotated vector fields, we still have a result regarding the bifurcation of limit cycles from a separatrix cycle similar to Theorem 3 in Section 4.6. We assume that (2) is a planar analytic system which has a separatrix cycle S_0 at a topological saddle $\mathbf{x}_0$. The separatrix cycle S_0 is said to be a *simple separatrix cycle* if the quantity

$$\sigma_0 \equiv \nabla \cdot \mathbf{f}(\mathbf{x}_0) \neq 0.$$

Otherwise, it is called a *multiple separatrix cycle*. A separatrix cycle S_0 is called stable or unstable if the displacement function $d(s)$ satisfies $d(s) < 0$ or $d(s) > 0$ respectively for all s in some neighborhood of $s = 0$ where $d(s)$ is defined; i.e., S_0 is stable (or unstable) if all of the trajectories in some inner or outer neighborhood of S_0 approach S_0 as $t \to \infty$ (or as $t \to -\infty$). The following theorem is proved using the displacement function as in Section 3.4 of Chapter 3; cf. [A–II], p. 304.

Theorem 1. *Let $\mathbf{x}_0$ be a topological saddle of the planar analytic system (2) and let S_0 be a simple separatrix cycle at $\mathbf{x}_0$. Then S_0 is stable iff $\sigma_0 < 0$.*

The following theorem, analogous to Theorem 1 in Section 4.5, is proved on pp. 309–312 in [A–II].

Theorem 2. *If S_0 is a simple separatrix cycle at a topological saddle $\mathbf{x}_0$ of the planar analytic system (2), then*

(i) *there is a $\delta > 0$ and an $\varepsilon > 0$ such that any system ε-close to (2) in the C^1-norm has at most one limit cycle in a δ-neighborhood of S_0, $N_\delta(S_0)$, and*

(ii) *for any $\delta > 0$ and $\varepsilon > 0$, there is an analytic system which is ε-close to (2) in the C^1-norm and has exactly one simple limit cycle in $N_\delta(S_0)$.*

Furthermore, if such a limit cycle exists, it is of the same stability as S_0; i.e., it is stable if $\sigma_0 < 0$ and unstable if $\sigma_0 > 0$.

Remark 1. If S_0 is a multiple separatrix cycle of (2), i.e., if $\sigma_0 = 0$, then it is shown on p. 319 in [A–II] that for all $\delta > 0$ and $\varepsilon > 0$, there is an analytic system which is ε-close to (2) in the C^1-norm which has at least two limit cycles in $N_\delta(S_0)$.

Example 1. The system

$$\dot{x} = y$$
$$\dot{y} = x + x^2 - xy + \mu y$$

defines a semicomplete family of rotated vector fields with parameter $\mu \in \mathbf{R}$. The Jacobian is

$$D\mathbf{f}(x, y) = \begin{bmatrix} 0 & 1 \\ 1 + 2x - y & -x + \mu \end{bmatrix}.$$

There is a saddle at the origin with $\sigma_0 = \mu$. There is a node or focus at $(-1, 0)$ and trace $D\mathbf{f}(-1, 0) = 1 + \mu$. Furthermore, for $\mu = -1$, if we translate the origin to $(-1, 0)$ and use equation (3') in Section 4.4, we find $\sigma = -3\pi/2 < 0$; i.e., there is a stable focus at $(-1, 0)$ at $\mu = -1$. It therefore follows from Theorem 5 and the table in Figure 1 in Section 4.6 that a unique stable limit cycle is generated in a supercritical Hopf bifurcation at $(-1, 0)$ at the bifurcation value $\mu = -1$. According to Theorems 1 and 4 in Section 4.6, this limit cycle expands monotonically with increasing μ until it intersects the saddle at the origin and forms a stable separatrix cycle S_0 at a bifurcation value $\mu = \mu_0$. Since the separatrix cycle S_0 is stable, it follows from Theorem 1 that $\sigma = \mu_0 \leq 0$. Numerical computation shows that $\mu_0 \simeq -.85$. The phase portraits for this system are shown in Figure 1.

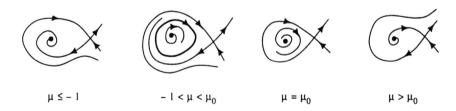

$\mu \leq -1$ $-1 < \mu < \mu_0$ $\mu = \mu_0$ $\mu > \mu_0$

Figure 1. The phase portraits for the system in Example 1.

The limit cycle for $\mu = -.9$ and the separatrix cycle for $\mu = \mu_0 \simeq -.85$ are shown in the numerical plots in Figure 2. The bifurcation diagram for this

4.8. Homoclinic Bifurcations

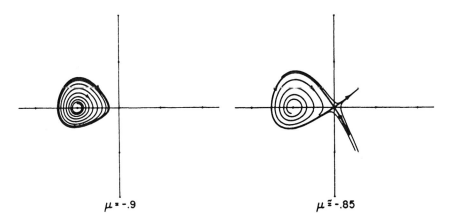

Figure 2. The limit cycle and the separatrix cycle for the system in Example 1.

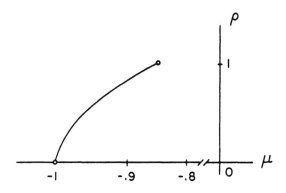

Figure 3. The bifurcation diagram for the system in Example 1.

system is shown in Figure 3 where ρ denotes the maximum distance of the limit cycle from the critical point at $(-1, 0)$. We see that a unique stable limit cycle bifurcates from the simple separatrix cycle S_0 as μ decreases from the bifurcation value $\mu = \mu_0$. And this is consistent with the table in Figure 1 in Section 4.6.

Remark 2. For planar systems, the same sort of results concerning the bifurcation of limit cycles from a graphic of (1) hold; an example is given in Problem 1.

The next example shows how a separatrix cycle can be obtained from a Hamiltonian system with a homoclinic loop. A limit cycle can then be made to bifurcate from the separatrix cycle by rotating the vector field in an appropriate sense.

Example 2. The system

$$\dot{x} = y$$
$$\dot{y} = x + x^2 \qquad (3)$$

was considered in Example 3 of Section 3.3 in Chapter 3. It is a Hamiltonian system with

$$H(x,y) = \frac{y^2}{2} - \frac{x^2}{2} - \frac{x^3}{3}.$$

It has a homoclinic loop S_0 at the saddle at the origin given by a motion on the curve

$$y^2 = x^2 + 2x^3/3.$$

There is a center inside S_0, the cycles being given by $H(x,y) = C$ with $C < 0$; cf. Figure 2 in Section 3.3 of Chapter 3. We next modify this system as follows:

$$\dot{x} = X(x,y,\alpha) \equiv y - \alpha(y^2 - x^2 - 2x^3/3)(x + x^2)$$
$$\dot{y} = Y(x,y,\alpha) \equiv x + x^2 + \alpha(y^2 - x^2 - 2x^3/3)y. \qquad (4)$$

Computing

$$\begin{vmatrix} X & Y \\ X_\alpha & Y_\alpha \end{vmatrix} = 2H(x,y)[y^2 + (x + x^2)^2],$$

we see that inside the homoclinic loop S_0, where $H(x,y) = C$ with $C < 0$, the vector field defined by (4) is rotated in the negative sense; i.e., for $\alpha > 0$ the trajectories of (4) cross the closed curves $H(x,y) = C$ with $C < 0$ from their exterior to their interior. The critical point $(-1, 0)$ of (4) is a stable focus. The homoclinic loop $H(x,y) = 0$ is preserved and we have the phase portrait shown in Figure 4 below.

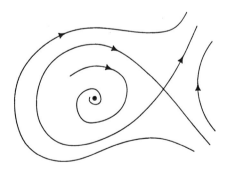

Figure 4. The phase portrait of the system (4) with an unstable separatrix cycle S_0.

4.8. Homoclinic Bifurcations

We now fix α at some positive value, say $\alpha = .1$, and rotate the vector field defined by (4) as in equation (5) in Section 4.6 in order to cause a limit cycle to bifurcate from the separatrix cycle S_0 in Figure 1. We obtain the one-parameter family of rotated vector fields

$$\dot{x} = P(x,y,\mu) \equiv X(x,y,.1)\cos\mu - Y(x,y,.1)\sin\mu \\ \dot{y} = Q(x,y,\mu) \equiv X(x,y,.1)\sin\mu + Y(x,y,.1)\cos\mu. \quad (5)$$

Since the separatrix cycle S_0 in Figure 1 is negatively oriented (with $\omega = -1$) and unstable on its interior (with the negative of the interior stability $\sigma = -1$; see the paragraph following Theorem 3 in Section 4.6), it follows from the table in Figure 1 in Section 4.6 that a limit cycle bifurcates from the separatrix cycle S_0 as μ increases from zero. The phase portraits for the system (5) are shown in Figure 5. The trace of the linear part of (5) at

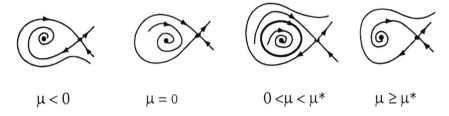

$\mu < 0$ \qquad $\mu = 0$ \qquad $0 < \mu < \mu^*$ \qquad $\mu \geq \mu^*$

Figure 5. The phase portraits for the system (5).

the critical point $(-1, 0)$ is easily computed. It is given by

$$\tau_\mu = \text{trace } Df(-1,0,\mu) = P_x(-1,0,\mu) + Q_y(-1,0,\mu) \\ = 2\left[\sin\mu - \frac{.1}{3}\cos\mu\right].$$

Clearly $\tau_\mu = 0$ at $\mu = \mu^* = \tan^{-1}(.1/3) \simeq .033$ and according to Theorem 5 and the table in Section 4.5 there is a subcritical Hopf bifurcation at the critical point $(-1, 0)$ at the bifurcation value $\mu = \mu^* \simeq .033$. Cf. Figure 5. Also, the limit cycle for system (5) with $\mu = .01$ is shown in Figure 6.

We see that for planar systems, one or more limit cycles can bifurcate from a separatrix cycle. For analytic systems, it follows from Dulac's Theorem in Section 3.3 of Chapter 3 that at most a finite number of limit cycles can bifurcate from a homoclinic loop. However, in higher dimensions, even in the analytic case, it is possible to have an infinite number of limit cycles or periodic orbits bifurcating from a homoclinic loop. This was seen to be the case for the Lorenz system discussed in Section 4.5. Indeed, it was the occurrence of a homoclinic loop, similar to that shown in Figure 7, that started the chain of events that led to the strange behavior encountered in the Lorenz system. We next give a brief discussion of transverse and

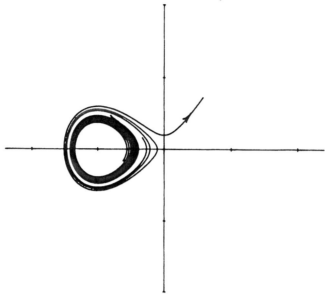

Figure 6. The limit cycle of the system (5) with $\mu = .01$ generated in a homoclinic loop bifurcation at $\mu = 0$.

tangential homoclinic orbits and point out that if the Poincaré map of an n-dimensional system with $n \geq 3$ has a transverse homoclinic orbit, then we typically get the strange kind of behavior encountered in Sparrow's numerical study of the Lorenz system; i.e., a kind of chaotic dynamics results when the Poincaré map has a transverse homoclinic orbit.

In Figure 7, the one-dimensional unstable manifold of the origin intersects the two-dimensional stable manifold of the origin tangentially and forms a homoclinic loop, Γ_0. Figure 8 shows a transverse heteroclinic orbit, Γ_0, where $W^s(\mathbf{x}_0)$ intersects $W^u(\mathbf{x}_1)$ transversally; i.e., at any point $\mathbf{q} \in \Gamma_0$, the sum of the tangent spaces $T_\mathbf{q} W^s(\mathbf{x}_0)$ and $T_\mathbf{q} W^u(\mathbf{x}_1)$ is equal to $\mathbf{R}^3$. It is not possible for a dynamical system defined by (2) to have a transverse homoclinic orbit since $\dim W^s(\mathbf{x}_0) + \dim W^u(\mathbf{x}_0) \leq n$ while transversality requires that $\dim W^s(\mathbf{x}_0) + \dim W^u(\mathbf{x}_0) > n$.

Even though dynamical systems do not have transverse homoclinic orbits, it is possible for the Poincaré map of a periodic orbit to have a transverse homoclinic orbit. And, as we shall see, this has some rather amazing consequences. In order to make these ideas more precise, we first give some basic definitions and then state the Stable Manifold Theorem for maps. In what follows we assume that the map $\mathbf{P} \colon \mathbf{R}^n \to \mathbf{R}^n$ is a C^1-diffeomorphism; i.e., a continuously differentiable map with a continuously differentiable inverse.

4.8. Homoclinic Bifurcations

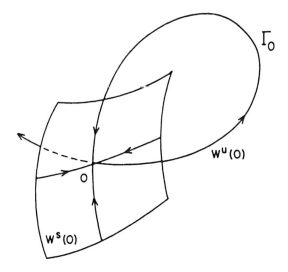

Figure 7. A tangential intersection of the unstable and stable manifolds of the origin which forms a homoclinic loop Γ_0.

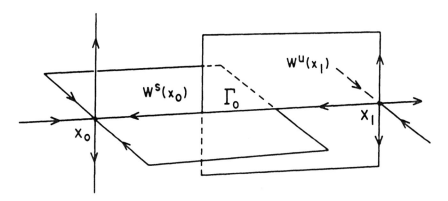

Figure 8. A transverse heteroclinic orbit Γ_0 of a dynamical system in $\mathbf{R}^3$.

Definition 1. A point $\mathbf{x}_0 \in \mathbf{R}^n$ is a *hyperbolic fixed point* of the diffeomorphism $\mathbf{P}\colon \mathbf{R}^n \to \mathbf{R}^n$ if $\mathbf{P}(\mathbf{x}_0) = \mathbf{x}_0$ and $D\mathbf{P}(\mathbf{x}_0)$ has no eigenvalue of unit modulus. An *orbit* of a map $\mathbf{P}\colon \mathbf{R}^n \to \mathbf{R}^n$ is a sequence of points $\{\mathbf{x}_j\}$ defined by $\mathbf{x}_{j+1} = \mathbf{P}(\mathbf{x}_j)$.

The eigenvalues and generalized eigenvectors of a linear map $\mathbf{x} \to A\mathbf{x}$ are defined as usual; cf. Chapter 1. The *stable, center and unstable subspaces of a linear map* $\mathbf{x} \to A\mathbf{x}$, where A is a real, $n \times n$ matrix, are defined as

$$E^s = \mathrm{Span}\{\mathbf{u}_j, \mathbf{v}_j \mid |\lambda_j| < 1\}$$

$$E^c = \text{Span}\{\mathbf{u}_j, \mathbf{v}_j \mid |\lambda_j| = 1\}$$
$$E^u = \text{Span}\{\mathbf{u}_j, \mathbf{v}_j \mid |\lambda_j| > 1\}$$

where we are using the same notation as in Definition 1 of Section 1.9 in Chapter 1. In the following theorem, we assume that the diffeomorphism $\mathbf{P}: \mathbf{R}^n \to \mathbf{R}^n$ has a hyperbolic fixed point which has been translated to the origin; cf. Theorem 1.4.2, p. 18 in [G/H] or Theorem 6.1, p. 27 in [Ru].

Theorem 3 (**The Stable Manifold Theorem for Maps**). *Let $\mathbf{P}: \mathbf{R}^n \to \mathbf{R}^n$ be a C^1-diffeomorphism with a hyperbolic fixed point $0 \in \mathbf{R}^n$. Then there exist local stable and unstable invariant manifolds S and U of class C^1 tangent to the stable and unstable subspaces E^s and E^u of $D\mathbf{P}(0)$ and of the same dimension such that for all $\mathbf{x} \in S$ and $n \geq 0$, $\mathbf{P}^n(\mathbf{x}) \in S$ and $\mathbf{P}^n(\mathbf{x}) \to 0$ as $n \to \infty$ and for all $\mathbf{x} \in U$ and $n \geq 0$, $\mathbf{P}^{-n}(\mathbf{x}) \in U$ and $\mathbf{P}^{-n}(\mathbf{x}) \to 0$ as $n \to \infty$.*

We define the global stable and unstable manifolds

$$W^s(0) = \bigcup_{n \geq 0} \mathbf{P}^{-n}(S)$$

and

$$W^u(0) = \bigcup_{n \geq 0} \mathbf{P}^n(U).$$

It can be shown that if $\mathbf{P}$ is a C^r-diffeomorphism, then $W^s(0)$ and $W^u(0)$ are invariant manifolds of class C^r.

Let us now assume that the C^1-diffeomorphism $\mathbf{P}: \mathbf{R}^n \to \mathbf{R}^n$ has a hyperbolic fixed point (which we assume has been translated to the origin) whose stable and unstable manifolds intersect transversally at a point $\mathbf{x}_0 \in \mathbf{R}^n$; cf. Figure 9.

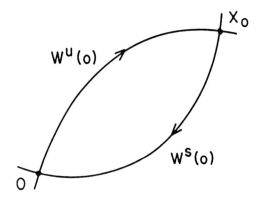

Figure 9. A transverse homoclinic point $\mathbf{x}_0$.

4.8. Homoclinic Bifurcations

Since $W^s(0)$ and $W^u(0)$ are invariant under $\mathbf{P}$, iterates of $\mathbf{x}_0, \mathbf{P}(\mathbf{x}_0)$, $\mathbf{P}^2(\mathbf{x}_0),\ldots$ as well as $\mathbf{P}^{-1}(\mathbf{x}_0), \mathbf{P}^{-2}(\mathbf{x}_0),\ldots$ also lie in $W^s(0) \cap W^u(0)$. We see that the existence of one transverse homoclinic point for $\mathbf{P}$ implies the existence of an infinite number of homoclinic points and it can be shown that they are all transverse homoclinic points which accumulate at 0; i.e., we have a transverse homoclinic orbit of $\mathbf{P}$. This leads to a "homoclinic tangle," part of which is shown in Figure 10, wherein $W^s(0)$ and $W^u(0)$ accumulate on themselves. These ideas are discussed more thoroughly in Chapter 3 of [Wi].

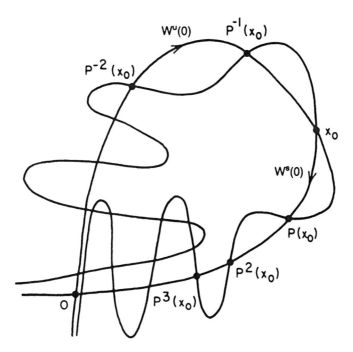

Figure 10. A transverse homoclinic orbit and the associated homoclinic tangle.

In a homoclinic tangle, a high enough iterate of $\mathbf{P}$ will lead to a horseshoe map as in Figure 11; cf. Figure 3.4.5, p. 318 in [Wi]. Furthermore, the existence of a horseshoe map results in "chaotic dynamics" as described below.

In the early sixties, Stephen Smale [31] described his now-famous horseshoe map and showed that it has a strange invariant set resulting in what is termed chaotic dynamics. We now give a brief description of (a variant of) the Smale horseshoe map. For more details, the reader should see Chapter 2 in [Wi] or Chapter 5 in [G/H]. The Smale horseshoe map $\mathbf{F}\colon D \to \mathbf{R}^2$ where $D = \{\mathbf{x} \in \mathbf{R}^2 \colon 0 \leq x \leq 1, 0 \leq y \leq 1\}$; geometrically, $\mathbf{F}$ contracts D

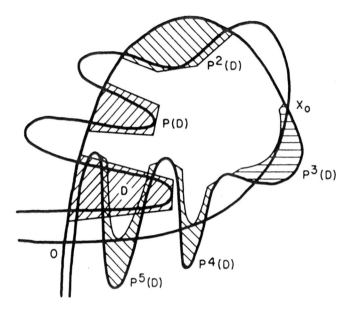

Figure 11. An iterate of **P** exhibiting a horseshoe map.

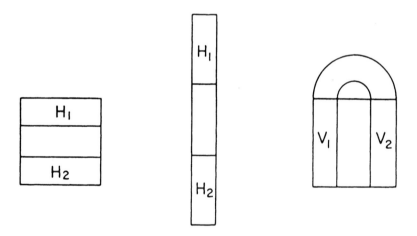

Figure 12. The Smale horseshoe map.

in the x-direction and expands D in the y-direction; and then **F** folds D back onto itself as shown in Figure 12 where $V_1 = \mathbf{F}(H_1)$ and $V_2 = \mathbf{F}(H_2)$. Furthermore, **F** is linear on the horizontal rectangles H_1 and H_2 shown in Figure 12.

4.8. Homoclinic Bifurcations

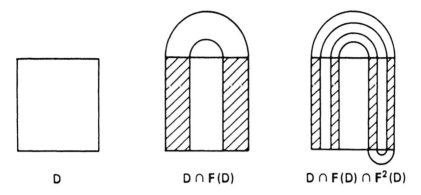

Figure 13. The action of $\mathbf{F}$ and $\mathbf{F}^2$ on D.

In order to begin to see the nature of the strange invariant set for $\mathbf{F}$, we next look at $\mathbf{F}^2(D)$. The action of $\mathbf{F}$ and $\mathbf{F}^2$ on D is shown in Figure 13. Similarly, the action of $\mathbf{F}^{-1}$ and $\mathbf{F}^{-2}$ on D is shown in Figure 14; cf. pp. 77–83 in [Wi].

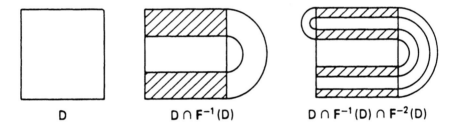

Figure 14. The action of $\mathbf{F}^{-1}$ and $\mathbf{F}^{-2}$ on D.

If we look at the sets

$$\Lambda_1 \equiv \mathbf{F}^{-1}(D) \cap D \cap \mathbf{F}(D)$$

and

$$\Lambda_2 \equiv \mathbf{F}^{-2}(D) \cap \mathbf{F}^{-1}(D) \cap D \cap \mathbf{F}(D) \cap \mathbf{F}^2(D)$$

pictured in Figure 15, we can begin to get an idea of the Cantor set structure of the strange invariant set

$$\Lambda \equiv \bigcap_{j=-\infty}^{\infty} \mathbf{F}^j(D)$$

of the Smale horseshoe map $\mathbf{F}$. The next theorem is due to Smale; cf. [G/H], p. 235, [31] and [32].

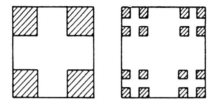

Figure 15. The sets Λ_1 and Λ_2 which lead to the invariant set Λ for **F**.

Theorem 4. *The horseshoe map* **F** *has an invariant Cantor set* Λ *such that:*

(i) Λ *contains a countable set of periodic orbits of* **F** *of arbitrarily long periods,*

(ii) Λ *contains an uncountable set of bounded nonperiodic orbits, and*

(iii) Λ *contains a dense orbit.*

Furthermore, any C^1-diffeomorphism **G** *which is sufficiently close to* **F** *in the C^1-norm has an invariant Cantor set Ω such that* **G** *restricted to Ω is topologically equivalent to* **F** *restricted to Λ.*

This theorem is proved by showing that the horseshoe map **F** is topologically equivalent to the shift map σ on bi-infinite sequences of zeros and ones:

$$\sigma(\ldots, s_{-n}, \ldots, s_{-1}; s_0, s_1, \ldots, s_n, \ldots)$$
$$= (\ldots, s_{-n}, \ldots, s_{-1}, s_0; s_1, \ldots, s_n, \ldots)$$

with $s_j \in \{0,1\}$ for $j = 0, \pm 1, \pm 2, \ldots$. Cf. Theorems 2.1.2–2.1.4 in [Wi]. Also, cf. Problem 6.

Whenever the Poincaré map of a hyperbolic periodic orbit has a transverse homoclinic point, we get the same type of chaotic dynamics exhibited by the Smale horseshoe map. This important result stated in the next section is known as the Smale–Birkhoff Homoclinic Theorem; cf. Theorem 5.3.5, p. 252 in [G/H]. The occurrence of transverse homoclinic points was first discovered by Poincaré [27] in 1890 in his studies of the three-body problem. Cf. [15]. Even at this early date he was well aware that very complicated dynamics would result from the occurrence of such points. In the next section, we briefly discuss Melnikov's Method which is one of the few analytical tools available for determining when the Poincaré map of a dynamical system has tangential or transverse homoclinic points. As was noted earlier, in dynamical systems, transverse homoclinic points and the resulting chaotic dynamics typically result from tangential homoclinic bifurcations.

4.8. Homoclinic Bifurcations

PROBLEM SET 8

1. Consider the system

$$\dot{x} = y + y(x^2 + y^2)$$
$$\dot{y} = x - x(x^2 - xy + y^2) + \mu y.$$

   (a) Show that this system is symmetric under the transformation $(x, y) \to (-x, -y)$ and that there is a saddle at the origin and foci at $(\pm 1, 0)$ for $|\mu| < 3$.

   (b) Show that for $y \neq 0$ this system defines a semicomplete one-parameter family of rotated vector fields with parameter $\mu \in \mathbf{R}$.

   (c) Use the results in Section 4.6 to show that a unique limit cycle is generated in a Hopf bifurcation at each of the critical points $(\pm 1, 0)$ at the bifurcation value $\mu = -1$ and that these limit cycles expand monotonically as μ increases until they intersect the origin and form a graphic S_0 at a bifurcation value $\mu = \mu_1 > -1$.

   (d) Use Theorem 3 in Section 4.6 to show that S_0 is stable; show that $\sigma_0 = \mu$ and deduce that $\mu_1 \leq 0$. (Numerical computation shows that $\mu_1 \simeq -.74$.)

   (e) Use the results in Section 4.6 to show that a unique limit cycle, surrounding both of the critical points $(\pm 1, 0)$, bifurcates from the graphic S_0 as μ increases beyond the bifurcation value μ_1. Can you determine how this one-parameter family of limit cycles terminates?

2. (a) Show that the system

$$\dot{x} = y$$
$$\dot{y} = x - x^3$$

   is a Hamiltonian system with $H(x, y) = y^2/2 - x^2/2 + x^4/4$, determine the critical points, and sketch the phase portrait for this system.

   (b) Sketch the phase portrait for the system

$$\dot{x} = X(x, y, \alpha) = y - \alpha H(x, y)(x - x^3)$$
$$\dot{y} = Y(x, y, \alpha) = x - x^3 + \alpha H(x, y)y$$

   for $\alpha > 0$. Note that this system defines a negatively rotated vector field inside the graphic $H(x, y) = 0$.

   (c) Now fix α at a positive value, say $\alpha = .1$, and embed the vector field defined in part (b) in a one-parameter family of rotated vector fields as in equation (5). For what range of μ will this system have a limit cycle i.e., determine the value of μ for which a limit cycle bifurcates from the graphic $H(x, y) = 0$ and the value of μ for which there is a Hopf bifurcation.

(d) What happens if you fix α at some negative value, say $\alpha = -.1$, in part (c)?

3. Carry out the same program as in Problem 2 for the Hamiltonian system
$$\dot{x} = 2y$$
$$\dot{y} = 12x - 3x^2.$$
Cf. [A–II], p. 306.

4. Find the stable, unstable and center subspaces, E^s, E^u and E^c for the linear maps $L(\mathbf{x}) = A\mathbf{x}$ where the matrix

(a) $A = \begin{bmatrix} 2 & 0 \\ 1 & -1 \end{bmatrix}$

(b) $A = \begin{bmatrix} 1/2 & 1 \\ 0 & 1/2 \end{bmatrix}$

(c) $A = \begin{bmatrix} 1 & -1 \\ 1 & 1 \end{bmatrix}$

(d) $A = \begin{bmatrix} 1 & 1 \\ 1 & 2 \end{bmatrix}.$

5. (a) Show that the map
$$\mathbf{P}(x, y) = (y, x - y - y^3)$$
is a diffeomorphism and find its inverse.

(b) Find the fixed points and the dimensions of the stable and unstable manifolds $W^s(\mathbf{x}_0)$ and $W^u(\mathbf{x}_0)$ at each of the fixed points $\mathbf{x}_0$ of $\mathbf{P}$.

6. (The Bernoulli Shift Map) Consider the mapping $F\colon [0, 1] \to [0, 1]$ defined by
$$F(x) = 2x \pmod 1.$$
Show that if $x \in [0, 1]$ has the binary expansion
$$x = .s_1 s_2 \cdots = \sum_{j=1}^{\infty} \frac{s_j}{2^j}$$
with $s_j \in \{0, 1\}$ for $j = 1, 2, 3, \ldots$, then $F(x) = .s_2 s_3 \ldots$ and for $n \geq 0$ $F^n(x) = .s_{n+1} s_{n+2} \ldots$. Show that the repeating sequences $.\overline{0}$ and $.\overline{1}$ are fixed points for F, that $.\overline{01}$ and $.\overline{10}$ are fixed points of F^2, and that $.\overline{001}, .\overline{010}$ and $.\overline{100}$ are fixed points of F^3. Fixed points of F^n are called periodic orbits of F of period n. Show that F has a countable number of periodic orbits of arbitrarily large period. Finally, show that if the sequences for x and y differ in the nth place, then $|F^{n-1}(x) - F^{n-1}(y)| \geq 1/2$. This illustrates the sensitive dependence of the map F on initial conditions.

4.9 Melnikov's Method

Melnikov's method gives us an analytic tool for establishing the existence of transverse homoclinic points of the Poincaré map for a periodic orbit of a perturbed dynamical system of the form

$$\dot{\mathbf{x}} = \mathbf{f}(\mathbf{x}) + \varepsilon \mathbf{g}(\mathbf{x}) \tag{1}$$

with $\mathbf{x} \in \mathbf{R}^n$ and $n \geq 3$. It can also be used to establish the existence of subharmonic periodic orbits of perturbed systems of the form (1). Furthermore, it can be used to show the existence of limit cycles and separatrix cycles of perturbed planar systems (1) with $\mathbf{x} \in \mathbf{R}^2$. This section is only intended as an introduction to Melnikov's method; however, more details are given in the next three sections for planar systems and the examples contained in this section serve to point out the versatility and power of this method. A general theory for planar systems is developed in the next section and some new results on second and higher order Melnikov theory are given in Sections 4.11 and 4.12. The reader should consult Chapters 4 in [G/H] and [Wi] for the general theory in higher dimensions $(n \geq 3)$.

We begin with a result for periodically perturbed planar systems of the form

$$\dot{\mathbf{x}} = \mathbf{f}(\mathbf{x}) + \varepsilon \mathbf{g}(\mathbf{x}, t) \tag{2}$$

where $\mathbf{x} \in \mathbf{R}^2$ and $\mathbf{g}$ is periodic of period T in t. Note that this system can be written as an autonomous system in $\mathbf{R}^3$ by defining $x_3 = t$. We assume that $\mathbf{f} \in C^1(\mathbf{R}^2)$ and that $\mathbf{g} \in C^1(\mathbf{R}^2 \times \mathbf{R})$. Following Guckenheimer and Holmes [G/H], pp. 184–188, we make the assumptions:

(A.1) For $\varepsilon = 0$ the system (2) has a homoclinic orbit

$$\Gamma_0: \mathbf{x} = \gamma_0(t), \quad -\infty < t < \infty,$$

at a hyperbolic saddle point $\mathbf{x}_0$ and

(A.2) For $\varepsilon = 0$ the system (2) has a one-parameter family of periodic orbits $\gamma_\alpha(t)$ of period T_α on the interior of Γ_0 with $\partial \gamma_\alpha(0)/\partial \alpha \neq 0$; cf. Figure 1.

The *Melnikov function*, $M(t_0)$, is then defined as

$$M(t_0) = \int_{-\infty}^{\infty} e^{-\int_{t_0}^{t} \nabla \cdot \mathbf{f}(\gamma_0(s))ds} \mathbf{f}(\gamma_0(t)) \wedge \mathbf{g}(\gamma_0(t), t+t_0) dt \tag{3}$$

where the wedge product of two vectors $\mathbf{u}$ and $\mathbf{v} \in \mathbf{R}^2$ is defined as $\mathbf{u} \wedge \mathbf{v} = u_1 v_2 - v_1 u_2$. Note that the Melnikov function $M(t_0)$ is proportional to the derivative of the Poincaré map with respect to the parameter ε in an

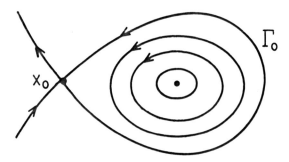

Figure 1. The phase portrait of the system (2) under the assumptions (A.1) and (A.2).

interior neighborhood of the separatrix cycle Γ_0 (or in a neighborhood of a cycle); cf. equation (6) in Section 4.5. Before stating the main result, established by Melnikov, concerning the existence of transverse homoclinic points of the Poincaré map, we need the following lemma which establishes the existence of a periodic orbit $\gamma_\varepsilon(t)$ of (2), and hence the existence of the Poincaré map $\mathbf{P}_\varepsilon$, for (2) with sufficiently small ε; cf. Lemma 4.5.1, p. 186 in [G/H].

Lemma 1. *Under assumptions* (A.1) *and* (A.2), *for ε sufficiently small, (2) has a unique hyperbolic periodic orbit $\gamma_\varepsilon(t) = \mathbf{x}_0 + 0(\varepsilon)$ of period T. Correspondingly, the Poincaré map $\mathbf{P}_\varepsilon$ has a unique hyperbolic fixed point of saddle type $\mathbf{x}_\varepsilon = \mathbf{x}_0 + 0(\varepsilon)$.*

Theorem 1. *Under the assumptions* (A.1) *and* (A.2), *if the Melnikov function $M(t_0)$ defined by (3) has a simple zero in $[0,T]$ then for all sufficiently small $\varepsilon \neq 0$ the stable and unstable manifolds $W^s(\mathbf{x}_\varepsilon)$ and $W^u(\mathbf{x}_\varepsilon)$ of the Poincaré map $\mathbf{P}_\varepsilon$ intersect transversally; i.e., $\mathbf{P}_\varepsilon$ has a transverse homoclinic point. And if $M(t_0) > 0$ (or < 0) for all t_0 then $W^s(\mathbf{x}_\varepsilon) \cap W^u(\mathbf{x}_\varepsilon) = \emptyset$.*

This theorem was established by Melnikov [21] in 1963. The idea of his proof is that $M(t_0)$ is a measure of the separation of the stable and unstable manifolds of the Poincaré map $\mathbf{P}_\varepsilon$; cf. [G/H], p. 188. Theorem 1 is an important result because it establishes the existence of a transverse homoclinic point for $\mathbf{P}_\varepsilon$. As was indicated in Section 4.8, this implies the existence of a strange invariant set Λ for some iterate, $\mathbf{F}_\varepsilon$, of $\mathbf{P}_\varepsilon$ and the same type of chaotic dynamics for (2) as for the Smale horseshoe map, according to the following theorem whose proof is outlined on p. 252 in [G/H] and on p. 108 in [Ru]:

Theorem (The Smale-Birkhoff Homoclinic Theorem). *Let $\mathbf{P}: \mathbf{R}^n \to \mathbf{R}^n$ be a diffeomorphism such that $\mathbf{P}$ has a hyperbolic fixed point of saddle type, $\mathbf{p}$, and a transverse homoclinic point $\mathbf{q} \in W^s(\mathbf{p}) \cap W^u(\mathbf{p})$.*

4.9. Melnikov's Method

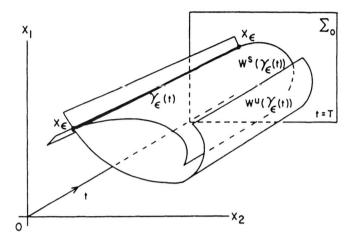

Figure 2. The hyperbolic fixed point x_ε of the Poincaré map P_ε in the section Σ_0 at $t = 0$ which is identified with the section at $t = T$.

Then there exists an integer N such that $\mathbf{F} = \mathbf{P}^N$ has a hyperbolic compact invariant Cantor set Λ on which $\mathbf{F}$ is topologically equivalent to a shift map on bi-infinite sequences of zeros and ones. The invariant set Λ

(i) contains a countable set of periodic orbits of $\mathbf{F}$ of arbitrarily long periods,

(ii) contains an uncountable set of bounded nonperiodic orbits, and

(iii) contains a dense orbit.

Another important result in the Melnikov theory is contained in the next theorem which establishes the existence of a homoclinic point of tangency in $W^s(\mathbf{x}_\varepsilon) \cap W^u(\mathbf{x}_\varepsilon)$; cf. Theorem 4.5.4, p. 190 in [G/H].

Theorem 2. *Assume that* (A.1) *and* (A.2) *hold for the system*

$$\dot{\mathbf{x}} = \mathbf{f}(\mathbf{x}) + \varepsilon \mathbf{g}(\mathbf{x}, t, \mu) \tag{4}$$

depending on a parameter $\mu \in \mathbf{R}$, *that* $\mathbf{g}$ *is* T-*periodic in* t, *that* $\mathbf{f} \in C^1(\mathbf{R}^2)$ *and that* $\mathbf{g} \in C^1(\mathbf{R}^2 \times \mathbf{R} \times \mathbf{R})$. *If the Melnikov function* $M(t_0, \mu)$ *defined by* (3) *has a quadratic zero at* $(t_0, \mu_0) \in [0, T] \times \mathbf{R}$, *i.e., if* $M(t_0, \mu_0) = \frac{\partial M}{\partial t_0}(t_0, \mu_0) = 0$, $\frac{\partial^2 M}{\partial t_0^2}(t_0, \mu_0) \neq 0$, *and* $\frac{\partial M}{\partial \mu}(t_0, \mu_0) \neq 0$, *then for all sufficiently small* $\varepsilon \neq 0$ *there is a bifurcation value* $\mu_\varepsilon = \mu_0 + 0(\varepsilon)$ *at which* $W^s(\mathbf{x}_\varepsilon)$ *and* $W^u(\mathbf{x}_\varepsilon)$ *intersect tangentially.*

Remark 1. If for $\varepsilon = 0$, (2) or (4) are Hamiltonian systems, i.e., if

$$\mathbf{f} = \left(\frac{\partial H}{\partial y}, -\frac{\partial H}{\partial x}\right)^T,$$

then $\nabla \cdot \mathbf{f} = 0$ and the Melnikov function has the simpler form

$$M(t_0) = \int_{-\infty}^{\infty} \mathbf{f}(\gamma_0(t)) \wedge \mathbf{g}(\gamma_0(t), t + t_0) dt. \tag{5}$$

Example 1. (Cf. [G/H], pp. 191–193.) Consider the periodically perturbed Duffing equation

$$\dot{x} = y$$
$$\dot{y} = x - x^3 + \varepsilon(\mu \cos t - 2.5y).$$

The parameter μ is the forcing amplitude. For $\varepsilon = 0$ we have a Hamiltonian system with Hamiltonian

$$H(x, y) = \frac{y^2}{2} - \frac{x^2}{2} + \frac{x^4}{4}.$$

For $\varepsilon = 0$ there is a saddle at the origin, centers at $(\pm 1, 0)$, and two homoclinic orbits

$$\Gamma_0^{\pm}: \gamma_0^{\pm}(t) = \pm(\sqrt{2}\operatorname{sech} t, -\sqrt{2}\operatorname{sech} t \tanh t)^T.$$

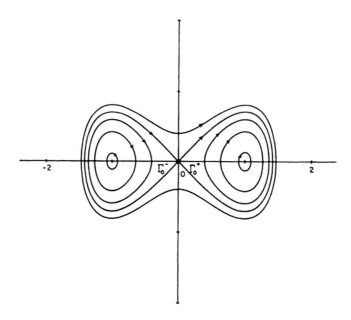

Figure 3. The phase portrait for the system in Example 1 with $\varepsilon = 0$.

We will compute the Melnikov function for $\gamma_0^+(t)$; the computation for $\gamma_0^-(t)$ is identical. From (5)

$$M(t_0) = \int_{-\infty}^{\infty} y_0(t)[\mu \cos(t + t_0) - 2.5y_0(t)] dt$$

4.9. Melnikov's Method

$$= -\sqrt{2}\mu \int_{-\infty}^{\infty} \operatorname{sech} t \tanh t \cos(t + t_0) dt$$

$$- 5 \int_{-\infty}^{\infty} \operatorname{sech}^2 t \tanh^2 t \, dt$$

$$= \sqrt{2}\mu\pi \operatorname{sech}\left(\frac{\pi}{2}\right) \sin t_0 - \frac{10}{3}.$$

The first integral can be found by the method of residues. See Problem 4. We therefore have

$$M(t_0) = \sqrt{2}\mu\pi \operatorname{sech}\left(\frac{\pi}{2}\right) \left[\sin t_0 - \frac{k_0}{\mu}\right]$$

where $k_0 = 10 \cosh(\pi/2)/(3\sqrt{2}\pi) \simeq 1.88$.

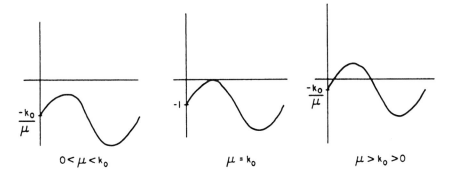

Figure 4. The function $\sin t_0 - k_0/\mu$.

We see in Figure 4 that if $0 < \mu < k_0$, $M(t_0)$ has no zeros and therefore by Theorem 1, $W^s(\mathbf{x}_\varepsilon) \cap W^u(\mathbf{x}_\varepsilon) = \emptyset$; if $\mu = k_0$ then $M(t_0)$ has a quadratic zero and by Theorem 2, there is a tangential intersection of $W^s(\mathbf{x}_\varepsilon)$ and $W^u(\mathbf{x}_\varepsilon)$ at some $\mu_\varepsilon = k_0 + 0(\varepsilon)$; and if $\mu > k_0 > 0$, $M(t_0)$ has a simple zero and by Theorem 1, $W^s(\mathbf{x}_\varepsilon)$ and $W^u(\mathbf{x}_\varepsilon)$ intersect transversally. This is borne out by the Poincaré maps for the perturbed Duffing equation of this example with $\varepsilon = .1$, computed by Udea in 1981, as shown in Figure 5 below; cf. Figure 4.5.3 in [G/H].

We would not expect the nonhyperbolic structure on the interior of the separatrix cycle Γ_0 to be preserved under small perturbations as in (2). However, under certain conditions, some of the periodic orbits $\gamma_\alpha(t)$ whose periods T_α are rational multiples of the period T of the perturbation function $\mathbf{g}$, i.e.,

$$T_\alpha = \frac{m}{n} T,$$

are preserved under small perturbations. We assume that $\mathbf{f}(\mathbf{x})$ is a Hamiltonian vector field on $\mathbf{R}^2$, that there is a one-parameter family of periodic

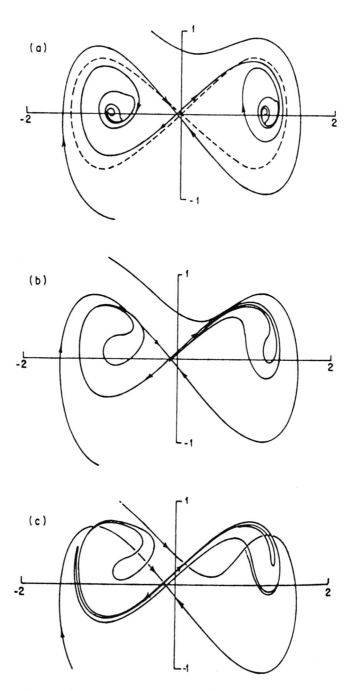

Figure 5. Poincaré maps for the perturbed Duffing equation in Example 1 with $\varepsilon = .1$ and (a) $\mu = 1.1$, (b) $\mu = 1.9$, and (c) $\mu = 3.0$. Reprinted with permission from Guckenheimer and Holmes (Ref. [G/H]).

4.9. Melnikov's Method

orbits $\gamma_\alpha(t)$ on the interior of the separatrix cycle Γ_0 as in Figure 1, and that the following assumption holds

$$\frac{\partial T_\alpha}{\partial h_u} \neq 0 \qquad (A.3)$$

where $h_\alpha = H(\gamma_\alpha(t))$.

Theorem 3. *If in equation* (2) $\mathbf{f}$ *is a Hamiltonian vector field, then under assumptions* (A.1)–(A.3), *if the subharmonic Melnikov function*

$$M_{m,n}(t_0) = \int_0^{mT} \mathbf{f}(\gamma_\alpha(t)) \wedge \mathbf{g}(\gamma_\alpha(t), t + t_0) dt \qquad (6)$$

along a subharmonic periodic orbit $\gamma_\alpha(t)$, *of period* mT/n, *has a simple zero in* $[0, mT]$ *then for all sufficiently small* $\varepsilon \neq 0$, *the system* (2) *has a subharmonic periodic orbit of period* mT *in an* ε-*neighborhood of* $\gamma_\alpha(t)$.

Results similar to Theorems 1–3 are given for Hamiltonian systems with two degrees of freedom in [G/H] on pp. 212–226 and in Chapter 4 of [Wi].

We next present some simple applications of the Melnikov theory for perturbed planar systems of the form

$$\dot{\mathbf{x}} = \mathbf{f}(\mathbf{x}) + \varepsilon \mathbf{g}(\mathbf{x}, \boldsymbol{\mu}) \qquad (7)$$

with $\mathbf{f} \in C^1(\mathbf{R}^2)$ and $\mathbf{g} \in C^1(\mathbf{R}^2 \times \mathbf{R}^m)$. For simplicity, we shall assume that $\mathbf{f}$ is a Hamiltonian vector field although this is not necessary if we use equation (3) for the Melnikov function. Similar to Theorems 1, 2 and 3 we have the following results established in the next section. Cf. [6, 37, 38].

Theorem 4. *Under the assumption* (A.1), *if there exists a* $\boldsymbol{\mu}_0 \in \mathbf{R}^m$ *such that the function*

$$M(\boldsymbol{\mu}) = \int_{-\infty}^\infty \mathbf{f}(\gamma_0(t)) \wedge \mathbf{g}(\gamma_0(t), \boldsymbol{\mu}) dt$$

satisfies

$$M(\boldsymbol{\mu}_0) = 0 \quad \text{and} \quad M_{\mu_1}(\boldsymbol{\mu}_0) \neq 0,$$

then for all sufficiently small $\varepsilon \neq 0$ *there is a* $\boldsymbol{\mu}_\varepsilon = \boldsymbol{\mu}_0 + 0(\varepsilon)$ *such that the system* (7) *with* $\boldsymbol{\mu} = \boldsymbol{\mu}_\varepsilon$ *has a unique homoclinic orbit in an* $0(\varepsilon)$ *neighborhood of the homoclinic orbit* Γ_0. *Furthermore, if* $M(\boldsymbol{\mu}_0) \neq 0$ *then for all sufficiently small* $\varepsilon \neq 0$ *and* $|\boldsymbol{\mu} - \boldsymbol{\mu}_0|$, *the system* (7) *has no separatrix cycle in an* $0(\varepsilon)$ *neighborhood of* $\Gamma_0 \cup \{\mathbf{x}_0\}$.

Theorem 5. *Under the assumption* (A.2), *if there exists a point* $(\boldsymbol{\mu}_0, \alpha_0) \in \mathbf{R}^{m+1}$ *such that the function*

$$M(\boldsymbol{\mu}, \alpha) = \int_0^{T_\alpha} \mathbf{f}(\gamma_\alpha(t)) \wedge \mathbf{g}(\gamma_\alpha(t), \boldsymbol{\mu}) dt$$

satisfies
$$M(\boldsymbol{\mu}_0, \alpha_0) = 0 \quad \text{and} \quad M_\alpha(\boldsymbol{\mu}_0, \alpha_0) \neq 0,$$
then for all sufficiently small $\varepsilon \neq 0$, the system (7) with $\boldsymbol{\mu} = \boldsymbol{\mu}_0$ has a unique hyperbolic limit cycle in an $0(\varepsilon)$ neighborhood of the cycle $\gamma_{\alpha_0}(t)$. If $M(\boldsymbol{\mu}_0, \alpha_0) \neq 0$, then for sufficiently small $\varepsilon \neq 0$ the system (6) with $\boldsymbol{\mu} = \boldsymbol{\mu}_0$ has no cycle in an $0(\varepsilon)$ neighborhood of the cycle $\gamma_{\alpha_0}(t)$.

Remark 2. Under the assumption (A.2), it can be shown that if there exists a point $(\boldsymbol{\mu}_0, \alpha_0) \in \mathbf{R}^{m+1}$ such that
$$M(\boldsymbol{\mu}_0, \alpha_0) = M_\alpha(\boldsymbol{\mu}_0, \alpha_0) = 0,$$
$$M_{\alpha\alpha}(\boldsymbol{\mu}_0, \alpha_0) \neq 0 \quad \text{and} \quad M_{\mu_1}(\boldsymbol{\mu}_0, \alpha_0) \neq 0,$$
then for all sufficiently small $\varepsilon \neq 0$ there exists a $\boldsymbol{\mu}_\varepsilon = \boldsymbol{\mu}_0 + 0(\varepsilon)$ such that the system (7) with $\boldsymbol{\mu} = \boldsymbol{\mu}_\varepsilon$ has a unique limit cycle of multiplicity two in an $0(\varepsilon)$ neighborhood of the cycle $\gamma_{\alpha_0}(t)$. This result as well as Theorems 4 and 5 are proved in the next section.

Example 2. Consider the Lienard equation
$$\dot{x} = y - \varepsilon[\mu_1 x + \mu_2 x^2 + \mu_3 x^3]$$
$$\dot{y} = -x$$
with $\boldsymbol{\mu} = (\mu_1, \mu_2, \mu_3)^T$. The unperturbed system with $\varepsilon = 0$ has a center at the origin with a one-parameter family of periodic orbits $\gamma_\alpha(t) = (\alpha \cos t, -\alpha \sin t)^T$ of period $T_\alpha = 2\pi$. It is easy to compute the Melnikov function for this example.

$$M(\boldsymbol{\mu}, \alpha) = \int_0^{T_\alpha} \mathbf{f}(\gamma_\alpha(t)) \wedge \mathbf{g}(\gamma_\alpha(t), \boldsymbol{\mu}) dt$$
$$= -\int_0^{2\pi} x_\alpha(t)[\mu_1 x_\alpha(t) + \mu_2 x_\alpha^2(t) + \mu_3 x_\alpha^3(t)] dt$$
$$= -\int_0^{2\pi} [\mu_1 \alpha^2 \cos^2 t + \mu_2 \alpha^3 \cos^3 t + \mu_3 \alpha^4 \cos^4 t] dt$$
$$= -2\pi\alpha^2 \left[\frac{\mu_1}{2} + \frac{3}{8}\mu_3 \alpha^2\right].$$

We see that the equation $M(\boldsymbol{\mu}, \alpha) = 0$ has a solution if and only if $\mu_1 \mu_3 < 0$. It follows from Theorem 5 that if $\mu_1 \mu_3 < 0$ then for all sufficiently small $\varepsilon \neq 0$ the Lienard system will have a unique limit cycle which is approximately a circle of radius

$$\alpha = \sqrt{\frac{4|\mu_1|}{3|\mu_3|}} + 0(\varepsilon).$$

Cf. Theorem 6 in Section 3.8 of Chapter 3.

4.9. Melnikov's Method

Example 3. Consider the perturbed Duffing equation

$$\dot{x} = y$$
$$\dot{y} = x - x^3 + \varepsilon[\alpha y + \beta x^2 y] \qquad (8)$$

of Exercise 4.6.4, p. 204 in [G/H], with the parameter $\mu = (\alpha, \beta)^T$. For $\varepsilon = 0$ this is a Hamiltonian system with Hamiltonian

$$H(x, y) = \frac{y^2}{2} - \frac{x^2}{2} + \frac{x^4}{4}.$$

The phase portrait for this system with $\varepsilon = 0$ is given in Figure 3. The graphic $\Gamma_0^- \cup \{0\} \cup \Gamma_0^+$ shown in Figure 3 corresponds to $H(x, y) = 0$; i.e., it is represented by motions on the curves defined by

$$y^2 = x^2 - \frac{x^4}{2}.$$

We compute the Melnikov function in Theorem 4 for this system using the fact that along trajectories of (8) $dt = dx/\dot{x} = dx/y$. We only compute $M(x)$ along the right-hand separatrix Γ_0^+; the computation along Γ_0^- is identical.

$$M(\mu) = \int_{-\infty}^{\infty} \mathbf{f}(\gamma_0^+(t)) \wedge \mathbf{g}(\gamma_0^+(t), \mu) dt$$
$$= \int_{-\infty}^{\infty} y_0^2(t)[\alpha + \beta x_0^2(t)] dt$$
$$= 2 \int_0^{\sqrt{2}} x\sqrt{1 - x^2/2}(\alpha + \beta x^2) dx$$
$$= \frac{4}{3}\alpha + \frac{16}{15}\beta.$$

We see that $M(\mu) = 0$ if and only if $\beta = -5\alpha/4$. Thus, according to Theorem 4, for all sufficiently small $\varepsilon \neq 0$, there is a $\mu_\varepsilon = \alpha(1, -5/4)^T + 0(\varepsilon)$ such that the system (8) with $\mu = \mu_\varepsilon$, i.e., with $\beta = -5\alpha/4 + 0(\varepsilon)$, has two homoclinic orbits $\Gamma_\varepsilon^\pm$ at the saddle at the origin in an ε-neighborhood of $\Gamma_0^\pm$. Also, for the vector field $\mathbf{F} = \mathbf{f} + \varepsilon \mathbf{g}$ in (8), we have

$$D\mathbf{F}(\mathbf{0}, \varepsilon, \mu) = \begin{bmatrix} 0 & 1 \\ 1 & \varepsilon\alpha \end{bmatrix}.$$

Thus, by Theorem 1 in Section 4.7, the separatrix cycles $\Gamma_\varepsilon^\pm \cup \{0\}$ are stable for $\sigma_0 = \varepsilon\alpha < 0$ and unstable for $\sigma_0 = \varepsilon\alpha > 0$. Cf. Figure 6 which shows the separatrix cycles of (8) for $\varepsilon\alpha < 0$.

We end this section by constructing the global phase portraits for the system in Example 3 for small $\varepsilon > 0, \beta > 0$ and $-\infty < \alpha < \infty$. The phase portraits for $\beta < 0$ can then be obtained by using the symmetry of the system (8) under the transformation $(t, x, y, \alpha, \beta) \to (-t, -x, -y, -\alpha, -\beta)$.

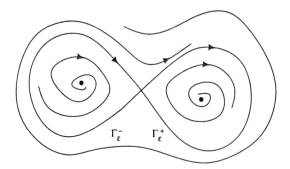

Figure 6. The graphic of the system (8) for small $\varepsilon \neq 0, \beta = -5\alpha/4 + 0(\varepsilon)$ and $\varepsilon\alpha < 0$.

The case when $\beta = 0$ is left as an exercise; cf. Problem 2. For fixed $\beta > 0$ and $\varepsilon > 0$, the system (8) defines a one-parameter family of rotated vector fields (mod $y = 0$) with parameter $\alpha \in \mathbf{R}$. Since the separatrix cycles $\Gamma_\varepsilon^\pm \cup \{0\}$, which exist for $\alpha = \alpha^* = -4\beta/5 + 0(\varepsilon)$ according to the results in Example 3, are stable on their interiors and are negatively oriented, it follows from Theorem 3 and the table in Figure 1 of Section 4.6 that a stable limit cycle is generated on the interior of each of the separatrix cycles $\Gamma_\varepsilon^\pm \cup \{0\}$ as α decreases from α^*. Also, since the graphic $\Gamma_\varepsilon^+ \cup \Gamma_\varepsilon^- \cup \{0\}$, shown in Figure 6, is stable on its exterior and negatively oriented, a stable limit cycle L_α^- is generated on the exterior of the graphic $\Gamma_\varepsilon^+ \cup \Gamma_\varepsilon^- \cup \{0\}$ as α increases from α^* according to Theorem 3 and the table in Section 4.6.

It follows from the equations (8) that the trace at the critical points at $(\pm 1, 0)$ is given by

$$\text{trace } D\mathbf{F}(\pm 1, 0, \varepsilon, \boldsymbol{\mu}) = \varepsilon(\alpha + \beta).$$

And using equation (3′) in Section 4.4, we find that for $\alpha = -\beta$ we have

$$\sigma = \frac{-3\pi\beta\varepsilon}{2^{1/2}} < 0.$$

Thus, for $\alpha = -\beta$ the weak foci at $(\pm 1, 0)$ are stable and they are negatively oriented. According to Theorem 5 and the table in Section 4.6, a unique stable limit cycle bifurcates from the critical points at $(\pm 1, 0)$ in a supercritical Hopf bifurcation at the bifurcation value $\alpha = -\beta$ (where we have fixed β at some positive value). And according to Theorems 1 and 6 in Section 4.6, these limit cycles expand monotonically with increasing α until they intersect the saddle at the origin and form the graphic $\Gamma_\varepsilon^+ \cup \Gamma_\varepsilon^- \cup \{0\}$ at $\alpha = \alpha^* = -4\beta/5 + 0(\varepsilon)$.

We next determine the behavior at infinity and draw the global phase portraits for (8). According to the theory in Section 3.10 of Chapter 3, the

4.9. Melnikov's Method

critical points at infinity are at $(0, \pm 1, 0)$ and $\pm(\beta, 1, 0)/\sqrt{1+\beta^2}$. According to Theorem 2 in Section 3.10 of Chapter 3, the behavior at these critical points is determined by the system

$$\pm \dot x = x^2 z^2 - x^4 + \alpha x z^2 + \beta x^3 - z^4$$
$$\pm \dot z = xz^3 - x^3 z + \alpha z^3 + \beta x^2 z.$$

The flow on the equator of the Poincaré sphere determines that the minus sign should be used in these equations. This system has two critical points at $(0,0)$ and at $(\beta, 0)$. According to Theorem 1 in Section 2.11 in Chapter 2, the critical point $(\beta, 0)$ is a saddle node. The origin is a higher degree critical point with two zero eigenvalues. Using a Liapunov function, it can be shown that it is a stable node as in Figure 7 below.

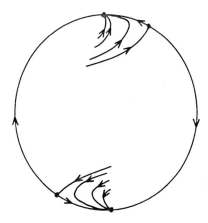

Figure 7. The behavior of the system (8) at infinity for $\varepsilon > 0$ and $\beta > 0$.

We now employ the Poincaré–Bendixson Theorem in Section 3.7 of Chapter 3 to deduce that, at least for the parameter range on α where (8) has a stable limit cycle L_α^- enclosing the three critical points $(0,0)$, $(\pm 1, 0)$, there must be an unstable limit cycle L_α^+ on the exterior of L_α^- which is the α-limit set of the two saddle separatrices shown in Figure 7. According to Theorem 1 and the table in Section 4.6, this unstable, negatively oriented limit cycle L_α^+ will contract as α increases; and similarly, the stable, negatively oriented limit cycle L_α^- will expand as α increases. The two limit cycles L_α^- and L_α^+ intersect in a saddle-node bifurcation at a semistable limit cycle at a bifurcation value $\alpha = \alpha^{**} > \alpha^*$. We have numerically determined that $\alpha^{**} = -k\beta + 0(\varepsilon)$ for some constant $k = .752\cdots$. Assuming that no other semistable limit cycles occur as α varies in $\mathbf{R}$, the global phase portraits for the system (8) with small $\varepsilon > 0$, any fixed $\beta > 0$ and $\alpha \in \mathbf{R}$ are shown in Figure 8. This assumption and the above-mentioned results concerning the behavior of the limit cycles of (8) are verified in

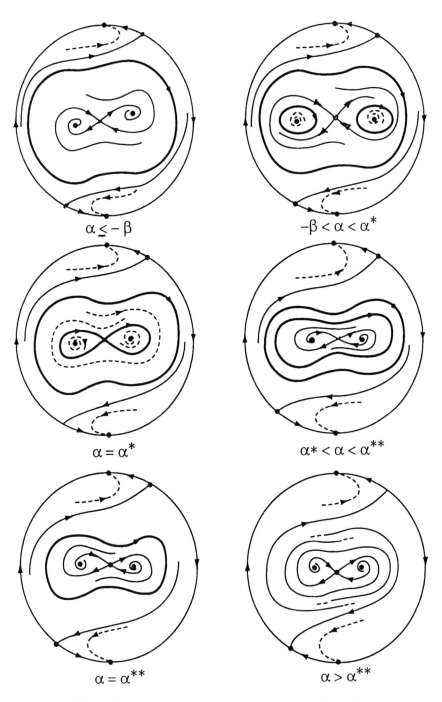

Figure 8. The global phase portraits for the system (8) with $\beta > 0$ and $0 < \varepsilon \ll 1$.

4.9. Melnikov's Method

the next section. The bifurcation diagram for the one-parameter family of limit cycles L_α generated at the critical point $(1, 0)$ in a supercritical Hopf bifurcation at $\alpha = -\beta$ and the bifurcation diagram for the one-parameter family of limit cycles $L_\rho = L^+_{\alpha(\rho)} \cup L^-_{\alpha(\rho)}$ generated on the exterior of the graphic as α increases from α^* are shown in Figure 9.

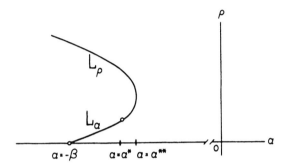

Figure 9. The bifurcation diagram for the two one-parameter families of limit cycles L_α and L_ρ of the system (8).

Numerical computation confirms the theoretical results described above. For $\varepsilon = .1$ and $\beta = 1$, the two stable limit cycles around the critical points $(\pm 1, 0)$ for $\alpha = -.9$, the separatrix cycle for $\alpha = \alpha^* \simeq -.8$, and the stable limit cycle L^-_α for $\alpha = -.77$ are shown in Figure 10. The unstable limit cycle L^+_α is also shown in each of the figures in Figure 10 for these values of α.

For $\alpha > \alpha^{**}$, there are an infinite number of bifurcation values which accumulate at $\alpha = \alpha^{**}$ and at which values there are trajectories connecting the saddle at the origin to the saddles at infinity; cf. [G/H], pp. 62–64. Let $\alpha^*_0 > \alpha^*_1 > \alpha^*_2 \cdots$ denote these bifurcation values. Then α^*_n approaches α^{**} from the right as $n \to \infty$ and the phase portraits for $\alpha = \alpha^*_0, \alpha^*_1, \ldots$ are shown in Figure 11. The serious student should determine the phase portraits for $\alpha^*_{n+1} < \alpha < \alpha^*_n$, $n = 0, 1, 2, \ldots$, in which cases no saddle connections exist. This completes our discussion of the perturbed Duffing equation (8).

Problem Set 9

1. Use Theorem 5 to show that if $\mu_3\mu_5 < 0$ and $0 < \mu_1\mu_5 < 9\mu_3^2/40$, then for all sufficiently small $\varepsilon \neq 0$ the Lienard equation

$$\dot{x} = y - \varepsilon[\mu_1 x + \mu_3 x^3 + \mu_5 x^5]$$
$$\dot{y} = -x$$

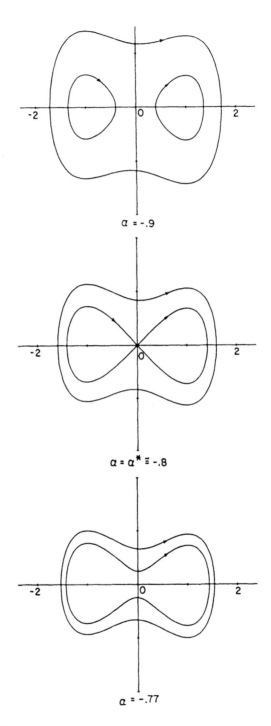

Figure 10. The limit cycles and the graphic of the system (8) with $\varepsilon = .1$ and $\beta = 1$.

4.9. Melnikov's Method

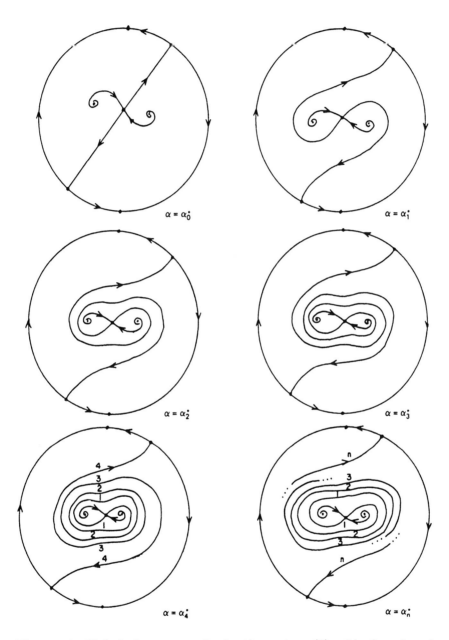

Figure 11. Global phase portraits for the system (8) with $\beta > 0$ and $0 < \varepsilon \ll 1$.

has exactly two limit cycles which are approximately circles of radii

$$\alpha = \sqrt{\frac{-3\mu_3 \pm \sqrt{9\mu_3^2 - 40\mu_1\mu_5}}{5\mu_5}}.$$

2. Construct the global phase portraits for the system (8) in Example 3 when $\beta = 0$ and $0 < \varepsilon \ll 1$. Use the fact that for $\alpha \neq 0$ and $\beta = 0$ the system (8) has a saddle-node at the critical points $(0, \pm 1, 0)$ at infinity. This can be obtained from Figure 7 when the critical points at $\pm(\beta, 1, 0)/\sqrt{1 + \beta^2}$ approach the critical points at $\pm(0, 1, 0)$ as $\beta \to 0^+$. How do the phase portraits in Figure 8 and in this problem for $\beta = 0$ change if $\varepsilon < 0$ (and $|\varepsilon| \ll 1$)? **Hint:** Use the symmetry $(t, x, y, \varepsilon, \alpha, \beta) \to (-t, x, -y, -\varepsilon, \alpha, \beta)$.

3. For $\varepsilon = 0$ the system

$$\dot{x} = y$$
$$\dot{y} = x + x^2 + \varepsilon y[\alpha + x]$$

is a Hamiltonian system which satisfies the assumptions (A.1) and (A.2); cf. Example 3 and Figure 2 in Section 3.3 of Chapter 3. Show that the homoclinic loop is given by a motion on the curve

$$\Gamma_0: y^2 = x^2 + \frac{2}{3}x^3.$$

Compute the Melnikov function along Γ_0 and use Theorem 4 to show that for all sufficiently small $\varepsilon \neq 0$ there is an $\alpha_\varepsilon = 6/7 + 0(\varepsilon)$ such that this system has a homoclinic loop at the origin in an $0(\varepsilon)$ neighborhood of Γ_0. Show that there is a subcritical Hopf bifurcation at the critical point $(-1, 0)$ at the bifurcation value $\alpha = 1$ and, assuming that there are no semistable limit cycles, sketch the phase portraits for this system for $-\infty < \alpha < \infty$.

4. Computation of $I \equiv \int_{-\infty}^{\infty} \operatorname{sech} t \cdot \tanh t \cdot \cos \omega_0(t + t_0) dt$ by the Method of Residues:

(a) Use integration by parts to show that

$$I = -\omega_0 \int_{-\infty}^{\infty} \operatorname{sech} t \cdot \sin \omega_0(t + t_0) dt$$

and then since $\operatorname{sech} t \cdot \sin \omega_0 t$ is an odd function, it follows that

$$I = -\omega_0 \sin \omega_0 t_0 \int_{-\infty}^{\infty} \operatorname{sech} t \cdot \cos \omega_0 t \, dt.$$

4.10. Global Bifurcations of Systems in R^2

(b) Let $z = t + i\tau$, $f(z) = \text{sech}\, z \cdot \cos\omega_0 z = \frac{\cos\omega_0 z}{\cosh z}$ and let C denote the contour shown below:

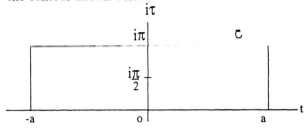

Then, by the Residue Theorem

$$\oint_C f(z)\,dz = 2\pi i \, \text{Res}_{z=i\pi/2}\, f(z).$$

Since $f(z)$ has a simple pole at $z = i\pi/2$ (and no other singularities inside or on C), show that $\text{Res}_{z=i\pi/2}\, f(z) = -i\cosh(\omega_0\pi/2)$.

(c) For the contour C given above, show that for $z = \pm a + i\tau$,

$$\left| \int_0^\pi \frac{\cos\omega_0(\pm a + i\tau)}{\cosh(\pm a + i\tau)}\, d\tau \right| \leq \pi e^{-a} \left(\frac{1 + e^{\omega_0 \pi}}{1 - e^{-2a}} \right)$$

which approaches zero as $a \to \infty$; and for $z = t + i\pi = -u + i\pi$, show that

$$\int_{a+i\pi}^{a-i\pi} f(z)\,dz = \cosh\omega_0\pi \int_{-a}^{a} \cos\omega_0 u \cdot \text{sech}\, u \, du.$$

(d) Finally, using the above results, show that

$$I = \frac{-\omega_0 \sin\omega_0 t_0 \cdot 2\pi \cosh(\omega_0\pi/2)}{1 + \cosh\omega_0\pi}$$
$$= -\pi\omega_0 \cdot \sin\omega_0 t_0 \cdot \text{sech}(\omega_0\pi/2)$$

since $1 + \cosh\omega_0\pi = 2\cosh^2(\omega_0\pi/2)$.

4.10 Global Bifurcations of Systems in R^2

Thus far in this chapter we have considered local bifurcations that occur at nonhyperbolic critical points and periodic orbits, and global bifurcations that occur at homoclinic loops and in a one-parameter family of periodic orbits. According to Peixoto's Theorem in Section 4.1, the only bifurcations that occur in a planar system are local bifurcations at a nonhyperbolic equilibrium point or limit cycle of the system, or global bifurcations that occur at a saddle–saddle connection or in a continuous band of cycles such as those surrounding a center of the system. Local bifurcations are well understood, and the theory has been developed earlier in this chapter. Global bifurcations are more difficult to understand; however, Melnikov's method gives us an excellent tool for studying global bifurcations that occur at homoclinic (or heteroclinic) loops or in a one-parameter family of periodic orbits of a perturbed dynamical system. In this section, we establish the first-order Melnikov theory for perturbed planar analytic systems of the form

$$\dot{\mathbf{x}} = \mathbf{f}(\mathbf{x}) + \varepsilon \mathbf{g}(\mathbf{x}, \varepsilon, \boldsymbol{\mu}) \tag{1_μ}$$

with $\mathbf{x} \in \mathbf{R}^2$ and $\boldsymbol{\mu} \in \mathbf{R}^m$, presented in the previous section, and give some additional examples of this theory. The Melnikov theory gives us an excellent tool for determining the parameter values for which a limit cycle bifurcates from a homoclinic (or heteroclinic) loop and for determining the number of limit cycles in a continuous band of cycles that are preserved under perturbation. In fact, the number, positions, and multiplicities of the limit cycles of (1_μ) for small $\varepsilon \neq 0$ are determined by the number, positions, and multiplicities of the zeros of the Melnikov function for (1_μ). If the Melnikov function is identically equal to zero across the continuous band of cycles, then a higher order analysis is necessary. This is presented in the next two sections.

We begin with a proof of the most basic result of the Melnikov theory for perturbed planar systems, Theorem 5 in the previous section. In order to give a proof of that theorem based on the Implicit Function Theorem, it is first necessary to determine the relationship between the displacement function and the Melnikov function in a neighborhood of a periodic orbit of (1_μ). The relationship is determined by integrating the first variation of (1_μ) with respect to ε along the periodic orbit. The details can be found in [7] or in the proof of Lemma 1.1 in [37]. Before giving that result, let us restate the assumptions (A.1) and (A.2) in the previous section for the system (1_μ) above. For convenience in presenting the theorems and proofs in this section, we shall assume that the parameter α is the arc length along an arc normal to the one-parameter family of periodic orbits $\gamma_\alpha(t)$ of (1_μ) with $\varepsilon = 0$ and that $\alpha \in I$, an interval of the real line. This assumption is not necessary as discussed in Remark 1 below.

(A.1) For $\varepsilon = 0$, the system (1_μ) has a homoclinic orbit

$$\Gamma_0 \colon \mathbf{x} = \gamma_0(t), \quad -\infty < t < \infty$$

at a hyperbolic saddle point $\mathbf{x}_0$.

(A.2) For $\varepsilon = 0$, the system (1_μ) has a one-parameter family of periodic orbits

$$\Gamma_\alpha \colon \mathbf{x} = \gamma_\alpha(t), \quad 0 \leq t \leq T_\alpha$$

4.10. Global Bifurcations of Systems in R^2

of period T_α with parameter $\alpha \in I \subset \mathbf{R}$ equal to the arc length along an arc Σ normal to the family Γ_α.

Lemma 1. *Assume that (A.2) holds for all $\alpha \in I$. Then the displacement function $d(\alpha, \varepsilon, \mu)$ of the analytic system (1_μ), defined in Section 3.4, is analytic in a neighborhood of $I \times \{0\} \times \mathbf{R}^m$, and in that neighborhood*

$$d(\alpha, \varepsilon, \mu) = \frac{-\varepsilon \omega_0 M(\alpha, \mu)}{|\mathbf{f}(\gamma_\alpha(0))|} + 0(\varepsilon^2) \qquad (2)$$

as $\varepsilon \to 0$, where $\omega_0 = \pm 1$ according to whether the one-parameter family of periodic orbits $\gamma_\alpha(t)$, with $\alpha \in I$, is positively or negatively oriented and $M(\alpha, \mu)$ is the Melnikov function for (1_μ) given by

$$M(\alpha, \mu) = \int_0^{T_\alpha} e^{-\int_0^t \nabla \cdot \mathbf{f}(\gamma_\alpha(s))ds} \mathbf{f}(\gamma_\alpha(t)) \wedge \mathbf{g}(\gamma_\alpha(t), 0, \mu) dt,$$

where T_α is the period of $\gamma_\alpha(t)$ for $\alpha \in I$.

Corollary 1. *Assume that (A.2) holds for all $\alpha \in I$. Then for $\alpha \in I$ and $\mu \in \mathbf{R}^m$*

$$d_\varepsilon(\alpha, 0, \mu) = \frac{-\omega_0}{|\mathbf{f}(\gamma_\alpha(0))|} M(\alpha, \mu).$$

And if for some $\alpha_0 \in I$ and $\mu_0 \in \mathbf{R}^m$, $M(\alpha_0, \mu_0) = 0$, it follows that

$$d_{\varepsilon\alpha}(\alpha_0, 0, \mu_0) = \frac{-\omega_0}{|\mathbf{f}(\gamma_{\alpha_0}(0))|} M_\alpha(\alpha_0, \mu_0).$$

Theorem 1. *Assume that (A.2) holds for all $\alpha \in I$. Then if there exists an $\alpha_0 \in I$ and a $\mu_0 \in \mathbf{R}^m$ such that*

$$M(\alpha_0, \mu_0) = 0 \quad \text{and} \quad M_\alpha(\alpha_0, \mu_0) \neq 0,$$

it follows that for all sufficiently small $\varepsilon \neq 0$, the analytic system (1_{μ_0}) has a unique, hyperbolic, limit cycle Γ_ε in an $0(\varepsilon)$ neighborhood of Γ_{α_0}; the limit cycle Γ_ε is stable if $\varepsilon\omega_0 M_\alpha(\alpha_0, \mu_0) > 0$ and unstable if $\varepsilon\omega_0 M_\alpha(\alpha_0, \mu_0) < 0$. Furthermore, if $M(\alpha_0, \mu_0) \neq 0$, then for all sufficiently small $\varepsilon \neq 0$, (1_{μ_0}) has no cycle in an $0(\varepsilon)$ neighborhood of Γ_{α_0}.

Proof. Under Assumption A.2, $d(\alpha, 0, \mu) \equiv 0$ for all $\alpha \in I$ and $\mu \in \mathbf{R}^m$. Define the function

$$F(\alpha, \varepsilon) = \begin{cases} \dfrac{d(\alpha, \varepsilon, \mu_0)}{\varepsilon} & \text{for } \varepsilon \neq 0 \\ d_\varepsilon(\alpha, 0, \mu_0) & \text{for } \varepsilon = 0. \end{cases}$$

Then $F(\alpha, \varepsilon)$ is analytic on an open set $U \subset \mathbf{R}^2$ containing $I \times \{0\}$, and by Corollary 1 and the above hypotheses,

$$F(\alpha_0, 0) = d_\varepsilon(\alpha_0, 0, \mu_0) = k_0 M(\alpha_0, \mu_0) = 0$$

and

$$F_\alpha(\alpha_0, 0) = d_{\varepsilon\alpha}(\alpha_0, 0, \mu_0) = k_0 M_\alpha(\alpha_0, \mu_0) \neq 0,$$

where the constant $k_0 = -\omega_0/|\mathbf{f}(\gamma_{\alpha_0}(0))|$. Thus, by the Implicit Function Theorem, cf. [R], there is a $\delta > 0$ and a unique function $\alpha(\varepsilon)$, defined and analytic for $|\varepsilon| < \delta$, such that $\alpha(0) = \alpha_0$ and $F(\alpha(\varepsilon), \varepsilon) = 0$ for all $|\varepsilon| < \delta$. It then follows from the above definition of $F(\alpha, \varepsilon)$ that for ε sufficiently small, $d(\alpha(\varepsilon), \varepsilon, \mu_0) = 0$ and for sufficiently small $\varepsilon \neq 0$, $d_\alpha(\alpha(\varepsilon), \varepsilon, \mu_0) \neq 0$. Thus, for sufficiently small $\varepsilon \neq 0$, there is a unique hyperbolic limit cycle Γ_ε of (1_μ) at a distance $\alpha(\varepsilon)$ along Σ. Since $\alpha(\varepsilon) = \alpha_0 + 0(\varepsilon)$, this limit cycle lies in an $0(\varepsilon)$ neighborhood of the cycle Γ_{α_0}. The stability of Γ_ε is determined by the sign of $d_\alpha(\alpha_0, 0, \mu_0)$; i.e., by the sign of $M_\alpha(\alpha_0, \mu_0)$, as was established in Section 3.4. Finally, if $M(\alpha_0, \mu_0) \neq 0$, then by Lemma 1 and the continuity of $F(\alpha, \mu)$, it follows for all sufficiently small $\varepsilon \neq 0$ and $|\alpha - \alpha_0| = (\varepsilon)$ that $d(\alpha, \varepsilon, \mu_0) \neq 0$; i.e., (1_{μ_0}) has no cycle in an $0(\varepsilon)$ neighborhood of the cycle Γ_{α_0}.

Remark 1. We note that it is not essential in Lemma 1 or in Theorem 1 for α to be the arc length ℓ along Σ, but only that α be a strictly monotone function of ℓ. In that case, the right-hand side of (2) must be multiplied by $\partial \ell / \partial \alpha$. Also, under the hypotheses of Theorem 1, it follows from the above proof that there is a unique one-parameter family of limit cycles Γ_ε of (1_{μ_0}) with parameter ε, which bifurcates from the cycle Γ_{α_0} of (1_{μ_0}) for $\varepsilon \neq 0$. Thus, as in Theorem 2 in Section 4.7 and the results in [25], Γ_ε can be extended to a unique, maximal, one-parameter family of limit cycles that is either cyclic or unbounded, or which terminates at a critical point or on a separatrix cycle of (1_μ). Note that Chicone and Jacob's example on p. 313 in [6] has one cyclic family and one unbounded family, and the examples at the end of this section have families that terminate at critical points and on separatrix cycles.

The next theorem establishes the relationship between zeros of multiplicity-two of the Melnikov function and the bifurcation of multiplicity-two limit cycles of (1_μ) for $\varepsilon \neq 0$ as noted in Remark 2 in the previous section. This result also holds for higher multiplicity zeros and limit cycles as in Theorem 1.3 in [37].

4.10. Global Bifurcations of Systems in R^2

Theorem 2. *Assume that* (A.2) *holds for all* $\alpha \in I$. *Then if there exists an* $\alpha_0 \in I$ *and a* $\mu_0 \in \mathbf{R}^m$ *such that*

$$M(\alpha_0, \mu_0) = M_\alpha(\alpha_0, \mu_0) = 0, \quad M_{\alpha\alpha}(\alpha_0, \mu_0) \neq 0,$$

and

$$M_{\mu_1}(\alpha_0, \mu_0) \neq 0,$$

it follows that for all sufficiently small ε *there is an analytic function* $\mu(\varepsilon) = \mu_0 + 0(\varepsilon)$ *such that for sufficiently small* $\varepsilon \neq 0$, *the analytic system* $(1_{\mu(\varepsilon)})$ *has a unique limit cycle of multiplicity-two in an* $0(\varepsilon)$ *neighborhood of the cycle* Γ_{α_0}.

Proof. Under assumption (A.2), $d(\alpha, 0, \mu) \equiv 0$ for all $\alpha \in I$ and $\mu \in \mathbf{R}^m$. Define the function $F(\alpha, \varepsilon, \mu)$ as in the proof of Theorem 1 with μ in place of μ_0. Then $F(\alpha, \varepsilon, \mu)$ is analytic in an open set containing $I \times \{0\} \times \mathbf{R}^m$. It then follows from Lemma 1 that, under the above hypotheses,

$$F(\alpha_0, 0, \mu_0) = d_\varepsilon(\alpha_0, 0, \mu_0) = k_0 M(\alpha_0, \mu_0) = 0,$$
$$F_\alpha(\alpha_0, 0, \mu_0) = d_{\varepsilon\alpha}(\alpha_0, 0, \mu_0) = k_0 M_\alpha(\alpha_0, \mu_0) = 0,$$
$$F_{\alpha\alpha}(\alpha_0, 0, \mu_0) = d_{\varepsilon\alpha\alpha}(\alpha_0, 0, \mu_0) = k_0 M_{\alpha\alpha}(\alpha_0, \mu_0) \neq 0,$$

and

$$F_{\mu_1}(\alpha_0, 0, \mu_0) = d_{\varepsilon\mu_1}(\alpha_0, 0, \mu_0) = k_0 M_{\mu_1}(\alpha_0, \mu_0) \neq 0.$$

Thus, by the Weierstrass Preparation Theorem for analytic functions, Theorem 69 on p. 388 in [A–II], there exists a $\delta > 0$ such that

$$F(\alpha, \varepsilon, \mu) = [(\alpha - \alpha_0)^2 + A_1(\varepsilon, \mu)(\alpha - \alpha_0) + A_2(\varepsilon, \mu)]\Phi(\alpha, \varepsilon, \mu),$$

where $A_1(\varepsilon, \mu)$, $A_2(\varepsilon, \mu)$, and $\Phi(\alpha, \varepsilon, \mu)$ are analytic for $|\varepsilon| < \delta$, $|\alpha - \alpha_0| < \delta$, and $|\mu - \mu_0| < \delta$; $A_1(0, \mu_0) = A_2(0, \mu_0) = 0$, $\Phi(\alpha_0, 0, \mu_0) \neq 0$, and $\partial A_2/\partial \mu_1(0, \mu_0) \neq 0$ since $F_{\mu_1}(\alpha_0, 0, \mu_0) \neq 0$. It follows from the above equation that

$$F_\alpha(\alpha_1, \varepsilon, \mu) = [2(\alpha - \alpha_0) + A_1(\varepsilon, \mu)]\Phi(\alpha, \varepsilon, \mu)$$
$$+ [(\alpha - \alpha_0)^2 + A_1(\varepsilon, \mu)(\alpha - \alpha_0) + A_2(\varepsilon, \mu)]\Phi_\alpha(\alpha, \varepsilon, \mu)$$

and

$$F_{\alpha\alpha}(\alpha_1, \varepsilon, \mu) = 2\Phi(\alpha, \varepsilon, \mu) + 2[2(\alpha - \alpha_0) + A_1(\varepsilon, \mu)]\Phi_\alpha(\alpha, \varepsilon, \mu)$$
$$+ [(\alpha - \alpha_0)^2 + A_1(\mu)(\alpha - \alpha_0) + A_2(\varepsilon, \mu)]\Phi_{\alpha\alpha}(\alpha, \varepsilon, \mu).$$

Therefore, if $2(\alpha - \alpha_0) + A_1(\varepsilon, \mu) = 0$ and $(\alpha - \alpha_0)^2 + A_1(\varepsilon, \mu)(\alpha - \alpha_0) + A_2(\varepsilon, \mu) = 0$, it follows from the above equations that (1_μ) has a multiplicity-two limit cycle. Thus, we set $\alpha = \alpha_0 - A_1(\varepsilon, \mu)/2$ and find from the first of the above equations that $F(\alpha_0 - A_1(\varepsilon, \mu)/2, \varepsilon, \mu) = 0$ if and only if the analytic function

$$G(\varepsilon, \mu) \equiv \frac{1}{4}A_1^2(\varepsilon, \mu) - A_2(\varepsilon, \mu) = 0$$

since by continuity $\Phi(\alpha, \varepsilon, \mu) \neq 0$ for small $|\varepsilon|, |\alpha - \alpha_0|$ and $|\mu - \mu_0|$. But $G(0, \mu_0) = 0$ since $A_1(0, \mu_0) = A_2(0, \mu_0) = 0$ and

$$\frac{\partial G}{\partial \mu_1}(0, \mu_0) = -\frac{\partial A_2}{\partial \mu_1}(0, \mu_0) \neq 0$$

since $F_{\mu_1}(\alpha_0, 0, \mu_0) \neq 0$. Let $\mu_0 = (\mu_1^{(0)}, \ldots, \mu_m^{(0)})$. Then it follows from the implicit function theorem, cf. [R], that there exists a $\delta > 0$ and a unique analytic function $g(\varepsilon, \mu_2, \ldots, \mu_m)$ such that $g(0, \mu_2^{(0)}, \ldots, \mu_m^{(0)}) = \mu_1^{(0)}$ and

$$G(\varepsilon, g(\varepsilon, \mu_2, \ldots, \mu_m), \mu_2, \ldots, \mu_m) = 0$$

for $|\varepsilon| < \delta, |\mu_2 - \mu_2^{(0)}| < \delta, \ldots, |\mu_m - \mu_m^{(0)}| < \delta$. For $|\varepsilon| < \delta$ we define $\mu(\varepsilon) = (g(\varepsilon, \mu_2^{(0)}, \ldots, \mu_m^{(0)}), \mu_2^{(0)}, \ldots, \mu_m^{(0)})$; then, $\mu(\varepsilon) = \mu_0 + 0(\varepsilon)$ and $(1_{\mu(\varepsilon)})$ has a unique multiplicity-two limit cycle Γ_ε through the point $\alpha(\varepsilon) = \alpha_0 - \frac{1}{2}A_1(\varepsilon, \mu(\varepsilon))$ on Σ; and by continuity with respect to initial conditions and parameters, it follows that Γ_ε lies in an $0(\varepsilon)$ neighborhood of the cycle Γ_{α_0} since $A_1(0, \mu_0) = 0$.

Remark 2. The proof of Theorem 2 actually establishes that there is an n-dimensional analytic surface $\mu_1 = g(\varepsilon, \mu_2, \ldots, \mu_m)$ through the point $(0, \mu_0) \in \mathbf{R}^{m+1}$ on which (1_μ) has a multiplicity-two limit cycle for $\varepsilon \neq 0$. On one side of this surface the system (1_μ) has two hyperbolic limit cycles, and on the other side (1_μ) has no limit cycle in an $0(\varepsilon)$ neighborhood of Γ_{α_0}; i.e., the system (1_μ) experiences a saddle-node bifurcation as we cross this surface. The side of the surface on which there are two limit cycles is determined by ω_0 and the sign of $M_{\mu_1}(\alpha_0, \mu_0)M_{\alpha\alpha}(\alpha_0, \mu_0)$. Cf. [38].

Remark 3. If $M(\alpha_0, \mu_0) = M_\alpha(\alpha_0, \mu_0) = \cdots = M_\alpha^{(k-1)}(\alpha_0, \mu_0) = 0$, but $M_\alpha^{(k)}(\alpha_0, \mu_0) \neq 0$ and $M_{\mu_1}(\alpha_0, \mu_0) \neq 0$, then it can be shown that for small $\epsilon \neq 0$ $(1_{\mu(\epsilon)})$ has a multiplicity-k limit cycle near Γ_{α_0}; however, if $M_\alpha^{(k)}(\alpha_0, \mu_0) = 0$ for all $k = 0, 1, 2, \ldots$, then $d_\varepsilon(\alpha, 0, \mu_0) \equiv 0$ for all $\alpha \in I$, and a higher order analysis in ε is necessary in order to determine the number, positions, and multiplicities of the limit cycles that bifurcate from the continuous band of cycles of (1_μ) for $\varepsilon \neq 0$. This type of higher order analysis is presented in the next two sections.

Besides the global bifurcation of limit cycles from a continuous band of cycles, there is another type of global bifurcation that occurs in planar systems, namely, the bifurcation of limit cycles from a separatrix cycle. The Melnikov theory for perturbed planar systems also gives us explicit information on this type of bifurcation; cf. Theorem 4 in the previous section. In order to prove Theorem 4 in Section 4.9, it is necessary to determine the relationship between the distance $d(\epsilon, \mu)$ between the saddle separatrices of (1_μ) along a normal line to the homoclinic orbit $\gamma_0(t)$ in (A.1) above at the point $\gamma_0(0)$. This is done by integrating the first variation of (1_μ) with respect to ε along the homoclinic orbit $\gamma_0(t)$. The details are carried out in Appendix I in [38]; cf. p. 358 in [37]:

4.10. Global Bifurcations of Systems in R^2

Lemma 2. *Under assumption* (A.1), *the distance between the saddle separatrices at the hyperbolic saddle point* x_0 *along the normal line to* $\gamma_0(t)$ *at* $\gamma_0(0)$ *is analytic in a neighborhood of* $\{0\} \times \mathbf{R}^m$ *and satisfies*

$$d(\varepsilon, \mu) = \frac{\varepsilon \omega_0 M(\mu)}{|\mathbf{f}(\gamma_0(0))|} + 0(\varepsilon^2) \tag{3}$$

as $\varepsilon \to 0$, *where* $M(\mu)$ *is the Melnikov function for* (1_μ) *along the homoclinic orbit* $\gamma_0(t)$ *given by*

$$M(\mu) = \int_{-\infty}^{\infty} e^{-\int_0^t \nabla \cdot \mathbf{f}(\gamma_0(s))ds} \mathbf{f}(\gamma_0(t)) \wedge \mathbf{g}(\gamma_0(t), 0, \mu) dt.$$

Theorem 3. *Under assumption* (A.1), *if there exists a* $\mu_0 \in \mathbf{R}^m$ *such that*

$$M(\mu_0) = 0 \quad \text{and} \quad M_{\mu_1}(\mu_0) \neq 0,$$

then for sufficiently small $\varepsilon \neq 0$ *there is an analytic function* $\mu(\varepsilon) = \mu_0 + 0(\varepsilon)$ *such that the analytic system* $(1_{\mu(\varepsilon)})$ *has a unique homoclinic orbit* Γ_ε *in an* $0(\varepsilon)$ *neighborhood of* Γ_0. *Furthermore, if* $M(\mu_0) \neq 0$, *then for all sufficiently small* $\varepsilon \neq 0$ *and* $|\mu - \mu_0|$ *the system* (1_μ) *has no separatrix cycle in an* $0(\varepsilon)$ *neighborhood of* Γ_0.

Proof. Under the hypotheses of Theorem 3, it follows from Lemma 2 that, for small ε, $d(\varepsilon, \mu)$ is an analytic function that satisfies $d(0, \mu_0) = 0$ and $d_{\mu_1}(0, \mu_1) \neq 0$. It therefore follows from the Implicit Function Theorem that there exists $\delta > 0$ and a unique analytic function $h(\varepsilon, \mu_1, \ldots, \mu_m)$ which satisfies $h(0, \mu_2^{(0)}, \ldots, \mu_m^{(0)}) = \mu_1^{(0)}$ and $d(\varepsilon, h(\varepsilon, \mu_2, \ldots, \mu_m), \mu_2, \ldots, \mu_m) = 0$ for $|\varepsilon| < \delta, |\mu_2 - \mu_2^{(0)}| < \delta, \ldots,$ and $|\mu_m - \mu_m^{(0)}| < \delta$. It therefore follows from the definition of the function $d(\varepsilon, \mu)$ that if we define $\mu(\varepsilon) = (h(\varepsilon, \mu_2^{(0)}, \ldots, \mu_m^{(0)}), \mu_2^{(0)}, \ldots, \mu_m^{(0)})$ for $|\varepsilon| < \delta$, then $\mu(\varepsilon) = \mu_0 + 0(\varepsilon)$ and $(1_{\mu(\varepsilon)})$ has a homoclinic orbit Γ_ε. It then follows from the uniqueness of solutions and from the continuity of solutions with respect to initial conditions, as on pp. 109–110 of [38], that Γ_ε is the only homoclinic orbit in an $0(\varepsilon)$ neighborhood of Γ_0 for $|\mu - \mu_0| < \delta$.

The next corollary, which is Theorem 3.2 in [37], determines the side of the homoclinic loop bifurcation surface $\mu_1 = h(\varepsilon, \mu_2, \ldots, \mu_m)$ through the point $(0, \mu_0) \in \mathbf{R}^{m+1}$ on which (1_μ) has a unique hyperbolic limit cycle in an $0(\varepsilon)$ neighborhood of Γ_0 for sufficiently small $\varepsilon \neq 0$. In the following corollary we let $\sigma_0 = -\text{sgn}[\nabla \cdot \mathbf{g}(x_0, 0, \mu_0)]$; then, under hypotheses (A.1) and (A.2), σ_0 determines the stability of the separatrix cycle Γ_ε of $(1_{\mu(\varepsilon)})$ in Theorem 3 on its interior and also the stability of the bifurcating limit cycle for small $\varepsilon \neq 0$. They are stable if $\sigma_0 > 0$ and unstable if $\sigma_0 < 0$. We note that under hypotheses (A.1) and (A.2), $\nabla \cdot \mathbf{f}(x_0) = 0$ and, in addition, if $\nabla \cdot \mathbf{g}(x_0, 0, \mu_0) = 0$, then more than one limit cycle can bifurcate from the homoclinic loop Γ_ε in Theorem 3 as μ varies from $\mu(\varepsilon)$; cf. Corollary 2 and the example on pp. 113–114 in [38].

438   4. Nonlinear Systems: Bifurcation Theory

Corollary 2. *Suppose that* (A.1) *and* (A.2) *are satisfied, that the one-parameter family of periodic orbits* Γ_α *lies on the interior of the homoclinic loop* Γ_0, *and that* $\Delta\mu_1 = \mu_1 - h(\varepsilon, \mu_2, \ldots, \mu_m)$ *with* h *defined in the proof of Theorem 3 above. Then for all sufficiently small* $\varepsilon \neq 0, |\mu_2 - \mu_2^{(0)}|, \ldots,$ *and* $|\mu_m - \mu_m^{(0)}|$,

(a) *the system* (1_μ) *has a unique hyperbolic limit cycle in an* $0(\varepsilon)$ *interior neighborhood of* Γ_0 *if* $\omega_0\sigma_0 M_{\mu_1}(\mu_0)\Delta\mu_1 > 0$;

(b) *the system* (1_μ) *has a unique separatrix cycle in an* $0(\varepsilon)$ *neighborhood of* Γ_0 *if and only if* $\mu_1 = h(\varepsilon, \mu_2, \ldots, \mu_m)$; *and*

(c) *the system* (1_μ) *has no limit cycle or separatrix cycle in an* $0(\varepsilon)$ *neighborhood of* Γ_0 *if* $\omega_0\sigma_0 M_{\mu_1}(\mu_0)\Delta\mu_1 < 0$.

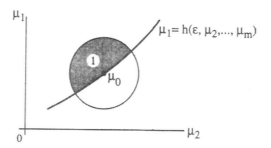

Figure 1. If $\omega_0\sigma_0 M_{\mu_1}(\mu_0) > 0$, then locally (1_μ) has a unique limit cycle if $\mu_1 > h(\varepsilon, \mu_2, \ldots, \mu_m)$ and no limit cycle if $h_1 < h(\varepsilon, \mu_2, \ldots, \mu_m)$ for all sufficiently small $\varepsilon \neq 0$.

We end this section with some examples illustrating the usefulness of the Melnikov theory in describing global bifurcations of perturbed planar systems. In the first example we establish Theorem 6 in Section 3.8.

Example 1. (Cf. [37], p. 348) Consider the perturbed harmonic oscillator

$$\dot{x} = -y + \varepsilon(\mu_1 x + \mu_2 x^2 + \cdots + \mu_{2n+1}x^{2n+1})$$
$$\dot{y} = x.$$

In this example, (A.2) is satisfied with $\gamma_\alpha(t) = (\alpha\cos t, \alpha\sin t)$, where the

4.10. Global Bifurcations of Systems in R^2

parameter $\alpha \in (0, \infty)$ is the distance along the x-axis. The Melnikov function is given by

$$M(\alpha, \mu) = \int_0^{2\pi} \mathbf{f}(\boldsymbol{\gamma}_\alpha(t)) \wedge \mathbf{g}(\boldsymbol{\gamma}_\alpha(t), 0, \mu) dt$$

$$= -\alpha^2 \int_0^{2\pi} (\mu_1 \cos^2 t + \cdots + \mu_{2n+1} \alpha^{2n} \cos^{2n+2} t) dt$$

$$= -2\pi\alpha^2 \left[\frac{\mu_1}{2} + \frac{3}{8}\mu_3 \alpha^2 + \cdots + \frac{\mu_{2n+1}}{2^{2n+2}} \binom{2n+2}{n+1} \alpha^{2n} \right].$$

From Theorems 1 and 2 we then obtain the following results:

Theorem 4. *For sufficiently small $\varepsilon \neq 0$, the above system has at most n limit cycles. Furthermore, for $\varepsilon \neq 0$ it has exactly n hyperbolic limit cycles asymptotic to circles of radii r_j, $j = 1, \ldots, n$ as $\varepsilon \to 0$ if and only if the nth degree equation in α^2*

$$\frac{\mu_1}{2} + \frac{3}{8}\mu_3 \alpha^2 + \cdots + \binom{2n+2}{n+1} \frac{\mu_{2n+1}}{2^{2n+2}} \alpha^{2n} = 0$$

has n positive roots $\alpha^2 = r_j^2, j = 1, \ldots, n$.

Corollary 3. *For $0 < \mu_1 < .3$ and all sufficiently small $\varepsilon \neq 0$, the system*

$$\dot{x} = -y + \varepsilon(\mu_1 x - 2x^3 + 3x^5)$$
$$\dot{y} = x$$

has exactly two limit cycles asymptotic to circles of radii

$$r = \sqrt{2/5}\sqrt{1 \pm \sqrt{1 - \mu_1/.3}}$$

as $\varepsilon \to 0$. Moreover, there exists a $\mu_1(\varepsilon) = .3 + 0(\varepsilon)$ such that for all sufficiently small $\varepsilon \neq 0$ this system with $\mu_1 = \mu_1(\varepsilon)$ has a limit cycle of multiplicity-two, asymptotic to the circle of radius $r = \sqrt{2/5}$, as $\varepsilon \to 0$. Finally, for $\mu_1 > .3$ and all sufficiently small $\varepsilon \neq 0$, this system has no limit cycles, and for $\mu_1 < 0$ it has exactly one hyperbolic limit cycle asymptotic to the circle of radius $\sqrt{2/5}\sqrt{1 + \sqrt{1 + |\mu_1|/.3}}$.

The results of this corollary are borne out by the numerical results shown in Figure 2, where the ratio of the displacement function to the radial distance, $d(r, \varepsilon, \mu_1)/r$, has been computed for $\varepsilon = .1$ and $\mu_1 = .28, .29, .30$, and .31.

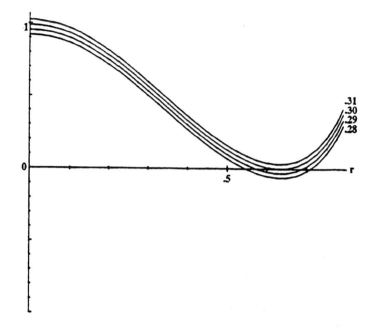

Figure 2. The function $d(r,\varepsilon,\mu_1)/r$ for the system in Corollary 3 with $\varepsilon = .1$ and $\mu_1 = .28, .29, .30,$ and $.31$.

Example 2. (Cf. [37], p. 365.) Consider the perturbed Duffing equation (8) in Section 4.9, which we can write in the form of the Lienard equation

$$\dot{x} = y + \varepsilon(\mu_1 x + \mu_2 x^3)$$
$$\dot{y} = x - x^3$$

with parameter $\mu = (\mu_1, \mu_2)$; cf. Problem 1, where it is shown that the parameters μ_1 and μ_2 are related to those in equation (8) in 4.9 by $\mu_1 = \alpha$ and $\mu_2 = \beta/3$. It was shown in Example 3 of Section 4.9 that the Melnikov function along the homoclinic loop Γ_0^+ shown in Figure 3 of Section 4.9 is given by

$$M(\mu) = \frac{4}{3}\alpha + \frac{16}{15}\beta = \frac{4}{3}\mu_1 + \frac{16}{5}\mu_2.$$

Thus, according to Theorem 3 above, there is an analytic function

$$\mu(\varepsilon) = \mu_1 \left(1, -\frac{5}{12}\right)^T + 0(\varepsilon)$$

such that for all sufficiently small $\varepsilon \neq 0$ the above system with $\mu = \mu(\varepsilon)$ has a homoclinic orbit Γ_ε^+ at the saddle at the origin in an $0(\varepsilon)$ neighborhood of Γ_0^+. We now turn to the more delicate question of the exact number and positions of the limit cycles that bifurcate from the one-parameter family of

4.10. Global Bifurcations of Systems in R^2

periodic orbits $\gamma_\alpha(t)$ for $\varepsilon \neq 0$. First, the one-parameter family of periodic orbits $\gamma_\alpha(t) = (x_\alpha(t), y_\alpha(t))^T$ can be expressed in terms of elliptic functions (cf. [G/H], p. 198 and Problem 4 below) as

$$x_\alpha(t) = \frac{\sqrt{2}}{\sqrt{2-\alpha^2}} dn\left(\frac{t}{\sqrt{2-\alpha^2}}, \alpha\right)$$

$$y_\alpha(t) = -\frac{\sqrt{2}\alpha^2}{2-\alpha^2} sn\left(\frac{t}{\sqrt{2-\alpha^2}}, \alpha\right) cn\left(\frac{t}{\sqrt{2-\alpha^2}}, \alpha\right)$$

for $0 \leq t \leq T_\alpha$, where the period $T_\alpha = 2K(\alpha)\sqrt{2-\alpha^2}$ and the parameter $\alpha \in (0,1)$; $K(\alpha)$ is the complete elliptic integral of the first kind. The parameter α is related to the distance along the x-axis by

$$x^2 = \frac{2}{2-\alpha^2} \quad \text{or} \quad \alpha^2 = \frac{2(x^2-1)}{x^2},$$

and we note that

$$\frac{\partial x}{\partial \alpha} = \frac{\sqrt{2}\alpha}{(2-\alpha^2)^{3/2}} > 0$$

for $\alpha \in (0,1)$ or, equivalently, for $x \in (1, \sqrt{2})$. For $\varepsilon = 0$, the above system is Hamiltonian, i.e., $\nabla \cdot \mathbf{f}(\mathbf{x}) = 0$. The Melnikov function along the periodic orbit $\gamma_\alpha(t)$ therefore is given by

$$M(\alpha, \boldsymbol{\mu}) = \int_0^{T_\alpha} \mathbf{f}(\gamma_\alpha(t)) \wedge \mathbf{g}(\gamma_\alpha(t), 0, \boldsymbol{\mu}) \, dt$$

$$= \int_0^{T_\alpha} [\mu_2 x_\alpha^6(t) + (\mu_1 - \mu_2)x_\alpha^4(t) - \mu_1 x_\alpha^2(t)] \, dt.$$

Then, substituting the above formula for $x_\alpha(t)$ and letting $u = t/\sqrt{2-\alpha^2}$, we find that

$$M(\alpha, \boldsymbol{\mu}) = \frac{1}{2}\int_0^{4K(\alpha)} \left[\mu_2 \left(\frac{2}{2-\alpha^2}\right)^3 dn^6 u + (\mu_1 - \mu_2)\left(\frac{2}{2-\alpha^2}\right)^2 dn^4 u \right.$$

$$\left. - \mu_1 \left(\frac{2}{2-\alpha^2}\right) dn^2 u\right] \sqrt{2-\alpha^2} \, du$$

$$= \frac{1}{(2-\alpha^2)^{5/2}} \int_0^{4K(\alpha)} [4\mu_2 dn^6 u + 2(\mu_1 - \mu_2)(2-\alpha^2)dn^4 u$$

$$- \mu_1(2-\alpha^2)^2 dn^2 u] du.$$

Using the formulas for the integrals of even powers of $dn(u)$ on p. 194 of [40],

$$\int_0^{4K(\alpha)} dn^2 u\, du = 4E(\alpha),$$

$$\int_0^{4K(\alpha)} dn^4 u\, du = \frac{4}{3}[(\alpha^2 - 1)K(\alpha) + 2(2 - \alpha^2)E(\alpha)],$$

and

$$\int_0^{4K(\alpha)} dn^6 u\, du = \frac{4}{15}[4(2-\alpha^2)(\alpha^2 - 1)K(\alpha) + (8\alpha^4 - 23\alpha^2 + 23)E(\alpha)],$$

where $K(\alpha)$ and $E(\alpha)$ are the complete elliptic integrals of the first and second kind, respectively, we find that

$$M(\alpha, \mu) = \frac{4}{(2-\alpha^2)^{5/2}} \left\{ \frac{4\mu_2}{15}[4(2-\alpha^2)(\alpha^2 - 1)K(\alpha) \right.$$
$$+ (8\alpha^4 - 23\alpha^2 + 23)E(\alpha)]$$
$$- \frac{2\mu_2}{3}(2-\alpha^2)[(\alpha^2 - 1)K(\alpha) + 2(2-\alpha^2)E(\alpha)]$$
$$+ \mu_1 \frac{2(2-\alpha^2)}{3}[(\alpha^2 - 1)K(\alpha) + 2(2-\alpha^2)E(\alpha)]$$
$$\left. - \mu_1(4 - 4\alpha^2 + \alpha^4)E(\alpha) \right\}.$$

After some algebraic simplification, this reduces to

$$M(\alpha, \mu) = \frac{4}{3(2-\alpha^2)^{5/2}} \left\{ \frac{-6\mu_2}{5}[(\alpha^4 - 3\alpha^2 + 2)K(\alpha) \right.$$
$$- 2(\alpha^4 - \alpha^2 + 1)E(\alpha)]$$
$$\left. + \mu_1[(\alpha^4 - 4\alpha^2 + 4)E(\alpha) - 2(\alpha^4 - 3\alpha^2 + 2)K(\alpha)] \right\}.$$

It therefore follows that $M(\alpha, \mu)$ has a simple zero iff

$$\frac{\mu_1}{\mu_2} = \frac{6[(\alpha^4 - 3\alpha^2 + 2)K(\alpha) - 2(\alpha^4 - \alpha^2 + 1)E(\alpha)]}{5[(\alpha^4 - 4\alpha^2 + 4)E(\alpha) - 2(\alpha^4 - 3\alpha^2 + 2)K(\alpha)]},$$

and that $M(\alpha, \mu) \equiv 0$ for $0 < \alpha < 1$ iff $\mu_1 = \mu_2 = 0$. Substituting $\alpha^2 = 2(x^2 - 1)/x^2$, the function μ_1/μ_2, given above, can be plotted, using Mathematica, as a function of x. The result for $1 < x < \sqrt{2}$ is given in Figure 3(a) and for $0 < x < 1$ in Figure 3(b). We note that the monotonicity of the function μ_1/μ_2 was established analytically in [37].

The next theorem then is an immediate consequence of Theorems 1 and 3 above. We recall that it was shown in the previous section that for $\mu_2 > 0$

4.10. Global Bifurcations of Systems in R^2

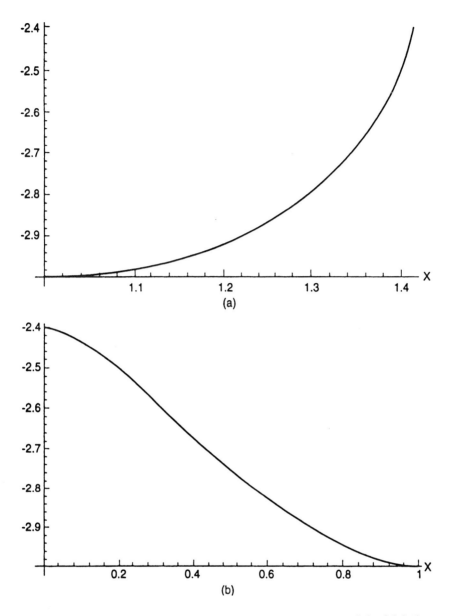

Figure 3. The values of μ_1/μ_2 that result in a simple zero of the Melnikov function for the system in Example 2 for (a) $1 < x < \sqrt{2}$ and (b) for $0 < x < 1$.

and for $\varepsilon > 0$ the system in this example has a supercritical Hopf bifurcation at the critical point $\pm(1, -\varepsilon(\mu_1 + \mu_2))$ at $\mu_1/\mu_2 = -3$. We state the following theorem for $\mu_2 > 0$ and $\varepsilon > 0$, although similar results hold for $\mu_2 < 0$ and/or $\varepsilon < 0$. We see that, even though the computation of the Melnikov function $M(\alpha, \mu)$ in this example is somewhat technical, the benefits are great: We determine the exact number, positions, and multiplicities of the limit cycles in this problem from the zeros of the Melnikov function. Considering that the single most difficult problem for planar dynamical systems is the determination of the number and positions of their limit cycles, the above-mentioned computation is indeed worthwhile.

Theorem 5. *For $\mu_2 > 0$ and for all sufficiently small $\varepsilon > 0$, the system in Example 2 with $-3\mu_2 < \mu_1 < \mu_1(\varepsilon) = -2.4\mu_2 + 0(\varepsilon)$ has a unique, hyperbolic, stable limit cycle around the critical point $(1, -\varepsilon(\mu_1 + \mu_2))$, born in a supercritical Hopf bifurcation at $\mu_1 = -3\mu_2$, which expands monotonically as μ_1 increases from $-3\mu_2$ to $\mu_1(\varepsilon) = -2.4\mu_2 + 0(\varepsilon)$; for $\mu_1 = \mu_1(\varepsilon)$, this system has a unique homoclinic loop around the critical point $(1, -\varepsilon(\mu_1 + \mu_2))$, stable on its interior in an $0(\varepsilon)$ neighborhood of the homoclinic loop Γ_0^+, defined in Section 4.9 and shown in Figure 3 of that section.*

Note that exactly the same statement follows from the symmetry of the equations in this example for the limit cycle and homoclinic loop around the critical point at $-(1, -\varepsilon(\mu_1 + \mu_2))$. In addition, the Melnikov function for the one-parameter family of periodic orbits on the exterior of the separatrix cycle $\Gamma_0^- \cup \{0\} \cup \Gamma_0^+$, shown in Figure 3 of Section 4.9, can be used to determine the exact number and positions of the limit cycles on the exterior of this separatrix cycle for sufficiently small $\varepsilon \neq 0$; cf. Problem 6.

We present one last example in this section, which also will serve as an excellent example for the second-order Melnikov theory presented in the next section.

Example 3. (Cf. [37], p. 362.) Consider the perturbed truncated pendulum equations

$$\dot{x} = y + \varepsilon(\mu_1 x + \mu_2 x^3)$$
$$\dot{y} = -x + x^3$$

with parameter $\boldsymbol{\mu} = (\mu_1, \mu_2)$. For $\varepsilon = 0$, this system has a pair of heteroclinic orbits connecting the saddles at $(\pm 1, 0)$. Cf. Figure 4.

It is not difficult to compute the Melnikov function along the heteroclinic orbits following what was done in Example 3 in Section 4.9. Cf. Problem 3. This results in

$$M(\boldsymbol{\mu}) = 2\sqrt{2}\left(\frac{\mu_1}{3} + \frac{\mu_2}{5}\right).$$

Thus, according to a slight variation of Theorem 3 above (cf. Remark 3.1 in [37]), there is an analytic function

$$\boldsymbol{\mu}(\varepsilon) = \mu_1 \left(1, -\frac{5}{3}\right)^T + 0(\varepsilon)$$

such that for all sufficiently small $\varepsilon \neq 0$ the above system with $\boldsymbol{\mu} = \boldsymbol{\mu}(\varepsilon)$ has a pair of heteroclinic orbits joining the saddles at $\pm(1, -\varepsilon(\mu_1 + \mu_2))$,

4.10. Global Bifurcations of Systems in R^2

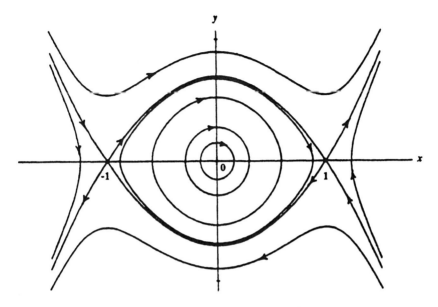

Figure 4. The phase portrait for the system in Example 3 with $\varepsilon = 0$.

i.e., a separatrix cycle, in an $0(\varepsilon)$ neighborhood of the heteroclinic orbits shown in Figure 4.

We next consider which of the cycles shown in Figure 4 are preserved under the above perturbation of the truncated pendulum when $\varepsilon \neq 0$. The one-parameter family of periodic orbits $\gamma_\alpha(t) = (x_\alpha(t), y_\alpha(t))^T$ can be expressed in terms of elliptic functions as follows (cf. Problem 5):

$$x_\alpha(t) = \frac{\sqrt{2}\alpha}{\sqrt{1+\alpha^2}} sn\left(\frac{t}{\sqrt{1+\alpha^2}}, \alpha\right)$$

$$y_\alpha(t) = \frac{\sqrt{2}\alpha}{1+\alpha^2} cn\left(\frac{t}{\sqrt{1+\alpha^2}}, \alpha\right) dn\left(\frac{t}{\sqrt{1+\alpha^2}}, \alpha\right)$$

for $0 \leq t \leq T_\alpha$, where the period $T_\alpha = 4K(\alpha)\sqrt{1+\alpha^2}$ and the parameter $\alpha \in (0,1)$; once again, $K(\alpha)$ denotes the complete elliptic integral of the first kind. The parameter α is related to the distance along the x-axis by

$$x^2 = \frac{2\alpha^2}{1+\alpha^2} \quad \text{or} \quad \alpha^2 = \frac{x^2}{2-x^2},$$

and we note that

$$\frac{\partial x}{\partial \alpha} = \frac{\sqrt{2}}{(1+\alpha^2)^{3/2}} > 0$$

for $\alpha \in (0,1)$ or, equivalently, for $x \in (0,1)$. For $\varepsilon = 0$, the above system is Hamiltonian, i.e., $\nabla \cdot \mathbf{f}(\mathbf{x}) = 0$. The Melnikov function along the periodic orbit $\gamma_\alpha(t)$ therefore is given by

$$M(\alpha,\mu) = \int_0^{T_\alpha} \mathbf{f}(\gamma_\alpha(t)) \wedge \mathbf{g}(\gamma_\alpha(t), 0, \mu) dt$$

$$= \int_0^{T_\alpha} [\mu_1 x_\alpha^2(t) + (\mu_2 - \mu_1) x_\alpha^4(t) - \mu_2 x_\alpha^6(t)] dt.$$

Then, substituting the above formula for $x_\alpha(t)$ and letting $u = t/\sqrt{1+\alpha^2}$, we get

$$M(\alpha,\mu) = \int_0^{4K(\alpha)} \left[\mu_1 \left(\frac{2\alpha^2}{1+\alpha^2}\right) sn^2 u + (\mu_2 - \mu_1)\left(\frac{2\alpha^2}{1+\alpha^2}\right)^2 sn^4 u \right.$$

$$\left. - \mu_2 \left(\frac{2\alpha^2}{1+\alpha^2}\right)^3 sn^6 u \right] \sqrt{1+\alpha^2} du$$

$$= \frac{2\alpha^2}{(1+\alpha)^{5/2}} \int_0^{4K(\alpha)} [\mu_1(1+\alpha^2)^2 sn^2 u$$

$$+ 2\alpha^2(\mu_2 - \mu_1)(1+\alpha^2) sn^4 u - 4\alpha^4 \mu_2 sn^6 u] du.$$

And then using the formulas for even powers of $sn(u)$ on p. 191 of [40],

$$\int_0^{4K(\alpha)} sn^2 u\, du = \frac{4}{\alpha^2}[K(\alpha) - E(\alpha)],$$

$$\int_0^{4K(\alpha)} sn^4 u\, du = \frac{4}{3\alpha^4}[(2+\alpha^2)K(\alpha) - 2(1+\alpha^2)E(\alpha)],$$

and

$$\int_0^{4K(\alpha)} sn^6 u\, du = \frac{4}{15\alpha^6}[(4\alpha^4 + 3\alpha^2 + 8)K(\alpha) - (8\alpha^4 + 7\alpha^2 + 8)E(\alpha)],$$

where $K(\alpha)$ and $E(\alpha)$ are the complete elliptic integrals of the first and second kind, respectively, we find that

$$M(\alpha,\mu) = \frac{8}{(1+\alpha^2)^{5/2}} \left\{ \mu_1(1+\alpha^2)^2[K(\alpha) - E(\alpha)] \right.$$

$$+ \frac{2}{3}(\mu_2 - \mu_1)(1+\alpha^2)[(2+\alpha^2)K(\alpha) - 2(1+\alpha^2)E(\alpha)]$$

$$\left. - \frac{4}{15}\mu_2[(4\alpha^4 + 3\alpha^2 + 8)K(\alpha) - (8\alpha^4 + 7\alpha^2 + 8)E(\alpha)] \right\}.$$

And after some algebraic simplification, this reduces to

$$M(\alpha,\mu) = \frac{-8}{3(1+\alpha^2)^{5/2}} \left\{ \frac{6\mu_2}{5}[(\alpha^4 - 3\alpha^2 + 2)K(\alpha) - 2(\alpha^4 - \alpha^2 + 1)E(\alpha)] \right.$$

$$\left. + \mu_1(\alpha^2 + 1)[(1-\alpha^2)K(\alpha) - (\alpha^2 + 1)E(\alpha)] \right\}.$$

4.10. Global Bifurcations of Systems in R^2

It follows that $M(\alpha, \mu)$ has a simple zero iff

$$\frac{\mu_1}{\mu_2} = \frac{-6[(\alpha^4 - 3\alpha^2 + 2)K(\alpha) - 2(\alpha^4 - \alpha^2 + 1)E(\alpha)]}{5(\alpha^2 + 1)[(1 - \alpha^2)K(\alpha) - (\alpha^2 + 1)E(\alpha)]}$$

and that $M(\alpha, \mu) \equiv 0$ for $0 < \alpha < 1$ iff $\mu_1 = \mu_2 = 0$. Substituting $\alpha^2 = x^2/(2-x^2)$, the ratio μ_1/μ_2, given above, can be plotted as a function of x (using Mathematica) for $0 < x < 1$; cf. Figure 5. We note that the

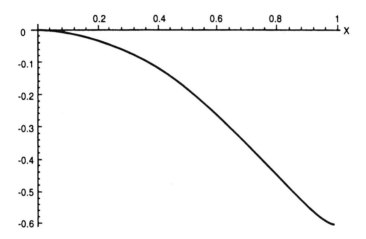

Figure 5. The values of μ_1/μ_2 that result in a simple zero of the Melnikov function for the system in Example 3 for $0 < x < 1$.

monotonicity of the function μ_1/μ_2 was established analytically in [37].

The next theorem then follows immediately from Theorems 1 and 3 in this section. It is easy to see that the system in this example has a Hopf bifurcation at the origin at $\mu_1 = 0$, and using equation (3') in Section 4.3 it follows that for $\mu_2 < 0$ and $\varepsilon > 0$ it is a supercritical Hopf bifurcation. We state the following for $\mu_2 < 0$ and $\varepsilon > 0$, although similar results hold for $\mu_2 > 0$ and/or $\varepsilon < 0$.

Theorem 6. *For $\mu_2 < 0$ and for all sufficiently small $\varepsilon > 0$, the system in Example 3 with $0 < \mu_1 < \mu_1(\varepsilon) = -.6\mu_2 + 0(\varepsilon)$ has a unique, hyperbolic, stable limit cycle around the origin, born in a supercritical Hopf bifurcation at $\mu_1 = 0$, which expands monotonically as μ_1 increases from 0 to $\mu_1(\varepsilon) = -.6\mu_2 + 0(\varepsilon)$; for $\mu_1 = \mu_1(\varepsilon)$, this system has a unique separatrix cycle, that is stable on its interior, in an $0(\varepsilon)$ neighborhood of the two heteroclinic orbits $\gamma_0^{\pm}(t) = \pm(\tanh t/\sqrt{2}, (1/\sqrt{2})\operatorname{sech}^2 t/\sqrt{2})$ shown in Figure 4.*

Problem Set 10

1. Show that the perturbed Duffing oscillator, equation (8) in Section 4.9,

$$\dot{x} = y$$
$$\dot{y} = x - x^3 + \varepsilon(\alpha y + \beta x^2 y),$$

can be written in the form

$$\ddot{x} - \varepsilon g(x, \mu)\dot{x} - x + x^3 = 0$$

with $g(x, \mu) = \alpha + \beta x^2$, and that this latter equation can be written in the form of a Lienard equation,

$$\dot{x} = y + \varepsilon(\mu_1 x + \mu_2 x^3)$$
$$\dot{y} = x - x^3,$$

where the parameter $\mu = (\mu_1, \mu_2) = (\alpha, \beta/3)$. Cf. Example 2.

2. Compute the Melnikov function $M(\alpha, \mu)$ for the quadratically perturbed harmonic oscillator in Bautin normal form,

$$\dot{x} = -y + \lambda_1 x - \lambda_3 x^2 + (2\lambda_2 + \lambda_5)xy + \lambda_6 y^2$$
$$\dot{y} = x + \lambda_1 y + \lambda_2 x^2 + (2\lambda_3 + \lambda_4)xy - \lambda_2 y^2,$$

with $\lambda_i = \sum_{j=1}^{\infty} \lambda_{i,j} \varepsilon^j$ for $i = 1, \ldots, 6$ and $\mu = (\lambda_1, \ldots, \lambda_6)$; and show that $M(\alpha, \mu) \equiv 0$ for all $\alpha > 0$ iff $\lambda_{11} = 0$. Cf. [6] and [37], p. 353.

3. Show that the system in Example 3 can be written in the form

$$\ddot{x} - g(x, \mu)\dot{x} + x - x^3 = 0$$

with $g(x, \mu) = \mu_1 + 3\mu_2 x^2$, and that this equation can be written in the form of the system

$$\dot{x} = y$$
$$\dot{y} = -x + x^3 + \varepsilon y(\mu_1 + 3\mu_2 x^2),$$

which for $\varepsilon = 0$ is Hamiltonian with

$$H(x, y) = \frac{y^2}{2} + \frac{x^2}{2} - \frac{x^4}{4}.$$

Follow the procedure in Example 3 of the previous section in order to compute the Melnikov function along the heteroclinic orbits, $H(x, y) = 1/4$, joining the saddles at $(\pm 1, 0)$. Cf. Figure 4.

4. Using the fact that the Jacobi elliptic function

$$y(u) = dn(u, \alpha)$$

satisfies the differential equation

$$y'' - (2 - \alpha^2)y + 2y^3 = 0,$$

4.10. Global Bifurcations of Systems in R^2

cf. [40], p. 25, show that the function $x_\alpha(t)$ given in Example 2 above satisfies the differential equation in that example with $\varepsilon = 0$, written in the form

$$\ddot{x} - x + x^3 = 0.$$

Also, using the fact that

$$dn'(u, \alpha) = -\alpha^2 sn(u, \alpha) cn(u, \alpha),$$

cf. [4], p. 25, show that $\dot{x}_\alpha(t) = y_\alpha(t)$ for the functions given in Example 2.

5. Using the fact that the Jacobi elliptic function

$$y(u) = sn(u, \alpha)$$

satisfies the differential equation

$$y'' + (1 + \alpha^2)y - 2\alpha^2 y^3 = 0,$$

cf. [40], p. 25, show that the function $x_\alpha(t)$ given in Example 3 above satisfies the differential equation in that example with $\varepsilon = 0$, written in the form

$$\ddot{x} + x - x^3 = 0.$$

Also, using the fact that

$$sn'(u, \alpha) = cn(u, \alpha) dn(u, \alpha),$$

cf. [40], p. 25, show that $\dot{x}_\alpha(t) = y_\alpha(t)$ for the functions given in Example 3.

6. The Exterior Duffing Problem (cf. [37], p. 366): For $\varepsilon = 0$ in the equations in Example 2 there is also a one-parameter family of periodic orbits on the exterior of the separatrix cycle shown in Figure 3 of the previous section. It is given by

$$x_\alpha(t) = \pm \frac{\sqrt{2}\alpha}{\sqrt{2\alpha^2 - 1}} cn\left(\frac{t}{\sqrt{2\alpha^2 - 1}}, \alpha\right)$$

$$y_\alpha(t) = \mp \frac{\sqrt{2}\alpha}{2\alpha^2 - 1} sn\left(\frac{t}{\sqrt{2\alpha^2 - 1}}, \alpha\right) dn\left(\frac{t}{\sqrt{2\alpha^2 - 1}}, \alpha\right),$$

where the parameter $\alpha \in (1/\sqrt{2}, 1)$. Compute the Melnikov function $M(\alpha, \mu)$ along the orbits of this family using the following formulas from p. 192 of [40]:

$$\int_0^{4K(\alpha)} cn^2 u\, du = \frac{4}{\alpha^2}[E(\alpha) - (1 - \alpha^2)K(\alpha)]$$

$$\int_0^{4K(\alpha)} cn^4 u\, du = \frac{4}{3\alpha^4}[2(2\alpha^2 - 1)E(\alpha) + (2 - 3\alpha^2)(1 - \alpha^2)K(\alpha)]$$

$$\int_0^{4K(\alpha)} cn^6 u\, du = \frac{4}{15\alpha^6}[(23\alpha^4 - 23\alpha^2 + 8)E(\alpha)$$
$$+ (1 - \alpha^2)(15\alpha^4 - 19\alpha^2 + 8)K(\alpha)].$$

Graph the values of μ_1/μ_2 that result in $M(\alpha, \mu) = 0$ and deduce that for $\mu_2 > 0$, all small $\varepsilon > 0$ and $\mu_1(\varepsilon) < \mu_1 < \mu_1^*(\varepsilon)$, where $\mu_1(\varepsilon) = -2.4\mu_2 + 0(\varepsilon)$ is given in Theorem 5 and $\mu_1^*(\varepsilon) \cong -2.256\mu_1 + 0(\varepsilon)$, the system in Example 2 has exactly two hyperbolic limit cycles surrounding all three of its critical points, a stable limit cycle on the interior of an unstable limit cycle, which respectively expand and contract monotonically with increasing μ_1 until they coalesce at $\mu_1 = \mu_1^*(\varepsilon)$ and form a multiplicity-two, semistable limit cycle. Cf. Figure 8 in Section 4.9. You should find that $M(\alpha, \mu) = 0$ if

$$\frac{\mu_1}{\mu_2} = \frac{6[(\alpha^4 - 3\alpha^2 + 2)K(\alpha) - 2(\alpha^4 - \alpha^2 + 1)E(\alpha)]}{5[(4\alpha^4 - 4\alpha^2 + 1)E(\alpha) - (\alpha^2 - 1)(2\alpha^2 - 1)K(\alpha)]}.$$

The graph of this function (using Mathematica) is given in Figure 6, where the maximum value of the function shown in the figure is $-2.256\ldots$, which occurs at $\alpha \cong .96$.

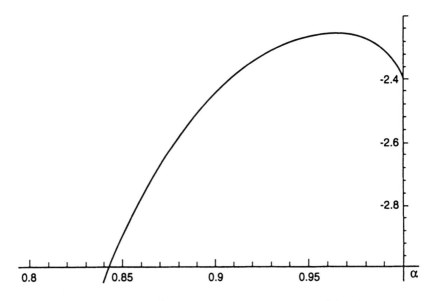

Figure 6. The values of μ_1/μ_2 that result in a zero of the Melnikov function for the exterior Duffing problem with $1/\sqrt{2} < \alpha < 1$ in Problem 6.

7. Consider the perturbed Duffing oscillator

$$\dot{x} = y + \varepsilon(\mu_1 x + \mu_2 x^2)$$
$$\dot{y} = x - x^3,$$

which, for $\varepsilon = 0$, has the one-parameter family of periodic orbits on the interior of the homoclinic loop Γ_0^+, shown in Figure 3 of the

4.10. Global Bifurcations of Systems in R^2

previous section, given in Example 2 above. Compute the Melnikov function $M(\alpha, \mu)$ along the orbits of this family using the following formulas from p. 194 of [40]:

$$\int_0^{4K(\alpha)} dn^3 u\, du = \pi(2 - \alpha^2)$$

$$\int_0^{4K(\alpha)} dn^5 u\, du = \frac{\pi}{4}(8 - 8\alpha^2 + 3\alpha^4).$$

Graph the values of μ_1/μ_2 that result in $M(\alpha, \mu) = 0$ and deduce that for $\mu_2 > 0$, all small $\varepsilon > 0$ and $-2\mu_2 < \mu_1 < \tilde{\mu}_1(\varepsilon) \cong -1.67\mu_2$, the above system has a unique, hyperbolic, stable limit cycle around the critical point $(1, -\varepsilon(\mu_1+\mu_2))$, born in a supercritical Hopf bifurcation at $\mu_1 = -2\mu_2$, which expands monotonically as μ_1 increases from $-2\mu_2$ to $\tilde{\mu}_1(\varepsilon) \cong -1.67\mu_2$, at which value this system has a unique homoclinic loop around the critical point at $(1, -\varepsilon(\mu_1 + \mu_2))$.

8. Show that the system in Problem 7 can be written in the form

$$\dot{x} = y$$
$$\dot{y} = x - x^3 + \varepsilon(\mu_1 y + 2\mu_2 xy),$$

and then follow the procedure in Example 3 of the previous section to compute the Melnikov function $M(\mu)$ along the homoclinic loop and show that $M(\mu) = 0$ iff $\mu_1 = -3\pi\mu_2/4\sqrt{2}$; i.e., according to Theorem 3, for all sufficiently small $\varepsilon \neq 0$ the above system has a homoclinic loop at $\mu_1 = -3\pi\mu_2/4\sqrt{2}+0(\varepsilon)$. Also, using the symmetry with respect to the y-axis, show that this system has a continuous band of cycles for $\mu_1 = 0$ and draw the phase portraits for $\mu_2 > 0$ and $\mu_1 \leq 0$.

9. Compute the Melnikov function $M(\alpha, \mu)$ for the Lienard equation

$$\dot{x} = -y + \varepsilon(\mu_1 x + \mu_3 x^3 + \mu_5 x^5 + \mu_7 x^7)$$
$$\dot{y} = x,$$

and show that for an appropriate choice of parameters it is possible to obtain three hyperbolic limit cycles or a multiplicity-three limit cycle for small $\varepsilon \neq 0$. **Hint:**

$$\frac{1}{2\pi}\int_0^{2\pi} \cos^2 t\, dt = \frac{1}{2}, \quad \frac{1}{2\pi}\int_0^{2\pi} \cos^4 t\, dt = \frac{3}{8}$$

$$\frac{1}{2\pi}\int_0^{2\pi} \cos^6 t\, dt = \frac{5}{16}, \quad \frac{1}{2\pi}\int_0^{2\pi} \cos^8 t\, dt = \frac{35}{128}.$$

4.11 Second and Higher Order Melnikov Theory

We next look at some recent developments in the second and higher order Melnikov theory. In particular, we present a theorem, proved in 1995 by Iliev [41], that gives a formula for the second-order Melnikov function for certain perturbed Hamiltonian systems in $\mathbf{R}^2$. We apply this formula to polynomial perturbations of the harmonic oscillator and to the perturbed truncated pendulum (Example 3 of the previous section). We then present Chicone and Jacob's higher order analysis of the quadratically perturbed harmonic oscillator [6], which shows that a quadratically perturbed harmonic oscillator can have at most three limit cycles. Two specific examples of quadratically perturbed harmonic oscillators are given, one with exactly three hyperbolic limit cycles and another with a multiplicity-three limit cycle.

Recall that, according to Lemma 1 in the previous section, under Assumption (A.2) in that section, the displacement function $d(\alpha, \varepsilon, \mu)$ of the analytic system (1_μ) is analytic in a neighborhood of $I \times \{0\} \times \mathbf{R}^m$, where I is an interval of the real axis. Since, under Assumption (A.2), $d(\alpha, 0, \mu) \equiv 0$ for all $\alpha \in I$ and $\mu \in \mathbf{R}^m$, it follows that there exists an $\varepsilon_0 > 0$ such that for all $|\varepsilon| < \varepsilon_0$, $\alpha \in I$, and $\mu_0 \in \mathbf{R}^m$,

$$d(\alpha, \varepsilon, \mu) = \varepsilon d_\varepsilon(\alpha, 0, \mu) + \frac{\varepsilon^2}{2!} d_{\varepsilon\varepsilon}(\alpha, 0, \mu) + \cdots$$
$$\equiv \varepsilon d_1(\alpha, \mu) + \varepsilon^2 d_2(\alpha, \mu) + \cdots.$$

As in the corollary to Lemma 1 in the previous section, $d_1(\alpha, \mu)$ is proportional to the first-order Melnikov function $M(\alpha, \mu)$, which we shall denote by $M_1(\alpha, \mu)$ in this section and which is given by equation (2) in the last section. The next theorem, proved by Iliev in [41] using Françoise's recursive algorithm for computing the higher order Melnikov functions [55], described in the next section, gives us a formula for $d_2(\alpha, \mu)$, i.e., for the second-order Melnikov function $M_2(\alpha, \mu)$ in terms of certain integrals along the periodic orbits Γ_α in Assumption (A.2) of the previous section. Iliev's Theorem applies to perturbed planar Hamiltonian (or Newtonian) systems of the form

$$\begin{aligned} \dot{x} &= y + \varepsilon f(x, y, \varepsilon, \mu) \\ \dot{y} &= U'(x) + \varepsilon g(x, y, \varepsilon, \mu), \end{aligned} \quad (1_\mu)$$

where $U(x)$ is a polynomial of degree two or more, and f and g are analytic functions. We note that for $\varepsilon = 0$ the system (1_μ) is a Hamiltonian (or Newtonian) system with

$$H(x, y) = \frac{y^2}{2} - U(x).$$

Before stating Iliev's Theorem, we restate Assumption (A.2) in the previous section with the energy $h = H(x, y)$ along the periodic orbit as the parameter:

4.11. Second and Higher Order Melnikov Theory

(A.2′) For $\varepsilon = 0$, the Hamiltonian system (1_μ) has a one-parameter family of periodic orbits

$$\Gamma_h: \mathbf{x} = \gamma_h(t), \qquad 0 \leq t \leq T_h,$$

of period T_h with parameter $h \in I \subset \mathbf{R}$ equal to the total energy along the orbit, i.e., $h = H(\gamma_h(0))$.

Theorem 1 (Iliev). *Under Assumption (A.2′), if $M_1(h, \mu) \equiv 0$ for all $h \in I$ and $\mu \in \mathbf{R}^m$, then the displacement function for the analytic system (1_μ)*

$$d(h, \varepsilon, \mu) = \frac{\varepsilon^2 M_2(h, \mu)}{|\mathbf{f}(\gamma_h(0))|} + 0(\varepsilon^3) \tag{2}$$

as $\varepsilon \to 0$, where the second-order Melnikov function is given by

$$M_2(h, \mu) = \oint_{\Gamma_h} [G_{1h}(x, y, \mu) P_2(x, h, \mu) - G_1(x, y, \mu) P_{2h}(x, h, \mu)] dx$$

$$- \oint_{\Gamma_h} \frac{F(x, y, \mu)}{y} [f_x(x, y, 0, \mu) + g_y(x, y, 0, \mu)] dx$$

$$+ \oint_{\Gamma_h} [g_\varepsilon(x, y, 0, \mu) dx - f_\varepsilon(x, y, 0, \mu) dy],$$

where

$$F(x, y, \mu) = \int_0^y f(x, s, 0, \mu) ds - \int_0^x g(s, 0, 0, \mu) ds,$$

$$G(x, y, \mu) = g(x, y, 0, \mu) + F_x(x, y, \mu),$$

$G_1(x, y, \mu)$ *denotes the odd part of $G(x, y, \mu)$ with respect to y, $G_2(x, y, \mu)$ denotes the even part of $G(x, y, \mu)$ with respect to y, $\tilde{G}_2(x, y^2, \mu) = G_2(x, y, \mu)$,*

$$G_{1h}(x, y, \mu) = \frac{\partial G_1}{\partial y}(x, y, \mu) \cdot \frac{1}{y},$$

$$P_2(x, h, \mu) = \int_0^x \tilde{G}_2(s, 2h + 2U(s), \mu) ds,$$

and $P_{2h}(x, h, \mu)$ denotes the partial of $P_2(x, h, \mu)$ with respect to h.

In view of formula (2) for the displacement function, it follows that if $M_1(h, \mu) \equiv 0$ for all $h \in I$ and $\mu \in \mathbf{R}^m$, then under Assumption (A.2′), Theorems 1 and 2 of Section 4.10 hold with $M_2(h, \mu)$ in place of $M(\alpha, \mu)$ and h in place of α in those theorems. The proofs in Section 4.10 remain valid as given. We therefore have a second-order Melnikov theory where the number, positions, and multiplicities of the limit cycles of (1_μ) are given by the number, positions, and multiplicities of the zeros of the second-order Melnikov function $M_2(h, \mu)$ of (1_μ). This idea is generalized to higher order in the following theorem, which also applies to the general system (1_μ) in Section 4.10 under Assumption (A.2); cf. Theorem 2.1 in [37].

Theorem 2. *Under Assumption* (A.2), *suppose that there exists a* $\mu_0 \in \mathbf{R}^m$ *such that for some* $k \geq 1$

$$d(\alpha, 0, \mu_0) \equiv d_1(\alpha, \mu_0) \equiv \cdots \equiv d_{k-1}(\alpha, \mu_0) \equiv 0$$

for all $\alpha \in I$, *and that there exists an* $\alpha_0 \in I$ *such that*

$$d_k(\alpha_0, \mu_0) = 0 \quad \text{and} \quad \frac{\partial d_k}{\partial \alpha}(\alpha_0, \mu_0) \neq 0;$$

then for all sufficiently small $\varepsilon \neq 0$, *the system* (1_{μ_0}) *has a unique hyperbolic limit cycle in an* $0(\varepsilon)$ *neighborhood of the cycle* Γ_{α_0}.

Remark 1. Theorem 2 in Section 4.10 also can be generalized to a higher order to establish the existence of a multiplicity-two limit cycle of (1_μ) in terms of a multiplicity-two zero of $M_2(\alpha, \mu)$ or, more generally, of $d_k(\alpha, \mu)$ with $k \geq 2$. As in Remark 3 in Section 4.10, we also can develop a higher order theory for higher multiplicity limit cycles; this is done in Theorem 2.2 in [37].

Our first example is a quadratically perturbed harmonic oscillator that has one limit cycle obtained from a second-order analysis. As you will see in Problem 1 and in Chicone and Jacobs' higher order analysis presented below, to second order any quadratically perturbed harmonic oscillator has at most one limit cycle.

Example 1. Consider the following quadratically perturbed harmonic oscillator:

$$\dot{x} = y + a(\varepsilon)x + b(\varepsilon)x^2 + c(\varepsilon)xy$$
$$\dot{y} = -x$$

with $a(\varepsilon) = \varepsilon a_1 + \varepsilon^2 a_2 + \cdots$, $b(\varepsilon) = \varepsilon b_1 + \varepsilon^2 b_2 + \cdots$, and $c(\varepsilon) = \varepsilon c_1 + \varepsilon^2 c_2 + \cdots$. For $\varepsilon = 0$, this is a Hamiltonian system with $H(x,y) = y^2/2 - U(x) = y^2/2 + x^2/2$; it has a one-parameter family of periodic orbits

$$x_h(t) = \sqrt{2h}\cos t, \qquad y_h(t) = \sqrt{2h}\sin t$$

with total energy $h \in (0, \infty)$. Using Lemma 1 in the previous section, it is easy to compute the first-order Melnikov function and find that

$$M_1(h, \mu) = 2\pi h a_1 \equiv 0$$

for all $h > 0$ iff $a_1 = 0$. Therefore, for a second-order analysis it is appropriate to consider the system

$$\dot{x} = y + \varepsilon(\varepsilon ax + bx^2 + cxy)$$
$$\dot{y} = -x,$$

4.11. Second and Higher Order Melnikov Theory

where we let $a_2 = a$, $b_1 = b$, and $c_1 = c$ for convenience in notation. In using Theorem 1 above to compute the second-order Melnikov function, we see that

$$f(x,y,\varepsilon,\mu) = \varepsilon ax + bx^2 + cxy, \qquad g(x,y,\varepsilon,\mu) = 0,$$
$$F(x,y,\mu) = bx^2y + cxy^2/2, \qquad G(x,y,\mu) = 2bxy + cy^2/2,$$
$$G_1(x,y,\mu) = 2bxy, \qquad G_2(x,y,\mu) = cy^2/2, \qquad G_{1h}(x,y,\mu) = 2bx/y,$$
$$P_2(x,h,\mu) = \int_0^x c(2h - s^2)/2 \, ds = c(hx - x^3/6).$$

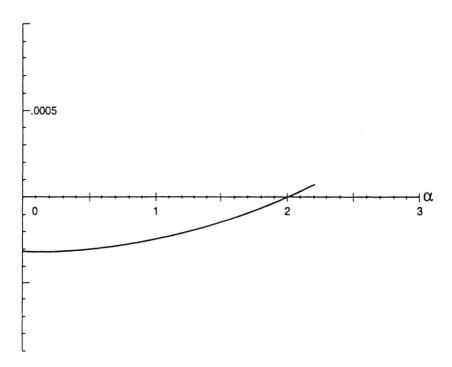

Figure 1. The function $d(\alpha,\varepsilon,\mu)/\alpha$ for the system in Example 1 with $\varepsilon = .01$, $a = c = -1$, and $b = 1$.

Thus,

$$M_2(h,\mu) = \oint_{\Gamma_h} \left[\frac{2bcx}{y}\left(hx - \frac{x^3}{6}\right) - 2bcx^2y\right] dx$$
$$- \oint_{\Gamma_h} \left(\frac{bx^2y + cxy^2/2}{y}\right)(2bx + cy)dx - \oint_{\Gamma_h} axdy.$$

Then, using $dx = ydt$ and $dy = xdt$ from the differential equations and the above formulas for $x_h(t)$ and $y_h(t)$, we find that

$$M_2(h, \mu) = 2bch \int_0^{2\pi} x_h^2(t)\,dt - \frac{bc}{3}\int_0^{2\pi} x_h^4(t)\,dt$$

$$- 4bc \int_0^{2\pi} x_h^2(t)y_h^2(t)\,dt + a \int_0^{2\pi} x_h^2(t)\,dt$$

$$= 2\pi h\left(a - \frac{bc}{2}h\right).$$

Thus, $M_2(h, \mu) = 0$ if $h = 2a/bc$. And since $h = \alpha^2/2$, where α is the radial distance, we see that for sufficiently small $\varepsilon \neq 0$ the above system has a limit cycle through the point $\alpha = 2\sqrt{a/bc} + 0(\varepsilon)$ on the x-axis iff $abc > 0$. For $\varepsilon = .01$, $a = c = -1$, and $b = 1$, Figure 1 shows a plot of $d(\alpha, \varepsilon, \mu)/\alpha$ for the above system and we see that a limit cycle occurs at $\alpha = 2 + 0(\varepsilon)$, as predicted.

Example 2. We next consider the perturbed truncated pendulum in Example 3 of the previous section. From the computation of the first-order Melnikov function in that example, we see that $M_1(h, \mu) \equiv 0$ iff $\mu_1 = \mu_2 = 0$. For a second-order analysis we therefore consider the system

$$\dot{x} = y + \varepsilon(\varepsilon ax + bx^2 + cxy)$$
$$\dot{y} = -x + x^3,$$

where we have taken $\mu_1 = \varepsilon a$ and $\mu_2 = 0$ in the system in Example 3 of the previous section, and we see that $M_1(h, \mu) \equiv 0$ for the above system. As was noted in the previous section, for $\varepsilon = 0$ there is a one-parameter family of periodic orbits $x_\alpha(t), y_\alpha(t)$ with parameter $\alpha \in (0, 1)$ that is related to the distance along the x-axis by $x^2 = 2\alpha^2/(1 + \alpha^2)$ and is related to the total energy by $h = \alpha^2/(1 + \alpha^2)^2$. This is obtained by substituting the functions $x_\alpha(t)$ and $y_\alpha(t)$, given in the previous section in terms of elliptic functions, into the Hamiltonian

$$H(x, y) = \frac{y^2}{2} + \frac{x^2}{2} - \frac{x^4}{4} = \frac{y^2}{2} - U(x)$$

for the above system with $\varepsilon = 0$. Let us now use Theorem 1 above to compute the second-order Melnikov function for this system. We have

$$f(x, y, \varepsilon, \mu) = \varepsilon ax + bx^2 + cxy, \qquad g(x, y, \varepsilon, \mu) = 0,$$

$$F(x, y, \mu) = bx^2 y + cxy^2/2, \qquad G(x, y, \mu) = 2bxy + cy^2/2,$$

$$G_1(x, y, \mu) = 2bxy, \qquad G_2(x, y, \mu) = cy^2/2, \qquad G_{1h}(x, y, \mu) = 2bx/y,$$

$$P_2(x, h, \mu) = \int_0^x c(2h - s^2 + s^4/2)/2\,ds = c(hx - x^3/6 + x^5/20),$$

4.11. Second and Higher Order Melnikov Theory

and $P_{2h}(x, h, \mu) = cx$. Thus it follows from Theorem 1 that

$$M_2(h, \mu) = 2bc \oint_{\Gamma_h} \left(\frac{hx^2}{y} - \frac{x^4}{6y} + \frac{x^6}{20y} - x^2 y \right) dx$$

$$- 2bc \oint_{\Gamma_h} x^2 y dx - a \oint_{\Gamma_h} x dy.$$

We then use $dx = y dt$ and $dy = (-x + x^3) dt$ from the above differential equations and substitute the formulas for $x_\alpha(t)$ and $y_\alpha(t)$ from Example 3 in the previous section, $u = t/\sqrt{1 + \alpha^2}$, and $h = \alpha^2/(1 + \alpha^2)^2$ into the above formula to get

$$M_2(\alpha, \mu) = 2bc \left\{ \frac{\alpha^2}{(1+\alpha^2)^2} \cdot \frac{2\alpha^2}{\sqrt{1+\alpha^2}} \int_0^{4K(\alpha)} sn^2 u\, du \right.$$

$$- \frac{1}{6} \cdot \frac{4\alpha^2}{(1+\alpha^2)^{3/2}} \int_0^{4K(\alpha)} sn^4 u\, du$$

$$+ \frac{8\alpha^6}{20(1+\alpha^2)^{5/2}} \int_0^{4K(\alpha)} sn^6 u\, du$$

$$- 2 \cdot \frac{4\alpha^4}{(1+\alpha^2)^{5/2}} \left[\int_0^{4K(\alpha)} sn^2 u\, du - (1+\alpha^2) \int_0^{4K(\alpha)} sn^4 u\, du \right.$$

$$\left. + \alpha^2 \int_0^{4K(\alpha)} sn^6 u\, du \right] \right\} + a \cdot \frac{2\alpha^2}{(1+\alpha^2)^{3/2}}$$

$$\cdot \left[(1+\alpha^2) \int_0^{4K(\alpha)} sn^2 u\, du - 2\alpha^2 \int_0^{4K(\alpha)} sn^4 u\, du \right],$$

where $K(\alpha)$ is the complete elliptic integral of the first kind and we have used the identities $cn^2 u = 1 - sn^2 u$ and $dn^2 u = 1 - \alpha^2 sn^2 u$ for the Jacobi elliptic functions. Note that $dh/d\alpha > 0$ for $\alpha \in (0, 1)$, and it therefore does not matter whether we use the energy h or α as our parameter. Now, using the formulas for the integrals of $sn^2 u$, $sn^4 u$, and $sn^6 u$ given in Example 3 in the last section, we find after some simplifications that

$$(1+\alpha^2)^{5/2} M_2(\alpha, \mu) = 2bc \left\{ -24\alpha^2 [K(\alpha) - E(\alpha)] \right.$$

$$+ \frac{88}{9}(1+\alpha^2)[(2+\alpha^2)K(\alpha) - 2(1+\alpha^2)E(\alpha)]$$

$$\left. - \frac{152}{75}[(4\alpha^4 + 3\alpha^2 + 8)K(\alpha) - (8\alpha^4 + 7\alpha^2 + 8)E(\alpha)] \right\}$$

$$+ 8a(1+\alpha^2) \left\{ (1+\alpha^2)[K(\alpha) - E(\alpha)] - \frac{2}{3}[(2+\alpha^2)K(\alpha) - 2(1+\alpha^2)E(\alpha)] \right\}.$$

It follows that $M_2(\alpha, \mu) = 0$ if

$$\frac{a}{bc} = \frac{-3[(94\alpha^4 - 42\alpha^2 + 188)K(\alpha) - (188\alpha^4 + 52\alpha^2 + 188)E(\alpha)]}{225(1+\alpha^2)[(\alpha^2 - 1)K(\alpha) + (\alpha^2 + 1)E(\alpha)]}.$$

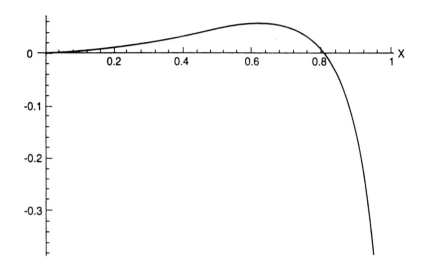

Figure 2. The values of a/bc that result in a zero of the second-order Melnikov function for the system in Example 2 for $0 < x < 1$.

Substituting $\alpha^2 = x^2/(2 - x^2)$ from Example 3 in the previous section, the above ratio can be plotted as a function of x (using Mathematica) for $0 < x < 1$. Cf. Figure 2.

Let us summarize our results for this example in a theorem that follows from Theorems 1 and 2 in the previous section (with $M_2(\alpha, \mu)$ in place of $M(\alpha, \mu)$ in those theorems) and from the fact that for $\varepsilon \neq 0$ the system in this example has a Hopf bifurcation at $a = 0$.

Theorem 3. *For $bc > 0$ and all sufficiently small $\varepsilon > 0$, the system in Example 2 with $a \leq 0$ has exactly one hyperbolic, unstable limit cycle, and for $0 < a < a(\varepsilon) \cong .0576bc + 0(\varepsilon)$ it has exactly two hyperbolic limit cycles, a stable limit cycle on the interior of an unstable limit cycle; these two limit cycles respectively expand and contract with increasing a until they coalesce at $a = a(\varepsilon)$ and form a multiplicity-two semistable limit cycle; there are no limit cycles for $a > a(\varepsilon)$. The stable limit cycle is born in a Hopf bifurcation at $a = 0$; the unstable limit cycle is born in a saddle-node bifurcation at the semistable limit cycle; it expands monotonically as a decreases from $a(\varepsilon)$, and it approaches an $0(\varepsilon)$ neighborhood of the heteroclinic orbits $\gamma_0^{\pm}(t)$, which are defined in Theorem 6 and shown in Figure 4 of Section 4.10, as a decreases to $-\infty$.*

The results of the Melnikov theory are borne out by the numerical results shown in Figure 3, where we have computed $d(x, \varepsilon, \mu)/x$ for the system in Example 2 with $\mu = (a, b, c)$, $b = c = 1$, $\varepsilon = .01$, and $a = .06, .01, -.01$, and $-.25$.

4.11. Second and Higher Order Melnikov Theory

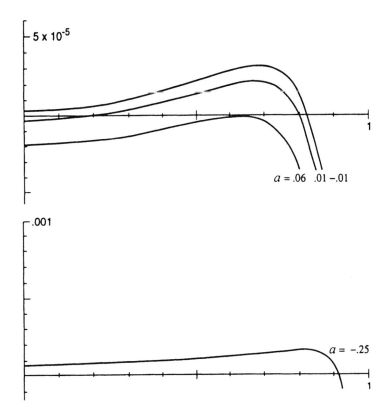

Figure 3. The displacement function $d(x, \varepsilon, \mu)/x$ for the system of Example 2 with $\varepsilon = .01$ and $b = c = 1$.

We next consider the quadratically perturbed harmonic oscillator
$$\begin{aligned} \dot{x} &= -y + \varepsilon[a_{10}(\varepsilon)x + a_{01}(\varepsilon)y + a_{20}(\varepsilon)x^2 + a_{11}(\varepsilon)xy + a_{02}(\varepsilon)y^2] \\ \dot{y} &= x + \varepsilon[b_{10}(\varepsilon)x + b_{01}(\varepsilon)y + b_{20}(\varepsilon)x^2 + b_{11}(\varepsilon)xy + b_{02}(\varepsilon)y^2]. \end{aligned} \quad (3)$$

Because of Bautin's Fundamental Lemma in [2] for this system, cf. Lemma 4.1 in [6], we are able to obtain very specific results for the system (3). The following result proved by Chicone and Jacobs in [6] is of fundamental importance in the theory of limit cycles.

Theorem 4. *For all sufficiently small $\varepsilon \neq 0$, the quadratically perturbed harmonic oscillator (3) has at most three limit cycles; moreover, for sufficiently small $\varepsilon \neq 0$, there exist coefficients $a_{ij}(\varepsilon)$ and $b_{ij}(\varepsilon)$ such that (3) has exactly three limit cycles.*

As in the proof of this theorem in [6], we obtain the following result from Bautin's Fundamental Lemma. This result also follows from Françoise's algorithm; cf. Problem 2 in Section 4.12.

Lemma 1. The displacement function $d(x, \varepsilon, \lambda)$ for the quadratically perturbed harmonic oscillator in the Bautin normal form

$$\dot{x} = -y + \lambda_1 x - \lambda_3 x^2 + (2\lambda_2 + \lambda_5)xy + \lambda_6 y^2$$
$$\dot{y} = x + \lambda_1 y + \lambda_2 x^2 + (2\lambda_3 + \lambda_4)xy - \lambda_2 y^2$$

with $\lambda_3 = 2\varepsilon$, $\lambda_6 = \varepsilon$, and $\lambda_i = \varepsilon\lambda_{i1} + \varepsilon^2\lambda_{i2} + \cdots$ for $i = 1, 2, 4, 5$, is given by

$$d(x, \varepsilon, \lambda) = \varepsilon d_1(x, \lambda) + \varepsilon^2 d_2(x, \lambda) + \cdots,$$

where

$d_1(x, \lambda) = 2\pi\lambda_{11}x$,

$d_2(x, \lambda) = 2\pi\lambda_{12}x - \dfrac{\pi}{4}\lambda_{51}x^3$,

$d_3(x, \lambda) = 2\pi\lambda_{13}x - \dfrac{\pi}{4}\lambda_{52}x^3$,

$d_4(x, \lambda) = 2\pi\lambda_{14}x - \dfrac{\pi}{4}\lambda_{53}x^3 + \dfrac{\pi}{24}\lambda_{21}\lambda_{41}(\lambda_{41} + 5)x^5$,

$d_5(x, \lambda) = 2\pi\lambda_{15}x - \dfrac{\pi}{4}\lambda_{54}x^3 + \dfrac{\pi}{24}[\lambda_{21}\lambda_{41}\lambda_{42} + (\lambda_{41} + 5)(\lambda_{21}\lambda_{42} + \lambda_{22}\lambda_{41})]x^5$,

and where, for $\lambda_4 = -5\varepsilon$,

$$d_6(x, \lambda) = 2\pi\lambda_{16}x - \frac{\pi}{4}\lambda_{55}x^3 + \frac{\pi}{24}(\lambda_{41}\lambda_{42}\lambda_{22}$$
$$+ \lambda_{42}^2\lambda_{21} + \lambda_{43}\lambda_{41}\lambda_{21})x^5 - \frac{\pi}{32}\lambda_{21}^3 x^7$$

for all $(x, \varepsilon, \lambda) \in U \times \mathbf{R}^6$, where U is some open subset of $\mathbf{R}^2$ that contains $(0, \infty) \times \{0\}$.

This lemma has the following corollaries, which give specific information about the number, location, and multiplicities of the limit cycles of quadratically perturbed harmonic oscillators.

Corollary 1. For $a > 0$ and all sufficiently small $\varepsilon \neq 0$, the quadratic system

$$\dot{x} = -y + \varepsilon^2 ax + \varepsilon(y^2 + 8xy - 2x^2)$$
$$\dot{y} = x + \varepsilon^2 ay + 4\varepsilon xy$$
(4)

has exactly one hyperbolic limit cycle asymptotic to the circle of radius $r = \sqrt{a}$ as $\varepsilon \to 0$.

This corollary is an immediate consequence of Theorem 2 and the above lemma, which implies that the displacement function for (4) is given by

$$d(x, \varepsilon, a) = 2\pi\varepsilon^2 x(a - x^2) + 0(\varepsilon^6).$$

Corollary 2. For $A > 0$, $B > 0$, and $A \neq B$, let $a = AB$, $b = A + B$, and $c = 1$. Then for all sufficiently small $\varepsilon \neq 0$, the quadratic system

$$\dot{x} = -y + \varepsilon^4 ax + 8\varepsilon^3 bxy + \varepsilon(y^2 - 24cxy - 2x^2),$$
$$\dot{y} = x + \varepsilon^4 ay + 12\varepsilon c(y^2 - x^2)$$
(5)

4.11. Second and Higher Order Melnikov Theory

has exactly two hyperbolic limit cycles asymptotic to circles of radii $r = \sqrt{A}$ and $r = \sqrt{B}$ as $\varepsilon \to 0$. For $ab > 0$ there exists a $c = (b^2/4a) + 0(\varepsilon)$ such that for all sufficiently small $\varepsilon \neq 0$ the system (5) has a unique semistable limit cycle of multiplicity two, asymptotic to the circle of radius $r = \sqrt{2|a|/|b|}$ as $\varepsilon \to 0$. For $c > b^2/4a$ and all sufficiently small $\varepsilon \neq 0$, the system (5) has no limit cycles.

The proof of this corollary is an immediate consequence of Theorem 2, Remark 1, and the above lemma, which implies that the displacement function for (5) is given by

$$d(x, \varepsilon, \mu) = 2\pi\varepsilon^4 x(a - bx^2 + cx^4) + 0(\varepsilon^6)$$

with $\mu = (a, b, c)$.

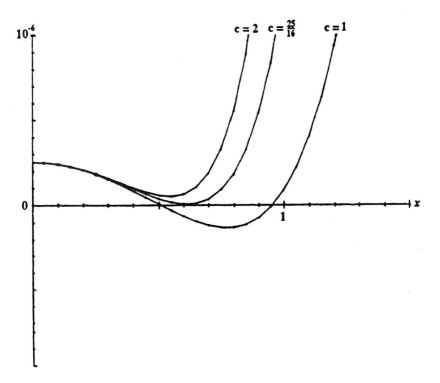

Figure 4. The function $d(x, \varepsilon, \mu)/x$ for the system (5) with $\varepsilon = .02$, $a = 1/4$, $b = 5/4$, and $c = 1$, $25/16$, and 2.

For example, according to Corollary 2, the quadratic system (5) with $a = 1/4$, $b = 5/4$ (i.e., $A = 1/4$, $B = 1$), and $c = 1$ has exactly two hyperbolic limit cycles asymptotic to circles of radii $r = 1/2$ and $r = 1$ as $\varepsilon \to 0$. Furthermore, for $a = 1/4$ and $b = 5/4$, there exists a $c = (25/16) + 0(\varepsilon)$ such that (5) has a semistable limit cycle of multiplicity two, asymptotic to the circle of radius $r = \sqrt{2/5} \cong .63$ as $\varepsilon \to 0$; and for $c > 25/16$, (5) has no limit cycles. This is borne out by the numerical results shown in Figure 4, where $d(x,\varepsilon,\mu)/x$ has been computed for $\varepsilon = .02$ and $c = 1, 25/16$, and 2.

Corollary 3. *For distinct positive constants A, B, and C, let $a = ABC$, $b = AB + AC + BC$, $c = A + B + C$, and $d = 1$. Then, for all sufficiently small $\varepsilon \neq 0$, the quadratic system*

$$\dot{x} = -y + \varepsilon^6 ax + 8\varepsilon^5 bxy + \varepsilon\left[y^2 + 8\left(\frac{d}{25}\right)^{1/3} xy - 2x^2\right],$$

$$\dot{y} = x + \varepsilon^6 ax - 12\varepsilon^3 \frac{3}{(5d)^{1/3}} xy + \varepsilon\left[4\left(\frac{d}{25}\right)^{1/3}(x^2 - y^2) - xy\right]$$

(6)

has exactly three hyperbolic limit cycles asymptotic to circles of radii $r = \sqrt{A}$, $\sqrt{B}$, and $\sqrt{C}$ as $\varepsilon \to 0$. Furthermore, for $ab > 0$ and $c = b^2/3a$, there exists a $d = (c^2/3b) + 0(\varepsilon)$ such that for all sufficiently small $\varepsilon \neq 0$, (6) has a unique limit cycle of multiplicity three, asymptotic to the circle of radius $r = \sqrt{3|a|/|b|}$ as $\varepsilon \to 0$.

The proof of this corollary is an immediate consequence of Theorem 2, Remark 1, and the above lemma, which implies that the displacement function for (6) is given by

$$d(x,\varepsilon,\mu) = 2\pi\varepsilon^6 x(a - bx^2 + cx^4 - dx^6) + 0(\varepsilon^7)$$

with $\mu = (a,b,c,d)$.

For example, according to Corollary 3, the quadratic system (6) with $a = 3/4$, $b = 11/4$, $c = 3$ (i.e., $A = 1/2$, $B = 1$, $C = 3/2$), and $d = 1$ has exactly three hyperbolic limit cycles asymptotic to circles of radii $r = \sqrt{1/2} \cong .71$, $r = 1$, and $r\sqrt{3/2} \cong 1.2$ as $\varepsilon \to 0$. This is borne out by the numerical result shown in Figure 5, where $d(x,\varepsilon,\mu)/x$ has been computed for $\varepsilon = .05$.

Furthermore, according to Corollary 3, for $a = 3/4$, $c = 3$, and $b = \sqrt{3ac} = 3\sqrt{3}/2$, there exists a $d = (2/\sqrt{3}) + 0(\varepsilon)$ such that the quadratic system (6) has a unique limit cycle of multiplicity three, asymptotic to the circle of radius $r = \sqrt{\sqrt{3/2}} \cong .93$ as $\varepsilon \to 0$. This is borne out by the numerical results in Figure 6, where $d(x,\varepsilon,\mu)/x$ has been computed for $\varepsilon = .05$.

We note that the formulas for $d_1(\alpha,\lambda)$ and $d_2(\alpha,\lambda)$ in Lemma 1 above follow from the first- and second-order Melnikov functions, respectively. Cf. Problem 2 in Section 4.10 and Problem 1 below. However, the remaining formulas in Lemma 1 depend on Bautin's Fundamental Lemma, Lemma 4.1 in [6], for their derivation. The results of the above lemma and its corollaries are tabulated in the second row of Table 1, and we note that, according to Theorem 4, a sixth- or higher-order analysis will produce at most three limit cycles in a quadratically perturbed harmonic oscillator. The first row

4.11. Second and Higher Order Melnikov Theory

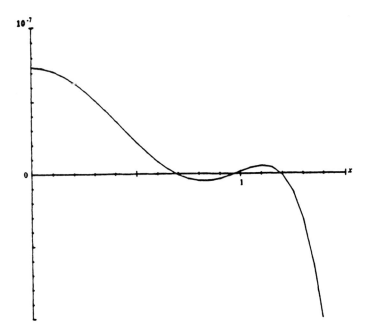

Figure 5. The function $d(x, \varepsilon, \mu)/x$ for the system (6) with $\varepsilon = .05$, $a = 3/4$, $b = 11/4$, $c = 3$, and $d = 1$.

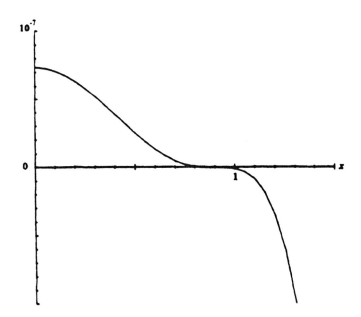

Figure 6. The function $d(x, \varepsilon, \mu)/x$ for the system (6) with $\varepsilon = .05$, $a = 3/4$, $b = 3\sqrt{3}/2$, $c = 3$, and $d = 2/\sqrt{3}$.

TABLE 4.1. The maximum number of limit cycles obtainable from a kth order analysis of an nth-degree polynomially perturbed harmonic oscillator.

	$k=1$	2	3	4	5	6
$n=1$	0	0	0	0	0	0
2	0	1	1	2	2	3
3	1	2				
4	1	3				
5	2	4				
$\vdots$	$\vdots$	$\vdots$				
$2m$	$m-1$	$2m-1$				
$2m+1$	m	$2m$				

in Table 1 simply reflects the fact that any linear system has no limit cycles. The first two columns in Table 1 summarize some recent work by the author and one of his REU students, J. Hunter Tart, during the summer of 1995; cf. Theorem 4 in Section 4.10 and Problems 4–7 below. The numbers in Table 1 are the maximum number of limit cycles that are possible (and that are obtained for some choice of coefficients) in a first-, second-, or kth-order analysis of a polynomially perturbed harmonic oscillator where the perturbation is of degree n. It would be an extremely interesting research project to complete some of the rows in this table; e.g., what is the maximum number of limit cycles possible in a cubically perturbed harmonic oscillator, and what order Melnikov theory is required to produce that number? For a cubically perturbed harmonic oscillator, it is known that this number is at least 11; cf. [42]. Also, it has been known for some time that for a cubically perturbed harmonic oscillator with homogenous cubic perturbations (and no quadratic perturbations), the maximum number of limit cycles is 5; cf. [43].

Remark 2. In a private communication, I.D. Iliev informed me that, using Françoise's recursive algorithm [55], he was able to show that the maximum number of limit cycles obtainable from a kth order analysis of an nth-degree polynomially perturbed harmonic oscillator is less than or equal to $[k(n-1)/2]$ where $[x]$ denotes the greatest integer in x. This work is discussed in the next section. From Table 1, we see that this upper bound is actually achieved for $k=1$ or 2 and all $n \geq 1$; and Iliev has also shown that it is achieved for $k=3$ and all $n \geq 1$. It follows from Theorem 4 and Table 1 that for $n=2$ this upper bound is also achieved for $1 \leq k \leq 7$, but that for $k \geq 6$, the maximum number of limit cycles of a quadratically perturbed harmonic oscillator is 3. And for $n=3$, Iliev has shown that this upper bound is achieved for $1 \leq k \leq 5$, but that for $k=6$, the maximum number of limit cycles obtainable from a 6th order analysis of a cubically perturbed harmonic oscillator is 5. Cf. Table 1 in Section 4.12. At this

4.11. Second and Higher Order Melnikov Theory

time, we do not know the maximum number of limit cycles possible for a cubically perturbed harmonic oscillator, only that it is greater than or equal to 11.

Problem Set 11

1. Use Theorem 1 to show that the second-order Melnikov function for the quadratically perturbed harmonic oscillator in Bautin normal form
$$\dot{x} = y + \varepsilon[\varepsilon\lambda_{12}x - 2x^2 - (2\lambda_{21} + \lambda_{51})xy + y^2]$$
$$\dot{y} = -x + \varepsilon[\varepsilon\lambda_{12}y - \lambda_{21}x^2 + (4 + \lambda_{41})xy + \lambda_{21}y^2]$$
is given by $M_2(h,\lambda) = 4\pi h(\lambda_{12} - \lambda_{51}h/4)$ or by $M_2(\alpha,\lambda) = 2\pi\lambda_{12}\alpha^2 - \pi\lambda_{51}\alpha^4/4$ since the radial distance $\alpha = \sqrt{2h}$ for the harmonic oscillator; cf. the formula for $d_2(\alpha,\lambda) = M_2(\alpha,\lambda)/\alpha$ in Lemma 1.

2. Use Theorem 1 to compute the second-order Melnikov function for the quadratically perturbed harmonic oscillator (4) in Corollary 1 above and compare your result with the formula for $d(x,\varepsilon,a)$ immediately following Corollary 1. (Be sure to let $y \to -y$ in order to transform (4) into the form (1_μ) to which Theorem 1 applies.)

3. Use Lemma 1, Theorem 2 and the results cited in Remark 1 to prove the corollaries in this section.

4. Use Theorem 1 to show that the second-order Melnikov function for the quartically perturbed harmonic oscillator
$$\dot{x} = y + \varepsilon(\varepsilon ax + bxy + \varepsilon dx^3 + ex^4)$$
$$\dot{y} = -x + \varepsilon cx^4$$
is given by $M_2(\alpha, \mu) = \pi\alpha^2(\frac{7}{16}c\alpha^6 + \frac{5}{24}be\alpha^4 + \frac{3}{4}d\alpha^2 + a)$, where α is the radial distance from the origin, and deduce that for sufficiently small $\varepsilon \neq 0$ this system can have three limit cycles for an appropriate choice of coefficients $\mu = (a,b,c,d,e)$.

5. Use Theorem 1 to show that the second-order Melnikov function for the harmonic oscillator with a sixth-degree perturbation of the form
$$\dot{x} = y + \varepsilon(\varepsilon ax + \varepsilon bx^3 + \varepsilon cx^5 + x^6)$$
$$\dot{y} = -x + \varepsilon(Ax^6 + Bx^4 + Cx^2)$$
is given by
$$M_2(\alpha) = \pi\alpha^2\left(\frac{99}{256}A\alpha^{10} + \frac{189}{320}B\alpha^8 + \frac{35}{32}C\alpha^6 + \frac{5}{8}c\alpha^4 + \frac{3}{4}b\alpha^2 + a\right)$$

and deduce that for sufficiently small $\varepsilon \neq 0$ this system can have five limit cycles for an appropriate choice of coefficients.

6. Use Theorem 1 to show that the second-order Melnikov function for the harmonic oscillator with a $2m$th degree perturbation of the form

$$\dot{x} = y + \varepsilon(\varepsilon a_1 x + \cdots + \varepsilon a_m x^{2m-1} + x^{2m})$$
$$\dot{y} = -x + \varepsilon(A_1 x^{2m} + \cdots + A_m x^2)$$

is given by $M_2(\alpha) = \pi \alpha^2 P_{2m-1}(\alpha^2)$ where $P_{2m-1}(\alpha^2)$ is a $(2m-1)$th degree polynomial in α^2, i.e., $P_{2m-1}(\alpha^2) = k_1 A_1 \alpha^{4m-2} + \cdots + a_1$ for some constant $k_1 \neq 0$. Then use Theorem 2 and Descarte's law of signs to deduce that for sufficiently small $\varepsilon \neq 0$, this system has $2m-1$ limit cycles for an appropriate choice of coefficients $a_1, \ldots, a_m$, $A_1, \ldots, A_m$.

7. Use Theorem 1 to show that the second order Melnikov function for the cubically perturbed harmonic oscillator

$$\dot{x} = y$$
$$\dot{y} = -x + \varepsilon(a + bx + xy + cx^3 + 3x^2 y - y^3)$$

is given by $M_2(h) = \pi h(2a + 3bh + 3ch^2)$ and hence (since $M_1(h) \equiv 0$), this system can have two limit cycles.

4.12 Françoise's Algorithm for Higher Order Melnikov Functions

In this section we describe J. P. Françoise's algorithm [55] for determining the first nonzero Melnikov function for a perturbed Hamiltonian system of the form

$$\dot{x} = H_y(x, y) + \varepsilon f(x, y, \varepsilon)$$
$$\dot{y} = -H_x(x, y) + \varepsilon g(x, y, \varepsilon) \qquad (1)$$

where f and g are nth-degree polynomials in x and y and analytic in ε. It is assumed that the unperturbed system, (1) with $\varepsilon = 0$, satisfies the assumption (A.2') in Section 4.11:

(A.2') For $\varepsilon = 0$ the Hamiltonian system (1) has a one-parameter family of periodic orbits

$$\Gamma_h: \mathbf{x} = \gamma_h(t), \quad 0 \leq t \leq T_h$$

of period T_h with parameter $h \in I \subset \mathbf{R}$ equal to the total energy along the orbit; i.e., $h = H(\gamma_h(0))$.

4.12. Françoise's Algorithm for Higher Order Melnikov Functions   467

Remark 1. It follows from Lemma 1 and Remark 1 in Section 4.10 for $k = 1$, from Theorem 1 in Section 4.11 for $k = 2$, or from Theorem 2.2 in [37] for general $k \geq 1$, that if we regard the displacement function $d(h, \varepsilon)$ for (1) as measuring the increment in the energy along a Poincaré section to the Hamiltonian flow (1), as was done in [57], then the first equation in the previous section becomes

$$d(h, \varepsilon) = \varepsilon d_1(h) + \varepsilon^2 d_2(h) + \cdots \quad (2)$$

where $d_k(h) = M_k(h)$, the kth-order Melnikov function for the system (1). We shall assume that this is the case throughout this section where we show how to use J. P. Françoise's algorithm to determine $d_k(h)$ when $d_1(h) = \cdots = d_{k-1}(h) \equiv 0$ for $h \in I$.

Since Françoise's algorithm is stated in terms of differential forms, we first make some basic remarks about differential one-forms. The student is already familiar with the idea of the total differential of a function. For example, the total differential of the second-degree polynomial $Q(x, y) = x^2 + 2xy - 3y^2 - x + y - 10$ is given by $dQ = (2x + 2y - 1)dx + (2x - 6y + 1)dy$. This total differential or exact differential is also called a polynomial one-form (of degree 1); however, not every polynomial one-form is an exact differential. But any **polynomial one-form of degree n** can be written as

$$\Omega = \sum_{i+j \leq n} a_{ij} x^i y^j \, dx + \sum_{i+j \leq n} b_{ij} x^i y^j \, dy. \quad (3)$$

Given a polynomial Hamiltonian function $H(x, y)$, it is of fundamental importance in applying Françoise's algorithm to the system (1) to know when we can decompose a polynomial one-form Ω into an exact differential dQ and a remainder depending on H as

$$\Omega = dQ + q dH \quad (4)$$

where $Q(x, y)$ and $q(x, y)$ are polynomials in x and y. Clearly, if $\Omega = dQ + q dH$, then since $dH = 0$ on any closed curve $H(x, y) = h$ and since the integral of an exact differential dQ around any closed curve is zero, it follows that

$$\int_{H=h} \Omega = \int_{H=h} dQ + \int_{H=h} q dH \equiv 0$$

for all $h \in I$. In order to apply Françoise's algorithm to the system (1), it is necessary that the converse of this statement hold; i.e., for a given polynomial function $H(x, y)$ we must be able to write any polynomial one-form Ω which satisfies $\int_{H=h} \Omega \equiv 0$ for all $h \in I$ as $\Omega = dQ + q dH$ for some polynomials $Q(x, y)$ and $q(x, y)$. If this is the case, then we say that the polynomial function $H(x, y)$ satisfies the condition (∗) in Definition 1 below. Before stating that definition, it is instructive to consider an example.

Example 1. Let $H(x,y) = (x^2+y^2)/2$ and consider a polynomial one-form of degree two

$$\Omega = (a_{20}x^2 + a_{11}xy + a_{02}y^2 + a_{10}x + a_{01}y + a_{00})dx$$
$$+ (b_{20}x^2 + b_{11}xy + b_{02}y^2 + b_{10}x + b_{01}y + b_{00})dy.$$

Integrating Ω around the closed curve $H(x,y) = h$, i.e. around a closed curve $\Gamma_h: \gamma_h(t) = \left(\sqrt{2h}\cos t, \sqrt{2h}\sin t\right)^T$ of the Hamiltonian system (1) with $\varepsilon = 0$, we find that

$$\int_{H=h} \Omega = 2h \int_0^{2\pi} (b_{10}\cos^2 t - a_{01}\sin^2 t)dt = 2\pi h(b_{10} - a_{01}).$$

Therefore, in view of the above remarks, the polynomial one-form Ω in this example cannot be written in the form (4) unless $b_{10} = a_{01}$. And if $b_{10} = a_{01}$, then it is shown in the proof of Lemma 1 below that there are specific polynomials $Q(x,y)$ and $q(x,y)$ of degrees 3 and 1 respectively such that Ω can be written in the form (4). In Problem 1 the student will see that any polynomial one-form of degree 3 or 4, as given in (3), can be written in the form (4) provided $\int_{H=h} \Omega \equiv 0$ for all $h \in I$, i.e., provided that $b_{10} = a_{01}$ and $b_{12} + 3b_{30} = a_{21} + 3a_{03}$. And in general, we shall see in Lemma 1 and Problem 1 below that for any polynomial one-form Ω of degree n, $\int_{H=h} \Omega \equiv 0$ for all $h \in I$ iff $[(n+1)/2]$ conditions on the coefficients a_{ij} and b_{ij} are satisfied (where $[x]$ denotes the greatest integer in x); and in that case Ω can be written in the form (4).

Definition 1. The polynomial function $H(x,y)$ is said to satisfy the condition (*) if for all polynomial one-forms Ω,

$$\int_{H=h} \Omega \equiv 0 \text{ for } h \in I \Rightarrow \exists \text{ polynomials}$$
$$Q(x,y) \text{ and } q(x,y) \text{ such that } \Omega = dQ + qdH. \quad (*)$$

Françoise [55] has shown that the Hamiltonian function for the harmonic oscillator, $H(x,y) = (x^2+y^2)/2$, satisfies the condition (*); cf. Lemma 1 below. And Iliev [59] has shown that the Bogdanov–Takens Hamiltonian, $H(x,y) = (x^2+y^2)/2 - x^3/3$, satisfies the condition (*); cf. Problem 3.

Before presenting Françoise's algorithm, we note that the system (1) can be written in differential form as

$$(H_x dx + H_y dy) + \varepsilon(fdy - gdx) = 0.$$

And if we expand the analytic functions f and g in Taylor series about $\varepsilon = 0$ as

$$f(x,y,\varepsilon) = f_1(x,y) + \varepsilon f_2(x,y) + \varepsilon^2 f_3(x,y) + \cdots$$

and

$$g(x,y,\varepsilon) = g_1(x,y) + \varepsilon g_2(x,y) + \varepsilon^2 g_3(x,y) + \cdots,$$

4.12. Françoise's Algorithm for Higher Order Melnikov Functions   469

then the above differential equation can be written as

$$dH - \varepsilon w_1 - \varepsilon^2 w_2 - \varepsilon^3 w_3 - \cdots = 0 \qquad (5)$$

where $dH = H_x dx + H_y dy$ and $w_j = g_j dx - f_j dy$ for $j = 1, 2, \ldots$.

The following theorem is J. P. Françoise's algorithm for determining the kth-order Melnikov function $M_k(h) = d_k(h)$ for the perturbed Hamiltonian system (3) modified, as in [41] or [57], to include the case when the functions f and g depend on ε.

Theorem 1 (Françoise). *Assume that the polynomial function $H(x,y)$ satisfies the condition (∗). Let $d(h, \varepsilon)$ be the displacement function for the differential equation (5), measuring the increment in energy along a Poincaré section to the Hamiltonian flow. Then if $d_1(h) = \cdots = d_{k-1}(h) \equiv 0$ for all $h \in I$ and for some integer $k \geq 2$, it follows that*

$$d_k(h) = \int_{H=h} \Omega_k$$

where

$$\Omega_1 = w_1, \quad \Omega_m = w_m + \sum_{i+j=m} q_i w_j$$

for $2 \leq m \leq k$, and the functions $q_i, i \leq 1 \leq k-1$, are determined successively from the formulas $\Omega_i = dQ_i + q_i dH$ for $i = 1, \ldots, k-1$.

Proof. This theorem is proved by induction on k as in [55]. For $k = 1$, we begin with the classical formula of Poincaré for the first-order Melnikov function

$$d_1(h) = \int_{H=h} \Omega_1 = \int_{H=h} w_1$$
$$= \int_{H=h} [g_1(x,y)dx - f_1(x,y)dy] = \int_0^{T_h} \mathbf{f}(\gamma_h(t)) \wedge \mathbf{g}(\gamma_h(t))dt$$

with $\mathbf{f} = (H_y, -H_x)^T$ and $\mathbf{g} = (f_1, g_1)^T$ as in Lemma 1 of Section 4.10 for the Hamiltonian system (1).

For $k = 2$, we see that if $d_1(h) \equiv 0$, i.e. if $\int_{H=h} \Omega_1 \equiv 0$ for all $h \in I$, then it follows from the fact that $H(x,y)$ satisfies the condition (∗) that

$$\Omega_1 = dQ_1 + q_1 dH$$

for some polynomials Q_1 and q_1. This allows us to define $\Omega_2 = w_2 + q_1 w_1$. And then multiplying the differential equation (5) by $(1 + \varepsilon q_1)$, we find

$$(1 + \varepsilon q_1) \left[dH - \varepsilon w_1 - \varepsilon^2 w_2 + 0\left(\varepsilon^3\right) \right] = 0$$

or

$$dH + \varepsilon(q_1 dH - w_1) - \varepsilon^2(w_2 + q_1 w_1) + 0\left(\varepsilon^3\right) = 0$$

or

$$dH - \varepsilon dQ_1 - \varepsilon^2 \Omega_2 + 0\left(\varepsilon^3\right) = 0$$

as $\varepsilon \to 0$. We then integrate this last equation around a trajectory γ of the differential equation (5) with $\gamma(0) = \gamma_h(0)$, using the fact that

$$\int_\gamma dH = d(h, \varepsilon)$$

and that $d(h, \varepsilon) = 0(\varepsilon^2)$ since $d_1(h) = 0$. This implies that $\int_\gamma dQ_1 = 0(\varepsilon^2)$ (which can be seen by using the mean value theorem) and that

$$\int_\gamma \Omega_2 = \int_{H=h} \Omega_2 + 0(\varepsilon^2).$$

This leads to

$$d(h, \varepsilon) - \varepsilon^2 \int_{H=h} \Omega_2 + 0\left(\varepsilon^3\right) = 0$$

or

$$\varepsilon^2 \left[d_2(h) - \int_{H=h} \Omega_2 \right] + 0\left(\varepsilon^3\right) = 0$$

as $\varepsilon \to 0$, which yields the desired result for $d_2(h)$.

Now let us assume that $d_1(h) = \cdots = d_{k-1}^{(h)} \equiv 0$ for $h \in I$ where $d_m(h) = \int_{H=h} \Omega_m$ for $1 \le m \le k-1$ and where we have defined

$$\Omega_1 = \omega_1 \quad \text{and} \quad \Omega_m = \omega_m + \sum_{i+j=m} q_i \omega_j \tag{6}$$

for $2 \le m \le k$, the functions q_i, $1 \le i \le k-1$, being determined successively from the formulas

$$\Omega_i = dQ_i + q_i dH \tag{7}$$

for $1 \le i \le k-1$ which follow from the assumption that $d_m(h) = \int_{H=h} \Omega_m \equiv 0$ for all $h \in I$ and for $1 \le m \le k-1$ since H satisfies the condition $(*)$. We then multiply the differential equation (5) by $(1 + \varepsilon q + \cdots + \varepsilon^{k-1} q_{k-1})$ to obtain

$$(1 + \varepsilon q_1 + \cdots + \varepsilon^{k-1} q_{k-1})[dH - \varepsilon \omega_1 - \cdots - \varepsilon^k \omega_k + 0(\varepsilon^{k+1})] = 0$$

or equivalently

$$dH + \varepsilon(q_1 dH - \omega_1) + \varepsilon^2(q_2 dH - q_1 \omega_1 - \omega_2) + \cdots$$
$$+ \varepsilon^{k-1}(q_{k-1} dH - q_{k-2}\omega_1 - \cdots - q_1 \omega_{k-2} - \omega_{k-1})$$
$$- \varepsilon^k(\omega_k + q_1 \omega_{k-1} + \cdots + q_{k-1}\omega_1) + 0(\varepsilon^{k+1}) = 0$$

which, according to (6) and (7), is equivalent to

$$dH - (\varepsilon dQ_1 + \cdots + \varepsilon^{k-1} dQ_{k-1}) - \varepsilon^k \Omega_k + 0(\varepsilon^{k+1}) = 0.$$

Integrating this last equation around the trajectory γ of (5), using the fact that

$$\int_\gamma dH = d(h, \varepsilon)$$

4.12. Françoise's Algorithm for Higher Order Melnikov Functions

and that $d(h,\varepsilon) = 0(\varepsilon^k)$ which implies that $\int_\gamma dQ_j = 0(\varepsilon^k)$ for $j = 1,\ldots,k-1$, we obtain

$$d(h,\varepsilon) - \varepsilon^k \int_\gamma \Omega_k + 0(\varepsilon^{k+1}) = 0$$

as $\varepsilon \to 0$. And since

$$\int_\gamma \Omega_k = \int_{H=h} \Omega_k + 0(\varepsilon^k)$$

we have

$$\varepsilon^k \left[d_k(h) - \int_{H=h} \Omega_k \right] + 0(\varepsilon^{k+1}) = 0$$

as $\varepsilon \to 0$, which implies that

$$d_k(h) = \int_{H=h} \Omega_k$$

and Theorem 1 is proved.

We now apply Françoise's algorithm to the polynomially perturbed harmonic oscillator, (1) with $H(x,y) = (x^2+y^2)/2$. The following lemma, proved by I. D. Iliev in [58], not only shows that the Hamiltonian $H(x,y) = (x^2+y^2)/2$ satisfies the condition (*) for any polynomial one-form Ω of degree n in x and y, but it also gives the degrees of the polynomials Q and q in (*) in terms of n. This lemma therefore makes it possible to obtain an upper bound on the number of limit cycles obtainable from a kth-order analysis of the nth-degree polynomially perturbed harmonic oscillator (1); cf. Table 1 below.

Lemma 1 (Iliev).*Any polynomial one-form Ω of degree n in x and y can be expressed as*

$$\Omega = dQ + qdH + \alpha(H)ydx$$

where $H(x,y) = (x^2+y^2)/2$, $Q(x,y)$ and $q(x,y)$ are polynomials of degrees $(n+1)$ and $(n-1)$ respectively and $\alpha(h)$ is a polynomial of degree $[\frac{1}{2}(n-1)]$ where $[x]$ denotes the greatest integer in x.

Remark 2. We note that it follows from Lemma 1 and the fact that

$$\int_{H=h} dQ = 0 \quad \text{and} \quad \int_{H=h} qdH = 0,$$

that

$$\int_{H=h} \Omega = \alpha(h) \int_{H=h} ydx = \alpha(h) \int_0^{2\pi} y_h^2(t) dt.$$

Thus, if $\int_{H=h} \Omega \equiv 0$ for all $h \in I$, it follows that $\alpha(h) \equiv 0$ for all $h \in I$ and hence, according to Lemma 1, $\Omega = dQ + qdH$; i.e., the Hamiltonian function $H(x,y) = (x^2+y^2)/2$ satisfies the condition (*).

Lemma 1 is proved by induction in [58]. We illustrate the idea of the proof by carrying out the first two steps in the induction proof and ask the student to do more in Problem 1 at the end of this section.

Proof (for $n = 1$ and 2). For $n = 1$ we wish to show that for any given polynomial one-form of degree 1,

$$\Omega = (a_{10}x + a_{01}y + a_{00})dx + (b_{10}x + b_{01}y + b_{00})dy,$$

we can find a second-degree polynomial

$$Q(x, y) = Q_{20}x^2 + Q_{11}xy + Q_{02}y^2 + Q_{10}x + Q_{01}y$$

and zeroth-degree polynomials $q(x, y) = q_{00}$ and $\alpha(h) = \alpha_0$ satisfying

$$\Omega = dQ + qdH + \alpha(H)ydx,$$

i.e., satisfying

$$(a_{10}x + a_{01}y + a_{00})dx + (b_{10}x + b_{01}y + b_{00})dy$$
$$= [(2Q_{20} + q_{00})x + (Q_{11} + \alpha_0)y + Q_{10}]dx$$
$$+ [Q_{11}x + (2Q_{02} + q_{00})y + Q_{01}]dy.$$

This is equivalent to $2Q_{20} + q_{00} = a_{10}, Q_{11} + \alpha_0 = a_{01}, Q_{10} = a_{00}; Q_{11} = b_{10}, 2Q_{02} + q_{00} = b_{01}$ and $Q_{01} = b_{00}$. There is one degree of freedom in this system of six equations in seven unknowns and we are free to choose $q_{00} = 0$. It will also be possible to choose $q_{00} = 0$ for any $n \geq 1$. We therefore arrive at

$$Q(x, y) = \frac{a_{10}}{2}x^2 + b_{10}xy + \frac{b_{01}}{2}y^2 + a_{00}x + b_{00}y,$$
$$q(x, y) = 0 \quad \text{and} \quad \alpha(h) = a_{01} - b_{10}.$$

For $n = 2$, we wish to show that for any given polynomial one-form of degree 2,

$$\Omega = (a_{20}x^2 + a_{11}xy + a_{02}y^2 + a_{10}x + a_{01}y + a_{00})dx$$
$$+ (b_{20}x^2 + b_{11}xy + b_{02}y^2 + b_{10}x + b_{01}y + b_{00})dy,$$

we can find a third-degree polynomial

$$Q(x, y) = Q_{30}x^3 + Q_{21}x^2y + Q_{12}xy^2 + Q_{03}y^3$$
$$+ Q_{20}x^2 + Q_{11}xy + Q_{02}y^2 + Q_{10}x + Q_{01}y,$$

a first-degree polynomial $q(x, y) = q_{10}x + q_{01}y$, and a zeroth-degree polynomial $\alpha(h) = \alpha_0$ satisfying

$$\Omega = dQ + qdH + \alpha(H)ydx,$$

4.12. Françoise's Algorithm for Higher Order Melnikov Functions 473

i.e., satisfying

$$\Omega = \big[(3Q_{30} + q_{10})x^2 + (2Q_{21} + q_{01})xy + Q_{12}y^2$$
$$+ 2Q_{20}x + (Q_{11} + \alpha_0)y + Q_{10}\big]\,dx$$
$$+ \big[Q_{21}x^2 + (2Q_{12} + q_{10})xy + (3Q_{03} + q_{01})y^2$$
$$Q_{11}x + 2Q_{02}y + Q_{01}\big]\,dy.$$

This is equivalent to $3Q_{30} + q_{10} = a_{20}, 2Q_{21} + q_{01} = a_{11}, Q_{12} = a_{02}, 2Q_{20} = a_{10}, Q_{11} + \alpha_0 = a_{01}, Q_{10} = a_{00}; Q_{21} = b_{20}, 2Q_{12} + q_{10} = b_{11}, 3Q_{03} + q_{01} = b_{02}, Q_1 = b_{10}, 2Q_{02} = b_{01}$, and $Q_{01} = b_0$. The solution of this system of twelve equations in twelve unknowns lead to

$$Q(x,y) = \frac{1}{3}(a_{20} - b_{11} + 2a_{02})x^3 + b_{20}x^2 y + a_{02}xy^2 + \frac{1}{3}(b_{02} - a_{11} + 2b_{20})y^3$$
$$+ \frac{1}{2}a_{10}x^2 + b_{10}xy + b_{01}y^2/2 + a_{00}x + b_{00}y,$$

$$q(x,y) = (b_{11} - 2a_{02})x + (a_{11} - 2b_{20})y \quad \text{and} \quad \alpha(h) = a_{01} - b_{10}.$$

The student is asked to provide the proof for $n = 3$ and for the highest-degree terms for $n = 4$ (noting that, as above, all of the other terms remain the same). This will allow the student to see how the general nth step in the induction proof should proceed.

Corollary 1. *For $H(x,y) = (x^2 + y^2)/2$, any integral $\int_{H=h} \Omega$ of a polynomial one-form of degree n has at most $[\frac{1}{2}(n-1)]$ isolated zeros in $(0, \infty)$.*

Proof. As in Remark 2, $\int_{H=h} \Omega = 0$ iff $\alpha(h) = 0$. And since $\alpha(h)$ is a polynomial of degree $[\frac{1}{2}(n-1)]$, according to Lemma 1, $\int_{H=h} \Omega$ has at most $[\frac{1}{2}(n-1)]$ isolated zeros in $(0, \infty)$.

The next corollary follows from Theorem 1 and Lemma 1.

Corollary 2. *If $H(x,y) = (x^2 + y^2)/2$ and $d_1(h) = \cdots = d_{k-1}(h) \equiv 0$ for $h \in I = (0, \infty)$, then Ω_k is a polynomial one-form of degree $k(n-1)+1$.*

Proof. The proof follows by induction on k. For $k = 1, \deg \Omega_1 = \deg w_1 = \deg (g_1 dx - f_1 dy) = n$ since f_1 and g_1 are polynomials of degree n in x and y. Then, according to Theorem 1, $\deg \Omega_2 = \deg (w_2 + q_1 w_1) = \deg (q_1 w_1) = (n-1) + n$ since $\deg \Omega_1 = n$ implies that $\deg q_1 = (n-1)$ by Lemma 1. Now assuming that $\deg \Omega_{k-1} = (k-1)(n-1)+1$, it follows from Lemma 1 that $\deg q_{k-1} = (k-1)(n-1)$ and then, according to Theorem 1,

$$\deg \Omega_k = \deg \left(w_k + \sum_{i+j=k} q_i w_j \right)$$
$$= \deg (q_{k-1} w_1) = (k-1)(n-1) + n = k(n-1) + 1$$

and this completes the proof of Corollary 2.

474   4. Nonlinear Systems: Bifurcation Theory

Theorem 2. *Assume that $H(x,y) = (x^2+y^2)/2$ and that $d_k(h)$ is the first Melnikov function for (1) that does not vanish identically. Then*

$$d_k(h) = \int_{H=h} \Omega_k$$

has at most $\left[\frac{1}{2}k(n-1)\right]$ zeros, counting their multiplicities, for $h \in (0,\infty)$.

Proof. By Corollary 2, Ω_k is a polynomial one-form of degree $k(n-1)+1$ and then by Corollary 1, $d_k(h) = \int_{H=h} \Omega_k$ has at most $\left[\frac{1}{2}k(n-1)\right]$ isolated zeros in $(0,\infty)$.

This theorem, together with Theorem 2 and Remark 1 in Section 4.11, then allows us to estimate the number of limit cycles obtainable from a kth-order analysis of an nth-degree polynomially perturbed harmonic oscillator. The results are displayed in Table 1 below. Cf. Table 1 at the end of Section 4.11. Iliev [58] has also shown that the first three columns of Table 1 are exact, i.e., that the maximum number of limit cycles given by Theorem 2 above for $k = 1, 2, 3$ is actually obtained for some perturbation, and that the numbers shown in the third row are exact except that for $n = 3$ and $k = 6$, the exact upper bound is 5 and not 6. We also know from Section 4.11 that the numbers in the first two rows of Table 1 are exact. Furthermore, Iliev [58] has conjectured that all of the numbers written explicitly in Table 1 are exact upper bounds except for $n = 3$ and $k = 6$, where the maximum number of limit cycles obtainable is 5 and not 6. The first row in Table 1 illustrates the fact that any linear system has no limit cycles and from Chicone and Jacob's result, Theorem 4 in Section 4.11, we know that the second row of Table 1 stabilizes at the value 3 for all $k \geq 6$; i.e., a quadratically perturbed harmonic oscillator has at most 3 limit cycles and it takes a 6th or higher order analysis in the small parameter ε to actually obtain 3 limit cycles. We believe that every row in Table 1 will stabilize at some value $N(n)$ for all $k \geq K(n)$. We know that $N(1) = 0, N(2) = 3$, that $N(3) \geq 11$, according to [42], and that $K(3) \geq 11$, according to Theorem 2.

TABLE 4.1. The maximum number of limit cycles obtainable from a kth-order analysis of an nth-degree polynomially perturbed harmonic oscillator.

	$k=1$	2	3	4	5	6	...
$n=1$	0	0	0	0	0	0	...
2	0	1	1	2	2	3	...
3	1	2	3	4	5	6	...
4	1	3	4	6	7	9	...
5	2	4	6	8	10	12	...
⋮	⋮	⋮	⋮	⋮	⋮	⋮	⋮
$2m$	$m-1$	$2m-1$	$3m-2$	$4m-2$	$5m-3$	$6m-3$	...
$2m+1$	m	$2m$	$3m$	$4m$	$5m$	$6m$	...

4.12. Françoise's Algorithm for Higher Order Melnikov Functions

Problem Set 12

1. (a) Carry out the third step in the induction proof of Lemma 1, i.e., for $n = 3$ determine polynomials

    $$Q(x,y) = Q_{40}x^4 + Q_{31}x^3y + Q_{22}x^2y^2 + Q_{13}xy^3 + Q_{04}y^4$$
    $$+ Q_{30}x^3 + Q_{21}x^2y + Q_{12}xy^2 + Q_{03}y^3 + Q_{20}x^2$$
    $$+ Q_{11}xy + Q_{02}y^2 + Q_{10}x + Q_{01}y,$$

    $q(x,y) = q_{20}x^2 + q_{11}xy + q_{02}y^2 + q_{10}x + q_{01}y$ and $\alpha(H(x,y)) = \alpha_1(x^2 + y^2)/2 + \alpha_0$ such that for any polynomial one-form Ω of degree 3 we have

    $$\Omega = dQ + qdH + \alpha(H)ydx.$$

    Note that all but the highest-degree terms in Q, q and α are exactly the same as those obtained in the proof of Lemma 1 for $n = 2$. Also, note that there are only four equations for the five unknowns $Q_{40}, Q_{22}, Q_{04}, q_{20}$ and q_{02}; so there is one degree of freedom and we can choose $q_{02} = 0$. In particular, you should find that $\alpha(h) = (a_{21} + 3a_{03} - b_{12} - 3b_{30})h/2 + (a_{01} - b_{10})$ and this implies that any polynomial one-form Ω of degree 3 satisfies $\int_{H=h} \Omega \equiv 0$ for all $h \in I$ iff $b_{10} = a_{01}$ and $b_{12} + 3b_{30} = a_{21} + 3a_{03}$.

    (b) Carry out the fourth step in the induction proof of Lemma 1 for the higher-degree terms not found in part (a); i.e., for $n = 4$, determine the highest-degree terms in $Q(x,y) = Q_{50}x^5 + Q_{41}x^4y + Q_{32}x^3y^2 + Q_{23}x^2y^3 + Q_{14}xy^4 + Q_{05}y^5$, and $q(x,y) = q_{30}x^3 + q_{21}x^2y + q_{12}xy^2 + q_{03}y^3$ such that $\Omega = dQ + qdH + \alpha(H)ydx$ is satisfied.

    (c) Carry out the nth step (separately for n even and for n odd) in the induction proof of Lemma 1 for the highest-degree terms in $Q(x,y), q(x,y)$ and $\alpha(H(x,y))$; cf. [58].

2. Consider the quadratically perturbed harmonic oscillator in the Bautin normal form given in Lemma 1 in Section 4.11:

    (a) Use Françoise's algorithm to show that if $d_1(h) = M_1(h) \equiv 0$ for all $h > 0$ (i.e., if $\lambda_{11} = 0$), then the second order Melnikov function, $M_2(h)$ or $d_2(h)$, for

    $$\dot{x} = y + \varepsilon[\varepsilon\lambda_{12}x - 2x^2 - (2\lambda_{21} + \lambda_{51})xy + y^2]$$
    $$\dot{y} = -x + \varepsilon[\varepsilon\lambda_{12}y - \lambda_{21}x^2 + (4 + \lambda_{41})xy + y^2]$$

    is given by $M_2(h) = 4\pi h\lambda_{12} - \pi h^2\lambda_{51}$ (as in Problem 1 in Section 4.11). Hint: As in the proof of Lemma 1, for $\Omega_1 = [-\lambda_{21}x^2 + (4+\lambda_{41})xy + \lambda_{21}y^2]dx + [2x^2 + (2\lambda_{21}+\lambda_{51})xy - y^2]dy$, you should find that $q_1(x,y) = \lambda_{51}x + \lambda_{41}y$.

(b) If $d_1(h) \equiv 0$ and $d_2(h) \equiv 0$ for all $h > 0$ (i.e., if $\lambda_{11} = 0$ and $\lambda_{12} = \lambda_{51} = 0$), use Françoise's algorithm to show that the third order Melnikov function, $M_3(h)$ or $d_3(h)$, for

$$\dot{x} = y + \varepsilon[\varepsilon^2 \lambda_{13} x - 2x^2 - (2\lambda_{21} + \varepsilon\lambda_{52})xy + y^2]$$
$$\dot{y} = -x + \varepsilon[\varepsilon^2 \lambda_{13} y - \lambda_{21} x^2 + (4 + \lambda_{41})xy + \lambda_{21} y^2]$$

is given by $M_3(h) = 4\pi h \lambda_{13} - \pi h^2 \lambda_{52}$ (as in Lemma 1 in Section 4.11 where $x = \alpha = \sqrt{2h}$ and $M_3(h) = \sqrt{2h} d_3(\sqrt{2h}, \lambda)$ with $d_3(x, \lambda)$ given in Lemma 1 of Section 4.11). **Hint:** You should find that $q_1(x, y) = \lambda_{41} y$ and, from Problem 1(a), that $q_2(x, y) = \lambda_{52} x - (2\lambda_{41} + \lambda_{41}^2)x^2 - \lambda_{41}(7\lambda_{21} - \lambda_{41} - 4)xy/4$.

(c) Use Françoise's algorithm to derive the formulas for $d_4(h), \ldots, d_6(h)$ in Lemma 1 in Section 4.11.

3. Consider the polynomially perturbed Hamiltonian system (1) with $H(x, y) = (x^2 + y^2)/2 - x^3/3$, that was treated by Iliev in [59]. Cf. Problem 3 in Section 4.13. In order to gain an understanding of the analysis carried out in [59], which parallels the analysis in this section for the perturbed harmonic oscillator, the student should carry out the first two steps (for $n = 1$ and 2) in the induction proof of the following lemma due to Iliev [59] where a polynomial function $p(x, y, H)$ is said to have weighted degree n if

$$p(x, y, H) = \sum_{i+j+2k \leq n} a_{ijk} x^i y^i H^k.$$

Lemma. For $H(x, y) = (x^2 + y^2)/2 - x^3/3$, any polynomial one-form Ω of degree n can be expressed as

$$\Omega = dQ + qdH + \alpha(H)ydx + \beta(H)xydx$$

where $Q(x, y, H)$ and $q(x, y, H)$ are polynomials of weighted degree $(n+1)$ and $(n-1)$ respectively and $\alpha(h)$ and $\beta(h)$ are polynomials of degrees $\left[\frac{1}{2}(n-1)\right]$ and $\left[\frac{1}{2}(n-2)\right]$ respectively.

Hint: In the proof of this lemma for $n = 2$ we have

$$Q(x, y, H) = Q_{30}x^3 + Q_{21}x^2 y + Q_{12}xy^2 + Q_{03}y^3$$
$$+ Q_{20}x^2 + Q_{11}xy + Q_{02}y^2 + Q_{10}x + Q_{01}y$$
$$+ (B_{10}x + B_{01}y)[(x^2 + y^2)/2 - x^3/3],$$

$q(x, y, H) = q_{10}x + q_{01}y$, $\alpha(h) = \alpha_0$ and $\beta(h) = \beta_0$. Note that it is possible to choose $B_{01} = 0$ and $q_{01} = 0$.

Based on this lemma, Iliev [59] was able to prove that for $k = 1$ and for $k = 2$ with n even, the polynomially perturbed Bogdanov–Takens system has at most $k(n - 1)$ limit cycles and for $k = 2$ with

4.13. The Takens–Bogdanov Bifurcation

n odd or for $k \geq 3$, it has at most $k(n-1) - 1$ limit cycles. This leads to the following table giving the maximum number of limit cycles obtainable in a kth-order analysis of an nth-degree polynomially perturbed Bogdanov–Takens Hamiltonian system. It is known that the first three columns are exact and that for $n = 2$ and $k \geq 2$ the maximum number of limit cycles is 2; cf. [59].

TABLE 4.2. The maximum number of limit cycles obtainable from a kth-order analysis of an nth-degree polynomially perturbed Bogdanov–Takens Hamiltonian system.

	$k=1$	2	3	4	5	...
$n=1$	0	0	0	0	0	...
2	1	2	2	2	2	...
3	2	3	5	7		...
4	3	6	8			...
5	4	7	11			...
$\vdots$	$\vdots$	$\vdots$	$\vdots$			...
$2m$	$2m-1$	$4m-2$	$6m-4$			...
$2m+1$	$2m$	$4m-1$	$6m-1$			...

4.13 The Takens–Bogdanov Bifurcation

In this section, we bring together several of the techniques and ideas developed in the previous sections in order to study the Takens–Bogdanov or double-zero eigenvalue bifurcation. This is the bifurcation that results from unfolding the normal form

$$\begin{aligned} \dot{x} &= y \\ \dot{y} &= x^2 \pm xy. \end{aligned} \tag{1}$$

In Section 2.13, we saw that this normal form is obtained from any system with a double zero eigenvalue of the form

$$\dot{\mathbf{x}} = \begin{bmatrix} 0 & 1 \\ 0 & 0 \end{bmatrix} \mathbf{x} + \begin{pmatrix} ax^2 + bxy + cy^2 \\ dx^2 + exy + fy^2 \end{pmatrix} + 0(|\mathbf{x}|^3)$$

when $d \neq 0$ and $e + 2a \neq 0$; cf. Remark 1 in Section 2.13. As was noted in that remark, a universal unfolding of the normal form (1) is given by

$$\begin{aligned} \dot{x} &= y \\ \dot{y} &= \mu_1 + \mu_2 y + x^2 \pm xy. \end{aligned}$$

This was proved independently by Takens [44] in 1974 and Bogdanov [45] in 1975. Since the above system with the plus sign is transformed into the above system with the minus sign under the linear transformation of coordinates $(x, y, t, \mu_1, \mu_2) \to (x, -y, -t, \mu_1, -\mu_2)$, it is only necessary to consider the above system with the plus sign,

$$\dot{x} = y$$
$$\dot{y} = \mu_1 + \mu_2 y + x^2 + xy, \quad (2)$$

in determining the different types of dynamical behavior that occur in unfoldings of (1). Cf. Problem 6. Since there are two parameters in the universal unfolding (2) of (1), we see that the Takens–Bogdanov bifurcation is a codimension-two bifurcation.

In order to determine the various types of dynamical behavior that occur for the system (2), let us begin by determining the location and the nature of the critical points of (2). For $\mu_1 = \mu_2 = 0$ the system (2) reduces to the system (1), which has a nonhyperbolic critical point at the origin, and it follows from Theorem 3 in Section 2.11 that the system (1) has a cusp at the origin. For $\mu_1 > 0$, the system (2) has no critical points; and for $\mu_1 = 0$ and $\mu_2 \neq 0$, (2) has one nonhyperbolic critical point at the origin. It follows from Theorem 1 in Section 2.11, or from the Center Manifold Theory in Section 2.12, that in this case (2) has a saddle-node at the origin. Cf. Problem 1. Finally, for $\mu_1 < 0$, the system (2) has two hyperbolic critical points,

$$\mathbf{x}_+ = (\sqrt{-\mu_1}, 0) \quad \text{and} \quad \mathbf{x}_- = (-\sqrt{-\mu_1}, 0).$$

Furthermore, the linear part of the vector field $\mathbf{f}$ in (2) at these critical points is determined by the matrix

$$D\mathbf{f}(\mathbf{x}_\pm) = \begin{bmatrix} 0 & 1 \\ \pm 2\sqrt{-\mu_1} & \mu_2 \pm \sqrt{-\mu_1} \end{bmatrix}.$$

It follows that for $\mu_1 < 0$, $\mathbf{x}_+$ is a saddle and that $\mathbf{x}_-$ is a source for $\mu_2 > \sqrt{-\mu_1}$ and a sink for $\mu_2 < \sqrt{-\mu_1}$. It is a weak focus for $\mu_2 = \sqrt{-\mu_1}$, and from equation (3') in Section 4.4 we find that

$$\sigma = \frac{3\pi}{2|2\mu_1|^{3/2}} > 0;$$

i.e., $\mathbf{x}_-$ is an unstable weak focus of multiplicity one for $\mu_1 < 0$ and $\mu_2 = \sqrt{-\mu_1}$.

We next investigate the bifurcations that take place in the system (2). For $\mu_1 = 0$ and $\mu_2 \neq 0$, the system (2) has a single-zero eigenvalue. According to Theorem 1 in Section 4.2, the system (2) experiences a saddle-node bifurcation at points on the μ_2-axis with $\mu_2 \neq 0$; i.e., there is a family of saddle-node bifurcation points along the μ_2-axis in the μ-plane:

$$SN: \mu_1 = 0, \quad \mu_2 \neq 0.$$

Cf. Problem 2. Next, according to Theorem 1 in Section 4.4, it follows that the system (2) experiences a Hopf bifurcation at the weak focus at $\mathbf{x}_-$ for

4.13. The Takens–Bogdanov Bifurcation

parameter values on the semiparabola $\mu_2 = \sqrt{-\mu_1}$ when $\mu_1 < 0$; i.e., for $\mu_1 < 0$, there is a curve of Hopf bifurcation points in the μ-plane given by

$$H: \mu_2 = \sqrt{-\mu_1} \quad \text{for } \mu_1 < 0.$$

Furthermore, since $\sigma > 0$, as was noted above, it follows from Theorem 1 in Section 4.4 that there is a subcritical Hopf bifurcation in which an unstable limit cycle bifurcates from $\mathbf{x}_-$ as μ_2 decreases from $\mu_2 = \sqrt{-\mu_1}$ for each $\mu_1 < 0$. This fact also follows from Theorem 5 in Section 4.6, since the system (2) forms a semicomplete family of rotated vector fields with parameter $\mu_2 \in \mathbf{R}$. Furthermore, it then follows from Theorems 1, 4, and 6 in Section 4.6 that for all $\mu_1 < 0$ the negatively oriented unstable limit cycle generated in the Hopf bifurcation at $\mu_2 = \sqrt{-\mu_1}$ expands monotonically as μ_2 decreases from $\sqrt{-\mu_1}$ until it intersects the saddle at $\mathbf{x}_+$ and forms a separatrix cycle or homoclinic loop at some parameter value $\mu_2 = h(\mu_1)$; i.e., there exists a homoclinic-loop bifurcation curve in the μ-plane given by

$$HL: \mu_2 = h(\mu_1) \quad \text{for } \mu_1 < 0.$$

It then follows from the results in [38] that $h(\mu_1)$ is analytic for all $\mu_1 < 0$. These are all of the bifurcations that occur in the system (2), and the bifurcation set in the μ-plane along with the various phase portraits that occur in the system (2) are shown in Figure 3 below.

In order to approximate the shape of the homoclinic-loop bifurcation curve HL, i.e., in order to approximate the function $h(\mu_1)$ for small μ_1, we use the rescaling of coordinates and parameters

$$x = \varepsilon^2 u, \quad y = \varepsilon^3 v, \quad \mu_1 = \varepsilon^4 \nu_1, \quad \text{and} \quad \mu_2 = \varepsilon^2 \nu_2 \tag{3}$$

with the time $t \to \varepsilon t$, given in [44], in order to reduce the system (2) to a perturbed system to which the Melnikov theory developed in Sections 4.9 and 4.10 applies. Substituting the rescaling transformation (3) into the system (2) yields

$$\begin{aligned} \dot{u} &= v \\ \dot{v} &= \nu_1 + u^2 + \varepsilon(\nu_2 v + uv). \end{aligned} \tag{4}$$

For $\varepsilon = 0$, this is a Hamiltonian system with Hamiltonian $H(u,v) = v^2/2 - u^3/3 - \nu_1 u$. The phase portrait for this Hamiltonian system (2) with $\varepsilon = 0$ is shown in Figure 1 for $\nu_1 < 0$, in which case there are two critical points at $(\pm\sqrt{-\nu_1}, 0)$. Cf. Figure 2 in Section 3.3.

The homoclinic loop Γ_0 in Figure 1 is given by

$$\gamma_0(t) = (u_0(t), v_0(t)) = \sqrt{-\nu_1}\left(1 - 3\operatorname{sech}^2\left(\frac{t}{\sqrt{2}}\right),\right.$$
$$\left. 3\sqrt{2}\operatorname{sech}^2\left(\frac{t}{\sqrt{2}}\right)\tanh\left(\frac{t}{\sqrt{2}}\right)\right).$$

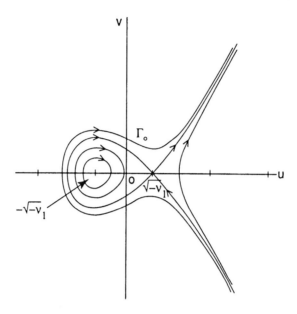

Figure 1. The phase portrait for the system (4) with $\varepsilon = 0$ and $\nu_1 < 0$.

Cf. p. 369 in [G/H]. Then, according to Theorem 4 in Section 4.9, the Melnikov function along the homoclinic loop Γ_0 is given by

$$M(\nu) = \int_{-\infty}^{\infty} \mathbf{f}(\gamma_0(t)) \wedge \mathbf{g}(\gamma_0(t), \nu) dt$$

$$= \int_{-\infty}^{\infty} v_0(t)[\nu_2 v_0(t) + u_0(t) v_0(t)] dt$$

$$= \frac{18|\nu_1|}{\sqrt{2}} \left[\nu_2 \int_{-\infty}^{\infty} \operatorname{sech}^4 \tau \tanh^2 \tau \, d\tau \right.$$

$$\left. + \sqrt{-\nu_1} \int_{-\infty}^{\infty} (1 - 3\operatorname{sech}^2 \tau) \operatorname{sech}^4 \tau \tanh^2 \tau \, d\tau \right]$$

$$= \frac{24|\nu_1|}{\sqrt{2}} \left[\frac{\nu_2}{5} - \frac{\sqrt{-\nu_1}}{7} \right],$$

where we have used the fact that $\operatorname{sech}^2 \tau = 1 - \tanh^2 \tau$ and

$$\int_{-\infty}^{\infty} \tanh^k \tau \operatorname{sech}^2 \tau \, d\tau = \frac{2}{k+1}.$$

Therefore, according to Theorem 4 in Section 4.9, the system (4) with $\nu_1 < 0$ has a homoclinic loop if

$$\nu_2 = \frac{5}{7}\sqrt{-\nu_1} + 0(\varepsilon).$$

4.13. The Takens–Bogdanov Bifurcation

It then follows from (3) that, for $\mu_1 < 0$, the system (2) has a homoclinic loop if

$$\mu_2 = h(\mu_1) = \frac{5}{7}\sqrt{-\mu_1} + 0(\mu_1)$$

as $\mu_1 \to 0$. Cf. Figure 3 below. We note that the above computation of the Melnikov function $M(\nu)$ along the homoclinic loop of the system (4) is equivalent to the computation carried out in Problem 3 in Section 4.9 since the system (4) can be reduced to the system in that problem by translating the origin to the saddle x_+ and by rescaling the coordinates so that the distance between the two critical points $x_\pm$ is equal to one. (Cf. Figure 1 above and Figure 2 in Section 3.3.)

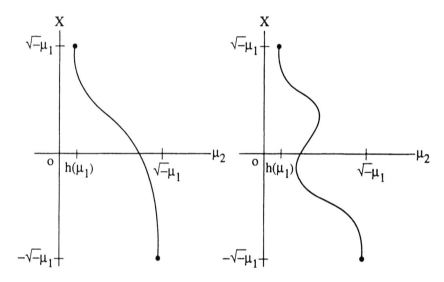

Figure 2. Possible variations of the x-intercept of the limit cycle of the system (2) for $\mu_1 < 0$.

As we have seen, the Melnikov theory, together with the rescaling transformation (3), allows us to approximate the shape of the homoclinic-loop bifurcation curve, HL, near the origin in the μ-plane; more importantly, however, it also allows us to show that, for small $\mu_1 < 0$, the system (2) has exactly one limit cycle for each value of $\mu_2 \in (h(\mu_1), \sqrt{-\mu_1})$. In other words, no semistable limit cycles occur in the one-parameter family of limit cycles generated in the Hopf bifurcation at the origin of system (2) when $\mu_2 = \sqrt{-\mu_1}$. This can happen as we saw in the one-parameter family of limit cycles L_ρ in Example 3 of Section 4.9; cf. Figure 9 in Section 4.9. In other words, for $\mu_1 < 0$ the bifurcation diagram for the one-parameter family of limit cycles generated in the Hopf bifurcation at the origin of (2) when $\mu_2 = \sqrt{-\mu_1}$ is given by Figure 2(a) and not 2(b).

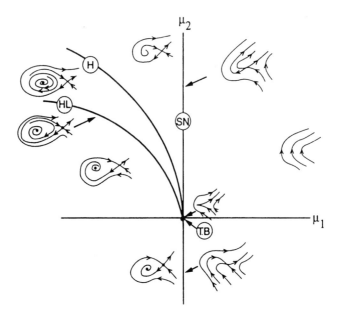

Figure 3. The bifurcation set and the corresponding phase portraits for the system (2).

The fact that there are no semistable limit cycles for the system (2) with $\mu_1 < 0$, such as those that occur at the turning points in Figure 2(b), follows from the fact that for $\mu_1 < 0$ and $h(\mu_1) < \mu_2 < \sqrt{-\mu_1}$ the system (2) has a unique limit cycle. This follows from Theorem 5 in Section 4.9, since the Melnikov function $M(\nu, \alpha)$ along the one-parameter family of periodic orbits of the rescaled system (4) has exactly one zero. The computation of the Melnikov function $M(\nu, \alpha)$, as well as establishing its monotonicity, was carried out by Blows and the author in Example 3.3 in [37]. Cf. Problem 3. The unfolding of the normal form (1), i.e., the Takens–Bogdanov bifurcation, is summarized in Figure 3, which shows the bifurcation set in the μ-plane consisting of the saddle-node, Hopf, and homoclinic-loop bifurcation curves SN, H, and HL, respectively, and the Takens–Bogdanov bifurcation point, TB, at the origin. The phase portraits for the system (2) with parameter values μ on the bifurcation set and in the components of the μ-plane in the complement of the bifurcation set also are shown in Figure 3.

In his 1974 paper [44], Takens also studies unfoldings of the normal form:
$$\dot{x} = y \\ \dot{y} = \pm x^3 - x^2 y. \tag{5}$$

Cf. Problems 2 and 3 in Section 2.13. He studies unfoldings of (5) that

4.13. The Takens–Bogdanov Bifurcation

preserve symmetries under rotations through π, i.e., unfoldings of the form

$$\begin{aligned}\dot{x} &= y \\ \dot{y} &= \mu_1 x + \mu_2 y \pm x^3 - x^2 y.\end{aligned} \quad (6)$$

Let us first of all consider the case with the plus sign, leaving most of the details for the student to do in Problem 4. For $\mu_1 \geq 0$ there is only the one critical point at the origin, and it is a topological saddle. For $\mu_1 < 0$ there are three critical points, a source at the origin for $\mu_2 > 0$ and a sink at the origin for $\mu_2 \leq 0$, as well as two saddles at $(\pm\sqrt{-\mu_1}, 0)$. There is a pitch-fork bifurcation at points on the μ_2-axis with $\mu_2 \neq 0$. There is a supercritical Hopf bifurcation at points on the μ_1-axis with $\mu_1 < 0$. And, using the rotated vector field theory in Section 4.6, it follows that for $\mu_1 < 0$ there is a homoclinic-loop bifurcation curve $\mu_2 = h(\mu_1) = -\mu_1/5 + 0(\mu_1^2)$. This approximation of the homoclinic-loop bifurcation curve follows from the Melnikov theory in Section 4.9 by making the rescaling transformation

$$x = \varepsilon u, \quad y = \varepsilon^2 v, \quad \mu_1 = \varepsilon^2 \nu_1, \quad \mu_2 = \varepsilon^2 \nu_2, \quad \text{and} \quad t \to \varepsilon t; \quad (7)$$

cf. [44]. Under this transformation, the system (6) with the plus sign assumes the form

$$\begin{aligned}\dot{u} &= v \\ \dot{v} &= -u + u^3 + \varepsilon v(\nu_2 - u^2)\end{aligned}$$

where we have set $\nu_1 = -1$, corresponding to $\mu_1 < 0$ as on p. 372 of [G/H]. But this is just Example 3 in Section 4.10 (with $\mu_1 = \nu_2$ and $\mu_2 = -1/3$); cf. Problem 3 in Section 4.10. From Theorem 6 in Section 4.10, it follows that for $\nu_1 = -1$ the homoclinic-loop bifurcation occurs at $\nu_2 = 1/5 + 0(\varepsilon)$. For the system (6) with the plus sign, this corresponds to the fact that the homoclinic-loop bifurcation curve in the $\boldsymbol{\mu}$-plane is given by

$$HL: \mu_2 = h(\mu_1) = -\mu_1/5 + 0(\mu_1^2) \quad \text{for } \mu_1 < 0$$

as $\mu_1 \to 0$. It is important to note that Theorem 6 in Section 4.10 also establishes the fact that there are no multiplicity-two limit cycle bifurcations for the system (6) with the plus sign. The student is asked to draw the bifurcation set in the $\boldsymbol{\mu}$-plane as well as the corresponding phase portraits in Problem 4 below; cf. Figure 7.3.5 in [G/H].

We next consider equation (6) with the minus sign, leaving most of the details for the student to do in Problem 5. For $\mu_1 < 0$ there is only one critical point at the origin, and it is a sink for $\mu_2 \leq 0$ and a source for $\mu_2 > 0$. For $\mu_1 > 0$ there are three critical points, a saddle at the origin and sinks at $(\pm\sqrt{\mu_1}, 0)$ for $\mu_2 > \mu_1$ and sources for $\mu_2 \leq \mu_1$. There is a pitch-fork bifurcation at points on the μ_2-axis with $\mu_2 \neq 0$. For $\mu_1 < 0$ there is a supercritical Hopf bifurcation at points on the μ_1-axis, and for $\mu_1 > 0$ there is a subcritical Hopf bifurcation at points on the line $\mu_2 = \mu_1$. Using the rotated vector field theory in Section 4.6, it follows that for $\mu_1 > 0$

there is a homoclinic-loop bifurcation curve $\mu_2 = h(\mu_1) = 4\mu_1/5 + 0(\varepsilon_1^2)$. This approximation of the homoclinic-loop bifurcation curve follows from the Melnikov theory in Section 4.10 by making the rescaling transformation (7) used by Takens in [44]. This transforms equation (6) with the minus sign into

$$\dot{u} = v$$
$$\dot{v} = u - u^3 + \varepsilon v(\nu_2 - u^2), \qquad (8)$$

where we have set $\nu_1 = +1$, corresponding to $\mu_1 > 0$ as on p. 373 of [G/H]. But this is just Example 2 in Section 4.10 (with $\mu_1 = \alpha = \nu_2$ and $\mu_2 = \beta/3 = -1/3$ as in Problem 1 in Section 4.10). Thus, from Theorem 5 in Section 4.10, it follows that for $\nu_1 = +1$ the homoclinic-loop bifurcation for the system (8) occurs at $\nu_2 = -2.4(-1/3) + 0(\varepsilon) = 4/5 + 0(\varepsilon)$; and for the system (6) with the minus sign this corresponds to $\mu_2 = h(\mu_1) = 4\mu_1/5 + 0(\mu_1^2)$ as $\mu_1 \to 0^+$. Theorem 5 in Section 4.10 also establishes that for $h(\mu_1) < \mu_2 < \mu_1$ there is exactly one limit cycle for this system. However, that is not the extent of the bifurcations that occur in the system (6) with the minus sign. There is also a curve of multiplicity-two limit cycles C_2 on which this system has a multiplicity-two semistable limit cycle. This follows from the computation of the Melnikov function for the exterior Duffing problem in Problem 6 of Section 4.10. It follows from the results of that problem that there is a multiplicity-two limit cycle bifurcation surface,

$$C_2: \ \mu_2 = C(\mu_1) \cong .752\mu_1 + 0(\mu_1^2)$$

as $\mu_1 \to 0^+$. The analyticity of the function $C(\mu_1)$ for $0 < \mu_1 < \delta$ and small $\delta > 0$ is established in [39]. Note that it is shown in Problem 1 in Section 4.10 that the system (8), the system of Example 2 in Section 4.10 and the system of Example 3 in Section 4.9 are all equivalent. Also this system was studied in Example 3.2 in [37]. It is left for the student to draw the bifurcation set in the μ-plane as well as the corresponding phase portraits in Problem 5 below; cf. Figure 7.3.9 in [G/H] and Figure 4 below.

We close this section with one final remark about Takens–Bogdanov or double-zero eigenvalue bifurcations. As in Remark 2 in Section 2.13, if $e + 2a = 0$ in the system at the beginning of this section, then the normal form for the double-zero eigenvalue is given by

$$\dot{x} = y$$
$$\dot{y} = ax^2 + bx^2y + cx^3y.$$

However, Dumortier et al. [46] show that, without loss of generality, the coefficient b of the x^2y-term can be taken as zero, and that it suffices to consider unfoldings of the normal form

$$\dot{x} = y$$
$$\dot{y} = x^2 \pm x^3y \qquad (9)$$

4.13. The Takens–Bogdanov Bifurcation

in this case; cf. Lemma 1, p. 384 in [46]. It is also shown in [46] that a universal unfolding of the normal form (9) is given by

$$\dot{x} = y$$
$$\dot{y} = \mu_1 + \mu_2 y + \mu_3 xy + x^2 \pm x^3 y, \qquad (10)$$

and this results in a codimension-three Takens–Bogdanov bifurcation, or a codimension-three cusp bifurcation if we choose to name the bifurcation after the type of critical point that occurs at the origin of (9), i.e., a cusp, as was done in [46]. The universal unfolding of (9), given by (10), is described in [46]. Cf. the bifurcation set in Figure 1 in Section 4.15.

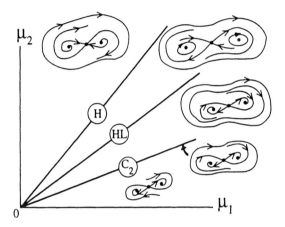

Figure 4. The bifurcation set in the first quadrant and the corresponding phase portraits for the system (6) with the minus sign.

PROBLEM SET 13

1. Show that the system (2) with $\mu_1 = 0$ and $\mu_2 \neq 0$,

   $$\dot{x} = y$$
   $$\dot{y} = \mu_2 y + x^2 + xy,$$

   has a center manifold approximated by $y = -x^2/\mu_2 + 0(x^3)$ as $x \to 0$, and that this results in a saddle-node at the origin.

2. Check that the conditions of Theorem 1 in Section 4.2 are satisfied by the system (2) with $\mathbf{v} = (1,0)^T$ and $\mathbf{w} = (\mu_2, -1)^T$, and show that for $\mu_1 = 0$ and $\mu_2 \neq 0$ the system (2) experiences a saddle-node bifurcation.

3. By translating the origin to the center shown in Figure 1 and normalizing the coordinates so that the distance between the critical points

in Figure 1 is one [or by translating the origin to the center and setting $\nu_1 = -1/4$ in (4)], the system (4) can be transformed into the system

$$\dot{x} = y$$
$$\dot{y} = -x + x^2 + \varepsilon(\alpha y + \beta xy).$$

Using the same procedure as in Problem 1 in Section 4.10, this system can be transformed into the system

$$\dot{x} = y + \varepsilon(ax + bx^2)$$
$$\dot{y} = -x + x^2$$

with $(a, b) = (\alpha, \beta/2)$. This is the system that was studied in [37]. It has a one-parameter family of periodic orbits given by

$$x_\alpha(t) = \frac{3\alpha^2}{2\sqrt{1 - \alpha^2 + \alpha^4}} sn^2\left(\frac{t}{2(1 - \alpha^2 + \alpha^4)^{1/4}}, \alpha\right)$$
$$+ \frac{1}{2}\left(1 - \frac{1 + \alpha^2}{\sqrt{1 - \alpha^2 + \alpha^4}}\right)$$

and

$$y_\alpha(t) = \frac{3\alpha^2}{2(1 - \alpha^2 + \alpha^4)^{3/4}} sn\left(\frac{t}{2(1 - \alpha^2 + \alpha^4)^{1/4}}, \alpha\right)$$
$$cn\left(\frac{t}{2(1 - \alpha^2 + \alpha^4)^{1/4}}, \alpha\right) dn\left(\frac{t}{2(1 - \alpha^2 + \alpha^4)^{1/2}}, \alpha\right)$$

for $0 < \alpha < 1$. Following the procedure used in Example 3 in Section 4.10, compute the Melnikov function $M(\alpha, a, b)$ along this one-parameter family of periodic orbits using the formulas

$$\int_0^{4K(\alpha)} sn^2 u\, du = \frac{4}{\alpha^2}[K(\alpha) - E(\alpha)]$$

$$\int_0^{4K(\alpha)} sn^4 u\, du = \frac{4}{3\alpha^4}[(2 + \alpha^2)K(\alpha) - 2(1 + \alpha^2)E(\alpha)]$$

given in Example 3 in Section 4.10 along with the fact that

$$\int_0^{4K(\alpha)} sn^m u\, du = 0$$

for any odd positive integer m; cf. [40], p. 191. Graph the values of a/b that result in $M(\alpha, a, b) = 0$, and deduce that for $b > 0$, all sufficiently small $\varepsilon > 0$, and $-2b/7 + 0(\varepsilon) < a < 0$, the above system has exactly one hyperbolic, unstable limit cycle, born in a subcritical Hopf bifurcation at $a = 0$, that expands monotonically as a decreases to the value $a(\varepsilon) = -2b/7 + 0(\varepsilon)$. [Note that for $\nu_1 = -1/4$, the system (4) has a homoclinic loop at $\nu_2 = 5/14 + 0(\varepsilon)$ according to the Melnikov computation in this section. Under the transformation of coordinates defined at the beginning of this problem, it can be shown that $\nu_2 = a + 1/2$ and $b = 1/2$ if $\nu_1 = -1/4$. Thus the Hopf bifurcation value $a = 0$ corresponds to $\nu_2 - 1/2 - \sqrt{-\nu_1}$, and the homoclinic-loop bifurcation value $a = -1/7 + 0(\varepsilon)$ corresponds to $\nu_2 = a + 1/2 = 5/14 + 0(\varepsilon)$, as given above.]

4. Consider the system (6) with the plus sign. Verify the statements made in this section concerning the critical points and bifurcations for that system, draw the bifurcation set in the μ-plane, and construct the phase portraits on the bifurcation set as well as for a point in each of the components in the complement of the bifurcation set.

5. Consider the system (6) with the minus sign. Verify the statements made in this section concerning the critical points and bifurcations for that system, draw the bifurcation set in the μ-plane, and construct the phase portraits on the bifurcation set as well as for a point in each of the components in the complement of the bifurcation set.

6. Draw the bifurcation set and corresponding phase portraits for the system

$$\dot{x} = y$$
$$\dot{y} = \mu_1 + \mu_2 y + x^2 - xy.$$

Hint: Apply the transformation of coordinates $(x, y, t, \mu_1, \mu_2) \to (x, -y, -t, \mu_1, -\mu_2)$ to the system (2) and to the results obtained for the system (2) at the beginning of this section.

7. Draw the bifurcation set and corresponding phase portraits for the system

$$\dot{x} = y$$
$$\dot{y} = \mu_1 + \mu_2 y + x^2.$$

Hint: For $\mu_2 = 0$, note that the system is symmetric with respect to the x-axis and apply Theorem 6 in Section 2.10.

4.14 Coppel's Problem for Bounded Quadratic Systems

In 1966, Coppel [47] posed the problem of determining all possible phase portraits for quadratic systems in $\mathbf{R}^2$ and classifying them by means of algebraic inequalities on the coefficients. This is a rather formidable problem and, in particular, a solution of Coppel's problem would include a solution of Hilbert's 16th problem for quadratic systems. It was pointed out by Dumortier and Fiddelaers that Coppel's problem as stated is insoluble; i.e., they pointed out that the phase portraits for quadratic systems cannot be classified by means of algebraic inequalities on the coefficients, but require analytic and even nonanalytic inequalities for their classification.

In 1968, Dickson and I initiated the study of bounded quadratic systems, i.e., quadratic systems that have all of their trajectories bounded for $t \geq 0$; cf. [48]. In that study we were looking for a subclass of the class of quadratic systems that was more amenable to solution and exhibited most of the interesting dynamical behavior found in the class of quadratic systems. In [48] we established necessary and sufficient conditions for a quadratic system to be bounded and determined all possible phase portraits for bounded quadratic systems with a partial classification of the phase portraits by means of algebraic inequalities on the coefficients.

This section contains a partial solution to Coppel's problem for bounded quadratic systems, modified to include analytic inequalities on the coefficients (i.e., inequalities involving analytic functions), under the assumption that any bounded quadratic system has at most two limit cycles. It is amazing that in 1900 Hilbert [49] was able to formulate the single most difficult problem for planar polynomial systems: to determine the maximum number and relative positions of the limit cycles for an nth-degree polynomial system. This is still an open problem for quadratic systems; however, we do know that there are quadratic systems with as many as four limit cycles (cf. [29]), and that any quadratic system has at most three local limit cycles (cf. [2]). The results on the number of limit cycles for bounded quadratic systems are somewhat more complete. First, it was shown in [48] that any bounded quadratic system has at most three critical points in $\mathbf{R}^2$. And in 1987 Coll et al. [50] were able to prove that any bounded quadratic system with one or two critical points in $\mathbf{R}^2$ has at most one limit cycle. It also was shown in [50] that any bounded quadratic system has at most two local limit cycles and that there are bounded quadratic systems with three critical points in $\mathbf{R}^2$ that have two local limit cycles. And just recently, Li et al. [51], using the global Melnikov method described in Section 4.10, were able to show that any bounded quadratic system that is near a center has at most two limit cycles. The conjecture, posed by the author, that any bounded quadratic system has at most two limit cycles is therefore reasonable, and it is consistent with all of the known results for bounded quadratic systems.

Since, according to [50], any bounded quadratic system with one or two critical points in $\mathbf{R}^2$ has at most one limit cycle, it was possible, using some of the author's results for establishing the existence and analyticity of homoclinic-loop bifurcation surfaces in [38], to solve Coppel's problem (as stated in [47]) for bounded quadratic systems with one critical point or for bounded quadratic systems with two critical points (if we allow analytic inequalities on the coefficients). These results are given below; they are closely related to Theorem C in [50]. The solution of Coppel's problem

4.14. Coppel's Problem for Bounded Quadratic Systems

for bounded quadratic systems (as stated below, where we allow analytic inequalities on the coefficients), under the assumption that any bounded quadratic system (with three critical points in $\mathbf{R}^2$) has at most two limit cycles, also is given in this section; however, it still remains to show exactly how the bifurcation surfaces in Theorem 5 below partition the parameter space of the system (5) in Theorem 5 into components for $\beta < 0$. This is accomplished for $\beta \geq 0$ in Figures 15 and 16 below; but for $\beta < 0$ it is still not completely clear how the codimension 3 Takens–Bogdanov bifurcation surface (described in Theorem 4 and Figure 1 in the next section) interact with the other bifurcation surfaces for bounded quadratic systems.

Coppel's Problem for Bounded Quadratic Systems. Determine all of the possible phase portraits for bounded quadratic systems in $\mathbf{R}^2$ and classify them by means of inequalities on the coefficients involving functions that are analytic on their domains of definition.

In this section we see that there is a rich structure in both the dynamics and the bifurcations that occur in the class of bounded quadratic systems: Their phase portraits exhibit multiple limit cycles, homoclinic loops, saddle connections on the Poincaré sphere, and two limit cycles in either the $(1, 1)$ or $(2, 0)$ configurations; while the evolution of their phase portraits includes Hopf, homoclinic-loop, multiple limit cycle, saddle-node and Takens–Bogdanov bifurcations. It will be seen that the families of multiplicity-two limit cycles that occur in the class of bounded quadratic systems terminate at either a homoclinic-loop bifurcation at "resonant eigenvalues" as described in [38, 39, 52], at a multiple Hopf bifurcation, or at a degenerate critical point as described in [39].

The main tools used to establish the results in this section are the theory of rotated vector fields developed in [7, 23] and presented in Section 4.6 and the results on homoclinic-loop bifurcation surfaces and multiple limit cycle bifurcation surfaces and their termination developed in [38, 39, 52]. The asymptotic results given in [51] serve as a nice check on the results of this section for bounded quadratic systems near a center, and they complement the numerical results in this section that describe the multiplicity-two limit cycle bifurcation surfaces whose existence follows from the theory of rotated vector fields as in [38]. In the problem set at the end of this section, the student is asked to determine the bifurcation surfaces that occur in the class of bounded quadratic systems. This serves as a nice application of the bifurcation theory that is developed in this chapter, and it allows the student to work on a problem of current research interest.

Let us begin by presenting some well-known results for BQS (i.e., for Bounded Quadratic Systems). The following theorem is Theorem 1 in [48]:

Theorem 1. *Any BQS is affinely equivalent to*

$$\dot{x} = a_{11}x$$
$$\dot{y} = a_{21}x + a_{22}y + xy \qquad (1)$$

with $a_{11} < 0$ and $a_{22} \leq 0$, or

$$\dot{x} = a_{11}x + a_{12}y + y^2$$
$$\dot{y} = a_{22}y \qquad (2)$$

with $a_{11} \leq 0, a_{22} \leq 0$, and $a_{11} + a_{22} < 0$; or

$$\dot{x} = a_{11}x + a_{12}y + y^2$$
$$\dot{y} = a_{21}x + a_{22}y - xy + cy^2 \qquad (3)$$

with $|c| < 2$ and either (i) $a_{11} < 0$, (ii) $a_{11} = 0$ and $a_{21} = 0$, or (iii) $a_{11} = 0$, $a_{21} \neq 0$, $a_{12} + a_{21} = 0$, and $ca_{21} + a_{22} \leq 0$.

As we shall see in the next two theorems, the BQS determined by (1) and (2), and also those determined by certain cases of (3) in Theorem 1, have either only one critical point or a continuum of critical points. The latter cases are integrable, and the cases with one critical point are easily treated using the results in [48] and [50]. The most interesting cases are those BQSs with two or three critical points which are determined by the remaining cases of system (3) in Theorem 1; cf. Theorems 4 and 5 below.

The solution to Coppel's problem for BQS1 and BQS2 (i.e., BQS with one or two critical points in $\mathbf{R}^2$, respectively) and for BQS with a continuum of critical points is given in the next three theorems, where we also give their global phase portraits. The first theorem follows from Lemmas 1, 3, 4, 15, 16, and 17 in [48] and the fact that any BQS1 has at most one limit cycle which was established in [50]. Note that it was pointed out in [50] that the phase portrait shown in Figure 1(b) was missing in [48].

Theorem 2. *The phase portrait of any BQS1 is determined by one of the separatrix configurations in Figure 1. Furthermore, the phase portrait of a quadratic system is given by Figure 1*

(a) *iff the quadratic system is affinely equivalent to (1) with $a_{11} < 0$ and $a_{22} < 0$;*

(b) *iff the quadratic system is affinely equivalent to (2) with $a_{11} < 2a_{22} < 0$;*

(c) *iff the quadratic system is affinely equivalent to (2) with $2a_{22} \leq a_{11} < 0$ or (3) with $|c| < 2$ and either*

  (i) $a_{11} = a_{12} + a_{21} = 0, a_{21} \neq 0$ and $a_{22} < \min(0, -ca_{21})$ or $a_{22} = 0 < -ca_{21}$,

  (ii) $a_{11} < 0, (a_{12} - a_{21} + ca_{11})^2 < 4(a_{11}a_{22} - a_{21}a_{12})$, and $a_{11} + a_{22} \leq 0$, or

  (iii) $a_{11} < 0$ and $(a_{12} - a_{21} + ca_{11}) = (a_{11}a_{22} - a_{21}a_{12}) = 0$;

(d) *iff the quadratic system is affinely equivalent to (3) with $|c| < 2$ and either*

4.14. Coppel's Problem for Bounded Quadratic Systems

(i) $a_{11} = a_{12} + a_{21} = 0$ and $0 < a_{22} < -ca_{21}$, or

(ii) $a_{11} < 0, a_{11} + a_{22} > 0$, and $(a_{12} - a_{21} + ca_{11})^2 < 4(a_{11}a_{22} - a_{12}a_{21})$.

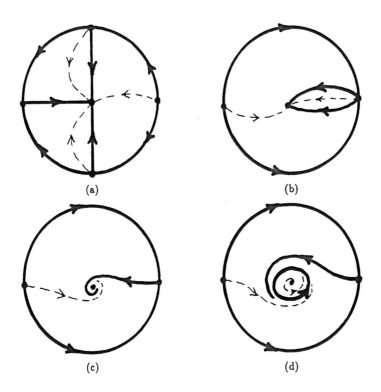

Figure 1. All possible phase portraits for BQS1.

We next give the results for BQS with a continuum of equilibrium points. As in [48], these cases all reduce to integrable systems; cf. Figure 10, 11, 12, 14, and 15, in [48].

Theorem 3. *The phase portrait of any BQS with a continuum of equilibrium points is determined by one of the configurations in Figure 2. Furthermore, the phase portrait of a quadratic system is given by Figure 2*

(a) *iff the quadratic system is affinely equivalent to (1) with $a_{11} < 0$ and $a_{22} = 0$;*

(b) *iff the quadratic system is affinely equivalent to (2) with $a_{11} = 0$ and $a_{22} < 0$;*

(c) *iff the quadratic system is affinely equivalent to (2) with $a_{11} < 0$ and $a_{22} = 0$;*

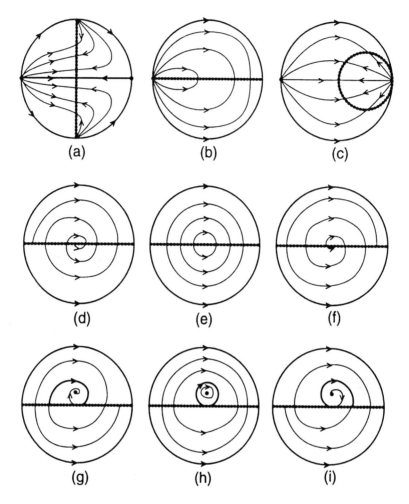

Figure 2. All possible phase portraits for a BQS with a continuum of equilibrium points.

(d) iff the quadratic system is affinely equivalent to (3) with $a_{11} = a_{21} = 0$ and $-2 < c < 0$;

(e) iff the quadratic system is affinely equivalent to (3) with $a_{11} = a_{21} = c = 0$;

(f) iff the quadratic system is affinely equivalent to (3) with $a_{11} = a_{21} = 0$ and $0 < c < 2$;

(g) iff the quadratic system is affinely equivalent to (3) with $a_{11} = a_{12} + a_{21} = ca_{21} + a_{22} = 0, a_{21} \neq 0$, and $-2 < c < 0$;

(h) iff the quadratic system is affinely equivalent to (3) with $a_{11} = a_{12} + a_{21} = a_{22} = c = 0$ and $a_{21} \neq 0$;

4.14. Coppel's Problem for Bounded Quadratic Systems

(i) iff the quadratic system is affinely equivalent to (3) with $a_{11} = a_{12} + a_{21} = ca_{21} + a_{22} = 0, a_{21} \neq 0$, and $0 < c < 2$.

Remark 1. Note that the classification of the phase portraits in Theorems 2 and 3 is determined by algebraic inequalities on the coefficients. Also, it follows from Theorem 3 and the results in [48] that any BQS with a center is affinely equivalent to (3) with $a_{11} = a_{12} + a_{21} = a_{22} = c = 0$ and $a_{21} \neq 0$, and that the corresponding phase portrait is determined by Figure 2(h).

We next present the solution to Coppel's problem for BQS2. In this case, there is a homoclinic-loop bifurcation surface whose analyticity follows from the results in [38]. Due to the existence of the homoclinic-loop bifurcation surface, the phase portraits of BQS2 cannot be classified by means of algebraic inequalities on the coefficients. However, any BQS2 has at most one limit cycle according to [50], and we therefore are able to solve Coppel's problem for BQS2 as stated at the beginning of this section.

First, note that it follows from Theorem 1 above and Lemma 8 in [48] that any BQS2 is affinely equivalent to (3) with $|c| < 2, a_{11} < 0, a_{12} - a_{21} + ca_{11} \neq 0$, and either $(a_{12} - a_{21} + ca_{11})^2 = 4(a_{11}a_{22} - a_{21}a_{12})$ or $(a_{11}a_{22} - a_{21}a_{12}) = 0$. As in [48] on p. 265, by translating the origin to the degenerate critical point of (3), it follows that any BQS2 is affinely equivalent to (3) with $|c| < 2, a_{11} < 0, a_{12} - a_{21} + ca_{11} \neq 0$ and $a_{11}a_{22} - a_{21}a_{12} = 0$. For $a_{11} < 0$, by making the linear transformation of coordinates $t \to |a_{11}|t, x \to |a_{11}|$ and $y \to y/|a_{11}|$, it follows that any BQS2 is affinely equivalent to

$$\dot{x} = -x + a_{12}y + y^2 \\ \dot{y} = a_{21}x + a_{22}y - xy + cy^2 \tag{3'}$$

with $|c| < 2, a_{21} - a_{12} + c \neq 0$, and $a_{22} = -a_{21}a_{12}$. Finally, by letting $\beta = a_{12}$ and $\alpha = a_{21} + c$, in which case $a_{22} = -a_{12}a_{21} = \beta c - \alpha\beta$ and $\alpha \neq \beta$, we obtain the following result:

Lemma 1. *Any BQS2 is affinely equivalent to the one-parameter family of rotated vector fields*

$$\dot{x} = -x + \beta y + y^2 \\ \dot{y} = \alpha x - \alpha\beta y - xy + c(-x + \beta y + y^2) \tag{4}$$

mod $x = \beta y + y^2$ with parameter $c \in (-2, 2)$ and $\alpha \neq \beta$. Furthermore, the system (4) is invariant under the transformation $(x, y, t, \alpha, \beta, c) \to (x - y, t, -\alpha, -\beta, -c)$, and it therefore suffices to consider $\alpha > \beta$. The critical points of (4) are at $0 = (0,0)$ and $P^+ = (x^+, y^+)$ with $x^+ = \alpha(\alpha - \beta)$ and $y^+ = \alpha - \beta$; P^+ is a node or focus, and 0 is a saddle-node or cusp; and 0 is a cusp iff $c = \alpha + 1/\beta$.

The last statement in Lemma 1 follows directly from the results in Lemma 8 in [48] regarding the critical points of (3). We next consider the bifurcations that take place in the three-dimensional parameter space of (4). By translating the origin to the critical point P^+ of (4), it is easy to show that there is a Hopf bifurcation at the critical point P^+ for any point (α, β, c) on the Hopf bifurcation surface

$$H^+ : c = \frac{1 + \alpha^2}{2\alpha - \beta}$$

in $\mathbf{R}^3$. This computation is carried out in Problem 1, where it also is shown by computing the Liapunov number at the weak focus of (4), σ, given by (3') in Section 4.4, that P^+ is a stable weak focus and that a supercritical Hopf bifurcation of multiplicity one occurs at points on H^+ as c increases. The Hopf bifurcation surface H^+ is shown in Figure 3. Note that for $\alpha > \beta$, the surface H^+ lies above the plane $c = 2$ for $\beta > 3/2$.

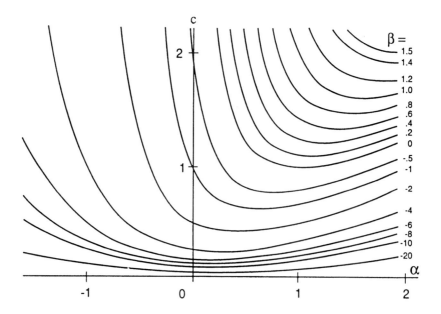

Figure 3. The Hopf bifurcation surface H^+.

According to the theory of rotated vector fields in Section 4.6, the stable (negatively oriented) limit cycle of the system (4), generated in the Hopf bifurcation at $c = (1 + \alpha^2)/(2\alpha - \beta)$, expands monotonically as c increases from this value until it intersects the saddle-node at 0 and forms a separatrix cycle around the critical point P^+; cf. Problem 6. This results in a homoclinic-loop bifurcation at a value of $c = h(\alpha, \beta)$, and, according to the results in [38], the function $h(\alpha, \beta)$ is an analytic function of α and β for all $(\alpha, \beta) \in \mathbf{R}^2$ with $2\alpha - \beta \neq 0$. Thus we have a homoclinic-loop bifurcation surface

$$HL^+ : c = h(\alpha, \beta).$$

4.14. Coppel's Problem for Bounded Quadratic Systems

A portion of the homoclinic-loop bifurcation surface HL^+, determined numerically by integrating trajectories of (4), is shown in Figure 4.

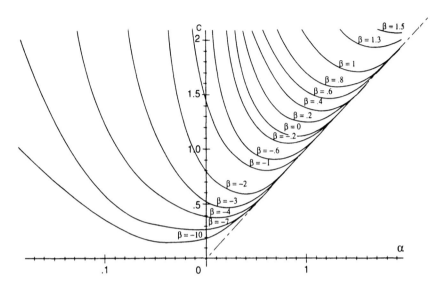

Figure 4. The homoclinic-loop bifurcation surface HL^+.

It follows from Lemma 11 in [48] that (4) has a saddle–saddle connection between the saddle-node at the origin and the saddle-node at the point $(1, 0, 0)$ on the equator of the Poincaré sphere iff $c = \alpha$. This plane in $\mathbf{R}^3$ determines the "saddle–saddle bifurcation surface"

$$SS: c = \alpha.$$

There is one other bifurcation that occurs in the class of BQS2 given by (4): A cusp or Takens–Bogdanov bifurcation occurs when both the determinant and the trace of the linear part of (4) are equal to zero (i.e., when the linear part of (4) has two zero eigenvalues); cf. Section 4.13. The Takens–Bogdanov bifurcation, for which (4) has a cusp at the origin, is derived in Problem 2 and it occurs for points on the surface

$$TB^0: c = \alpha + \frac{1}{\beta}.$$

These are the only bifurcations that occur in the class BQS2 according to Peixoto's theorem in Section 4.1. We shall refer to the plane $\alpha = \beta$ as a saddle-node bifurcation surface since $P^+ \to 0$ as $\alpha \to \beta$ in (4); i.e., as shown in Problem 3, there is a saddle-node bifurcation surface

$$SN: \alpha = \beta.$$

For $\alpha = \beta$ the system (4) has only one critical point which is located at the origin; and the phase portrait for (4) with $\alpha = \beta$ is given by Figure 1(c).

The relationship of the bifurcation surfaces described above for $|c| < 2$ is determined by Figure 6 below, which shows the bifurcation surfaces in this section for various values of β in the intervals $(-\infty, -2), [-2, -1/2),$ $[-1/2, 0), [0, \beta^*), [\beta^*, 3/2), [3/2, 2),$ and $[2, \infty)$. The number $\beta^* \cong 1.43$; this is the value of β at which the homoclinic-loop bifurcation surface HL^+ leaves the region in $\mathbf{R}^3$ where $|c| < 2$; cf. Figure 4. The phase portraits for (4) with (α, β, c) in the various regions shown in the "charts" in Figure 6 follow from the results in [48] for BQS2 and from the fact that any BQS2 has at most one limit cycle, which was established in [50]. These results then lead to the following theorem. Note that the configuration (a'), referred to in Figure 6, is obtained from the configuration (a), shown in Figure 5, by rotating that configuration through π radians about the x-axis. Similar statements hold for the configurations (b'), ..., (j').

Theorem 4. *The phase portrait for any BQS2 is determined by one of the separatrix configurations in Figure 5. Furthermore, there exists a homoclinic-loop bifurcation function $h(\alpha, \beta)$ that is defined and analytic for all $\alpha \neq \beta/2$. The bifurcation surfaces*

$$H^+: c = \frac{1+\alpha^2}{2\alpha - \beta},$$
$$HL^+: c = h(\alpha, \beta),$$
$$SS: c = \alpha,$$

and

$$TB^0: c = \alpha + \frac{1}{\beta}$$

partition the region

$$R = \{(\alpha, \beta, c) \in \mathbf{R}^3 \mid \alpha > \beta, |c| < 2\}$$

of parameters for the system (4) into components, the specific phase portrait that occurs for the system (4) with (α, β, c) in any one of these components being determined by the charts in Figure 6.

Remark 2. The components of the region R defined in Theorem 4 and described by the charts in Figure 6 also can be described by analytic inequalities on the coefficients $\alpha, \beta,$ and c in (4). In fact, it can be seen from the last three charts shown in Figure 6 that for $\beta < 0, \alpha > \beta,$ and $|c| < 2$, the system (4) has the phase portrait determined by Figure 5

(a) iff $c = \alpha$ and $c \leq (1+\alpha^2)/(2\alpha - \beta)$.

(b) iff $c = \alpha$ and $c > (1+\alpha^2)/(2\alpha - \beta)$.

(c') iff $\alpha + 1/\beta < c < \alpha$ and $c \leq (1+\alpha^2)/(2\alpha - \beta)$.

(c_1) iff $\alpha < c \leq (1+\alpha^2)/(2\alpha - \beta)$.

4.14. Coppel's Problem for Bounded Quadratic Systems

(d) iff $c > \alpha$ and $(1+\alpha^2)/(2\alpha - \beta) < c < h(\alpha, \beta)$.

(e) iff $c = h(\alpha, \beta)$.

(f') iff $(1+\alpha^2)/(2\alpha - \beta) < c < \alpha + 1/\beta$.

(g') iff $\alpha + 1/\beta < c < \alpha$ and $c > (1+\alpha^2)/(2\alpha - \beta)$.

(h$'_1$) iff $c \leq (1+\alpha^2)/(2\alpha - \beta)$ and $c < \alpha + 1/\beta$.

(h) iff $c > h(\alpha, \beta)$.

(i') iff $c = \alpha + 1/\beta$ and $c \leq (1+\alpha^2)/(2\alpha - \beta)$.

(j') iff $c = \alpha + 1/\beta$ and $c > (1+\alpha^2)/(2\alpha - \beta)$.

Similar results follow from the first four charts in Figure 6 for $\beta \geq 0$. In fact, the configurations f', h$'_1$, i', and j' do not occur for $\beta \geq 0$; the inequalities necessary and sufficient for c' or g' are $c < \alpha$ and $c \leq (1+\alpha^2)/(2\alpha - \beta)$ or $c > (1+\alpha^2)/(2\alpha - \beta)$, respectively; and the above inequalities necessary and sufficient for the configurations a, b, c$_1$, d, e, and h remain unchanged for $\beta \geq 0$.

Note that the system (4) with $\alpha > \beta \geq 2$ and $|c| < 2$ has the single phase portrait determined by the separatrix configuration in Figure 5(c'). Also note that, due to the symmetry of the system (4) cited in Lemma 1, if for a given $c \in (-2, 2), \beta \in (-\infty, \infty)$, and $\alpha > \beta$ the system (4) has one of the configurations a, b, c', c$_1$, d, e, f', g', h', h$_1$, i' or j' described in Figure 5, then the system (4) with $-\alpha$ and $-\beta$ in place of α and β (and $-\alpha < -\beta$), and with $-c$ in place of c, will have the corresponding configuration a', b', c, c$'_1$, d', e', f, g, h, h$'_1$, i, or j, respectively. We see that every one of the configurations in Figure 5 (or one of these configurations rotated about the x-axis) is realized for some parameter values in the last chart in Figure 6. The labeling of the phase portraits in Figure 5 was chosen to correspond to the labeling of the phase portraits for BQS3 in Figure 8 below.

It is instructive to see how the results of Theorem 4, together with the algebraic formula for the surface H^+ given in Theorem 4 and with the analytic surface HL^+ approximated by the numerical results in Figure 4, can be used to determine the specific phase portrait that occurs for a given BQS2 of the form (4).

Example 1. Consider the BQS2 given by (4) with $\alpha = 1$ and $\beta = .4 \in (0, \beta^*)$, i.e., by

$$\begin{aligned} \dot{x} &= -x + .4y + y^2 \\ \dot{y} &= x - .4y - xy + c(-x + .4y + y^2), \end{aligned} \quad (4')$$

where we let $c = 1.2, 1.3, 1.375$, and 1.5. Cf. the fourth chart in Figure 6. It then follows from Theorem 4 and Figures 3 and 4 that this BQS2 with

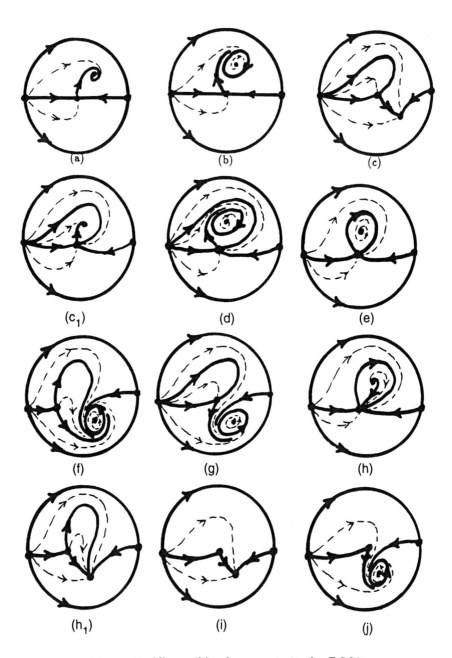

Figure 5. All possible phase portraits for BQS2.

4.14. Coppel's Problem for Bounded Quadratic Systems

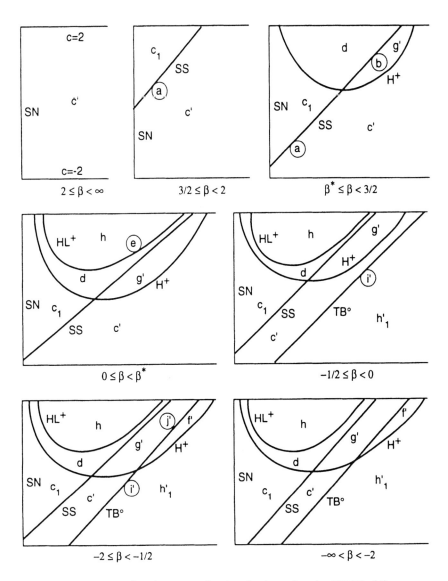

Figure 6. The charts in the (α, c)-plane for the BQS2, (4).

$c = 1.2, 1.3, 1.375$, and 1.5 has its phase portrait determined by Figures 5 (c_1), (d), (e), and (h) respectively. This is borne out by the numerical results shown in Figure 7. Of course, the computer-drawn phase portrait for $c = 1.375$ only approximates the homoclinic loop that occurs at $c = 1.375\cdots$.

We next present the solution to Coppel's problem for BQS3 under the assumption that any BQS3 has at most two limit cycles. It has been shown in [51] that any BQS3 which is near a center has at most two limit cycles, and, because of the results presented in this section and in [53], we believe

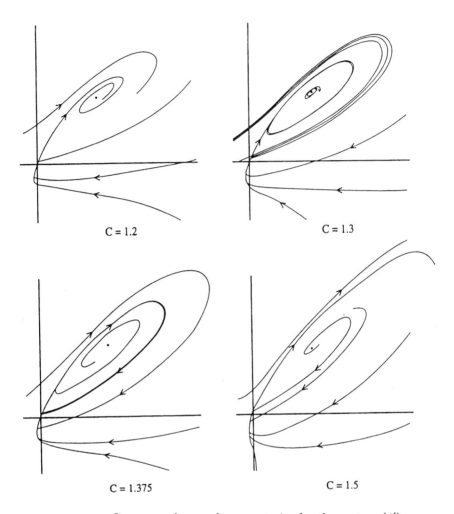

Figure 7. Computer-drawn phase portraits for the system (4′).

that it is true in general that any BQS3 (and therefore any BQS) has at most two limit cycles.

As we shall see, the class BQS3 has both homoclinic-loop and multiplicity-two limit cycle bifurcation surfaces that are described by functions whose analyticity follows from the results in [38]. Thus, just as in the case of BQS2, it is once again necessary to allow inequalities involving analytic functions in the solution of Coppel's problem for BQS3.

It follows from Theorem 1 above and Lemma 8 in [48] that any BQS3 is affinely equivalent to (3) with $|c| < 2, a_{11} < 0$, and $(a_{12} - a_{21} + ca_{11})^2 > 4(a_{11}a_{22} - a_{21}a_{12}) \neq 0$. It was shown in [48] that for any BQS3 of the form (3), the middle critical point (ordered according to the size of the

4.14. Coppel's Problem for Bounded Quadratic Systems

y-component of the critical point) is a saddle. Thus, by translating the origin to the lower critical point and by making the linear transformation of coordinates $t \to |a_{11}|t, x \to x/|a_{11}|$ and $y \to y/|a_{11}|$ in (3) with $a_{11} < 0$, it follows that any BQS3 is affinely equivalent to (3') above with $|c| < 2$ and $(a_{21} \quad a_{12} + c)^2 > 4(-a_{22} - a_{12}a_{21}) > 0$. Therefore, by letting $\beta = a_{12}, \alpha = a_{21} + c$, and $\gamma^2 = -a_{22} - a_{21}a_{12}$, a positive quantity, it follows from the above inequalities that $|\alpha - \beta| > 2|\gamma| > 0$, and we obtain the following result:

Lemma 2. *Any BQS3 is affinely equivalent to the one-parameter family of rotated vector fields*

$$\begin{aligned}\dot{x} &= -x + \beta y + y^2 \\ \dot{y} &= \alpha x - (\alpha\beta + \gamma^2)y - xy + c(-x + \beta y + y^2)\end{aligned} \quad (5)$$

mod $x = \beta y + y^2$ *with parameter* $c \in (-2, 2)$ *and* $|\alpha - \beta| > 2|\gamma| > 0$. *Furthermore, the system is invariant under the transformation* $(x, y, t, \alpha, \beta, \gamma, c) \to (x, -y, t, -\alpha, -\beta, -\gamma, -c)$, *and it therefore suffices to consider* $\alpha - \beta > 2\gamma > 0$. *The critical points of (5) are at* $0 = (0, 0)$, $P^+ = (x^+, y^+)$, *and* $P^- = (x^-, y^-)$ *with* $x^\pm = (\beta + y^\pm)y^\pm$ *and* $2y^\pm = \alpha - \beta \pm [(\alpha - \beta)^2 - 4\gamma^2]^{1/2}$. *The origin and* P^+ *are nodes or foci, and* P^- *is a saddle. The y-components of* $0, P^-$, *and* P^+ *satisfy* $0 < y^- < y^+$; *i.e.,* $0, P^-$, *and* P^+ *are in the relative positions shown in the following diagram:*

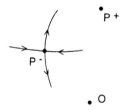

The last statements in Lemma 2 follow directly from the results in Lemma 8 in [48] regarding the critical points of (3). The bifurcations that take place in the four-dimensional parameter space of (5) are derived in the problems at the end of this section. Hopf and homoclinic-loop bifurcations occur at both 0 and P^+; these bifurcation surfaces are denoted by H^+, H^0, HL^+, and HL^0; cf. Problems 1, 5, and 6. There are also multiplicity-two Hopf bifurcations that occur at points in H^+ and H^0; these surfaces are denoted by H_2^+ and H_2^0; cf. Problems 1 and 5. Note that it was shown in Proposition C5 in [50] that there are no multiplicity-two Hopf bifurcations for BQS1 and BQS2, and that there are no multiplicity-three Hopf bifurcations for BQS3. There are multiplicity-two homoclinic-loop bifurcations that occur in HL^+, as is shown in Problem 7 at the end of this section, and this surface is denoted by HL_2^+. Also, just as in the class BQS2, it follows from Lemma 11 in [48] that the class BQS3 has a saddle–saddle bifurcation surface

$$SS: c = (\alpha + \beta + S)/2,$$

where $S = \sqrt{(\alpha - \beta)^2 - 4\gamma^2}$, and for $(\alpha, \beta, \gamma, c) \in SS$ the system (5) has a saddle–saddle connection between the saddle P^- and the saddle-node at the point $(1, 0, 0)$ on the equator of the Poincaré sphere. There is also a saddle-node bifurcation that occurs as $\alpha \to \beta + 2\gamma$; i.e., as $P^+ \to P^-$, and this results in the following saddle-node bifurcation surface for (5):

$$SN: \alpha = \beta + 2\gamma.$$

Cf. Problem 3.

Next we point out that there is a Takens–Bogdanov (or cusp) bifurcation surface TB^+ that occurs at points where H^+ intersects HL^+ on SN; i.e., as in Figure 3 in Section 4.13, $TB^+ = H^+ \cap HL^+ \cap SN$. As is shown in Problem 4, it is given by

$$TB^+: c = \frac{1}{\beta + 2\gamma} + \beta + \gamma \quad \text{and} \quad \alpha = \beta + 2\gamma.$$

It is shown in Problem 8 that there is a transcritical bifurcation that occurs as $\gamma \to 0$; i.e., as $0 \to P^-$, and this results in the following transcritical bifurcation surface for (5):

$$TC: \gamma = 0.$$

Finally, as was noted earlier, there is the Takens–Bogdanov (or cusp) bifurcation surface TB^0 that occurs at points where H^0 intersects HL^0 on TC; i.e., $TB^0 = H^0 \cap HL^0 \cap TC$. Cf. Problems 2 and 3. It is given by

$$TB^0: c = \alpha + \frac{1}{\beta};$$

cf. Theorem 4 above. All of these bifurcations are derived in the problem set at the end of this section, including the multiplicity-two limit cycle bifurcation surfaces C_2^+ and C_2^0 whose existence and analyticity follow from the results in [38]; cf. Problem 6. These bifurcation surfaces for BQS3 are listed in the next theorem, where they are described by either algebraic or analytic functions of the parameters $\alpha, \beta, \gamma,$ and c that appear in (5). Furthermore, these are the only bifurcations that occur in the class BQS3, according to Peixoto's theorem.

The relative positions of the bifurcation surfaces described above and in Theorem 5 below are determined by the atlas and charts for the system (5) given in Figures $\mathcal{A}$ and $\mathcal{C}$ in [53] and shown in Figures 15 and 16 below for $\beta \geq 0$. The phase portraits for (5) with $(\alpha, \beta, \gamma, c)$ in the various components of the region R, defined in Theorem 5 below and determined by the atlas and charts, in [53], follow from the results in [48] for BQS3 under the assumption that any BQS3 has at most two limit cycles. These results are summarized in the following theorem.

4.14. Coppel's Problem for Bounded Quadratic Systems

Theorem 5. *Under the assumption that any BQS3 has at most two limit cycles, the phase portrait for any BQS3 is determined by one of the separatrix configurations in Figure 8. Furthermore, there exist homoclinic-loop and multiplicity-two limit cycle bifurcation functions $h(\alpha, \beta, \gamma), h_0(\alpha, \beta, \gamma), f(\alpha, \beta, \gamma),$ and $f_0(\alpha, \beta, \gamma)$, analytic on their domains of definition, such that the bifurcation surfaces*

$$H^+ : c = \frac{1 + \alpha(\alpha + \beta + S)/2}{\alpha + S},$$

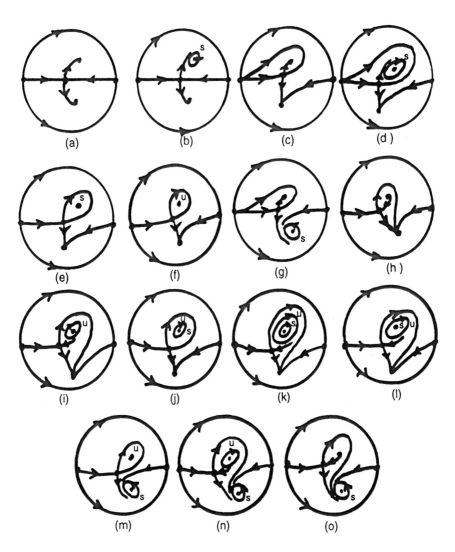

Figure 8. All possible phase portraits for BQS3.

504   4. Nonlinear Systems: Bifurcation Theory

$$H_2^+: c = \frac{-b + \sqrt{b^2 - 4ad}}{2a},$$

$$H^0: c = \alpha + \frac{1+\gamma^2}{\beta},$$

$$H_2^0: c = \frac{\alpha\beta - 2\alpha^2 - 1 + \sqrt{(\alpha\beta - 2\alpha^2 - 1)^2 - 4(\alpha - \beta)(\beta - 2\alpha)}}{2(\beta - 2\alpha)},$$

$$HL^+: c = h(\alpha, \beta, \gamma),$$

$$HL_2^+: c = \frac{1 + \alpha(\alpha + \beta - S)/2}{\alpha - S},$$

$$HL^0: c = h_0(\alpha, \beta, \gamma),$$

$$SS: c = (\alpha + \beta + S)/2 \text{ or } \alpha = c + \gamma^2/(c - \beta),$$

$$C_2^+: c = f(\alpha, \beta, \gamma),$$

and

$$C_2^0: c = f_0(\alpha, \beta, \gamma)$$

with $S = \sqrt{(\alpha - \beta)^2 - 4\gamma^2}$, $a = 2(2S - \beta)$, $b = (\alpha + \beta - S)(\beta - 2S) + 2$, and $d = \beta - \alpha - 3S$ partition the region

$$R = \{(\alpha, \beta, \gamma, c) \in \mathbf{R}^4 \mid \alpha > \beta + 2\gamma, \gamma > 0, |c| < 2\}$$

of parameters for the system (5) *into components, the specific phase portrait that occurs for the system* (5) *with* $(\alpha, \beta, \gamma, c)$ *in any one of these components being determined by the atlas and charts in Figures* $\mathcal{A}$ *and* $\mathcal{C}$ *in* [53], *which are shown in Figures* 15 *and* 16 *below for* $\beta \geq 0$.

The purpose of the atlas and charts presented in [53] and derived below for $\beta \geq 0$ is to show how the bifurcation surfaces defined in Theorem 5 partition the region of parameters for the system (5),

$$R = \{(\alpha, \beta, \gamma, c) \in \mathbf{R}^4 \mid \alpha > \beta + 2\gamma, \gamma > 0, |c| < 2\},$$

into components and to specify which phase portrait in Figure 8 corresponds to each of these components.

The "atlas," shown in Figure $\mathcal{A}$ in [53] and in Figure 15 below for $\beta \geq 0$, gives a partition of the upper half of the (β, γ)-plane into components together with a chart for each of these components. The charts are specified by the numbers in the atlas in Figure $\mathcal{A}$. Each of the charts 1–5 in Figure 16 determines a partition of the region

$$E = \{(\alpha, c) \in \mathbf{R}^2 \mid \alpha > \beta + 2\gamma, |c| < 2\}$$

in the (α, c)-plane into components (determined by the bifurcation surfaces $H^+, \ldots, C_2^0$ in Theorem 5) together with the phase portrait from Figure 8 that corresponds to each of these components. The phase portraits are denoted by a–o or a'–o' in the charts in Figure $\mathcal{C}$ in [53] and in Figure 16 below. As was mentioned earlier, the phase portraits a'–o' are obtained by

4.14. Coppel's Problem for Bounded Quadratic Systems

rotating the corresponding phase portraits a–o throughout π radians about the x-axis.

In the atlas in Figure 15, each of the curves $\Gamma_1, \ldots, \Gamma_4$ that partition the first quadrant of the (β, γ)-plane into components defines a fairly simple event that takes place regarding the relative positions of the bifurcation surfaces defined in Theorem 5. For example, the saddle–saddle connection bifurcation surface SS, defined in Theorem 5, intersects the region E in the (α, c)-plane iff $\beta + \gamma < 2$. (This fact is derived below.) Cf. Charts 1 and 2 in Figure 16 where we see that for all $(\alpha, c) \in E$ and $\beta + \gamma > 2$ the system (5) has the single phase portrait c' determined by Figure 8.

In what follows, we describe each of the curves $\Gamma_1, \ldots, \Gamma_4$ that appear in the atlas in Figure 15 as well as what happens to the bifurcation surfaces in Theorem 5 as we cross these curves.

A. $\Gamma_1^\pm$: SS INTERSECTS SN ON $c = \pm 2$

From Theorem 5, the bifurcation surfaces SS and SN are given by

$$SS: c = \frac{\alpha + \beta + S}{2}$$

and

$$SN: \alpha = \beta + 2\gamma,$$

respectively, where $S = \sqrt{(\alpha - \beta)^2 - 4\gamma^2}$. Substituting $\alpha = \beta + 2\gamma$ into the equation for S shows that $S = 0$ on SN; substituting those quantities into the SS equation shows that SS intersects SN at the point

$$SS \cap SN: \alpha = \beta + 2\gamma, \quad c = \beta + \gamma.$$

Thus SS intersects SN on $c = \pm 2$ iff $(\beta, \gamma) \in \Gamma_1^\pm$, where

$$\Gamma_1^\pm: \beta + \gamma = \pm 2.$$

It follows that in the (α, c)-plane the SS and SN curves have the relative positions shown in Figure 9.

B. $\Gamma_4^\pm$: TB^+ INTERSECTS SN ON $c = \pm 2$

As was determined in Problem 4, a Takens–Bogdanov bifurcation occurs at the critical point P^+ of the system (5), given in Lemma 2, for points on the Takens–Bogdanov surface

$$TB^+: c = \frac{1}{\beta + 2\gamma} + \beta + \gamma.$$

Setting $c = \pm 2$ in this equation determines the curves

$$\Gamma_4^\pm: \frac{1}{\beta + 2\gamma} + \beta + \gamma = \pm 2,$$

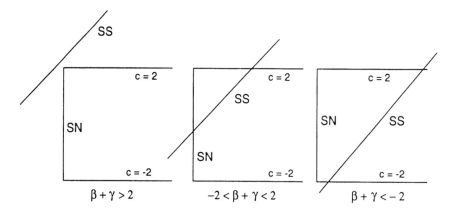

Figure 9. The position of the SS curve in the (α, c)-plane.

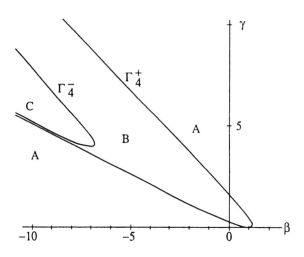

Figure 10. The curves $\Gamma_4^\pm$ in the (β, γ)-plane.

where the point $p \in TB^+ \cap SN$ enters and leaves the region $|c| < 2$ in the (α, c)-plane, respectively. The curves $\Gamma_4^\pm$ are shown in Figure 10. For points (β, γ) in between these two curves we have a Takens–Bogdanov bifurcation point TB^+ in the closure of the region E in the (α, c)-plane; cf. Figure 11 and charts 4 and 5 in Figure 16.

4.14. Coppel's Problem for Bounded Quadratic Systems

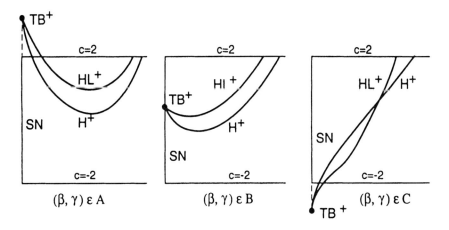

Figure 11. The position of the TB^+ point in the (α, c)-plane.

C. Γ_2: H^+ INTERSECTS $c = 2$

First, consider the case when $\gamma = 0$. In this case, it follows from Theorem 4 that
$$H^+ : c = \frac{1 + \alpha^2}{2\alpha - \beta}.$$

It then follows that $\partial c/\partial \alpha = 0$ iff $\alpha^2 - \alpha\beta - 1 = 0$ and that for any $\beta \in \mathbf{R}$, the H^+ curve has a minimum at $\alpha = (\beta + \sqrt{\beta^2 + 4})/2$. This minimum point occurs at the intersection of the SS and H^+ curves in the (α, c)-plane. (This follows since for $\gamma = 0, c = \alpha$ on the SS curve, and substituting $c = \alpha$ into the above formula for H^+ yields $\alpha^2 - \alpha\beta - 1 = 0$.) Thus, for $\gamma = 0$, the minimum point on the H^+ curve intersects the horizontal line $c = 2$ at the point $c = \alpha = (\beta = \sqrt{\beta^2 + 4})/2 = 2$, which implies that $\beta = 3/2$; cf. Figure 6.

Next consider the case when $\gamma > 0$. In this case, it follows from Theorem 5 that
$$H^+ : c = \frac{1 + \alpha(\alpha + \beta + S)/2}{\alpha + S},$$

where $S = \sqrt{(\alpha - \beta)^2 - 4\gamma^2}$. Once again, we set $\partial c/\partial \alpha = 0$ to find the minimum point on the H^+ curve (when it exists). This yields

$$(2\alpha + \beta)S^2 + [\alpha^2 - 2 + (\alpha - \beta)^2 - 4\gamma^2]S - (\alpha - \beta)(2 + \alpha\beta) = 0. \quad (*)$$

And setting $c = 2$ in the H^+ equation yields

$$2 + (\alpha - 4)S + \alpha(\alpha + \beta - 4) = 0. \quad (**)$$

Eliminating α between the two equations $(*)$ and $(**)$ then yields the curve

$$\Gamma_2 : \gamma = \gamma_2(\beta)$$

508   4. Nonlinear Systems: Bifurcation Theory

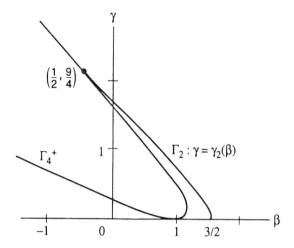

Figure 12. The curve Γ_2 in the (β, γ)-plane.

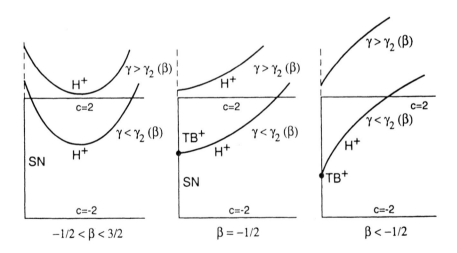

Figure 13. The position of the H^+ curve in the (α, c)-plane.

in the (β, γ)-plane, where H^+ first intersects the horizontal line $c = 2$ in the (α, c)-plane; cf. Figures 12 and 13. The curve Γ_2 was determined numerically; it is shown in Figure 12 and in the atlas in Figure 15 below.

4.14. Coppel's Problem for Bounded Quadratic Systems

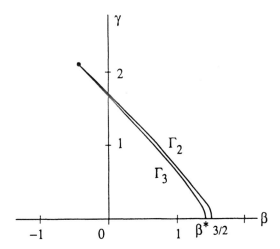

Figure 14. The curve Γ_3 in the (β, γ)-plane.

D. $\Gamma_3 \colon HL^+$ Intersects $c = 2$

For $\gamma = 0$, it was noted just prior to Theorem 4 above that the homoclinic loop bifurcation surface HL^+ intersects the region $|c| < 2$ iff $\beta < \beta^*$, where $\beta^* \cong 1.43$ was determined numerically. Cf. Figure 4.

For $\gamma > 0$, it has been determined numerically, by integrating trajectories of (5), that for $-\frac{1}{2} < \beta < \beta^*$, the HL^+ curve intersects the region $|c| < 2$ iff the point (β, γ) lies below the curve

$$\Gamma_3 \colon \gamma = \gamma_3(\beta)$$

in the (β, γ)-plane. Cf. Figure 14, where we see that the curve Γ_3 parallels the Γ_2 curve in the (β, γ)-plane, going from the point $(\beta^*, 0)$ to the point $(-\frac{1}{2}, 2\frac{1}{4})$ common to Γ_2, Γ_3, and Γ_4^+. This is not surprising since the HL^+ curves "parallel" the H^+ and SS curves in the (α, c)-plane. The Γ_2 and Γ_3 curves are shown in Figure 14.

The Atlas in the First Quadrant

At this point we can determine exactly which phase portrait occurs in the system (5) for $(\alpha, \beta, \gamma, c) \in \mathbf{R}$ with $\beta \geq 0$. The curves $\Gamma_1^+, \Gamma_2, \Gamma_3$, and Γ_4^+, discussed above, are shown in Figure 15 together with the chart numbers 1–5 that correspond to each of the components in the first quadrant of the (β, γ)-plane that are determined by the curves Γ_1^+–Γ_4^+. We also show charts 1–5 in Figure 16 and the phase portraits that occur in these charts

510   4. Nonlinear Systems: Bifurcation Theory

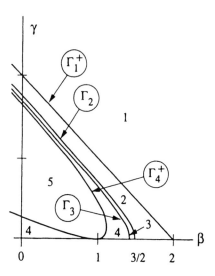

Figure 15. The atlas $\mathcal{A}$ in the first quadrant of the (β, γ)-plane.

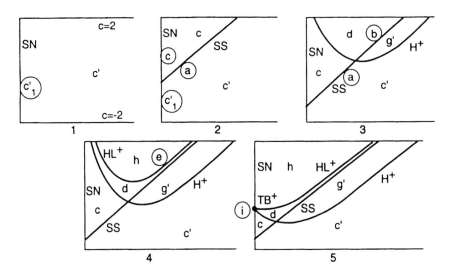

Figure 16. The charts in the (α, c)-plane, that appear in the first quadrant of the atlas $\mathcal{A}$ shown in Figure 15.

4.14. Coppel's Problem for Bounded Quadratic Systems

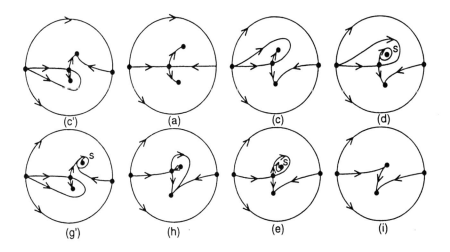

Figure 17. The phase portraits that occur in the charts shown in Figure 16; cf. Figure 8 (and Figure 5 for c, c'_1 and i on SN).

in Figure 17. This should give the student a very good idea of how the results in this section allow us to determine the phase portrait of any BQS of the form (5) with $\beta \geq 0$ and $\gamma \geq 0$. It also should be clear that as $\gamma \to 0$, the first four charts shown in Figure 16 reduce to the first four charts in Figure 6, and the phase portraits shown in Figure 17 reduce to the corresponding phase portraits in Figure 5. Note that all of the phase portraits shown in Figure 17 occur in chart 5 shown in Figure 16.

E. THE SURFACE $\beta = 0$

As we cross the plane $\beta = 0$, the bifurcation surfaces H^0 and HL^0 enter into the region R; i.e., for $\beta \geq 0$ (and $\gamma \geq 0$), the H^0 and HL^0 curves do not intersect the region

$$E = \{(\alpha, c) \in \mathbf{R}^2 \mid \alpha > \beta + 2\gamma, |c| < 2\}$$

in the (α, c)-plane; and for $\beta < 0$ (and $\gamma > 0$), they do. Also, for $\gamma = 0$ and $\beta \geq 0$, there is no Takens–Bogdanov curve TB^0 in the region E in the (α, c)-plane, while for $\beta < 0$ there is; cf. Problem 8. Figure 18 depicts what happens as we cross the plane $\beta = 0$; cf. Problem 9.

It is instructive at this point to look at some examples of how Theorem 5, together with the atlas an charts in [53] can be used to determine the phase portrait of a given BQS3 of the form (5). The atlas and charts determine which phase portrait in Figure 8 occurs for a specific BQS3 of the form (5), provided that we use the algebraic formulas given in Theorem 5 and/or the numerical results given in [53] for the various bifurcation surfaces listed in Theorem 5. We consider the system (5) with $\beta = -10$ in the following examples because some interesting bifurcations occur for large negative

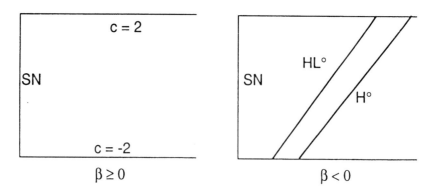

Figure 18. The appearance of the H^0 and HL^0 curves in the (α, c)-plane for $\beta < 0$.

values of β, and also because we can compare the results for $\beta = -10$ with the asymptotic results given in [51] and in Theorem 6 below for large negative β. This is done in Example 4 below.

Example 2. Consider the system (5) with $\beta = -10$ and $\gamma = 3.5$. The bifurcation curves for $\beta = -10$ and $\gamma = 3.5$ are shown in [53] and in Figure 19. The bifurcation curves H^+, HL^+, H^0, HL^0, SS, C_2^+, and C_2^0 partition the region $\alpha > -3$ and $|c| < 2$ into various components. The phase portrait for the system (5) with $\beta = -10, \gamma = 3.5$, and (α, c) in any one of these components is determined by Figure 8 above. Note that every one of the configurations a–o or a'–o' in Figure 8 occurs in Figure 19.

Also note that the multiplicity-two limit cycle bifurcation curve C_2^+ has two branches, one of them going from the left-hand point H_2^+ on the curve H^+ to the point HL_2^+ on the curve HL^+, and the other branch going from the right-hand point H_2^+ to infinity, asymptotic to the SS curve, as $\alpha \to \infty$. Cf. the termination principle for one-parameter families of multiple limit cycles in [39]. A similar comment holds for the multiplicity-two limit cycle bifurcation curve C_2^0 shown in Figure 19. The region of the (α, c)-plane containing the two branches of the C_2^+ curve is shown on an expanded scale in Figure 20. The system (5) with $\beta = -10$, $\gamma = 3.5$, and (α, c) in the shaded regions in Figure 20 has two limit cycles around the critical point P^+; the phase portrait for these parameter values is determined by the configuration (k) in Figure 8 above.

The bifurcation curves HL^+, HL^0, C_2^+, and C_2^0; i.e., the graphs of the functions $c = h(\alpha, -10, 3.5)$, $c = h_0(\alpha, -10, 3.5)$, $c = f(\alpha, -10, 3.5)$, and $c = f_0(\alpha, -10, 3.5)$, respectively, were determined numerically. The most efficient and accurate way of doing this is to compute the Poincaré map $P(r)$ along a ray through the critical point P^+ in order to determine the HL^+ and C_2^+ curves (or through the critical point 0 in order to determine

4.14. Coppel's Problem for Bounded Quadratic Systems

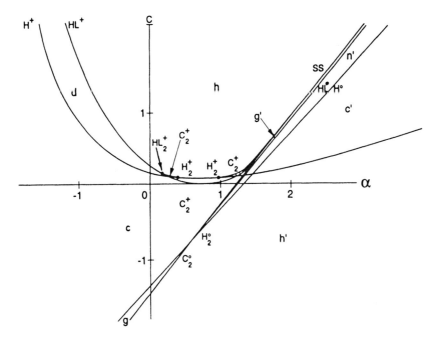

Figure 19. The bifurcation curves H^+, HL^+, H^0, HL^0, SS, C_2^+, and C_2^0 for the system (5) with $\beta = -10$ and $\gamma = 3.5$.

the HL^0 and C_2^0 curves). The displacement function $d(r) \equiv P(r) - r$ divided by r, i.e., $d(r)/r$, along the ray $\theta = \pi/6$ through the point P^+ for the system (5) with $\beta = -10$, $\gamma = 3.5$, and $\alpha = 1.1$ is shown in Figure 21 for various values of c. In Figure 21(a) we see that for $\alpha = 1.1$, a homoclinic loop occurs at $c \cong .04$, i.e., $(1.1, .04 \cdots)$ is a point on the homoclinic loop bifurcation curve HL^+ for $\beta = -10$ and $\gamma = 3.5$, as shown in Figure 20. Also, the displacement function curve $d(r)/r$ shown in Figure 21(a) is tangent to the r-axis (which is equivalent to saying that the curve $d(r)$ is tangent to the r-axis) at $c \cong .09$. The blow-up of some of these curves, given in Figure 21(b), shows that the displacement function $d(r)$ is tangent to the r-axis at $c \cong .0885$; i.e., $(1.1, .0885 \cdots)$ is a point on the right-hand branch of the multiplicity-two cycle bifurcation curve C_2^+ for $\beta = -10$ and $\gamma = 3.5$, as shown in Figure 20. It also can be seen in Figure 21(b) that the system (5) with $\beta = -10$, $\gamma = 3.5$, $\alpha = 1.1$, and $c = .088$ has two limit cycles at distances $r \cong 4.9$ and $r \cong 7.7$ along the ray $\theta = \pi/6$ through the critical point P^+; and for $c = .087$ there are two limit cycles at distances $r \cong 1.7$ and $r \cong 8.4$ along the ray $\theta = \pi/6$ through the critical point P^+. Cf. Figure 8(k).

Figure 22 shows a blow-up of the region in Figure 19 where the curves SS, H^0, and HL^0 intersect and where the curve C_2^0 emerges from the

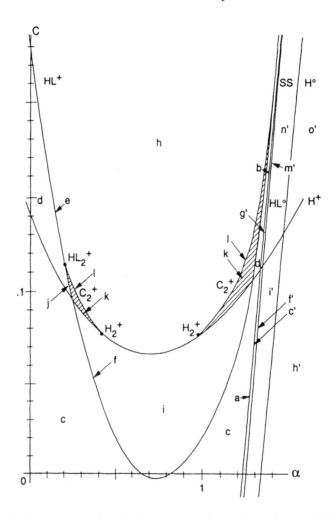

Figure 20. The regions in which (5) with $\beta = -10$ and $\gamma = 3.5$ has two limit cycles around the critical point P^+.

point H_2^0 on the H^0 curve. The curve C_2^0 is tangent to H^0 at H_2^0, and it is asymptotic to the curve HL^0 as α or c decrease without bound.

Example 3. Once again consider the system (5) with $\beta = -10$, but this time with $\gamma = 3$. The bifurcation curves for this case are shown in Figure 23. We see that the bifurcation curve C_2^+ only has one branch, which goes from the point HL_2^+ on the curve HL^+ to infinity along the SS curve as $\alpha \to \infty$. The reason why there can be one or two branches of the bifurcation curve C_2^+ in the (α, c)-plane for various values of β and γ is discussed in [53]. Once again, the points on the bifurcation curves HL^+ and C_2^+ were computed

4.14. Coppel's Problem for Bounded Quadratic Systems

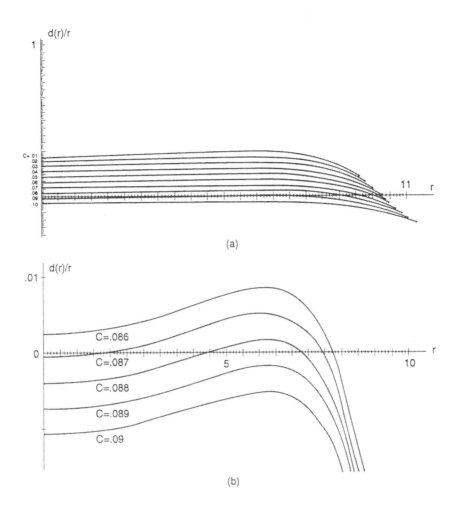

Figure 21. The displacement function $d(r)/r$ for the system (5) with $\beta = -10$, $\gamma = 3.5$, and $\alpha = 1.1$.

using the Poincaré map as described in the previous example. The point HL_2^+ and the bifurcation curves H^+, H^0, and SS follow from the algebraic formulas in Theorem 5.

We next compare the results of Theorem 5 with the asymptotic results in [51], where Li et al. study the unfolding of the center for a BQS given in Remark 1 above. They study the system

$$\begin{aligned} \dot{x} &= -\delta x - ay + y^2 \\ \dot{y} &= \delta\nu_2 x + by - xy + \delta\nu_3 y^2 \end{aligned} \quad (6)$$

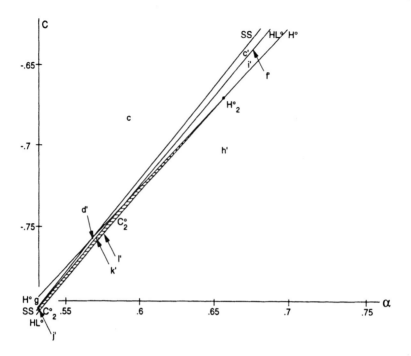

Figure 22. A blow-up of the region in Figure 12 where SS, H^0, and HL^0 cross.

for $a > 0$, $b < 0$, $0 < \delta \ll 1$, and $|\delta \nu_3| < 2$; cf. equation (1.1) and Theorem C in [51]. The system (6) with $\delta = 0$ is affinely equivalent to the BQS2 with a center given in Remark 1. We note that there is a removable parameter in the system (6); i.e., for $a > 0$ the transformation of coordinates $t \to at$, $x \to x/a$ and $y \to y/a$ reduces to (6) to

$$\dot{x} = -\delta x - y + y^2$$
$$\dot{y} = \delta\nu_2 x + by - xy + \delta\nu_3 y^2 \tag{6'}$$

with $b < 0$. For $\delta > 0$, the linear transformation of coordinates $t \to \delta t$, $x \to x/\delta$, and $y \to y/\delta$ transforms (6') into

$$\dot{x} = -x - \frac{1}{\delta}y + y^2$$
$$\dot{y} = \nu_2 x + \frac{b}{\delta}y - xy + \delta\nu_3 y^2 \tag{6''}$$

with $b < 0$. Comparing (6'') to the system (5), we see that they are identical with the parameters relayed by

$$\beta = -\frac{1}{\delta} \qquad \alpha = \nu_2 + \delta\nu_3$$
$$c = \delta\nu_3 \qquad \gamma = \sqrt{(\nu_2 - b)/\delta} \tag{7}$$

4.14. Coppel's Problem for Bounded Quadratic Systems

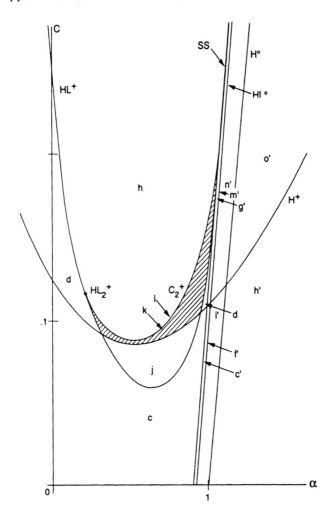

Figure 23. The bifurcation curves H^+, HL^+, H^0, SS, and C_2^+ for the system (5) with $\beta = -10$ and $\gamma = 3$, and the shaded region in which (5) has two limit cycles around the critical point P^+.

for $\delta > 0$ and $\nu_2 \geq b$. Note that the transcritical bifurcation surface $\gamma = 0$ corresponds to $\nu_2 = b$ in (7). Since the Jacobian of the (nonlinear) transformation defined by (7),

$$\left| \frac{\partial(\alpha, \beta, \gamma, c)}{\partial(\delta, \nu_2, \nu_3, b)} \right| = \frac{1}{\delta\sqrt{\delta(\nu_2 - b)}},$$

it follows that (7) defines a one-to-one transformation of the region

$$\{(\alpha, \beta, \gamma, c) \in \mathbf{R}^4 \mid \beta < 0, \gamma > 0\}$$

onto the region
$$\{(\delta, \nu_2, \nu_3, b) \in \mathbf{R}^4 \mid \delta > 0, \nu_2 > b\}.$$
For $0 < \delta \ll 1$, the asymptotic formulas for the bifurcation surfaces H^+, H_2^+, HL^+, HL_2^+, and C_2^+ (denoted by H, A_1, hℓ, A_2 and dℓ) in [51] can be compared to the corresponding bifurcation surfaces in Theorem 5 above with $\beta = -1/\delta \ll -1$. Substituting the parameters defined by (7) into Theorem C in [51], or letting $\beta \to -\infty$ in Theorem 5 above, leads to the same asymptotic formulas for the bifurcation surfaces H^+, H_2^+, and HL_2^+. These formulas are given in Theorem 6 below. This serves as a nice check on our work. In addition, we obtain a bonus from the results in [51]. Namely, an asymptotic formula for the bifurcation surface C_2^+; this does not follow from Theorem 5, since only the existence of the function $f(\alpha, \beta, \gamma)$ is given in Theorem 5. This asymptotic formula for C_2^+ follows from the Melnikov theory in [51]; cf. Section 4.10. The last statement in Theorem 6 follows from the results in [38] and [52].

Theorem 6. *For $\beta = -1/\delta \ll -1$ and $\gamma^2 = |\beta|\Gamma^2$ in (5), it follows that*
$$H^+: c = (1 - \Gamma^2\alpha + \alpha^2)\delta + 0(\delta^2),$$
$$H_2^+: c = \frac{2}{3}\delta + 0(\delta^2), \quad 2\alpha = \Gamma^2 \pm \sqrt{\Gamma^4 = 4/3} + 0(\delta) \quad \text{for } \Gamma \geq \sqrt[4]{4/3},$$
$$HL_2^+: c = 2\delta + 0(\delta^2), \quad \alpha = 3\delta + 0(\delta^2),$$

and
$$C_2^+: c = -2\delta\left[2\alpha^2 - 2\Gamma^2\alpha - 1 + \sqrt{(2\alpha^2 - 2\Gamma^2\alpha - 1)^2 - 1}\right] + 0(\delta^2)$$

as $\delta \to 0$. Furthermore, for each fixed $\beta \ll -1$ and $\gamma^2 = |\beta|\Gamma^2$ with $\Gamma > \sqrt[4]{4/3}$, the multiplicity-two limit cycle bifurcation curve C_2^+ is tangent to the H^+ curve at the point(s) H_2^+, and it has a flat contact with the HL^+ curve at the point HL_2^+.

Remark 3. The result for the homoclinic-loop bifurcation surface HL^+ given in [51], namely that $\nu_2 = 0(\delta)$, does not add any significant new result to Theorem 5. However, just as Li et al. give the tangent line to C_2^+ at HL_2^+; i.e., $\nu_3 = -2\nu_2/\delta + 4 + 0(\delta)$, as the linear approximation to HL^+ at HL_2^+ in Figure 1.4 in [51], we also give the linear approximation to HL^+ at HL_2^+:
$$c = -\frac{2}{3}\alpha + 4\delta + 0(\delta^2)$$

as $\delta \to 0$. This is simply the equation of the tangent line to C_2^+ at HL_2^+ for $\beta = -1/\delta \ll -1$ and $\gamma^2 = |\beta|\Gamma^2$, and it provides a local approximation for the bifurcation surface HL^+ near HL_2^+ for small $\delta > 0$. It can be shown using this linear approximation for HL^+ at HL_2^+ and the asymptotic approximation for H^+ and C_2^+ given in Theorem 6 that, for any fixed $\Gamma \geq (4/3)^{1/4}$, the branch of C_2^+ from H_2^+ to HL_2^+ lies in an $0(\delta)$ neighborhood of $H^+ \cup HL^+$ above H^+ and HL^+.

4.14. Coppel's Problem for Bounded Quadratic Systems

We also obtain the following asymptotic formulas for small $\delta > 0$ from Theorem 5 (where $\beta = -1/\delta$). Note that the first formula for the Hopf bifurcation surface H^0 is exact.

$$H^0: c = \alpha - \Gamma^2 - \delta,$$

$$H_2^0: c = -\frac{1}{\Gamma^2} + 0(\delta), \quad \alpha = \Gamma^2 - \frac{1}{\Gamma^2} + 0(\delta),$$

$$SS: c = \alpha - \Gamma^2 + 0(\delta^2) \quad \text{for } \alpha = \Gamma^2 + 0(\delta),$$

$$SS \cap H^0: c = -\frac{1}{\Gamma^2} + 0(\delta), \quad \alpha = \Gamma^2 - \frac{1}{\Gamma^2} + 0(\delta).$$

It also follows from Theorem 5 that the surfaces SS crosses the plane $c = 0$ at $\alpha = |\beta|\gamma^2 = \Gamma^2$ for all $\beta < 0$.

Let us compare the results in Examples 2 and 3 above with the asymptotic results given above and in Theorem 6.

Example 4. Figure 24 shows the bifurcation curves H^+, HL^+, H^0, SS, and C_2^+ as well as the points H_2^+ and HL_2^+ on H^+ and HL^+, respectively, given by Theorem 5 for $\beta = -10$ and $\gamma = 3.5$. Cf. Figure 20. It also shows the approximations $\sim H^+, \sim H^0, \sim SS$, and $\sim C_2^+$ to these curves as dashed curves (and the approximation $\sim H_2^+$ to H_2^+) given by the asymptotic formulas in Theorem 6 and the above formulas for $\beta = -10$ and $\gamma = 3.5$. The approximation is seen to be reasonably good for this reasonably large negative value of $\beta = -10$. (For larger negative values of β, the approximation is even better, as is to be expected, and as is illustrated in Example 5 below.) In Figure 24, we see that the approximation of H^+ by the asymptotic formula in Theorem 6 is particularly good for $1 < \alpha < 1.5$ but not as good for α near zero; however, the difference between the H^+ and $\sim H^+$ curves at $\alpha = 0, .04 = 0(\delta^2)$ for $\delta = -1/\beta = .1$ in this case.

Figure 25 shows the same type of comparison for $\beta = -10$ and $\gamma = 3$. We note that both $C_2^+ \cap H^+ = \emptyset$ and $(\sim C_2^+) \cap (\sim H^+) = \emptyset$; i.e., there are no H_2^+ nor $\sim H_2^+$ points on H^+ or $\sim H^+$, respectively. Thus, the asymptotic formulas in Theorem 6 also yield some qualitative information about the bifurcation curves H^+ and C_2^+ for $\beta \ll -1$.

Example 5. We give one last example to show just how good the asymptotic approximations in Theorem 6 are for large negative β. We consider the case with $\beta = -100$ and $\gamma = 12$, in which case $\Gamma = \delta\gamma^2 = 1.44 > \sqrt[4]{4/3}$ and, according to the asymptotic formula in Theorem 6 for H_2^+, there will be two points H_2^+ on the curve H^+. Since the bifurcation curves given by Theorem 5 and their asymptotic approximations given by Theorem 6 (and the formulas following Theorem 6) are so close, especially for $.3 < \alpha < 2$, we first show just the approximations $\sim H^+, \sim C_2^+, \sim SS, \sim H^0$, and $\sim H_2^+$ in Figure 26. These same curves are shown as dashed curves in Figure 27 along with the exact bifurcation curves given by Theorem 5. The comparison is seen to be excellent. In particular, H^+ and $\sim H^+$ as well as H^0 and $\sim H^0$, and SS and $\sim SS$ are indistinguishable (on this scale) for $.3 < \alpha < 2$. For

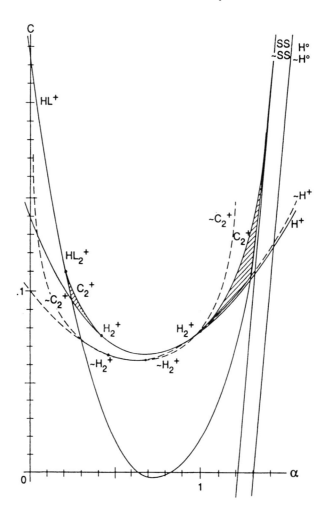

Figure 24. A comparison of the bifurcation curves given by Theorem 5 with their asymptotic approximations given by Theorem 6 for $\beta = -10$ and $\gamma = 3.5$.

α near zero, the approximation of H^+ by $\sim H^+$ is within $.0003 = 0(\delta^2)$ for $\delta = -1/\beta = 1/100$ in this case. One final comment: In Figures 26 and 27, we see that there is a portion of $\sim C_2^+$ between the two points $\sim H_2^+$ on $\sim H^+$. However, this portion of $\sim C_2^+$ (for $\Gamma > \sqrt[4]{4/3}$) has no counterpart on C_2^+, since dynamics tells us that there are no limit cycles for parameter values in the region above H^+ in this case. Cf. Remark 10 in [38].

We end this section with a theorem summarizing the solution of Coppel's problem for BQS, as stated in the introduction, modulo the solution to Hilbert's 16th problem for BQS3:

4.14. Coppel's Problem for Bounded Quadratic Systems

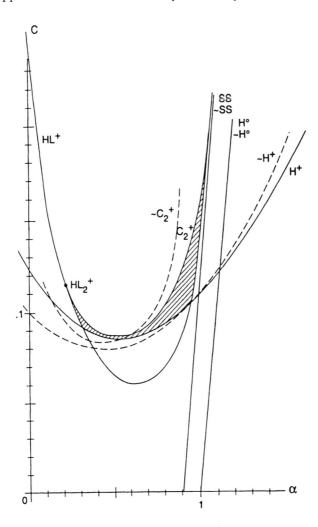

Figure 25. A comparison of the bifurcation curves given by Theorem 5 with their asymptotic approximations given by Theorem 6 for $\beta = -10$ and $\gamma = 3$.

Theorem 7. *Under the assumption that any BQS3 has at most two limit cycles, the phase portrait of any BQS is determined by one of the configurations in Figures 1, 2, 5, or 8. Furthermore, any BQS is affinely equivalent to one of the systems (1)–(5) with the algebraic inequalities on the coefficients given in Theorem 2 or 3 or in Lemma 1 or 2, the specific phase portrait that occurs for any one of these systems being determined by the algebraic inequalities given in Theorem 2 or 3, or by the partition of the regions in Theorem 4 or 5 described by the analytic inequalities defined by the charts in Figure 6 or by the atlas and charts in Figures A and C in [53], which are shown in Figures 15 and 16 for $\beta \geq 0$.*

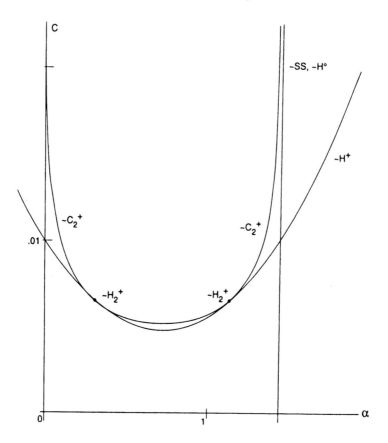

Figure 26. The asymptotic approximations for the bifurcation curves H^+, H^0, SS, and C_2^+ given by Theorem 6 for $\beta = -100$ and $\gamma = 12$.

Corollary 1. *There is a BQS with two limit cycles in the $(1,1)$ configuration, and, under the assumption that any BQS3 has at most two limit cycles, the phase portrait for any BQS with two limit cycles in the $(1,1)$ configuration is determined by the separatrix configuration in Figure 8(n).*

Corollary 2. *There is a BQS with two limit cycles in the $(2,0)$ configuration, and, under the assumption that any BQS3 has at most two limit cycles, the phase portrait for any BQS with two limit cycles in the $(2,0)$ configuration is determined by the separatrix configuration in Figure 8(k).*

Remark 4. The termination of any one-parameter family of multiplicity-m limit cycles of a planar, analytic system is described by the termination principle in [39]. We note that, as predicted by the above-mentioned termination principle, the one-parameter families of simple or multiplicity-two limit cycles whose existence is established by Theorem 5 (several of which are exhibited in Examples 2–5) terminate either

(i) as the parameter or the limit cycles become unbounded, or

4.14. Coppel's Problem for Bounded Quadratic Systems

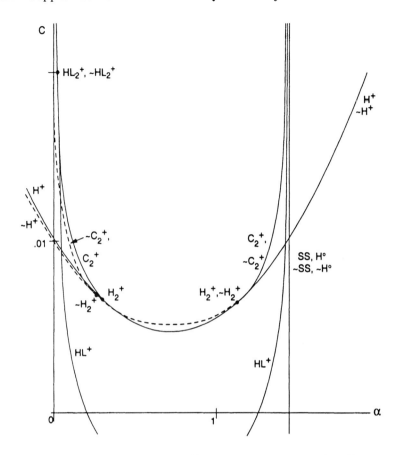

Figure 27. A comparison of the bifurcation curves given by Theorems 5 and 6 for $\beta = -100$ and $\gamma = 12$.

(ii) at a critical point in a Hopf bifurcation of order $k = 1$ or 2, or

(iii) on a graphic or separatrix cycle in a homoclinic loop bifurcation of order $k = 1$ or 2, or

(iv) at a degenerate critical point (i.e., a cusp) in a Takens–Bogdanov bifurcation.

Problem Set 14

In this problem set, the student is asked to determine the bifurcations that occur in the BQS2 or BQS3 given by

$$\dot{x} = -x + \beta y + y^2$$
$$\dot{y} = \alpha x - (\alpha\beta + \gamma^2)y - xy + c(-x + \beta y + y^2) \quad (5)$$

with $\alpha - \beta \geq 2\gamma \geq 0$; cf. Lemmas 1 and 2. According to Lemma 2, the critical points of (5) are at $0 = (0,0)$ and $P^\pm = (x^\pm, y^\pm)$ with

$$x^\pm = (\beta + y^\pm)y^\pm$$

and

$$y^\pm = \frac{\alpha - \beta \pm \sqrt{(\alpha-\beta)^2 - 4\gamma^2}}{2}. \tag{8}$$

If we let $\mathbf{f}(x,y)$ denote the vector field defined by the right-hand side of (5), it follows that

$$D\mathbf{f}(0,0) = \begin{bmatrix} -1 & \beta \\ \alpha - c & c\beta - \alpha\beta - \gamma^2 \end{bmatrix},$$

and that

$$D\mathbf{f}(x^\pm, y^\pm) = \begin{bmatrix} -1 & \beta + 2y^\pm \\ \alpha - c - y^\pm & \beta(c-\alpha) + (2c-\alpha)y^\pm \end{bmatrix}$$

$$= \begin{bmatrix} -1 & \alpha \pm S \\ \dfrac{\alpha + \beta - 2c \mp S}{2} & \dfrac{(c-\alpha)(\alpha+\beta) + c(\alpha-\beta) \pm (2c-\alpha)S}{2} \end{bmatrix},$$

where $S = \sqrt{(\alpha-\beta)^2 - 4\gamma^2}$.

If we use $\delta(x,y)$ for the determinant and $\tau(x,y)$ for the trace of $D\mathbf{f}(x,y)$, then it follows from the above formulas that

$$\delta(0,0) = \det D\mathbf{f}(0,0) = \gamma^2 \geq 0,$$
$$\tau(0,0) = \operatorname{tr} D\mathbf{f}(0,0) = -1 + c\beta - \alpha\beta - \gamma^2,$$
$$\delta(x^\pm, y^\pm) = \det D\mathbf{f}(x^\pm, y^\pm) = \pm S y^\pm,$$

and

$$\tau(x^\pm, y^\pm) = \operatorname{tr} D\mathbf{f}(x^\pm, y^\pm) = -1 + \beta(c-\alpha) + (2c-\alpha)y^\pm.$$

These formulas will be used throughout this problem set in deriving the formulas for the bifurcation surfaces listed in Theorem 5 (which reduce to those in Theorem 4 for $\gamma = 0$).

1. (a) Show that for $\alpha \neq \beta + 2\gamma$ there is a Hopf bifurcation at the critical point P^+ of (5) for parameter values on the Hopf bifurcation surface

$$H^+: c = \frac{1 + \alpha(\alpha + \beta + S)/2}{\alpha + S},$$

where $S = \sqrt{(\alpha-\beta)^2 - 4\gamma^2}$ and that, for $\gamma = 0$ and $\alpha > \beta$ as in Lemma 1, this reduces to the Hopf bifurcation surface for (4) given by

$$H^+: c = \frac{1+\alpha^2}{2\alpha - \beta}.$$

Furthermore, using formula (3′) in Section 4.4, show that for points on the surface H^+, P^+ is a stable weak focus (of multiplicity one) of the system (4), and that a supercritical Hopf bifurcation occurs at points on H^+ as c increases. Cf. Theorem 5′ in Section 4.15.

(b) Use equation (3′) in Section 4.4 and the fact that a BQS3 cannot have a weak focus of multiplicity $m \geq 3$ proved in [50] to show that the system (5) has a weak focus of multiplicity two at P^+ for parameter values $(\alpha, \beta, \gamma, c) \in H^+$ that lie on the multiplicity-two Hopf bifurcation surface

$$H_2^+ : c = \frac{-b + \sqrt{b^2 - 4ad}}{2a},$$

where $a = 2(2S - \beta)$, $b = (\alpha + \beta - S)(\beta - 2S) + 2$, and $d = \beta - \alpha - 3S$ with S given above. Note that the quantity σ, given by equation (3′) in Section 4.4, determines whether we have a supercritical or a subcritical Hopf bifurcation, and that σ changes sign at points on H_2^+; cf. Figure 20. Cf. Theorem 6′ in Section 4.15.

2. Show that there is a Takens–Bogdanov bifurcation at the origin of the system (5) for parameter values on the Takens–Bogdanov bifurcation surface

$$TB^0 : c = \alpha + \frac{1}{\beta} \quad \text{and} \quad \gamma = 0$$

for $\alpha \neq \beta$; cf. Theorems 3 and 4 in Section 4.15. Note that the system (5) reduces to the system (4) for $\gamma = 0$. Also, cf. Problem 8 below.

3. Note that for $\alpha = \beta + 2\gamma$, the quantity $S = \sqrt{(\alpha - \beta)^2 - 4\gamma^2} = 0$. This implies that $x^+ = x^-$ and $y^+ = y^-$; i.e., as $\alpha \to \beta + 2\gamma$, $P^+ \to P^-$. Show that for $\alpha = \beta + 2\gamma$, $\delta(x^\pm, y^\pm) = 0$ and $\tau(x^\pm, y^\pm) \neq 0$ if $c \neq 1/(\beta + 2\gamma) + \beta + \gamma$; i.e., $D\mathbf{f}(x^\pm, y^\pm)$ has one zero eigenvalue in this case. Check that the conditions of Theorem 1 in Section 4.2 are satisfied, i.e, show that the system (5) has a saddle-node bifurcation surface given by

$$SN : \alpha = \beta + 2\gamma.$$

Cf. Theorem 1′ and Problem 1 in Section 4.15. Note that this equation reduces to $\alpha = \beta$ for the system (4), where $\gamma = 0$ and as Theorem 2 in the next section shows, in this case we have a saddle-node or cusp bifurcation of codimension two.

4. Show that for $\alpha = \beta + 2\gamma$ and $c = 1/(\beta + 2\gamma) + \beta + \gamma$, the matrix

$$A \equiv D\mathbf{f}(x^\pm, y^\pm) = \begin{bmatrix} -1 & \alpha \\ -\frac{1}{\alpha} & 1 \end{bmatrix},$$

and that $\delta(x^{\pm}, y^{\pm}) = \tau(x^{\pm}, y^{\pm}) = 0$, where, as was noted in Problem 3, $(x^+, y^+) = (x^-, y^-)$ for $\alpha = \beta + 2\gamma$. Since the matrix $A \neq 0$ has two zero eigenvalues in this case, it follows from the results in Section 4.13 that the quadratic system (5) experiences a Takens–Bogdanov bifurcation for parameter values on the Takens–Bogdanov surface

$$TB^+ : c = \frac{1}{\beta + 2\gamma} + \beta + \gamma \quad \text{and} \quad \alpha = \beta + 2\gamma$$

for $\gamma \neq 0$; cf. Theorems 3' and 4' in Section 4.15. Note that it was shown earlier in this section that for the TB^+ points to lie in the region $|c| < 2$, it was necessary that the point (β, γ) lie in the region between the curves $\Gamma_4^{\pm}$ in Figure 10; this implies that $\beta + 2\gamma > 0$, i.e., that $\alpha > 0$ in this case. It should also be noted that a codimension two Takens–Bogdanov bifurcation occurs at points on the above TB^+ curve for $\gamma \neq -(\beta^2 + 2)/2\beta$; however, for $\beta < 0$ and $\gamma = -(\beta^2 + 2)/2\beta$, a codimension three Takens–Bogdanov bifurcation occurs on the TB^+ curve defined above. Cf. the remark at the end of Section 4.13, reference [46] and Theorem 4' in the next section. Also, it can be shown that there are no codimension four bifurcations that occur in the class of bounded quadratic systems.

5. Similar to what was done in Problem 1, for $\gamma \neq 0$ and $\beta \neq 0$ set $\tau(0,0) = 0$ to find the Hopf bifurcation surface

$$H^0 : c = \alpha + \frac{1 + \gamma^2}{\beta}$$

for the critical point at the origin of (5). Cf. Theorem 5 in Section 4.15. Then, using equation (3') in Section 4.4 and the result in [50] cited in Problem 1, show that for parameter values on H^0 and on

$$H_2^0 : c = \frac{-b + \sqrt{b^2 - 4ad}}{2a}$$

with $b = 1 + 2\alpha^2 - \alpha\beta$, $a = \beta - 2\alpha$, and $d = \alpha - \beta$, the system (5) has a multiplicity-two weak focus at the origin. Cf. Theorem 6 in Section 4.15.

6. Use the fact that the system (5) forms a semi-complete family of rotated vector fields mod $x = \beta y + y^2$ with parameter $c \in \mathbf{R}$ and the results of the rotated vector field theory in Section 4.6 to show that there exists a function $h(\alpha, \beta, \gamma)$ defining the homoclinic-loop bifurcation surface

$$HL^+ : c = h(\alpha, \beta, \gamma),$$

for which the system (5) has a homoclinic loop at the saddle point P^- that encloses P^+. This is exactly the same procedure that was used in Section 4.13 in establishing the existence of the homoclinic-loop

4.14. Coppel's Problem for Bounded Quadratic Systems

bifurcation surface for the system (2) in that section. The analyticity of the function $h(\alpha, \beta, \gamma)$ follows from the results in [38]. Carry out a similar analysis, based on the rotated vector field theory in Section 4.6, to establish the existence (and analyticity) of the surfaces HL^0, C_2^1, and C_2^0. Remark 10 in [38] is helpful in establishing the existence of the C_2^+ and C_2^0 surfaces, and their analyticity also follows from the results in [38]. Cf. Remarks 2 and 3 in Section 4.15.

7. Use Theorem 1 and Remark 1 in Section 4.8 to show that for points on the surface HL^+, the system (5) has a multiplicity-two homoclinic-loop bifurcation surface given by

$$HL_2^+ : c = \frac{1 + \alpha(\alpha + \beta - S)/2}{\alpha - S}$$

with S given above. Note that under the assumption that (5) has at most two limit cycles, there can be no higher multiplicity homoclinic loops.

8. Note that as $\gamma \to 0$, the critical point $P^- \to 0$. Show that for $\gamma = 0, \delta(0,0) = 0$ and that $\tau(0,0) \neq 0$ for $c \neq 1/\beta + \alpha$; i.e., $Df(0,0)$ has one zero eigenvalue in this case. Check that the conditions in equation (3) in Section 4.2 are satisfied in this case, i.e., show that the system (5) has a transcritical bifurcation for parameter values on the transcritical bifurcation surface

$$TC : \gamma = 0.$$

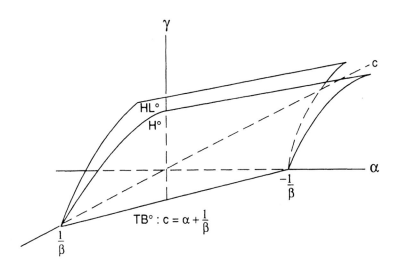

Figure 28. The Takens–Bogdanov bifurcation surface $TB^0 = H^0 \cap HL^0 \cap TC$ in the (α, c)-plane for a fixed $\beta < 0$.

Note that the H^0 and HL^0 surfaces intersect in a cusp on the $\gamma = 0$ plane as is shown in Figure 28.

9. Re-draw the charts in Figure 16 for $-1 \ll \beta < 0$. **Hint:** As in Figure 18, the HL^0 and H^0 curves enter the region E for $\beta < 0$, and for points on the HL^0 curve we have the phase portrait (f') in Figure 8, etc.

4.15 Finite Codimension Bifurcations in the Class of Bounded Quadratic Systems

In this final section of the book, we consider the finite codimension bifurcations that occur in the class of bounded quadratic systems (BQS), i.e., in the BQS (5) in Section 4.14:

$$\begin{aligned} \dot{x} &= -x + \beta y + y^2 \\ \dot{y} &= \alpha x - (\alpha\beta + \gamma^2)y - xy + c(-x + \beta y + y^2) \end{aligned} \quad (1)$$

with $\alpha \geq \beta + 2\gamma, \gamma \geq 0$ and $|c| < 2$. As in Lemma 2 of Section 4.14, the system (1) defines a one-parameter family of rotated vector fields mod $x = \beta y + y^2$ with parameter c and it has three critical points $0, P^\pm$ with a saddle at P^- and nodes or foci at 0 and P^+. The coordinates $(x^\pm, y^\pm)$ of $P^\pm$ are given in Lemma 2 of Section 4.14.

We consider saddle-node bifurcations at critical points with a single-zero eigenvalue, Takens–Bogdanov bifurcations at a critical point with a double-zero eigenvalue, and Hopf or Hopf–Takens bifurcations at a weak focus. Unfortunately, there is no universally accepted terminology for naming bifurcations. Consequently, the saddle-node bifurcation of codimension two referred to in Theorem 3.4 in [60], i.e., in Theorem 2 below, is also called a cusp bifurcation of codimension two in Section 4.3 of this book and in [G/S]; however, once the codimension of the bifurcation is given and the bifurcation diagram is described, the bifurcation is uniquely determined and no confusion should arise concerning what bifurcation is taking place, no matter what name is used to label the bifurcation.

In this section we see that the only finite-codimension bifurcations that occur at a critical point of a BQS are the saddle-node (SN) bifurcations of codimension 1 and 2, the Takens–Bogdanov (TB) bifurcations of codimension 2 and 3, and the Hopf (H) or Hopf–Takens bifurcations of codimension 1 and 2 and that whenever one of these bifurcations occurs at a critical point of the BQS (1), a universal unfolding of the vector field (1) exists in the class of BQS. We use a subscript on the label of a bifurcation to denote its codimension and a superscript to denote the critical point at which it occurs: for example, SN_2^0 will denote a codimension-2, saddle-node bifurcation at the origin, as in Theorem 2 below.

4.15. Finite Codimension Bifurcations

Most of the results in this section are established in the recent work of Dumortier, Herssens and the author [60]. This section, along with the work in [60], serves as a nice application of the bifurcation theory, normal form theory, and center-manifold theory presented earlier in this book. In presenting the results in [60], we use the definition of the codimension of a critical point given in Definition 3.1.7 on p. 295 in [Wi-II]. The codimension of a critical point measures the degree of degeneracy of the critical point. For example, the saddle-node at the origin of the system in Example 4 of Section 4.2 for $\mu = 0$ has codimension 1, the node at the origin of the system in Example 1 in Section 4.3 has codimension 2 and the cusp at the origin of the system (1) in Section 4.13 has codimension 2. We begin this section with the results for the single-zero-eigenvalue or saddle-node bifurcations that occur in the BQS (1).

A. SADDLE-NODE BIFURCATIONS

First of all, note that as $\gamma \to 0$ in the system (1), the critical point $P^- \to 0$ and the linear part of (1) at $(0,0)$ has a single-zero eigenvalue for $\beta(c-\alpha) \neq 1$; cf. Problem 1. The next theorem, which is Theorem 3.2 in [60], describes the codimension-1, saddle-node bifurcation that occurs at the origin of the system (1).

Theorem 1 (SN_1^0). *For $\gamma = 0, \alpha \neq \beta$ and $\beta(c-\alpha) \neq 1$, the system (1) has a saddle-node of codimension 1 at the origin and*

$$\begin{aligned} \dot{x} &= -x + \beta y + y^2 \\ \dot{y} &= \mu + \alpha x - \alpha\beta y - xy + c(-x + \beta y + y^2) \end{aligned} \quad (2)$$

is a universal unfolding of (1), in the class of BQS for $|c| < 2$, which has a saddle-node bifurcation of codimension 1 at $\mu = 0$. The bifurcation diagram for this bifurcation is given by Figure 2 in Section 4.2.

The proofs of all of the theorems in this section follow the same pattern: We reduce the system (1) to normal form, determine the resulting flow on the center manifold, and use known results to deduce the appropriate universal unfolding of this flow. We illustrate these ideas by outlining the proof of Theorem 1. Cf. the proof of Theorem 3.2 in [60].

The system (1) under the linear transformation of coordinates

$$\begin{aligned} x &= u + \beta v \\ y &= (c-\alpha)u + v, \end{aligned}$$

which reduces the linear part of (1) at the origin to its Jordan normal form, becomes

$$\begin{aligned} \dot{u} &= u + a_{20}u^2 + a_{11}uv + a_{02}v^2 \\ \dot{v} &= b_{20}u^2 + b_{11}uv + b_{02}v^2 \end{aligned} \quad (3)$$

where $a_{20} = (\alpha - c)(\alpha\beta c - \alpha + \beta - \beta c^2 + c)/[\beta(c-\alpha) - 1]^2, \ldots, b_{02} = (\beta - \alpha)/[\beta(c-\alpha) - 1]^2$, cf. Problem 2 or [60], and where we have also let $t \to [\beta(c-\alpha) - 1]t$. On the center manifold,

$$u = -a_{02}v^2 + 0(v^3),$$

of (3) we have a flow defined by

$$\dot{v} = b_{02}v^2 + 0(v^3)$$

with $b_{02} \neq 0$ since $\alpha \neq \beta$. Thus, there is a saddle-node (of codimension 1) at the origin of (3). Furthermore, the system obtained from (2) under the above linear transformation of coordinates, together with $t \to -[\beta(c-\alpha) - 1]^2 t$, has a flow on its center manifold defined by

$$\dot{v} = \mu + (\beta - \alpha)v^2 + 0(v, v^3, \mu^2, \ldots). \tag{4}$$

As in Section 4.3, the $0(v)$ terms can be eliminated by translating the origin and, as in equation (4) in Section 4.3, we see that the above differential equation is a universal unfolding of the corresponding normal form (4) with $\mu = 0$; i.e., the system (2) is a universal unfolding of the system (1) in this case. Furthermore, by translating the origin to the $0(\mu)$ critical point of (2), the system (2) can be put into the form of system (1) which is a BQS for $|c| < 2$.

Remark 1. The unfolding (2), with parameter μ, of the system (1) with $\gamma = 0, \alpha \neq \beta$ and $\beta(c-\alpha) \neq 1$, gives us the generic saddle-node, codimension-1 bifurcation described in Sotomayor's Theorem 1 in Section 4.2 (Cf. Problem 1), while the unfolding (1) with parameter γ gives us the transcritical bifurcation, labeled TC in Section 4.14.

We next note that as $\alpha \to \beta + 2\gamma$ in the system (1), the critical point $P^- \to P^+$ and the linear part of (1) at P^+ has a single-zero eigenvalue for $c \neq \beta + \gamma + 1/(\beta + 2\gamma)$; cf. Problem 3 in Section 4.14. The next theorem gives the result corresponding to Theorem 1 for the codimension-1, saddle-node bifurcation that occurs at the critical point P^+ of the system (1). This bifurcation was labeled SN in Section 4.14.

Theorem 1' (SN_1^+). *For $\alpha = \beta + 2\gamma, \gamma \neq 0$ and $(\beta + 2\gamma)(c - \alpha + \gamma) \neq 1$, the system (1) has a saddle-node of codimension 1 at $P^+ = (x^+, y^+)$ and*

$$\begin{aligned}\dot{x} &= -x + \beta y + y^2 \\ \dot{y} &= \mu + (\beta + 2\gamma)x - (\beta + \gamma)^2 y - xy + c(-x + \beta y + y^2)\end{aligned} \tag{5}$$

is a universal unfolding of (1), in the class of BQS for $|c| < 2$, which has a saddle-node bifurcation of codimension 1 at $\mu = 0$. The bifurcation diagram for this bifurcation is given by Figure 2 in Section 4.2

If both $\gamma \to 0$ and $\alpha \to \beta + 2\gamma$ in (1), then both $P^\pm \to 0$ and the linear part of (1) still has a single-zero eigenvalue for $\beta(c-\alpha) \neq 1$; cf. Problem 1.

4.15. Finite Codimension Bifurcations

The next theorem, which is Theorem 3.4 in [60], cf. Remark 3.5 in [60], describes the codimension-2, saddle-node bifurcation that occurs at the origin of the system (1) which, according to the center manifold reduction in [60], is a node of codimension 2. The fact that (6) below is a BQS for $|c| < 2$ follows, as in [60], by showing that (6) has a saddle-node at infinity.

Theorem 2 (SN_2^0). *For $\gamma = 0, \alpha = \beta$ and $\beta(c - \alpha) \neq 1$, the system (1) has a node of codimension 2 at the origin and*

$$\begin{aligned} \dot{x} &= -x + \beta y + y^2 \\ \dot{y} &= \mu_1 + \beta x - (\beta^2 + \mu_2)y - xy + c(-x + \beta y + y^2) \end{aligned} \tag{6}$$

is a universal unfolding of (1), in the class of BQS for $|c| < 2$, which has a saddle-node (or cusp) bifurcation of codimension 2 at $\boldsymbol{\mu} = \mathbf{0}$. The bifurcation diagram for this bifurcation is given by Figure 2 in Section 4.3.

In the proof of Theorem 2, or of Theorem 3.4 in [60], we use a center manifold reduction to show that the system (1), under the conditions listed in Theorem 2, reduces to the normal form (5) in Section 4.3 whose universal unfolding is given by (6) in Section 4.3, i.e., by (6) above; cf. Problem 2.

B. TAKENS–BOGDANOV BIFURCATIONS

As in paragraph A above, as $\gamma \to 0, P^- \to 0$; however, the linear part of (1) at the origin has a double-zero eigenvalue for $\beta(c - \alpha) = 1$; cf. Problem 1. The next theorem, which follows from Theorem 3.8 in [60], describes the codimension-2, Takens–Bogdanov bifurcation that occurs at the origin of the system (1) which, according to the results in [60], is a cusp of codimension 2. The fact that the system (7) below is a BQS for $|c| < 2$ and $\mu_2 \sim 0$ follows, as in [60], by looking at the behavior of (7) on the equator of the Poincaré sphere where there is a saddle-node.

Theorem 3 (TB_2^0). *For $\gamma = 0$, $\alpha \neq \beta$, $\beta(c - \alpha) = 1$ and $\beta \neq 2c$, the system (1) has a cusp of codimension 2 at the origin and*

$$\begin{aligned} \dot{x} &= -x + \beta y + y^2 \\ \dot{y} &= \mu_1 + \alpha x - (\alpha\beta + \mu_2)y - xy + c(-x + \beta y + y^2) \end{aligned} \tag{7}$$

is a universal unfolding of (1), in the class of BQS for $|c| < 2$ and $\mu_2 \sim 0$, which has a Takens–Bogdanov bifurcation of codimension 2 at $\boldsymbol{\mu} = \mathbf{0}$. The bifurcation diagram for this bifurcation is given by Figure 3 in Section 4.13.

In the proof of Theorem 3, or of Theorem 3.8 in [60], we show that the system (1), under the conditions listed in Theorem 3, reduces to the normal form (1) in Section 4.13 whose universal unfolding is given by (2) in Section 4.13, i.e., by (7) above; cf. Problem 3.

The next theorem gives the result corresponding to Theorem 3 for the codimension-2, Takens–Bogdanov bifurcation that occurs at the critical point P^+ of the system (1).

Theorem 3' (TB_2^+). *For $\alpha = \beta + 2\gamma$, $\gamma \neq 0$, $(\beta + 2\gamma)(c - \alpha + \gamma) = 1$ and $\beta^2 + 2\beta\gamma + 2 \neq 0$, the system (1) has a cusp of codimension 2 at the critical point P^+ and*

$$\dot{x} = -x + \beta y + y^2$$
$$\dot{y} = \mu_1 + (\beta + 2\gamma)x + [(\beta + \gamma)^2 + \mu_2]y - xy + c(-x + \beta y + y^2) \quad (8)$$

is a universal unfolding of (1), in the class of BQS for $|c| < 2$ and $\mu_2 \sim 0$, which has a Takens–Bogdanov bifurcation of codimension 2 at $\boldsymbol{\mu} = \boldsymbol{0}$. The bifurcation diagram for this bifurcation is given by Figure 3 in Section 4.13.

The next theorem, which follows from Theorem 3.9 in [60], describes the codimension-3, Takens–Bogdanov bifurcation that occurs at the origin of the system (1), which, according to the results in [61], is a cusp of codimension 3; cf. Remarks 1 and 2 in Section 2.13. The fact that the system (9) below is a BQS for $|c| < 2, \mu_2 \sim 0$ and $\mu_3 \sim 0$, follows as in [60], by showing that (9) has a saddle-node at infinity.

Theorem 4 (TB_3^0). *For $\gamma = 0$, $\alpha \neq \beta$, $\beta(c - \alpha) = 1$ and $\beta = 2c$, the system (1) has a cusp of codimension 3 at the origin and*

$$\dot{x} = -x + \beta y + y^2$$
$$\dot{y} = \mu_1 + \alpha x - (\alpha\beta + \mu_2)y - (1 + \mu_3)xy + c(-x + \beta y + y^2) \quad (9)$$

is a universal unfolding of (1), in the class of BQS for $|c| < 2, \mu_2 \sim 0$ and $\mu_3 \sim 0$, which has a Takens–Bogdanov bifurcation of codimension 3 at $\boldsymbol{\mu} = \boldsymbol{0}$. The bifurcation diagram for this bifurcation is given by Figure 1 below.

In proving this theorem, we show that the system (1), under the conditions listed in Theorem 4, reduces to the normal form (9) in Section 4.13 whose universal unfolding is given by (10) in Section 4.13, i.e. by (9) above; cf. [46] and the proof of Theorem 3.9 in [60].

The next theorem describes the Takens–Bogdanov bifurcation TB_3^+ that occurs at the critical point P^+ of the system (1).

Theorem 4' (TB_3^+). *For $\alpha = \beta + 2\gamma$, $\gamma \neq 0$, $(\beta + 2\gamma)(c - \alpha + \gamma) = 1$ and $\beta^2 + 2\beta\gamma + 2 = 0$, the system (1) has a cusp of codimension 3 at the critical point P^+ and*

$$\dot{x} = -x + \beta y + y^2$$
$$\dot{y} = \mu_1 + (\beta + 2\gamma)x + [(\beta + \gamma)^2 + \mu_2]y - (1 + \mu_3)xy + c(-x + \beta y + y^2) \quad (10)$$

is a universal unfolding of (1), in the class of BQS for $|c| < 2$, $\mu_2 \sim 0$ and $\mu_3 \sim 0$, which has a Takens–Bogdanov bifurcation of codimension 3 at

4.15. Finite Codimension Bifurcations

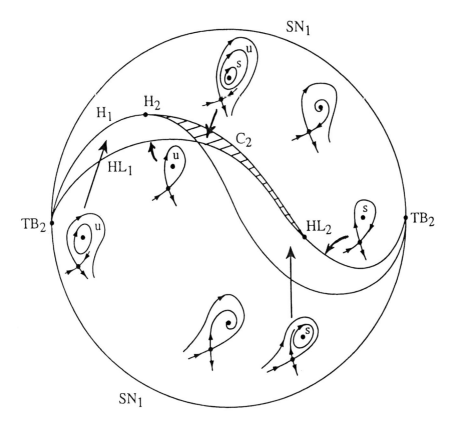

Figure 1. The bifurcation set and the corresponding phase portraits for the codimension-3 Takens–Bogdanov bifurcation (where s and u denote stable and unstable limit cycles or separatrix cycles respectively).

$\mu = 0$. *The bifurcation diagram for this bifurcation is described in Figure 1 above.*

It was shown in [46] and in [61] that the bifurcation diagram for the system (9), which has a Takens–Bogdanov bifurcation of codimension 3 at $\mu = 0$, is a cone with its vertex at the origin of the three-dimensional parameter space (μ_1, μ_2, μ_3). The intersection of this cone with any small sphere centered at the origin can be projected on the plane and, as in [46] and [61], this results in the bifurcation diagram (or bifurcation set) for the system (9) or for the system (10) shown in Figure 1 above. The bifurcation diagram in a neighborhood of either of the TB_2 points is shown in detail in Figure 3 of Section 4.13. The Hopf and homoclinic-loop bifurcations of codimension 1 and 2, H_1, H_2, HL_1, and HL_2 were defined in Theorem 5 in Section 4.14 and are discussed further in the next paragraph. Also, in

Figure 1 we have deleted the superscripts on the labels for the bifurcations since Figure 1 applies to either (9) or (10).

C. Hopf or Hopf–Takens Bifurcations

As in Problem 5 in Section 4.14, the system (1) has a weak focus of multiplicity 1 (or of codimension 1) at the origin if $c = \alpha + (1+\gamma^2)/\beta$ and $c \neq h_2^0(\alpha,\beta)$ where

$$h_2^0(\alpha,\beta) = \left[\alpha\beta - 2\alpha^2 - 1 + \sqrt{(\alpha\beta - 2\alpha^2 - 1)^2 - 4(\alpha-\beta)(\beta-2\alpha)}\right] / (2\beta - 4\alpha).$$

The next theorem follows from Theorem 3.16 in [60].

Theorem 5 (H_1^0). *For $\gamma \neq 0$, $\beta \neq 0$, $c = \alpha + (1+\gamma^2)/\beta$ and $c \neq h_2^0(\alpha,\beta)$, the system (1) has a weak focus of codimension 1 at the origin and the rotated vector field*

$$\begin{aligned}\dot{x} &= -x + \beta y + y^2 \\ \dot{y} &= \alpha x - (\alpha\beta + \gamma^2)y - xy + (c+\mu)(-x + \beta y + y^2)\end{aligned} \quad (11)$$

with parameter $\mu \in \mathbf{R}$ is a universal unfolding of (1), in the class of BQS for $|c| < 2$ and $\mu \sim 0$, which has a Hopf bifurcation of codimension 1 at $\mu = 0$. The bifurcation diagram for this bifurcation is given by Figure 2 in Section 4.4.

The idea of the proof of Theorem 5 is that under the above conditions, the system (1) can be brought into the normal form in Problem 1(b) in Section 4.4 and, as in Theorem 5 and Problem 1(b) in Section 4.6, a rotation of the vector field then serves as a universal unfolding of the system. In [60] we used the normal form for a weak focus of a BQS given in [50] together with a rotation of the vector field to obtain a universal unfolding.

The next theorem treats the Hopf bifurcation at the critical point P^+ and, as in Theorem 5 or Problem 1 in Section 4.14, we define the function $h_2^+(\alpha,\beta,\gamma) = \left(-b + \sqrt{b^2 - 4ad}\right)/2a$ with $a = 2(2S - \beta)$, $b = (\alpha + \beta - S)(\beta - 2S) + 2$, $d = \beta - \alpha - 3S$ and $S = \sqrt{(\alpha - \beta)^2 - 4\gamma^2}$.

Theorem 5' (H_1^+). *For $\alpha \neq \beta + 2\gamma$, $\beta^2 - 2\alpha\beta - 4\gamma^2 \neq 0$, $c = [1 + \alpha(\alpha + \beta + S)/2]/(\alpha + S)$ and $c \neq h_2^+(\alpha,\beta,\gamma)$, the system (1) has a weak focus of codimension 1 at P^+ and the rotated vector field (11) with parameter $\mu \in \mathbf{R}$ is a universal unfolding of (1), in the class of BQS for $|c| < 2$ and $\mu \sim 0$, which has a Hopf bifurcation of codimension 1 at $\mu = 0$. The bifurcation diagram for this bifurcation is given by Figure 2 in Section 4.4.*

The next theorem, describing the Hopf-Takens bifurcation of codimension 2 that occurs at the origin of the system (1) follows from Theorem 3.20 in [60]. The details of the proof of that theorem are beyond the scope of

4.15. Finite Codimension Bifurcations 535

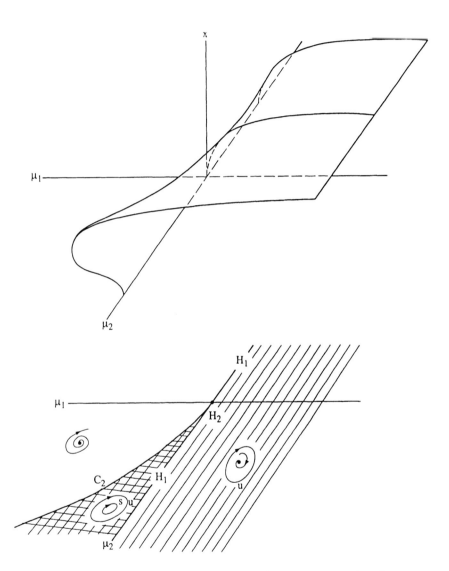

Figure 2. The bifurcation diagram and the bifurcation set (in the μ_1, μ_2 plane) for the codimension-2 Hopf–Takens bifurcation. Note that at $\mu_1 = \mu_2 = 0$ the phase portrait has an unstable focus (and no limit cycles) according to Theorem 4 in Section 4.4. Cf. Figure 6.1 in [G/S].

this book; however, after reducing the system (12) to the normal form for a BQS with a weak focus in [50], we can use Theorem 4 in Section 4.4 and the theory of rotated vector fields in Section 4.6 to analyze the codimension-2, Hopf–Takens bifurcation and draw the corresponding bifurcation set shown in Figure 2 above. Cf. Problem 4. The fact that (12) is a universal unfolding for the Hopf–Takens bifurcation of codimension 2 follows from the results of Kuznetsov [64], as in [60]; cf. Remark 4 below. The results for the Hopf–Takens bifurcation of codimension-2 that occurs at the critical point P^+ of the system (1) are given in Theorem 6' below. Recall that it follows from the results in [50] that a BQS cannot have a weak focus of multiplicity (or codimension) greater than two.

Theorem 6 (H_2^0). For $\gamma \neq 0, \beta \neq 0, c = \alpha + (1+\gamma^2)/\beta$ and $c = h_2^0(\alpha, \beta)$, the system (1) has a weak focus of codimension 2 at the origin and

$$\begin{aligned} \dot{x} &= -x + \beta y + y^2 \\ \dot{y} &= \alpha x - (\alpha\beta + \gamma^2)y - (1+\mu_2)xy + (c+\mu_1)(-x + \beta y + y^2) \end{aligned} \qquad (12)$$

is a universal unfolding of (1), in the class of BQS for $|c| < 2$, $\mu_1 \sim 0$ and $\mu_2 \sim 0$, which has a Hopf–Takens bifurcation of codimension 2 at $\boldsymbol{\mu} = \mathbf{0}$. The bifurcation diagram for this bifurcation is given by Figure 2 above.

Theorem 6' (H_2^+). For $\alpha \neq \beta + 2\gamma$, $\beta^2 - 2\alpha\beta - 4\gamma^2 \neq 0$, $c = [1 + \alpha(\alpha + \beta + S)/2]/(\alpha + S)$ and $c = h_2^+(\alpha, \beta, \gamma)$, the system (1) has a weak focus of codimension 2 at P^+ and the system (12) is a universal unfolding of (1), in the class of BQS for $|c| < 2$, $\mu_1 \sim 0$, and $\mu_2 \sim 0$, which has a Hopf–Takens bifurcation of codimension 2 at $\boldsymbol{\mu} = \mathbf{0}$. The bifurcation diagram for this bifurcation is given by Figure 2 above.

We conclude this section with a few remarks concerning the other finite-codimension bifurcations that occur in the class of BQS.

Remark 2. It follows from Theorem 6 above and the theory of rotated vector fields that there exist multiplicity-2 limit cycles in the class of BQS. (This also follows as in Theorem 5 and Problem 6 in Section 4.14.) The BQS (1) with parameter values on the multiplicity-2 limit cycle bifurcation surfaces C_2^0 or C_2^+ in Theorem 5 of Section 4.14 has a universal unfolding given by the rotated vector field (11), in the class of BQS for $|c| < 2$ and $\mu \sim 0$, which, in either of these cases, has a codimension-1, saddle-node bifurcation at a semi-stable limit cycle (as described in Theorem 1 of Section 4.5) at $\mu = 0$. The bifurcation diagram for this bifurcation is given by Figure 2 in Section 4.5.

Remark 3. As in Theorem 5 in Section 4.14, there exist homoclinic loops of multiplicity 1 and also homoclinic loops of multiplicity 2 in the class of BQS. And under the assumption that any BQS has at most two limit cycles, there are no homoclinic loops of higher codimension; however, Hilbert's

4.15. Finite Codimension Bifurcations

16th Problem for the class of BQS is still an open problem; cf. Research Problem 2 below. The BQS (1) with parameter values on the homoclinic-loop bifurcation surfaces HL^0 and HL^+ (or on the SS bifurcation surface) in Theorem 5 of Section 4.14 has a universal unfolding given by the rotated vector field (11), in the class of BQS for $|c| < 2$ and $\mu \sim 0$, which in either of these cases has a homoclinic-loop bifurcation of codimension 1 at $\mu = 0$. For parameter values on the bifurcation surface $HL^+ \cap HL_2^+$ in Theorem 5 of Section 4.14, it is conjectured that the BQS (1) has a universal unfolding given by the system (12), in the class of BQS for $|c| < 2, \mu_1 \sim 0$ and $\mu_2 \sim 0$, which has a homoclinic-loop bifurcation of codimension 2 at $\mu = 0$, the bifurcation diagram being given by Figure 8 (or Figure 10) in [38]; cf. Theorem 3 and Remark 10 in [38]. Also, cf. Figure 1 above, Figure 20 in Section 4.14 and Figure 7 (or Figure 12) in [38]. Finally, for parameter values on the homoclinic-loop bifurcation surface HL^+ (or on the SS bifurcation surface) in Theorem 4 in Section 4.14, which has a saddle-node at the origin, it is conjectured that the BQS (1) has a universal unfolding given by a rotation of the vector field (1), as in equation (11), together with the addition of a parameter μ_1, as in equation (7), to unfold the saddle node at the origin; this will result in a codimension-2 bifurcation which splits both the saddle-node and the homoclinic loop (or saddle–saddle connection).

Remark 4. In this section we have considered the finite codimension bifurcations that occur in the class of bounded quadratic systems. In this context, it is worth citing some recent results regarding two of the higher codimension bifurcations that occur at critical points of planar systems:

A. The single-zero eigenvalue or saddle node bifurcation of codimension m, SN_m: In this case, the planar system can be put into the normal form

$$\dot{x} = -x^{m+1} + O(|\mathbf{x}|^{m+2})$$
$$\dot{y} = -y + O(|\mathbf{x}|^{m+2})$$

and a universal unfolding of this normal form is given by

$$\dot{x} = \mu_1 + \mu_2 x + \cdots + \mu_m x^{m-1} - x^{m+1}$$
$$\dot{y} = -y.$$

cf. Section 4.3.

B. A pair of pure imaginary eigenvalues, the Hopf-Takens bifurcation of codimension m, H_m: It has recently been shown by Kuznetsov [64] that any planar C^1-system

$$\dot{\mathbf{x}} = \mathbf{f}(\mathbf{x}, \mu) \tag{13}$$

which has a weak focus of multiplicity one at the origin for $\mu = 0$, with the eigenvalues of $D\mathbf{f}(\mathbf{x}_\mu, \mu)$ crossing the imaginary axis at $\mu = 0$, can be transformed into the normal form in Theorem 2 in Section 4.4 with $b = 0$ and $a = \pm 1$ by smooth invertible coordinate and parameter transformations and a reparameterization of time, a universal unfolding of that normal form being given by the universal unfolding in Theorem 2 in Section 4.4 with $b = 0$ and $a = \pm 1$ (the plus sign corresponding to a subcritical Hopf bifurcation and the minus sign corresponding to a supercritical Hopf bifurcation). Furthermore, Kuznetsov [64] showed that any planar C^1-system (13) which has a weak focus of multiplicity two at the origin for $\boldsymbol{\mu} = \mathbf{0}$ and which satisfies certain regularity conditions can be transformed into the following normal form with $\boldsymbol{\mu} = \mathbf{0}$ which has a universal unfolding given by

$$\dot{x} = \mu_1 x - y + \mu_2 x |\mathbf{x}|^2 \pm x|\mathbf{x}|^4 + O(|\mathbf{x}|^6)$$
$$\dot{y} = x + \mu_1 y + \mu_2 y |\mathbf{x}|^2 \pm y|\mathbf{x}|^4 + O(|\mathbf{x}|^6).$$

Finally, it is conjectured that any planar C^1-system (13) which has a weak focus of multiplicity m at the origin for $\boldsymbol{\mu} = \mathbf{0}$ and which satisfies certain regularity conditions can be transformed into the following normal form with $\boldsymbol{\mu} = \mathbf{0}$ which has a universal unfolding given by

$$\dot{x} = \mu_1 x - y + \mu_2 x|\mathbf{x}|^2 + \cdots + \mu_m x|\mathbf{x}|^{2(m-1)} \pm x|\mathbf{x}|^{2m} + O(|\mathbf{x}|^{2(m+1)})$$
$$\dot{y} = x + \mu_1 y + \mu_2 y|\mathbf{x}|^2 + \cdots + \mu_m y|\mathbf{x}|^{2(m-1)} \pm y|\mathbf{x}|^{2m} + O(|\mathbf{x}|^{2(m+1)}).$$

PROBLEM SET 15

1. Show that as $\gamma \to 0$ the critical point P^- of (1) approaches the origin, that $\delta(0,0) = 0$ and that $\tau(0,0) = -1 + \beta(c - \alpha)$. Cf. the formulas for δ and τ in Problem Set 14. Also, show that the conditions of Sotomayor's Theorem 1 in Section 4.2 are satisfied by the system (2) for $(c - \alpha)\beta \neq 1, \alpha \neq \beta$ and $\gamma = 0$ and by the system (5) for $\gamma \neq 0, (\beta + 2\gamma)(c - \alpha - \gamma) \neq 1$ and $\alpha = \beta + 2\gamma$.

2. Use the linear transformation following Theorem 1 to reduce the system (1) with $\gamma = 0$, $\alpha = \beta$ and $\beta(c - \alpha) \neq 1$ to the system (3) with $b_{02} = 0$ and show that on the center manifold, $u = -a_{02}v^2 + 0(v^3)$, of (3) we have a flow determined by $\dot{v} = -v^3 + 0(v^4)$, after an appropriate rescaling of time. And then, using the same linear transformation (and rescaling the time), show that the flow on the center manifold of the system obtained from (6) is determined by $\dot{v} = \mu_1 + \mu_2 v - v^3 + 0(\mu_1 v, \mu_1^2, \mu_2^2, v^4, \ldots)$. Cf. equation (16) and Problem 6(b) in Section 4.3.

3. Show that under the linear transformation of coordinates $x = (u - v)/(\alpha c - \alpha^2 - 1)$, $y = (c - \alpha)u/(\alpha c - \alpha^2 - 1)$, the system (1) with $\gamma = 0, \alpha \neq \beta, \beta(c - \alpha) = 1$ and $\beta \neq 2c$ reduces to

$$\dot{u} = v + au^2 + buv$$
$$\dot{v} = u^2 + euv$$

4.15. Finite Codimension Bifurcations

with $a = (c^2 - \alpha c - 1)/(\alpha c - \alpha^2 - 1)$ and $b = e = 1/(\alpha c - \alpha^2 - 1)$ and note that $e + 2a = -(1 - 2c^2 + 2\alpha c)/(\alpha^2 - \alpha c + 1) \neq 0$ iff $(1 - 2c^2 + 2\alpha c) \neq 0$ or equivalently iff $\beta \neq 2c$. As in Remark 1 in Section 2.13, the normal form (1) in Section 4.13 results from any system of the above form if $e + 2a \neq 0$ and, as was shown by Takens [44] and Bogdanov [45], the universal unfolding of that normal form is given by (2) in Section 4.13; and this leads to the universal unfolding (7) of the system (1) in Theorem 3.

4. (a) Use the results of Theorem 4 (or Problem 8b) in Section 4.4 to show that the system

$$\dot{x} = \mu x - y + x^2 + xy$$
$$\dot{y} = x + \mu y + x^2 + mxy$$

has a weak focus of multiplicity 2 for $\mu = 0$ and $m = -1$. (Also, note that from Theorem 4 in Section 4.4, $W_2 = -8 < 0$ for $\mu = 0$, $m = -1$, $n = 0$ and $a = b = \ell = 1$.) Show that this system defines a family of negatively rotated vector fields with parameter μ in a neighborhood of the origin and use the results of Section 4.6 and Theorem 2 in Section 4.1 to establish that this system has a bifurcation set in a neighborhood of the point $(0, -1)$ in the (μ, m) plane given by the bifurcation set in Figure 2 above (the orientations and stabilities being opposite those in Figure 2.)

(b) In the case of a perturbed system with a weak focus of multiplicity 2 such as the one in Example 3 of Section 4.4,

$$\dot{x} = y - \varepsilon[\mu x + a_3 x^3 + \frac{16}{5} x^5]$$
$$\dot{y} = -x,$$

(where we have set $a_5 = 16/5$) we can be more specific about the shape of the bifurcation curve C_2 near the origin in Figure 2: For $\mu = a_3 = 0$, use equation (3') in Section 4.4 to show that this system has a weak focus at the origin of multiplicity $m \geq 2$ and note that by Theorem 5 in Section 3.8, the multiplicity $m \leq 2$. Also, show that for $\mu = a_3 = 0$ and $\varepsilon > 0$, the origin is a stable focus since $\dot{r} < 0$ for $x \neq 0$. For $\mu = 0$, $\varepsilon > 0$ and $a_3 \neq 0$, use equation (3') in Section 4.4 to find σ. Then show that for $\varepsilon > 0$ this system defines a system of negatively rotated vector fields (mod $x = 0$) with parameter μ and use the results of Section 4.6 and Theorem 2 in Section 4.1 to establish that this system has a bifurcation set in a neighborhood of the origin in the (μ, a_3) plane given by the bifurcation set in Figure 2 above (the stabilities of the limit cycles being opposite those in Figure 2). Finally, show that the Melnikov function for this system is given by

$$M(\alpha, \mu) = -\pi\alpha^2(\mu + \frac{3}{4}a_3\alpha^2 + 2\alpha^4)$$

and, using Theorem 2 in Section 4.10, deduce that for sufficiently small $\varepsilon > 0$ this system has a multiplicity-2 limit cycle in an $O(\varepsilon)$ neighborhood of the circle of radius $r = \sqrt[4]{\mu/2}$ iff $a_3 = -8\sqrt{2\mu}/3$; i.e., the bifurcation curve C_2 in Figure 2 above is given by $a_3 = -8\sqrt{2\mu}/3 + O(\varepsilon)$ for sufficiently small $\varepsilon > 0$ in this system. Note that the curve $a_3 = -4\sqrt{\mu}$ for which the system in Example 3 in Section 4.4 has two limit cycles asymptotic to circles of radii $r = \sqrt[4]{\mu}$ and $r = \sqrt[4]{\mu/4}$ as $\varepsilon \to 0$, lies in the region bounded by the curve C_2 and the a_3 axis where this system has two limit cycles.

Research Problems

We list the major research problems that remain to be solved in order to complete our understanding of the dynamics in the class of BQS. These are essentially the same open problems that were listed at the end of our paper [60].

1. Complete the study of the finite codimension bifurcations that occur in the class of BQS, i.e., those bifurcations discussed in Remarks 2 and 3 above.

2. Prove that any BQS has at most two limit cycles, i.e., solve Hilbert's 16th Problem for BQS.

3. Completely describe the dynamics in the class of BQS. One approach to this problem would be to complete the solution of Coppel's problem for BQS3 by first solving Research Problem 2 above and then completing the description of how the bifurcation surfaces described in Sections 4.14 and 4.15 partition the parameter space for the BQS (1), with $\beta < 0$ and $\gamma > 0$, into components (as well as determining which phase portrait in Figure 8 of the previous section corresponds to each component.)

References

[A-I] A.A. Andronov, E.A. Leontovich, I.I. Gordon and A.G. Maier, *Qualitative Theory of Second-Order Dynamical Systems*, John Wiley and Sons, New York, 1973.

[A-II] A.A. Andronov, et al., "Theory of Bifurcations of Dynamical Systems on a Plane," *Israel Program for Scientific Translations*, Jerusalem, 1971.

[B] I. Bendixson, "Sur les courbes définies par des équations différentielles," *Acta Math.*, **24** (1901), 1–88.

[C] H.S. Carslaw, *Theory of Fourier Series and Integrals*, Dover Publications, Inc., New York, 1930.

[Ca] J. Carr, *Applications of Center manifold Theory*, Springer-Verlag, New York, 1981.

[C/H] S.N. Chow and J.K. Hale, *Methods of Bifurcation Theory*, Springer-Verlag, New York, 1982.

[C/L] E.A. Coddington and N. Levinson, *Theory of Ordinary Differential Equations*, McGraw Hill, New York, 1955.

[Cu] C.W. Curtis, *Linear Algebra*, Allyn and Bacon Inc., Boston, 1974.

[D] H. Dulac, "Sur les cycles limites," *Bull Soc. Math. France*, **51** (1923), 45–188.

[G] L.M. Graves, *The Theory of Functions of Real Variables*, McGraw Hill, New York, 1956.

[G/G] M. Golubitsky and V. Guillemin, *Stable Mappings and their Singularities*, Springer-Verlag, New York, 1973.

[G/H] J. Guckenheimer and P. Holmes, *Nonlinear Oscillations, Dynamical Systems, and Bifurcations of Vector Fields*, Springer-Verlag, New York, 1983.

[G/S] M. Golubitsky and D.G. Schaeffer, *Singularities and Groups in Bifurcation Theory*, Springer-Verlag, New York, 1985.

[H] P. Hartman, *Ordinary Differential Equations*, John Wiley and Sons, New York, 1964.

[H/S] M.W. Hirsch and S. Smale, *Differential Equations, Dynamical Systems and Linear Algebra*, Academic Press, New York, 1974.

[L]   S. Lefschetz, *Differential Equations: Geometric Theory*, Interscience, New York, 1962.

[Lo]   F. Lowenthal, *Linear Algebra with Linear Differential Equations*, John Wiley and Sons, New York, 1975.

[N/S]   V.V. Nemytskii and V.V. Stepanov, *Qualitative Theory of Differential Equations*, Princeton University Press, Princeton, 1960.

[P/d]   J. Palais Jr. and W. deMelo, *Geometric Theory of Dynamical Systems*, Springer-Verlag, New York, 1982.

[P]   H. Poincaré, "Mémoire sur les courbes définies par une equation différentielle," *J. Mathématiques*, **7** (1881), 375–422; Ouevre (1880–1890), Gauthier-Villar, Paris.

[R]   W. Rudin, *Principles of Mathematical Analysis*, McGraw Hill, New York, 1964.

[Ru]   D. Ruelle, *Elements of Differentiable Dynamics and Bifurcation Theory*, Academic Press, New York, 1989.

[S]   C. Sparrow, *The Lorenz Equations: Bifurcations, Chaos, and Strange Attractors*, Springer-Verlag, New York, 1982.

[W]   P. Waltman, A *Second Course in Elementary Differential Equations*, Academic Press, New York, 1986.

[Wi]   S. Wiggins, *Global Bifurcations and Chaos*, Springer-Verlag, New York, 1988.

[Wi-II] S. Wiggins, *Introduction to Applied Nonlinear Dynamical Systems and Chaos*, Springer-Verlag, New York, 1990.

[Y]   Ye Yan-Qian, *Theory of Limit Cycles*, Translations of Mathematical Monographs, Vol. 66, American Mathematical Society, Rhode Island, 1986.

Additional References

[1] K. Alligood, J. Mallet-Paret and J. Yorke, An index for the global continuation of relatively isolated sets of periodic orbits, *Lecture Notes in Mathematics*, **1007**, Springer-Verlag (1983), 1–21.

[2] N. Bautin, On the number of limit cycles which appear with a variation of coefficients from an equilibrium position of focus or center type, *Mat. Sb.*, **30** (1952), 181–196; A.M.S. Translation No. 100 (1954), 3–19.

[3] T. Blows and N. Lloyd, The number of small amplitude limit cycles of Lienard equations, *Math Proc. Cambridge Phil. Soc.*, **95** (1984), 751–758.

[4] L. Cherkas and L. Zhilevich, Some criteria for the absence of limit cycles and for the existence of a single limit cycle, *Differential Equations*, **13** (1977), 529–547.

[5] L.Cherkas, Estimation of the number of limit cycles of autonomous systems, *Differential Equations*, **13** (1977), 529–547.

[6] C. Chicone and M. Jacobs, Bifurcation of limit cycles from quadratic isochrones, *J. Diff. Eq.*, **91** (1991), 268–326.

[7] G. Duff, Limit cycles and rotated vector fields, *Ann. Math.*, **57** (1953), 15–31.

[8] J. Ecalle, Finitude des cycles limites et accelero-sommation de l'application de retour, *Lecture Notes in Mathematics*, **1455**, Springer-Verlag (1990), 74–159.

[9] D. Grobman, Homeomorphisms of systems of differential equations, *Dokl, Akad. Nauk SSSR*, **128** (1959) 880–881.

[10] J. Hadamard, Sur l'iteration et les solutions asymptotiques des equations differentielles, *Bull. Soc. Math. France*, **29** (1901), 224–228.

[11] J. Hale, integral manifolds for perturbed differentiable systems, *Ann. Math.*, **73** (1961), 496–531.

[12] P. Hartman, A lemma in the theory of structural stability of differential equations. *Proc. A.M.S.*, **11** (1960), 610–620.

[13] P. Hartman, On local homeomorphisms of Euclidean spaces, *Bol. Soc. Math. Mexicana*, **5** (1960), 220–241.

[14] M. Hirsch, C. Pugh and M. Shub, Invariant manifolds, *Lectures Notes in Mathematics*, **583** Springer-Verlag (1977).

[15] P. Holmes, Poincaré, celestial mechanics, dynamical systems theory, and chaos, *Physics Report*, **193** (1990), 137–163.

[16] J. Kotus, M. Kyrch and Z. Nitecki, Global structural stability of flows on open surfaces, *Memoirs A.M.S.*, **37** (1982), 1–108.

[17] A. Liapunov, Probleme general de la stabilitie du mouvement, *Ann. Fac. Sci. Univ. Toulouse*, **9** (1907), 203–475; *Ann. of Math. Studies*, Princeton University Press, **17** (1947).

[18] A. Lins, W. de Melo and C. Pugh, On Lienard's equation, *Lecture Notes in Mathematics*, **597**, Springer-Verlag (1977), 335–357.

[19] L. Lyagina, The integral curves of the equation $y' = (ax^2 + bxy + cy^2)/(dx^2) + exy + fy^2)$, *Uspehi Mat. Nauk*, **6** (1951), 171–183.

[20] L. Markus, Global structure of ordinary differential equations in the plane, *Trans. A.M.S.*, **76** (1954), 127–148.

[21] V. Melinokov, On the stability of the center for time periodic perturbations, *Trans. Moscow Math. Soc.*, **12** (1963). 1–57.

[22] M. Peixoto, Structural stability on two-dimensional manifolds, *Topology*, **1** (1962), 101–120.

[23] L. Perko, Rotated vector field and the global behavior of limit cycles for a class of quadratic systems in the plane, *J. Diff. Eq.*, **18** (1975), 63–86.

[24] L. Perko, On the accumulation of limit cycles, *Proc. A.M.S.*, **99** (1987), 515–526.

[25] L. Perko, Global families of limit cycles of planar analytic systems, *Trans. A.M.S.*, **322** (1990), 627–656.

[26] O. Perron, Uber Stabiltat und asymptotisches verhalten der Integrale von Differentialgleichungssystem, *Math. Z.*, **29** (1928), 129–160.

[27] H. Poincaré, Sur les equations de la dynamique et le probleme des trois corps, *Acta math.*, **13** (1890), 1–270.

[28] H. Poincaré, Sur les proprietes des fonctions definies par les equations aux differences partielles, *Oeuvres*, **1**, Gauthier-Villars, paris (1929).

[29] Shi Songling, A concrete example of the existence of four limit cycles for plane quadratic systems, *Sci Sinica*, **23** (1980), 153–158.

[30] S. Smale, Stable manifolds for differential equations and diffeomorphisms, *Ann. Scuola Norm. Sup. Pisa*, **17** (1963), 97–116.

[31] S. Smale, Diffeomorphisms with many periodic points, *Differential and Combinatorial Topology*, Princeton University Press (1963), 63–80.

[32] S. Smale, Differentiable dynamical systems, *Bull. A.M.S.*, **73** (1967), 747–817.

[33] S. Sternberg, On local C^n contractions of the real line, *Duke Math. J.*, **24** (1957), 97–102.

[34] A. Wintner, Beweis des E. Stromgrenschen dynamischen Abschulusprinzips der preiodischen Bahngruppen im restringeirten Dreikorpenproblem, *Math. Z.*, **34** (1931), 321–349.

[35] Zhang Zhifen, On the uniqueness of limit cycles of certain equations of nonlinear oscillations, *Dokl. Akad. Nauk SSSR*, **19** (1958), 659–662.

[36] Zhang Zhifen, On the existence of exactly two limit cycles for the Lienard equation, *Acta Math. Sinica*, **24** (1981), 710–716.

[37] T.R. Blows and L.M. Perko, Bifurcation of limit cycles from centers and separatrix cycles of planar analytic systems, *SIAM Review*, **36** (1994), 341–376.

[38] L.M. Perko, Homoclinic loop and multiple limit cycle bifurcation surfaces, *Trans. A.M.S.*, **344** (1994), 101–130.

[39] L.M. Perko, Multiple limit cycle bifurcation surfaces and global families of multiple limit cycles, *J. Diff. Eq.*, **122** (1995), 89–113.

[40] P.B. Byrd and M.D. Friedman, *Handbook of Elliptic Integrals for Scientists and Engineers*, Springer-Verlag, New York, 1971.

[41] I.D. Iliev, On second order bifurcations of limit cycles, *J. London Math. Soc*, **58** (1998), 353–366.

[42] H. Zoladek, Eleven small limit cycles in a cubic vector field, *Nonlinearity*, **8** (1995), 843–860.

[43] K.S. Sibriskii, On the number of limit cycles in the neighborhood of a singular point, *Differential Equations*, **1** (1965), 36–47.

[44] F. Takens, Forced oscillations and bifurcations, Applications of Global Analysis I, *Comm. Math. Inst. Rijksuniversitat Utrecht*, **3** (1974), 1–59.

[45] R.I. Bogdanov, Versal deformations of a singular point on the plane in the case of zero eigenvalues, *Functional Analysis and Its Applications*, **9** (1975), 144–145.

[46] F. Dumortier, R. Roussarie, and J. Sotomayor, Generic 3-parameter families of vector fields on the plane, unfolding a singularity with nilpotent linear part. The cusp case of codimension 3, *Ergod. Th. and Dynam. Sys.*, **7** (1987), 375–413.

[47] W.A. Coppel, A survey of quadratic systems, *J. Diff. Eq.*, **2** (1966), 293–304.

[48] R.J. Dickson and L.M. Perko, Bounded quadratic systems in the plane, *J. Diff. Eq.*, **7** (1970), 251–273.

[49] D. Hilbert, Mathematical problems, *Bull A.M.S.*, **8** (1902), 437–479.

[50] B. Coll, A. Gasull, and J. Lilibre, Some theorems on the existence uniqueness and nonexistence of limit cycles for quadratic systems, *J. Diff. Eq.*, **67** (1987), 372–399.

[51] C. Li, J. Lilibre, and Z. Zhang, Weak foci, limit cycles and bifurcations for bounded quadratic systems, *J. Diff. Eq.*, **115** (1995), 193-233.

[52] S.N. Chow, B. Deng, and B. Fiedler, Homoclinic bifurcation at resonant eigenvalues, *J. Dynamics and Diff. Eq.*, **2** (1990), 177–244.

[53] L.M. Perko, Coppel's problem for bounded quadratic systems, *N.A.U. Research Report*, **M-94** (1994), 1–218.

[54] N. Markley, The Poincaré-Bendixson theorem for the Klein bottle, *Trans. Amer. Math. Soc.*, **135** (1969), 17–34.

[55] J.P. Françoise, Successive derivatives of the first return map, application to the study of quadratic vector fields, *Ergod. Theory Dynam. Syst.*, **16** (1996), 87–96.

[56] D.S. Shafer, Structural stability and generic properties of planar polynomial vector fields, *Revista Matematica Iberoamericana*, **3** (1987), 337-355.

[57] I.D. Iliev and L.M. Perko, Higher order bifurcations of limit cycles, *J. Diff Eq.*, **154** (1999), 339–363.

[58] I.D. Iliev, The number of limit cycles due to polynomial perturbations of the harmonic oscillator, *Math. Proc. Cambridge Phil Soc.*, **127** (1999), 317–322.

Additional References

[59] I.D. Iliev, On the number of limit cycles available from polynomial perturbations of the Bogdanov-Takens Hamiltonian, *Israel J. Math.*, **115** (2000), 269–284.

[60] F. Dumortier, C. Herssens and L.M. Perko, Local bifurcations and a survey of bounded quadratic systems, *J. Diff. Eq.*, **164** (2000), 430–467.

[61] F. Dumoritier and P Fiddelaers, Quadratic models for general local 3-parameter bifurcations on the plane, *Trans. Am. Math. Soc.*, **326** (1991), 101–126.

[62] C. Li, Two problems of planar quadratic systems, *Sci. Sinica, Ser. A*, **26** (1983), 471–481.

[63] L.M. Perko, Rotated vector fields, *J. Diff. Eq.*, **103** (1993), 127–145.

[64] Y.A. Kuznetsov, *Elements of Applied Bifurcation Theory*, Applied Math. Sciences, Vol. 112, Springer-Verlag, Berlin, 1995.

Index

Italic page numbers indicate where a term is defined.

α-limit cycle, *204*
α-limit point, *192*
α-limit set, *192*
Analytic function, 69
Analytic manifold, 107
Annular region, 294
Antipodal points, 269
Asymptotic stability, *129*, 131
Asymptotically stable periodic orbits, *202*
Atlas, *107*, 118, 244
Attracting set, *194*, 196
Attractor, *194*, 195
Autonomous system, 65

Bautin's lemma, 460
Behavior at infinity, 267, 272
Bendixson sphere, 235, 268, 292
Bendixson's criteria, 264
Bendixson's index theorem, 305
Bedixson's theorem, 140
Bifurcation
  homoclinic, 374, 387, *401*, 405, 416, 438
  Hopf, 350, 352, 353, 381, 389, 395
  period doubling, 362, 371
  pitchfork, 336, 337, 341, 344, 368, 369, 371, 380
  saddle connection, 324, 328, 381
  saddle node, 334, 338, 344, 364, 369, 379, 387, 495, 502, 529
  transcritical, 331, 338, 340, 366, 369
  value, 296, *334*

Bifurcation at a nonhyperbolic equilibrium point, 334
Bifurcation at a nonhyperbolic periodic orbit, 362, 371, 372
Bifurcation from a center, 422, 433, 434, 454, 474
Bifurcation from a multiple focus, 356
Bifurcation from a multiple limit cycle, 371
Bifurcation from a multiple separatrix cycle, 401
Bifurcation from a simple separatrix cycle, 401
Bifurcation set, 315
Bifurcation theory, 315
Bifurcation value, 104, 296, *334*
Blowing up, *268*, 291
Bogdanov-Takens bifurcation, 477
Bounded quadratic systems, 487, 488, 503, 528
Bounded trajectory, 246

$C(E)$, 68
$C^1(E)$, *68*, 316
C^1 diffeomorphism, 127, *190*, 213, 408
C^1 function, 68
C^1 norm, *316*, 318
C^1 vector field, 96, 284
$C^k(E)$, *69*
C^k conjugate vector fields, 191
C^k equivalent vector fields, 190
C^k function, 69
C^k manifold, 107
C^k norm, 355
Canonical region, 295

Cauchy sequence, 73
Center, 23, 24, *139*, 143
Center focus, *139*, 143
Center manifold of a periodic orbit, 228
Center manifold of an equilibrium point, 116, 154, 161, 343, 349
Center manifold theorem, 116, 155, 161
Center manifold theorem for periodic orbits, 228
Center subspace, 5, 9, *51*, 55
Center subspace of a map, 407
Center subspace of a periodic orbit, 226
Central projection, 268
Characteristic exponent, 222, 223
Characteristic multiplier, 222, 223
Chart, 107
Cherkas' theorem, 265
Chicone and Jacobs' theorem, 459
Chillingworth's theorem, 189
Circle at infinity, 269
Closed orbit, 202
Codimension of a bifurcation, *343*, 344–347, 359, 371, 478, 485
Competing species, *298*
Complete family of rotated vector fields, 384
Complete normed linear space, 73, 316
Complex eigenvalues, 28, 36
Compound separatrix cycle, 208, *245*
Conservation of energy, 172
Continuation of solutions, 90
Continuity with respect to initial conditions, 10, 20, 80
Continuous function, 68
Continuously differentiable function, 68
Contraction mapping principle, 78
Convergence of operators, 11
Coppel's problem, 487, 489, 521
Critical point, *102*
Critical point of multiplicity m, 337
Critical points at infinity, 271, 277
Cusp, 150, 151, 174
Cusp bifurcation, 345, 347

Cycle, *202*
Cyclic family of periodic orbits, 398
Cylindrical coordinates, 95

$D\mathbf{f}$, 67
$D^k\mathbf{f}$, 69
Deficiency indices, 42
Degenerate critical point, 23, 173, 313
Degenerate equilibrium point, 23, 173, 313
Derivative, *67*, 69
Derivative of the Poincaré map, 214, 216, 221, 223, 225, 362
Derivative of the Poincaré map with respect to a parameter, 370, 415
Diagonal matrix, 6
Diagonalization, 6
Diffeomorphism, 127, 182, 213
Differentiability with respect to initial conditions, 80
Differentiability with respect to parameters, 84
Differentiable, 67
Differentiable manifold, *107*, 118
Differentiable one-form, 467
Discrete dynamical system, 191
Displacement function, 215, 364, 396, 433
Duffing's equation, 418, 423, 440, 447, 449
Dulac's criteria, 265
Dulac's theorem, 206, 217
Dynamical system, 2, 181, *182*, 187, 191
Dynamical system defined by differential equation, 183, 184, 187

Eigenvalues
  complex, 28, 36
  distinct, 6
  pure imaginary, 23
  repeated, 33
Elementary Jordan blocks, 40, 49
Elliptic domain, 148, 151
Elliptic functions, 442, 445, 448
Elliptic region, 294
Elliptic sector, 147

Index

Equilibrium point, 2, 65, *102*
Escape to infinity, 246
Euler-Poincaré characteristic of a surface, 299, *306*
Existence uniqueness theorem, 74
Exponential of an operator, *12*, 13, 15, 17

f_σ, 284
Fixed point, 102, 406
Floquet's theorem, 221
flow
    of a differential equation, 96
    of a linear system, 54
    of a vector field, 96
    on a manifold, 284
    on S^2, 271, 274, 326
    on a torus, 200, 238, 311, 312, 325
Focus, 22, 24, 25, *139*, 143
Françoise's algorithm, 469
Fundamental existence uniqueness theorem, 74
Fundamental matrix solution, *60*, 77, 83, 85, 224
Fundamental theorem for linear systems, 17

Gauss' model, 298
General solution, 1
Generalized eigenvector, *33*, 51
Generalized Poincaré Bendixson theorem, 245
Generic property, 325, 331
Genus, 306, 307
Global behavior of limit cycles and periodic orbits, 389, 390, 395
Global bifurcations, 431
Global existence theorem, 184, *187*, 188, 189
Global Lipschitz condition, 188
Global phase portrait, 280, 283, 287
Global stability, 202
Global stable and unstable manifolds, 113, 203, 408
Gradient system, *176*, 178
Graphic, 207, *245*, 333, 388
Gronwall's inequality, 79

Hamiltonian system, *171*, 178, 210, 234
Harmonic oscillator, 171
Hartman Grobman theorem, 120
Hartman's theorem, 127
Heteroclinic orbit, 207
Higher order Melnikov method, 452, 453, 466, 469
Hilbert's 16th problem, 262
Homeomorphism, 107
Homoclinic bifurcation, 374, 387, 401, 405, 438, 494, 501, 536
Homoclinic explosion, 374
Homoclinic orbit, 207, 375
Homoclinic tangle, 409
Hopf bifurcation, 296, 350, 352, 353, 376, 381, 389, 405, 494, 503, 534
Horseshoe map, 409, 412
Hyperbolic equilibrium point, 102
Hyperbolic fixed point of a map, 407
Hyperbolic flow, 54
Hyperbolic periodic orbit, 226
Hyperbolic region, 294
Hyperbolic sector, 147

$I_\mathbf{f}(C)$, 299
$I_\mathbf{f}(\mathbf{x}_0)$, 302, 306
Iliev's lemma, 471
Iliev's theorem, 453
Implicit function theorem, 213, 362, 434, 436, 437
Improper node, 21
Index
    of a critical point, *302*, 306
    of a Jordan curve, 299
    of a saddle, node, focus or center, 305
    of a separatrix cycle, 303
    of a surface, 299, *306*, 307
Index theory, 299
Initial conditions, 1, 71, 80
Initial value problem, 16, 29, 71, 74, 76, 78
Invariant manifolds, 107, 111, 114, 226, 241, 408
Invariant subset, *99*, 194
Invariant subspace, 16, 20, 54

Jacobian matrix, 67
Jordan block, 40, 42, 49
Jordan canonical form, 39, 47
Jordan curve, 204, 247, *299*
Jordan curve theorem, 204

Kernel of a linear operator, 42
Klein bottle, 307, 313

$L(\mathbf{R}^n)$, 10
Left maximal interval, 91
Level curves, 177
Liapunov function, 131
Liapunov number, 218, 352, 353
Liapunov theorem, 131
Lienard equation, 136, 253, 440, 442
Lienard system, 253
Lienard's theorem, 254
Limit cycle, 195, *204*
Limit cycle of multiplicity k, 216
Limit orbit, 194
Limit set, 192
Linear
    approximation, 102
    flow, 54
    subspace, 51
    system, 1, 20
    transformation, 7, 20
Linearization about a periodic orbit, 221
Linearization of a differential equation, *102*, 221
Liouville's theorem, 86, 232
Lipschitz condition, 71
Local bifurcations, 315
Local center manifold theorem, 155, 161
Local limit cycle, 260
Local stable and unstable manifolds, 114, 203
Locally Lipschitz, 71
Lorenz system, 104, 198, 201, *373*

Manifold, 107
    center, 115, 155, 160
    differentiable, *107*, 118
    global stable and unstable, 113, 201, 398, 408
    invariant, 107, 111, 113, 201, 223, 241, 406
    local stable and unstable, 113, 203, 408
Maps, 211, 380, 382, 407
Markus' theorem, 295
Maximal family of periodic orbits, 400
Maximal interval of existence, 65, 67, 87, *90*, 94
Melnikov function, 415, 418, 421, 433, 437, 453, 467, 469
Melnikov's method, 316, 415, 416, 421, 433, 435, 452, 466
Method of residues, 430
Morse-Smale system, 331
Multiple eigenvalues, 33
Multiple focus, *218*, 356
Multiple limit cycle, *216*, 371
Multiple separatrix cycle, 401
Multiplicity of a critical point, *337*
Multiplicity of a focus, 218
Multiplicity of a limit cycle, 216

Negative half-trajectory, 192
Negatively invariant set, 99
Neighborhood of a set, 194
Newtonian system, 173, 180
Nilpotent matrix, *33*, 50
Node, 22, 24, 25, *139*, 143
Nonautonomous linear system, 77, 86
Nonautonomous system, 63, 66, 77
Nondegenerate critical point, 173
Nonhomogenous linear system, 60
Nonhyperbolic equilibrium point, 102, 147
Nonlinear systems, 65
Nonwandering point, 324
Nonwandering set, *324*, 331
Norm
    C^1-norm, 316, 318
    C^k-norm, 355
    Euclidian, 11
    matrix, 10, 15
    operator, 10, 15
    uniform, 73
Normal form, 163, 168, 170
Number of limit cycles, 254, 260, 262

ω-limit cycle, 203, *204*

Index

ω-limit point, 192
ω-limit set, 192
Operator norm, 10
Orbit, 191, 195, 200, 407
Orbit of a map, 407
Ordinary differential equation, 1
Orientable manifold, *107*, 118
Orthogonal systems of differential equations, 177

Parabolic region, 294
Parabolic sector, 147
Peixoto's theorem, 325
Pendulum, 174
Period, 202
Period doubling bifurcation, 371, 376
Period doubling cascade, 378
Periodic orbit, 202
Periodic orbit of saddle type, 204
Periodic solution, 202
Perko's planar termination principle, 400
Perturbed Duffing equation, 418, 423, 440, 447, 449
Perturbed dynamical systems, 415, 421, 444
Perturbed harmonic oscillator, 260, 438, 448, 454, 459, 464, 474
Perturbed truncated pendulum, 444, 455
Phase plane, 2
Phase portrait, 2, 9, 20
Picard's method of successive approximations, 72
Pitchfork bifurcation, 336, 337, 339, 341, 344, 368, 374, 378
Poincaré-Bendixson theorem, 245
Poincaré-Bendixson theorem for two-dimensional manifolds, 250
Poincaré index theorem, 307
Poincaré map, 211, *213*, 218, 362, 370, 372
Poincaré map for a focus, 218
Poincaré sphere, 268, 274
Polar coordinates, 28, 137, 144, 382
Polynomial one-form,
Positive half trajectory, 192
Positively invariant set, 99

Predator prey problem, 298
Projective geometry, 235, 268
Projective plane, 269, 306, 309, 311
Proper node, 21, *140*
Pure imaginary eigenvalues, 23
Putzer algorithm, 39

Real distinct eigenvalues, 6
Recurrent trajectory, 250
Regular point, 246
Rest point or equilibrium point, 102
Right maximal interval, 91
Rotated vector fields, 384
Rotated vector field (mod $G = 0$), 392

Saddle, 21, 24, 25, 102, *140*, 142
Saddle at infinity (SAI), 327
Saddle connection, 324, 327, 328, 388
Saddle-node, 149, 150
Saddle-node bifurcation, 334, 338, 339, 344, 364, 369, 387, 436, 478, 495, 502, 529
Second order Melnikov function, 453
Sector, 147, 293
Semicomplete family of rotated vector fields, 384
Semisimple matrix, 50
Semi-stable limit cycle, 202, 216, 220, 387
Separatrix, 21, 28, 140, *293*
Separatrix configuration, 276, *295*
Separatrix cycle, 206, 207, *244*, 387, 404
Shift map, 412, 414
Simple limit cycle, 216
Simple separatrix cycle, 401
Singular point, 102
Sink, 26, 56, *102*, 130
Smale-Birkhoff homoclinic theorem, 412, *416*
Smale horseshoe map, 380, 409, 412
Smale's theorem, 412
Smooth curve, 284
Solution curve, 2, 96, 191
Solution of a differential equation, 71
Solution of an initial value problem, 71
Sotomayor's theorem, 338

Source, 26, 56, *102*
Spherical pendulum, 171, 237
Spiral region, 294
Stability theory, 51, 129
Stable equilibrium point, 129
Stable focus, 22, *139*
Stable limit cycle, 202, 215
Stable manifold of an equilibrium point, 113
Stable manifold of a periodic orbit, 203, 225
Stable manifold theorem, 107
Stable manifold theorem for maps, 408
Stable manifold theorem for periodic orbits, 225
Stable node, 22, *139*
Stable periodic orbit, *200*
Stable separatrix cycle, 401
Stable subspace, 5, 9, *51*, 55, 58
Stable subspace of a map, 407
Stable subspace of a periodic orbit, 225
Stereographic projection, 235
Strange attractor, 198, 376
Strip region, 294
Strong C^1-perturbations, 326
Structural stability, 315, *317*, 318, 325, 326, 327
Structural stability on $\mathbf{R}^2$, 327
Structural stable dynamical system, 317
Structurally stable vector field, *317*, 318, 325
Subcritical Hopf bifurcation, *352*
Subharmonic Melnikov function, 421
Subharmonic periodic orbit, 421
Subspaces, 5, 9, *51*, 59
Successive approximations, 73, 77, 111, 122, 125
Supercritical Hopf bifurcation, *352*
Surface, 306
Swallow tail bifurcation, 346
Symmetric system, 145
System of differential equations, 1, 65, 181

$T_\mathbf{p} S^2$, 283
$T_\mathbf{p} M$, 284

Takens-Bogdanov bifurcation, 477, 482, 495, 502, 531
Tangent bundle, 284
Tangent plane, 283
Tangent space, 284
Tangent vector, 283
Tangential homoclinic bifurcation, 401, 412, *417*, 419
Topological saddle, *140*, 141, 151
Topologically conjugate, 119, 184
Topologically equivalent, 107, 119, *183*, 184, 187, 295, 318
Trajectory, 96, *191*, 192, 201
Transcritical bifurcation, 336, 338, 340, 366, 369
Transversal, 212, *246*
Transversal intersection of manifolds, 212, *331*, 408, 416
Transverse homoclinic orbit, 316, 401, 406, 416, 419
Transverse homoclinic point, 408, 412, 416, 419
Triangulation of a surface, 306, 313
Two-dimensional surface, 306

Unbounded oscillation, 246
Uncoupled linear systems, 17
Unfolding of a vector field, *343*
Uniform continuity, 78
Uniform convergence, 73, 92
Uniform norm, 73
Uniqueness of limit cycles, 254, 257
Uniqueness of solutions, 66, 74
Universal unfolding, *343*, 348, 359
Unstable equilibrium point, 129
Unstable focus, 22, *139*
Unstable limit cycle, 202, 215
Unstable manifold of an equilibrium point, 113
Unstable manifold of a periodic orbit, 203, 225
Unstable node, 22, *140*
Unstable periodic orbit, 202
Unstable separatrix cycle, 401
Unstable subspace, 4, 9, *51*, 55, 58
Unstable subspace of a map, 407
Unstable subspace of a periodic orbit, 226
Upper Jordan canonical form, 40, 48

Index

Van der Pol equation, 136, 254, 257, 263
Variation of parameters, 62
Variational equation or linearization of a differential equation, *102*, 221
Vector field, 3, 96, 102, 183, 190, 283, 305, 317
Vector field on a manifold, *283*, 284, 288, 306, 325, 326

$W^c(\mathbf{0})$, 116, 155, 160
$W^s(\mathbf{0})$, 113, 408
$W^u(\mathbf{0})$, 113, 408
$W^c(\Gamma)$, 228
$W^s(\Gamma)$, 203, 225
$W^u(\Gamma)$, 203, 225
Weak focus or multiple focus, *218*, 356
Wedge product, *370*, 384
Weierstrass preparation theorem, 435
Whitney's theorem, 284
Whitney topology, 326
Wintner's principle of natural termination, 399

Zero eigenvalues, 150, 154, 164, 168, 477, 482, 531
Zero of a vector field, 102
Zhang's theorem, 257, 259

Texts in Applied Mathematics

(continued from page ii)

31. *Brémaud:* Markov Chains: Gibbs Fields, Monte Carlo Simulation, and Queues.
32. *Durran:* Numerical Methods for Wave Equations in Geophysical Fluid Dynamics.
33. *Thomas:* Numerical Partial Differential Equations: Conservation Laws and Elliptic Equations.
34. *Chicone:* Ordinary Differential Equations with Applications.
35. *Kevorkian:* Partial Differential Equations: Analytical Solution Techniques, 2nd ed.
36. *Dullerud/Paganini:* A Course in Robust Control Theory: A Convex Approach.
37. *Quarteroni/Sacco/Saleri:* Numerical Mathematics.
38. *Gallier:* Geometric Methods and Applications: For Computer Science and Engineering.
39. *Atkinson/Han:* Theoretical Numerical Analysis: A Functional Analysis Framework.
40. *Brauer/Castillo-Chávez:* Mathematical Models in Population Biology and Epidemiology.
41. *Davies:* Integral Transforms and Their Applications, 3rd ed.
42. *Deuflhard/Bornemann:* Scientific Computing with Ordinary Differential Equations.
43. *Deuflhard/Hohmann:* Numerical Analysis in Modern Scientific Computing: An Introduction, 2nd ed.
44. *Knabner/Angermann:* Numerical Methods for Elliptic and Parabolic Partial Differential Equations.
45. *Larsson/Thomée:* Partial Differential Equations with Numerical Methods.
46. *Pedregal:* Introduction to Optimization.

LaVergne, TN USA
02 April 2010
177996LV00001B/12/P